T0213986

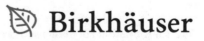

Foundations for Undergraduate Research in Mathematics

Series Editor
Aaron Wootton
University of Portland, Portland, USA

More information about this series at http://www.springer.com/series/15561

Hannah Callender Highlander • Alex Capaldi
Carrie Diaz Eaton

Editors

An Introduction to Undergraduate Research in Computational and Mathematical Biology

From Birdsongs to Viscosities

 Birkhäuser

Editors
Hannah Callender Highlander
Department of Mathematics
University of Portland
Portland, OR, USA

Alex Capaldi
Mathematics & Statistics
Valparaiso University
Valparaiso, IN, USA

Carrie Diaz Eaton
Bates College
Lewiston, ME, USA

ISSN 2520-1212   ISSN 2520-1220   (electronic)
Foundations for Undergraduate Research in Mathematics
ISBN 978-3-030-33647-9   ISBN 978-3-030-33645-5   (eBook)
https://doi.org/10.1007/978-3-030-33645-5

Mathematics Subject Classification: 92B05, 92C20, 92C42, 93A30, 76Z05, 37N25, 90B15, 60H30

This book is published under the imprint Birkhäuser, www.birkhauser-science.com by the registered company Springer Nature Switzerland AG.
The registered company address is: Gewerbestrasse 11, 6330 Cham, Switzerland

Series Preface

Research experience has become an increasingly important aspect of undergraduate programs in mathematics. Students fortunate enough to take part in such research, either through their home institution or via an external program, are exposed to the heart of the discipline. These students learn valuable skills and habits of mind that reach beyond what is typically addressed by the undergraduate curriculum, and are often more attractive to graduate programs and future employers than peers without research experience.

Despite their growing value in the community, research experiences for undergraduate students in mathematics are still the exception rather than the rule. The time commitment required to run or partake in a successful program can be prohibitive, and support for students and mentors is limited. For the faculty member, establishing such a program often requires taking a sophisticated topic outside the scope of the typical undergraduate curriculum and translating it in order to make it accessible to undergraduates with limited backgrounds. It also requires identifying problems or projects that are amenable to undergraduate exploration yet still relevant and interesting to the wider mathematical community. For the undergraduate, pursuing research can often mean reviewing extensive articles and technical texts while meeting regularly with a faculty member. This is no easy feat for the modern undergraduate with a heavy class load, or one who counts on a summer job to help offset academic costs.

The primary goal of the *Foundations for Undergraduate Research in Mathematics* series is to provide faculty and undergraduates with the tools they need to pursue collaborative or independent undergraduate research without the burden it often requires. In order to attain this goal, each volume in the series is a collection of chapters by researchers who have worked extensively with undergraduates to a great degree of success. Each chapter will typically include the following:

- A list of classes from the standard undergraduate curriculum that serve as prerequisites for a full understanding of the chapter
- An expository treatment of a topic in mathematics, written so a student with the stated prerequisites can understand it with limited guidance

- Exercises and Challenge Problems to help students grasp the technical concepts introduced
- A list of specific open problems that could serve as projects for interested undergraduates
- An extensive bibliography and carefully chosen citations intended to provide undergraduates and faculty mentors with a keen interest in the topic with appropriate further reading

On reading a chapter and doing the recommended exercises, the intention is that a student is now ready to pursue research, either collaboratively or independently, in that area. Moreover, that student has a number of open research problems at hand on which they can immediately get to work.

Undergraduate research programs are prevalent in the sciences, technology and engineering and their tremendous benefits are well-documented. Though far less common, the benefits of undergraduate research in mathematics are equally valuable as their scientific counterparts, and increased participation is strongly supported by all the major professional societies in mathematics. As the pioneering series of its type, *Foundations for Undergraduate Research in Mathematics* will take the lead in making undergraduate research in mathematics significantly more accessible to students from all types of backgrounds and with a wide range of interests and abilities.

Aaron Wootton
University of Portland

Introduction

While each volume of the *Foundations for Undergraduate Research in Mathematics* (FURM) series will be characterized by a commitment to undergraduate education, and representative of a particular subdiscipline within mathematics, we designed our volume to bring some unique features to the series, in particular in the range of types of mathematics, the depth of partner discipline understanding, and special highlights on mentoring and the process of research. This project is dear to the authors because we have been active in guiding undergraduate research projects for many years, and the unique applications in this volume are a reflection of our own philosophy of engagement. We are inspired by our students' passions. Some students are passionate about a specific area within biology and bring with them an enthusiasm for asking questions. Some students come with a love for mathematics and a desire to use their mathematical prowess to solve pressing problems of our day. We then have the privilege of helping students discover the right mathematics for the question or, if constrained by our mathematical area of training, the right question for the mathematics. Most importantly, we encourage mentors to take the backseat and let students own the process as much as possible.

Because the journey sometimes involves finding the right math, the content in this volume spans the breadth of mathematical and computational biology. This includes many topics that we believe to be appropriate for student–faculty exploration that are not normally addressed by the undergraduate curriculum (e.g., Hidden Markov Models, bifurcation diagrams, convolutional neural networks, agent-based models). The diversity of areas also allows multiple options for students of varying mathematical exposure to start research in mathematical and computational biology.

Equally important is the emphasis of this volume on applications. While we endeavor to help make any new math accessible, the examples of research included in the text also explain the biology needed to be a researcher in this area. They explain the fundamental questions our disciplinary friends have and then show how mathematics gives new understanding to these biological challenges. While each chapter has a different application focus, some chapters include a variety of applications, all centered around that chapter's primary theme. The volume thus provides a wide variety of application areas to inspire students.

Finally, we have solicited from individuals who have won awards for their mentoring and whose students have won national research awards—individuals like Dr. Castillo-Chavez and Dr. Prieto Lagarica. As such, some of our authors also take the time to talk about the productive engagement in undergraduate research, from both the student's and the mentor's point of view.

Our intended audience is broad: undergraduate mathematics faculty, with a particular emphasis on faculty interested in (but not necessarily experienced in) mathematical and computational biology, and students with sophomore- to junior-level coursework as a background and likely studying in a STEM field. Undergraduate faculty wishing to direct research will benefit from the many project ideas suggested by the authors herein, as will faculty simply wishing to expand their own research repertoire in a new direction. Undergraduate STEM students will appreciate the accessible yet rigorous treatment of topics previously relegated to the graduate curriculum, some of these are with fairly minimal prerequisite assumptions (e.g., calculus and linear algebra). The benefits of undergraduate research, both for the mentee and the mentor, are myriad. One in particular, though, is contributing to the scientific community writ large. We hope that if this book inspires you to advance scientific knowledge in the field of computational biology through research, then you will consider submitting your work for publication to one of many undergraduate research journals serving the field such as *Involve*, *SIAM Undergraduate Research Online*, or *Spora: A Journal of Biomathematics*.

Chapters 1–4 focus on ordinary differential equations from modeling to analysis. Chapters 5–7 focus on agent-based and network models. Chapters 8–10 focus on other computational approaches such as parallel computing, neural networks, and computational fluid dynamics. Chapters are ordered by prerequisite knowledge in and between these sets.

In Chapter 1, "Building New Models: Rethinking and Revising ODE Model Assumptions," the commonly adopted assumptions of ordinary differential equations models, where the applications are endless (disease dynamics is a major area), are revisited. The reader is introduced to an array of techniques of relaxing such assumptions, with an emphasis on investigating the resulting implications. Prerequisites for interested readers include some familiarity with basic ideas from calculus, differential equations, calculus-based probability, and linear algebra.

Chapter 2, "A Tour of the Basic Reproductive Number and the Next Generation of Researchers," then takes readers through a set of techniques to analyze SIR disease models accessible to anyone who has taken differential equations-oriented modeling course. They also describe a number of projects, such as student–teacher ratios and co-infection, that have utilized these techniques in the context of the well-known undergraduate summer research program at Arizona State University. The authors also provide a section on mentoring for faculty working with undergraduate researchers.

Next, Chapter 3, "The Effect of External Perturbations on Ecological Oscillators," introduces readers to the world of oscillatory behaviors that are present in a host of real-world applications, from seasonal outbreaks of childhood diseases to action potentials in neurons. Students with a foundation in ordinary differential equations and some experience in modeling and simulation will find this chapter to be an exciting adventure into analyzing oscillating systems.

Chapter 4, "Exploring Modeling by Programming: Insights from Numerical Experimentation," then moves on to look at protein regulation, parasite transmission in cats, and treatment costs in an influenza outbreak. They use numerical methods and experimentation with their ordinary differential equations models to conduct their research. A first course in differential equations and introductory programming would be a helpful background.

The focus of Chapter 5, "Simulating Bacterial Growth, Competition, and Resistance with Agent-Based Models and Laboratory Experiments," is centered on bacteria: their growth, competition, and adaptation, including antibiotic resistance. These ideas are modeled using individual-based models as opposed to structured population models seen in many of the other chapters of this text. The reader will note that a very brief introduction to the free software package NetLogo would be helpful before reading this chapter, but otherwise, it is rather accessible to any undergraduate student eager to learn about this method of modeling.

Chapter 6, "Agent-Based Modeling in Mathematical Biology: A Few Examples," is unique in its presentation style. This chapter is intended to guide readers through an inquiry-based independent study in agent-based models, designed for students that have no prior modeling or programming experience. However, to more quickly move through the modules, programming and graph theory are recommended. A research project in infection response to serve as an illustration of the types of projects, along with student co-author reflections ongoing through the independent study and conducting this research project, is included at the end. Also summaries of additional application areas along with invasive species, biofilms, and cellular dynamics are included.

Next, Chapter 7, "Network Structure and Dynamics of Biological Systems," introduces the basic concepts of networks in biological systems and provides the reader with a foundation on how to assess the effects of network structure on system dynamics. This chapter illustrates well the fact that networks are ubiquitous in biology, with applications to gene regulatory networks, food webs, and the spread of disease through social contact networks, just to name a few. A first course in calculus-based probability is recommended, and some familiarity with graph theory or network theory would be helpful for students interested in this area.

Chapter 8, "What Are the Chances?—Hidden Markov Models," presents the concept of Hidden Markov Models to study a wide variety of applications, including facial recognition, language translation, animal movement, and gene discovery. Students familiar with probability and programming will find this chapter readily accessible. For the student with experience in parallel computing, optional research projects are provided to give students a taste of how parallel computing can increase efficiency in Hidden Markov Modeling.

Next, in Chapter 9, "Using Neural Networks to Identify Bird Species from Birdsong Samples," the authors analyze scalograms (visualizations of sound) to identify which bird species made the sound. They use the mathematical construct of a wavelet transform on the scalograms and then feed them into the computational tool of a convolutional neural network to perform this classification. Students should expect linear algebra and multivariable calculus and introductory programming (preferably in Python) to assist them in understanding.

Finally, Chapter 10, "Using Regularized Singularities to Model Stokes Flow: A Study of Fluid Dynamics Induced by Metachronal Ciliary Waves," illustrates some current work in computational fluid dynamics as applied to pulmonary cilia (hair-like fibers within the lung). The authors also provide plenty of suggestions for additional extensions. The required prerequisites are vector calculus and programming; however, it is beneficial if the reader also has an understanding of cellular biology, physics, partial differential equations, and computational fluid dynamics.

Contents

Building New Models: Rethinking and Revising ODE Model Assumptions

Paul J. Hurtado

Abstract Ordinary Differential Equation (ODE) models are ubiquitous throughout the sciences, and form the backbone of the branch of mathematics known as applied dynamical systems. However, despite their utility and their analytical and computational tractability, modelers must make certain simplifying assumptions when modeling a system using ODEs. Relaxing (or otherwise changing) these assumptions may lead to the derivation of new ODE or non-ODE models and sometimes these new models can yield results that differ meaningfully in the context of a given application. The goal of this chapter is to explore some approaches to relaxing these ODE model assumptions to derive models which can then be analyzed in ways that parallel or build upon an existing ODE model analysis. To accomplish this, the first part of this chapter (Sect. 2) reviews some common methods for the application and analysis of ODE models. The next section (Sect. 3) explores various ways of deriving new models by modifying the assumptions of existing ODE models. This allows investigators to explore the extent to which ODE model results are robust to changes in model assumptions, and to answer questions that are better addressed using non-ODE models. The last part of this chapter suggests a few specific project ideas (Sect. 4) and encourages undergraduate researchers to share their results through presentations and publications (Sect. 5).

Suggested Prerequisites *Knowledge of topics from a standard calculus sequence, basic linear algebra, an introductory course in probability, and some familiarity with very basic differential equations models are assumed. Readers would also benefit from having some basic programming experience (in R, Python, MATLAB, or a similar high-level scientific programming language),*

Electronic supplementary material The online version of this chapter (https://doi.org/10.1007/978-3-030-33645-5_1) contains supplementary material, which is available to authorized users.

P. J. Hurtado (✉)
Department of Mathematics and Statistics, University of Nevada, Reno (UNR), Reno, NV, USA
e-mail: phurtado@unr.edu

© Springer Nature Switzerland AG 2020   1
H. Callender Highlander et al. (eds.), *An Introduction to Undergraduate Research in Computational and Mathematical Biology*, Foundations for Undergraduate Research in Mathematics, https://doi.org/10.1007/978-3-030-33645-5_1

and familiarity with calculus-based probability, stochastic processes (e.g., homogeneous and non-homogeneous Poisson processes), statistical theory of estimators, biology, or experience with mathematical modeling in an applied context.

1 Introduction

Ordinary Differential Equations (ODEs) models are ubiquitous throughout the sciences, and are a cornerstone of the branch of applied mathematics known as applied dynamical systems. Examples include the exponential and logistic growth models, Newton's laws of motion, Ohm's law, chemical kinetics models, infectious disease models for the control and treatment of diseases ranging from the common cold to HIV and Ebola, and models for large scale ecosystem dynamics.

The utility of ODEs as mathematical models is partly a reflection of their analytical and computational tractability, and also the relative ease of formulating ODE model equations from assumptions about the system being modeled. Modelers also have access to a well-developed set of mathematical approaches and computational tools for analyzing ODEs [4, 10, 39, 41, 45, 50, 55, 56, 84, 94–96, 119, 126, 127, 149, 155, 160, 165]. If there is a practically important or scientifically interesting real-world system that changes over time, odds are very good that someone has modeled it with one or more ODEs. However, despite their utility, modelers must also make certain simplifying assumptions when using ODEs to model a real-world system (e.g., that systems are well mixed, individuals behave identically, etc.), and sometimes alternative assumptions can lead to different outcomes that are very relevant in the context of the motivating application.

The goal of this chapter is to explore some approaches to relaxing these ODE model assumptions to come up with new models that can then be used in ways that parallel or build upon an existing ODE model application. This allows investigators to explore the extent to which results obtained from the analysis of an ODE model are robust to changes in certain model assumptions, and/or to address a broader range of questions, some of which might be better answered through the analysis of a similar ODE or non-ODE model (e.g., using a stochastic model of population growth and decline to assess the distribution of time to extinction under different population management scenarios [105]). Undergraduate students reading this chapter are encouraged to initially focus on understanding the broader context of the technical details presented in this chapter, and then later attempt to more fully understand specific details, with the help of a research mentor or advisor, as they work on a project.

A great recipe for an undergraduate research project is to (a) find an interesting paper in a reputable, peer-reviewed journal that includes some analysis of an ODE model, (b) repeat one or more analyses to confirm a specific published result, and then (c) answer similar questions via the analysis of a new model derived by modifying one or more assumptions of that original ODE model. Mathematically, these projects often go one of two ways: Changing ODE model assumptions will sometimes result in a new set of ODEs, while changing other assumptions can lead to the formulation of a new non-ODE model, e.g., a model that is defined using partial differential equations (PDEs), stochastic differential equations (SDEs), or discrete-time difference equations.

This chapter is divided into two main sections. First, we review some common approaches to using ODEs models in Sect. 2, including methods of mathematical, computational, and statistical analysis. Second, in Sect. 3, we address some of the different ways that ODE model assumptions can be modified to derive new ODE or non-ODE models, and additional project suggestions are presented in Sect. 4.

Examples are used throughout to illustrate the process of deriving a new model, and some guidance is provided for analyzing or simulating these derived models. Some examples include computer code written in the R programming language [137]. R is a popular, free computing platform that is widely used, easy to install, and well supported in terms of free resources to get started programming in R. For readers that have no prior programming experience, or want to learn R, relevant resources for getting started with R are provided in the appendix. To help readers who are proficient in another language (e.g., Python, MATLAB, or Mathematica) but who are unfamiliar with R, the R code in this chapter is annotated with comments so that it can be more easily ported to other languages.

Lastly, many sections below include *exercises* that the readers should work through as they read the sections. These are intended to help solidify the topics being discussed, and should not require consulting outside resources. The *challenge problems* are more involved, and are intended to push the reader to dig deeper into the topic, or perhaps inspire a research project, by exploring additional resources including software tools and other peer-reviewed publications.

2 Techniques for Analyzing ODE Models

Before we explore some of the ways in which we can alter ODE model assumptions to obtain new (sometimes non-ODE) models, it is helpful to consider the kinds of questions investigators address using ODE models, and some related methods of ODE model analysis. This is done in Sects. 2.2–2.7. But first, for context, it helps to review how ODEs can often (but not always, e.g., see Newton's laws of motion) be viewed as *mean field* models of a stochastic process, as this is often the natural context in which to critically evaluate and revise the assumptions of biological models.

2.1 ODEs as Mean Field Models

First order systems of ODEs are systems of equations that can be written as

$$\frac{d\mathbf{x}}{dt} = f(t, \mathbf{x}, \theta), \quad \mathbf{x}(0) = \mathbf{x}_0, \tag{1}$$

where the *state variable* vector $\mathbf{x}(t) = (x_1(t), \ldots, x_n(t))^\mathsf{T} \in \mathbb{R}^n$, the *initial condition* vector $\mathbf{x}_0 \in \mathbb{R}^n$, the parameter vector $\theta = (\theta_1, \ldots, \theta_p) \in \mathbb{R}^p$, and the function[1] $f : \mathbb{R}^n \mapsto \mathbb{R}^n$. We call such systems *autonomous* when the right hand side f only depends on time through its dependence on the state variables, i.e., when the right side of the system of equations can be written $f(\mathbf{x}, \theta)$. While these are deterministic equations, in the context of mathematical models we often think of ODEs as *mean field* approximations of an underlying stochastic model. More specifically, biological and other physical systems made up of large numbers of discrete individuals can often be modeled as continuous-time, discrete-state stochastic models (e.g., where the model tracks $X(t)$, the integer number of individuals alive at time t, which changes stochastically to reflect births and deaths in the population). Thus, mean field models attempt to capture a particular kind of average behavior of a stochastic model.

As detailed below, mean field ODEs that approximate these stochastic models can often be derived from a stochastic model in two main steps: First, the continuous-time model is approximated with a discrete-time model (step size Δt), and the stochastic rule that describes how state variables change over one time step is replaced with a deterministic rule obtained by simply averaging over that stochastic model. Second, letting $\Delta t \to 0$, the discrete-time model converges to an ODE (e.g., see [5, 20, 92]) and the similar approach to derive stochastic differential equations (SDEs) in [3]).

For example, consider a simple exponential growth model of the growing number of bacteria, $x(t)$, with *per capita* growth rate r, in a Petri dish that was inoculated with $x_0 > 0$ bacteria at time $t = 0$, given by

$$\frac{dx(t)}{dt} = r\, x(t), \quad x(0) = x_0. \tag{2}$$

Note that this is a continuous-state (i.e., $x(t) \in \mathbb{R}$) deterministic model, despite the fact that we envision the actual biological process as a integer-valued number of bacterial cells dividing and dying stochastically over time. *This dual perspective is very important in that it provides a foundation for thinking about model assumptions, and in some cases a starting point for deriving new models.*

To clarify this link between stochastic and deterministic approaches, consider the following derivation of Eq. (2) from the following stochastic model[2] of bacterial population growth. Let $X(t)$ be the number of cells at time t. Assume that, over a sufficiently small time period of duration $\Delta t > 0$, the probability p of each cell dividing to give rise to a new cell is assumed to be identical for all cells, independent of time, and given by

---

[1] This function is often referred to as the *right hand side*, abbreviated RHS, of the ODE.

[2] It is worth mentioning here that multiple different stochastic models can yield the same mean field ODE model.

$$p = r\,\Delta t, \quad r > 0. \tag{3}$$

A *stochastic, discrete-time map* can be used to model this process, as follows. If there are $X(t)$ cells at time t, then Δt time units later, at time $t + \Delta t$, there are

$$X(t + \Delta t) = X(t) + U(X(t), r\,\Delta t) \tag{4}$$

cells present, where $U(n, p)$ is a binomially distributed random variable where for each of the $n = X(t)$ cells there is a probability $p = r\,\Delta t$ that a given cell will give rise to a new cell by time $t + \Delta t$.

To derive the ODE model of exponential growth Eq. (2) as an approximation of the above stochastic model, we follow the steps described above: The discrete-time stochastic model is approximated by a *mean field*[3] deterministic model, and then we let the time step $\Delta t \to 0$ to yield the desired ODE.

In this first step, the number of new cells U is replaced by the expected value $E(U)$ as follows. Assume that the number of cells $X(t)$ is sufficiently large, and that we use an appropriately small choice of Δt so that U is well approximated by a binomial distribution. The expected value of this binomial random variable, where $n = X(t)$ trials and probability of success $p = r\,\Delta t$, is the product $n\,p = X(t)\,r\,\Delta t$, thus we make the approximation $U(X(t), r\,\Delta t) \approx E(U(n = X(t), p = r\,\Delta t)) = X(t)\,r\,\Delta t$. This yields the *mean field* discrete-time model

$$x(t + \Delta t) = x(t) + r\,x(t)\Delta t. \tag{5}$$

Note that we distinguish between these new state variable quantities $x(t)$ (which are $\mathbb{R}$-valued expectations) and their integer-valued counterparts $X(t)$, even though we interpret both as representing the number of bacteria present at time t. Lastly, rearranging this deterministic equation, and taking the limit as $\Delta t \to 0$, yields

$$\frac{x(t + \Delta t) - x(t)}{\Delta t} \quad \to \quad \frac{dx(t)}{dt} = r\,x(t). \tag{6}$$

There are a few remarks on the above derivation that are worth mentioning. First, a more careful derivation, that explicitly tracks error terms in the above approximations, would yield the same mean field result, but those details were omitted for brevity and clarity (for a more detailed example, see [92]). Second, mean field models locally average how state variable change in time, and do not average

---

[3]The deterministic equation derived in this first step is called a *mean field* model because, for a given point $X(t)$ in state space, we take the distribution of possible steps the system might take on the next time step, and define a new model by replacing that random quantity with its mean—in this case, the mean of the binomial random variable $U(n = X(t), p = r\,\Delta t)$. One could visualize this on a grid of state space values with vectors pointing from $x(t)$ to $x(t + \Delta t)$, forming a vector field.

over full trajectories of a stochastic model. Thus, mean field model solutions can differ markedly from the average over multiple stochastic model trajectories. Third, for readers who are familiar with Poisson processes [145], this is a very useful context for thinking about these underlying stochastic processes associated with a mean field ODE. In fact, the approach above (i.e., discretizing time to derive a mean field model) very much parallels the typical approach to deriving properties of homogeneous and non-homogeneous Poisson processes by discretizing time then taking the continuum limit (e.g., see the book [40], the standard derivation of the binomial approximation of the Poisson distribution [106], and [92] for a some relevant aspects of Poisson processes). In the context of Poisson processes, what has been assumed in the stochastic model above can be rephrased as follows: each individual cell replicates itself according to a Poisson process with rate r where the events generated by the Poisson process mark the times at which replication events occur.

Exercise 1 It is often appropriate to make simplifying assumptions when formulating mathematical models of real-world phenomena. The art of modeling is striking a balance between simplicity, mathematical tractability, and capturing the important mechanism that drive real-world processes. Often, these simplifying assumptions are not explicitly stated as an exhaustive list as part of the model description, for example, above we have assumed all cells are basically identical in terms of their replication rates. What other *implicit* assumptions have been made in the above example? *Hint: What was assumed about resource limitation, cell mortality, whether or not cells influence one another in terms of death and cell division, and so on?*

Challenge Problem 1 Formulate a similar stochastic model to Eq. (4), but replace replication rate symbol r with b and include a mortality rate d. Show this yields an identical mean field model to Eq. (2) where $r = b - d$. *Hint: Show $E(\Delta X) \approx X b \Delta t - X d \Delta t$.*

This ability to reframe the stochastic model assumptions in terms of Poisson processes is important because the ODE model Eq. (2) reflects these underlying exponential distributions through the per capita growth rate r. Importantly, this pattern generalizes: when the appearance or loss of individuals in a given state follows *i.i.d.* Poisson process first event time distributions with rate $r(t)$ (i.e., exponential distributions if r is constant) then the per capita loss or growth rates in corresponding mean field ODEs are the coefficients $-r(t)$ or $r(t)$, respectively. This is illustrated by the following theorem, which is stated here without a proof (the proof follows from more general theorems in [92]).

Theorem 1 *Consider a continuous-time, stochastic, state transition model in which individuals spend independent and identically distributed amounts of time in a given state (X) and then transition to subsequent state Y. Suppose that individual dwell times are either exponentially distributed with constant rate r, or (more generally) follow the first event time distribution under a non-homogeneous Poisson process with rate $r(t)$. Further assume individuals enter X at rate $\Lambda_X(t)$ and the dwell time in state Y follows the first event time under a (non-homogeneous) Poisson process* *with rate $\mu(t)$. Then, if $x(t)$ and $y(t)$ are the expected number of individuals in each state X and Y, respectively, the mean field ODE model for this case is*

$$\frac{dx}{dt} = \Lambda_X(t) - r(t)\,x(t) \tag{7a}$$

$$\frac{dy}{dt} = r(t)\,x(t) - \mu(t)\,y(t). \tag{7b}$$

To further illustrate how these underlying Poisson process rates are reflected in ODE model terms, consider the well-known SIR model of infectious disease transmission [98] (for more on SIR-type epidemic models, see [4, 7, 22, 23, 43, 98]). In this model, a closed population of size N is assumed to be made up of individuals who are either *susceptible* to an infectious disease, currently *infected* (and *infectious*), or *recovered* from infection and immune to reinfection by the pathogen. Accordingly, we let state variables $S(t)$, $I(t)$, and $R(t)$ correspond to the number of susceptible, infectious, and recovered individuals at time t. The SIR model is a simple ODE model of how these numbers of individuals in each state change over time, and is given by the equations

$$\frac{d}{dt}S(t) = -\lambda(t)\,S(t) \tag{8a}$$

$$\frac{d}{dt}I(t) = \lambda(t)\,S(t) - \gamma\,I(t) \tag{8b}$$

$$\frac{d}{dt}R(t) = \gamma\,I(t) \tag{8c}$$

where $\lambda(t) = \beta\,I(t)$ is the per capita infection rate (also called the *force of infection* [7]), and γ is the per capita recovery rate. Note that, by the assumed dependence of $\lambda(t)$ on $I(t)$ this is a nonlinear system of ODEs. This model can also be viewed as the mean field model for an underlying stochastic state transition model of a large but finite number of individuals that each transition from state S to I following "event times" (the time spent in a given state; also called *dwell times* or *residence times*) that obey a non-homogeneous Poisson process first event time distribution with rate $\lambda(t)$. Each individual similarly transitions from I to R following a homogeneous Poisson process first event time distributions with rate γ, hence the time an individual spends in state I is exponentially distributed with mean $1/\gamma$ (see [98] for a derivation, and see [9, 13], and references therein for examples of the convergence of stochastic models to mean field ODEs).

Altering the assumptions of ODE models like the SIR model above is often best done when the modeler has adopted this mean field perspective. For example, if we would like to relax the assumption that individuals immediately become infectious and instead assume that they experience a *latent period* where they are infected but not yet infectious for a time period that is exponentially distributed with mean $1/\nu$, then we can introduce an *exposed* class and let $E(t)$ be the number in this class at time t. This yields the well-known SEIR model:

$$\frac{d}{dt} S(t) = -\lambda(t) \, S(t) \tag{9a}$$

$$\frac{d}{dt} E(t) = \lambda(t) \, S(t) - \nu \, E(t) \tag{9b}$$

$$\frac{d}{dt} I(t) = \nu \, E(t)) - \gamma \, I(t) \tag{9c}$$

$$\frac{d}{dt} R(t) = \gamma \, I(t). \tag{9d}$$

Exercise 2 Under the SIR model, the time spent infected and infectious is the same, and follows an exponential distribution with rate γ (i.e., with mean duration $1/\gamma$). This is not the case under the SEIR model. Describe and compare the mean time infected (i.e., between entering the exposed state and entering the recovered state) for both the SIR and SEIR models, in terms of the mean time spent in the exposed state and the mean time spent in the infectious state.

In summary, systems of ODEs that model individuals transitioning among different states are often best thought of as a mean field model for some underlying (often unspecified) continuous-time stochastic process. While in general it is true that multiple different stochastic models can yield the same mean field ODE model, it is still often very useful to associate a given ODE with the continuous-time stochastic state transition model defined in terms of Poisson processes where the Poisson process rates are given by the overall rates (i.e., ODE model terms) or the per capita rates (i.e., coefficients on those terms) that appear in the right hand side of the ODE model. We will explore one way of constructing such a model from a system of ODEs using the Stochastic Simulation Algorithm in Sect. 3.3.1, but throughout the remainder of this chapter we will often draw upon the intuition associated with this underlying stochastic model implied by a given mean field ODE model.

With this perspective in mind, we next review some standard ways of using ODE models in applications.

2.2   Equilibrium Stability Analysis

Many questions about dynamical systems can be answered by asking *What is the eventual state of* $\mathbf{x}(t)$ *as* $t \to \infty$? For example, Does the disease epidemic end with a local extinction of the pathogen, or does the disease become *endemic* and persist by establishing itself at low levels in the population? Will new harvest regulations in a fishery maintain a sustainable population of fish, or will the population be pushed to extinction? Sometimes knowing where a system is going helps inform the path it takes to get there, so these asymptotic results as $t \to \infty$ can also be useful for understanding short-term dynamics. Since many questions can be addressed by mathematical and computational analyses that reveal the long-term behavior of a system, this is often a focus of an introductory course in mathematical biology [4, 50], and the main focus of courses in applied dynamical systems [10, 84, 160] and bifurcation theory [104, 165]. Here we review the basics of equilibrium stability analysis, and in the next section we discuss analyses that build on these concepts to further illuminate the behavior of ODE models in a dynamical systems context.

Let us revisit the simple exponential growth or decay model Eq. (2) where we assume a positive initial value, i.e., $x(0) = x_0 > 0$, and that $r \in \mathbb{R}$. Solutions to this ODE have the form

$$x(t) = x_0 \, e^{r \, t} \tag{10}$$

thus we can see that the sign of r determines whether solutions $x(t)$ diverge towards ∞ ($r > 0$) or converge towards zero ($r < 0$). Here we focus on solutions like the latter, where the long-term behavior of the system is to settle on a steady-state value, more commonly referred to as an *equilibrium point*.

Formally, an equilibrium point of a model $d\mathbf{x}/dt = f(\mathbf{x})$ is a state $\mathbf{x}_*$ such that $f(\mathbf{x}_*) = 0$. Starting at such points yields solutions that remain there for all time, which we refer to as *equilibrium solutions*.

An *equilibrium stability analysis* aims to accomplish two goals: First, to determine how many different *equilibrium points* exist and find expressions for those state values as a function of the model parameters, and second, to determine the *(local) asymptotic stability* of those equilibria. Identifying equilibria is straightforward: simply set the system of equations equal to zero, and solve for the state variable vector $\mathbf{x}_*$ that satisfies $f(\mathbf{x}_*) = 0$. The second step in a stability analyses requires us to first clarify what is meant by an equilibrium point being *asymptotically stable*.

In short, we call an equilibrium point *stable* if small perturbations away from that point in state space yield trajectories that converge back towards said equilibrium point, and *unstable* if small perturbations yield trajectories that diverge away from the equilibrium point. More precisely, we can use the formal definition of equilibrium stability given in [160]: If, for an equilibrium point $\mathbf{x}_*$, there is a $\delta > 0$ such that any trajectory $\mathbf{x}(t)$ that starts within a distance $< \delta$ from $\mathbf{x}_*$ eventually converges to $\mathbf{x}_*$, then we call $\mathbf{x}_*$ *attracting*. If for $\epsilon > 0$ there is a $\delta_\epsilon > 0$ such that whenever a trajectory $\mathbf{x}(t)$ starts closer than δ_ϵ to $\mathbf{x}_*$ the trajectory remains within a distance ϵ of $\mathbf{x}_*$, then we call $\mathbf{x}_*$ *Lyapunov stable*. If $\mathbf{x}_*$ is both attracting and Lyapunov stable, we say $\mathbf{x}_*$ is (locally) *asymptotically stable* and this is what is meant below when an equilibrium point $\mathbf{x}_*$ is called *stable*, i.e., for sufficiently small perturbations away from that point, trajectories $\mathbf{x}(t)$ stay within a small neighborhood of that point and asymptotically converge towards $\mathbf{x}_*$ (i.e., $\mathbf{x}(t) \to \mathbf{x}_*$) as $t \to \infty$.

In practice, determining the stability of an equilibrium point is fairly straight-forward; however, having an intuitive understanding of why the standard approach works requires some explanation. Because the *vector field*[4] $f(\mathbf{x})$ is usually smooth (i.e., because $f(\mathbf{x})$ is differentiable), zooming in very close to a given point (e.g., an equilibrium point) the vector field is increasingly well approximated by a linear vector field. Mathematically, this is a straightforward consequence of Taylor's Theorem: Recall that Taylor's Theorem for functions of one variable [112] states that if a function is differentiable at a then for x values in a neighborhood of a the function can be approximated by

$$f(x) \approx f(a) + f'(a)(x - a) + \ldots + \frac{1}{k!} f^{(k)}(a)(x - a)^k \tag{11}$$

where the difference between $f(x)$ and the approximation above vanishes as $x \to a$. This gives the linear approximation

$$f(x) \approx f(a) + f'(a)(x - a) \tag{12}$$

for values of x in a small neighborhood of a. Similarly, Taylor's Theorem for multivariate functions[5] $f : \mathbb{R}^n \to \mathbb{R}^n$ gives the linear approximation

$$f(\mathbf{x}) \approx f(\mathbf{a}) + \mathbf{J}(\mathbf{x} - \mathbf{a}) \tag{13}$$

where $\mathbf{J}$ is the $n \times n$ *Jacobian* matrix whose entries are the partial derivatives of f with respect to each component of $\mathbf{x}$ evaluated at $\mathbf{x} = \mathbf{a}$, i.e., if $f(\mathbf{x}) = (f_1(\mathbf{x}), \ldots, f_n(\mathbf{x}))^\mathsf{T}$ then

$$J_{ij} = \frac{\partial f_i}{\partial x_j}\bigg|_{\mathbf{x}=\mathbf{a}} \tag{14}$$

Since $f(\mathbf{x}_*) = 0$ (i.e., if $\mathbf{x}_*$ is an equilibrium point), Taylor's Theorem implies that the linear approximation near $\mathbf{x}_*$ is

$$\frac{d\mathbf{x}}{dt} \approx \mathbf{J}(\mathbf{x} - \mathbf{x}_*). \tag{15}$$

Moreover, if we let $\mathbf{u} = \mathbf{x} - \mathbf{x}_*$ be the perturbation of $\mathbf{x}$ away from $\mathbf{x}_*$, then

$$\frac{d\mathbf{u}}{dt} \approx \mathbf{J}\mathbf{u} \tag{16}$$

and thus the deviations ($\mathbf{u}$) of trajectories very near $\mathbf{x}_*$ are approximately solutions to this linear system of first order differential equations.

---

[4]For each point in state space, one can think of the derivative $\frac{d\mathbf{x}}{dt}$ as a vector pointing in a direction that is tangent to the trajectory passing through that point.

[5]See upper level dynamical systems texts and online resources for the Taylor expansion of multivariate functions.

The importance of this linear approximation in the context of determining the stability of an equilibrium point $\mathbf{x}_*$ is that the eigenvalues [159] of matrix $\mathbf{J}$ can tell us whether or not $\mathbf{x}_*$ is stable (e.g., see linear stability analysis sections in [45, 84, 160, 165] or related texts on nonlinear dynamics). Why? Small perturbations off of the equilibrium point will yield trajectories that are approximately $\mathbf{x}(t) \approx \mathbf{u}(t) + \mathbf{x}_*$ where $\mathbf{u}(t)$ is a solution to the above linearization Eq. (16). Importantly, solutions to linear ODEs can be written explicitly in terms of the eigenvalues and eigenvectors of the coefficient matrix $\mathbf{J}$ as shown for the 2-dimensional case in [45] and for higher dimensions in [84]. Specifically, these solutions look like sums of exponential growth and decay terms (sometimes multiplied by sine and cosine terms if $\mathbf{J}$ has complex eigenvalues [84]) along the different directions determined by the eigenvectors of $\mathbf{J}$, where the growth and decay rates along these eigenvectors are the *real parts of the corresponding eigenvalues of* $\mathbf{J}$. Thus, if the eigenvalues of $\mathbf{J}$ all have negative real parts, small perturbations off of $\mathbf{x}_*$ will exponentially decay back to that point, and if any eigenvalue has a positive real part, trajectories that start as perturbations off of $\mathbf{x}_*$ will diverge away from that point. The following stability theorem formally summarizes this intuition (for further details, see [84, 165] or similar texts):

Theorem 2 (Equilibrium Stability Criteria) *Suppose* $\mathbf{x}_*$ *is an equilibrium point of* $d\mathbf{x}/dt = f(\mathbf{x})$ *and* $\mathbf{J}$ *is the Jacobian evaluated at* $\mathbf{x}_*$. *Further assume the real parts of the eigenvalues of* $\mathbf{J}$ *are all non-zero (i.e.,* $Re(\lambda_i) \neq 0$ *for* $i = 1, \ldots, n$). *Then the equilibrium point is stable if all* $Re(\lambda_i) < 0$, *and unstable if any* $Re(\lambda_i) > 0$. *If any eigenvalues have* $Re(\lambda_i) = 0$ *the stability of* $\mathbf{x}_*$ *is determined by higher order terms in the Taylor expansion of* f *about* $\mathbf{x} = \mathbf{x}_*$. *If the system is one dimensional (i.e., if* $f : \mathbb{R} \to \mathbb{R}$) *then* x_* *is stable if* $f'(x_*)$ *is negative (and unstable if* $f'(x_*)$ *is positive).*

In computational applications, where it suffices to check stability of an equilibrium point for a specific set of parameter values, the above criteria are very practical. It is usually straightforward to compute the Jacobian and use standard approaches to find its eigenvalues (e.g., using the eigen() function in R). However, when looking to obtain analytical stability conditions, it is often cumbersome to find general expressions for eigenvalues and to find conditions under which they have a negative real part. This is especially true for higher dimensional nonlinear systems.

Importantly, we do not need to find these eigenvalues explicitly to determine if an equilibrium point is stable. The Routh–Hurwitz criteria (Theorem 3 below) provide a sometimes simpler way of checking whether or not all eigenvalues of Jacobian $\mathbf{J}$, evaluated at equilibrium point $\mathbf{x}_*$, have negative real part by checking a set of criteria that involve the coefficients of the Jacobian's characteristic polynomial[6] [159],

---

[6]This definition is used instead of the more common $p(\lambda) = \det(\mathbf{J} - \lambda\mathbf{I})$ since it ensures that the roots are the eigenvalues of $\mathbf{J}$ and that the polynomial is *monic, i.e., has a leading coefficient of 1 on the* λ^n *term.*

$$p(\lambda) = \det(\lambda \mathbf{I} - \mathbf{J}). \tag{17}$$

In short, since the eigenvalues are the roots of the characteristic polynomial of the Jacobian, satisfying the Routh–Hurwitz criteria implies that each of those eigenvalues has negative real part, and thus, the equilibrium point is stable. This is often the preferred approach when looking for general stability criteria (e.g., an inequality describing parameter value relationships that yield a stable equilibrium point) because in practice it typically yields meaningful stability criteria in fewer intermediate steps compared to finding the eigenvalues, and their real parts, directly.

The following statement of the Routh–Hurwitz criteria was adapted from Ch. 4 in [4][7] and pg 233–234 of [45] (see also [117]).

Theorem 3 (Routh–Hurwitz Criteria) *Consider the monic polynomial*

$$p(\lambda) = \lambda^n + a_1 \lambda^{n-1} + \cdots + a_{n-1} \lambda + a_n \tag{18}$$

with real coefficients a_i. All roots of Eq. (18) have negative real part if and only if all n principal minors of H_n (where $a_j = 0$ for $j > n$) have positive determinants.

$$H_n = \begin{bmatrix} a_1 & 1 & 0 & 0 & \cdots & 0 \\ a_3 & a_2 & a_1 & 1 & \cdots & 0 \\ a_5 & a_4 & a_3 & a_2 & \cdots & 0 \\ \vdots & \vdots & \vdots & \vdots & \cdots & \vdots \\ a_{2n-1} & a_{2n-2} & a_{2n-3} & a_{2n-4} & \cdots & a_n \end{bmatrix} \tag{19}$$

The principal minors H_k, $k = 1, \ldots, n$, are

$$H_1 = a_1, \quad H_2 = \begin{bmatrix} a_1 & 1 \\ a_3 & a_2 \end{bmatrix}, \quad \ldots \tag{20}$$

Equivalently, the real parts of the roots of the characteristic polynomial (i.e., the real parts of each eigenvalue) all have negative real part if and only if the following hold (similar criteria can be derived using the matrices above for $n \geq 6$):

$n = 2$: $a_2 > 0$, *and* $a_1 > 0$.

$n = 3$: $a_1 > 0$, $a_3 > 0$ *and* $a_1 a_2 > a_3$.

$n = 4$: $a_1 > 0$, $a_3 > 0$, $a_4 > 0$, *and* $a_1 a_2 a_3 > a_3^2 + a_1^2 a_4$.

$n = 5$: $a_i > 0$ $(i = 1, \ldots, 5)$, $a_1 a_5 + a_1 a_2 a_3 > a_3^2 + a_1^2 a_4$ *and*

$$(a_1 a_4 - a_5)(a_1 a_5 + a_1 a_2 a_3 - a_3^2 - a_1^2 a_4) > a_5 (a_1 a_2 - a_3)^2.$$

---

[7]Here we give the correct $n = 5$ case, which is missing a few terms in [4].

Challenge Problem 2 Look up the analogues of Theorems 2 and 3 for determining the stability of *fixed points* for discrete maps.

Here we have only briefly touched upon equilibrium stability analysis. Interested readers are encouraged to consult standard applied dynamical systems texts, such as [4, 45, 160], for additional details and related topics like the asymptotic stability analysis of equilibria (also called *fixed points*) in discrete-time systems.

The following exercises illustrate the application of Theorems 2 and 3.

Exercise 3 The well-known logistic equation, where $r, K > 0$ and $x \geq 0$, is given by

$$\frac{dx}{dt} = r x \left(1 - \frac{x}{K} \right) \tag{21}$$

and has equilibria $x = 0$ and $x = K$. Use Theorem 2 to show that $x = 0$ is unstable and $x = K$ is stable. *Hint: What is the derivative of $f(x)$ with respect to x?*

Exercise 4 The following extension of the logistic equation includes a strong Allee effect: below a certain threshold ($\alpha > 0$) the population declines towards 0.

$$\frac{dx}{dt} = r x \left(1 - \frac{x}{K} \right) \left(\frac{x}{\alpha} - 1 \right) \tag{22}$$

Find all three equilibria of this model, and determine their stability.

The next few exercises deal with the Rosenzweig–MacArthur predator–prey model (for background on this and related models, see [118, 125])

$$\frac{dN}{dt} = r N \left(1 - \frac{N}{K} \right) - \frac{a P}{k + N} N \tag{23a}$$

$$\frac{dP}{dt} = \chi \frac{a N}{k + N} P - \mu P \tag{23b}$$

where N and P are the prey and predator densities, respectively, $r > 0$ and $K > 0$ are the logistic growth rate and carrying capacity for the prey, $a > 0$ is the maximum per-predator rate of predation, k is the half-saturation value of the prey population in the Holling type-2 predation rate term (when $N = k$ the per-predator predation rate is $a/2$), $\chi > 0$ is a conversion factor between the number of prey consumed and predators born, and $\mu > 0$ is the predator mortality rate corresponding to having assumed exponentially distributed predator lifetimes with a mean lifespan $1/\mu$.

Exercise 5 Find equilibrium points of the Rosenzweig–MacArthur model above. Note that it may be helpful to write these equilibrium values where one state variable value is given as a function of parameter values, and the other is written as a function of parameter values as well as the equilibrium value of the other state variable. *Hint:*

There are three non-negative equilibria: one with zero organisms, one with only prey, and one with both predators and prey coexisting.

Exercise 6 Find the Jacobian for this system (recall Eq. (14)), and evaluate it at each equilibrium point. *Hint: The Jacobian for the case with zero organisms is*

$$\begin{bmatrix} r & 0 \\ 0 & -\mu \end{bmatrix}. \tag{24}$$

Challenge Problem 3 Find the characteristic polynomial of each Jacobian above and apply the Routh–Hurwitz criteria to determine their stability. *Hint: For the case with zero organisms, the characteristic polynomial is*

$$p(\lambda) = \lambda^2 + (\mu - r)\lambda - r\,\mu. \tag{25}$$

Challenge Problem 4 Repeat the above stability analysis for the Rosenzweig–MacArthur model using a *computer algebra system* like the free software wxMaxima [163] (an improved user interface for Maxima [148]), Sage, or commercial software such as Maple or Mathematica.

Challenge Problem 5 What are some other kinds of *attractors* besides equilibrium points (e.g., limit cycles) and how do we determine their stability?

2.3 Bifurcation Analysis

An equilibrium stability analysis often reveals parameter space *thresholds* that mark important qualitative transitions in the long-term behavior of solutions. These typically manifest as inequalities involving model parameters and equilibrium values.

For example, in our simple exponential decay equation, the coefficient $a = 0$ divides *parameter space* into cases of exponential decay ($a < 0$) and those leading to exponential growth ($a > 0$). In infectious disease models, one can often find an expression for the *basic reproduction number* of the pathogen (commonly denoted as $\mathcal{R}_0$ and interpreted as the expected number of new infections per infectious individual over the average duration of infectiousness when the number of susceptible individuals is at the disease-free equilibrium). For example, for the SIR model given by Eqs. (8) the number susceptible at the disease-free steady state is N, hence the rate of new infections per infectious individual, per unit time, is βN and the mean duration of infectiousness is $1/\gamma$. Thus[8] (multiplying rate times time) we have $\mathcal{R}_0 = \beta N/\gamma$. It is commonly the case that no epidemic will occur if $\mathcal{R}_0 < 1$

---

[8]These basic reproduction numbers are best derived from stability criteria, which often require some rearrangement using a similar interpretation to arrive at the proper form of $\mathcal{R}_0$.

(i.e., the disease-free equilibrium is stable) but an epidemic will occur if $\mathscr{R}_0 > 1$ (i.e., when infectious individuals more than replace themselves, on average). For more on $\mathscr{R}_0$ calculations and its role in epidemic thresholds, see [22, 23, 35, 43].

In addition to changes in the stability of equilibria (or in some cases, their appearance or disappearance) threshold phenomena also arise in non-equilibrium dynamics. For example, the transition between steady-state and oscillatory behavior is quite common in applications, e.g., it is exhibited by the Rosenzweig–MacArthur model mentioned above in which there is a transition from predator and prey coexisting at equilibrium to asymptotically periodic solutions where the predator and prey populations oscillate together (this example will be discussed in more detail below). In some higher dimensional models, these can include much more complex dynamics, including transitions to chaos.

Bifurcation theory deals with the mathematical study of these transitions [73, 84, 104, 160, 165]. Formally, *bifurcations* are the thresholds in parameter space that indicate where there are changes in the topological structure of the model behavior (i.e., in the vector field defined by f). For example, we can consider the set of all parameters for which predators and prey coexist at equilibrium as *topologically equivalent*,[9] and topologically distinct from the case where the predators are unable to live on the prey and go extinct. In essence, bifurcation thresholds partition parameter space into regions that yield the same qualitative behavior, making them an excellent tool for studying model dynamics in an applied setting.

The goals of a *bifurcation analysis* are to characterize where bifurcations occur in parameter space, and to identify the types of bifurcation(s) involved. Importantly, identifying the different types of bifurcations informs us of the long-term behavior of solutions for parameters on either side of these bifurcation thresholds. These boundaries are, fortunately, usually very well defined, e.g., they often involve an equilibrium point where one of the real parts of its eigenvalues passes through zero (recall that our stability criteria above say that an equilibrium point is stable if the eigenvalues of the corresponding Jacobian have negative real part). More complex bifurcations involving other types of attractors (e.g., limit cycles or chaotic attractors) can be similarly characterized by extending the above notions of equilibria and their stability to these other sets[10] of points, e.g., limit cycles, and their stability. Importantly, there are a small number of named bifurcations that we often see arise in applications, e.g., *Hopf* bifurcations (where an isolated equilibrium point changes stability and gives rise to a *limit cycle*—an isolated closed curve in state space that is a solution of the model—i.e., a change from steady-state to oscillatory dynamics), *transcritical* bifurcations (two equilibrium points pass through one another and exchange stability), *saddle-node* bifurcations (where

---

[9]For readers unfamiliar with topological equivalence, this essentially means that the gross qualitative features of solutions are the same across *equivalent* parameter sets, e.g., the number of equilibrium points and their stability will be the same, even though their exact numerical values may change. See [160, pg. 156], or more advanced treatments of topological equivalence in the texts mentioned above.

[10]See the definition of an *invariant set*, and different types of attractors, in texts like [73, 104].

two equilibria collide and vanish). For a more detailed treatment of the different bifurcations that commonly arise in applications, consult standard texts such as [73, 84, 104, 160, 165].

In practice, the first step of a bifurcation analysis is to conduct an equilibrium stability analysis. From there, stability thresholds can be identified and these can sometimes be useful for identifying specific types of bifurcations. This will often reveal equilibrium bifurcations like saddle-node and transcritical bifurcations, but this can also reveal bifurcations involving non-equilibrium attractors like Hopf bifurcations that involve limit cycles. For example, there is a very useful result [74, 75] that states that, when using the Routh–Hurwitz criteria to determine equilibrium stability, the loss of stability via the failure of the criterion listed last, for each n value considered at the end of Theorem 3, coincides with the loss of equilibrium stability via a Hopf bifurcation.

Computational methods also exist for conducting bifurcation analyses (e.g., see [41, 55, 56, 155]), and can be quite useful for exploring model dynamics for parameter values near a specific parameter set. A less sophisticated approach is to simply use brute force to simulate solutions over a range of parameter values, and analyze those simulation results to infer bifurcations (see Sect. 2.6 below for more on simulating solutions to ODEs on the computer). For example, Fig. 1 shows simulated asymptotic states of the Rosenzweig–MacArthur model over a range of maximum predation rates, a.

The preferred approach, however, is to use computational methods that are specifically designed to find bifurcations in parameter space. The reason? First, the above simulation-based approach will typically miss part of the picture in situations where different initial conditions lead to different attractors (i.e., cases of *bistability* or *multistability*), or in cases where a bifurcation involves the creation or disappearance of either an unstable equilibrium point or unstable limit cycle (which can be practically impossible to find by running simulations unless you know ahead of time where or how to look for them). Simulation based approaches also rely on the investigator to infer what bifurcations might or might not be involved, whereas software designed for bifurcation analysis includes rigorous mathematical criteria that are automatically checked to notify the investigator when certain types of bifurcations are found. Two popular software packages for conducting bifurcation analyses are MatCont [41], which is freely available software that runs in Matlab, and the software AUTO which is available in different forms including through the XPPAUT simulation software [55, 56] which includes an interface for AUTO. An introduction to using such software is beyond the scope of this book; however, various tutorials and examples can easily be found online (see also the list of software available on the SIAM Dynamical Systems activity group website [155]: https://dsweb.siam.org/Software).

Challenge Problem 6 Improve upon the bifurcation diagram in Fig. 1 with an equilibrium stability analysis and by using bifurcation continuation software such as MatCont or XPPAUT to do a computational bifurcation analysis of the model given by Eqs. (23). Use the parameter values given in the figure caption. *Hint: Are there any missing equilibria or other stable or unstable attractors not shown in Fig. 1?*

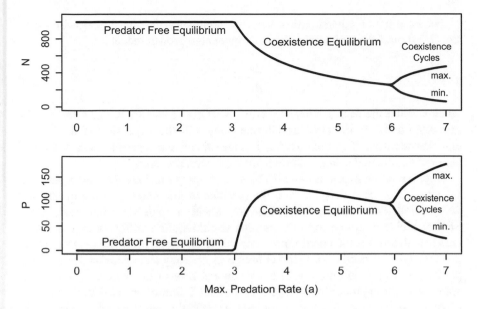

Fig. 1 A "brute force" bifurcation diagram for the Rosenzweig–MacArthur model, generated by numerically simulating solutions over very long time periods then plotting the long-term state variable values as a function of parameter a. For solutions that converge to limit cycle oscillations, the min and max values are plotted. Parameter values used: $r = 1$, $K = 1000$, $a = 5$, $k = 500$, $\chi - 0.5$, $\mu - 1$, and initial condition $(N_0, P_0) = (1000, 10)$. These results suggest a transcritical bifurcation near $a = 3$ and a Hopf bifurcation near $a = 6$

2.4 A Few Comments on Approximation

In applications, some questions cannot be appropriately answered using the asymptotic analyses described above, but might instead be addressed using careful approximations or simulations (numerical solutions to ODEs are discussed in Sect. 2.6).

Approximation methods can be very useful for gaining insight into specific phenomena. For example, in Sect. 2.2 linear approximation via Taylor's Theorem was used to clarify the behavior of trajectories near equilibria. Similar techniques exist that use other expansions to make local approximations, e.g., Fourier expansion can be used to approximate periodic solutions to quantify period–amplitude relationships (for more on investigating oscillations in nonlinear dynamical systems, see [140]). Whatever the technique, a good presentation of approximation-based results should balance intuition and mathematical rigor, and hopefully make clear the conditions for which the approximation is a good one.

For example,[11] consider this heuristic argument: the logistic growth model Eq. (21) should approximately obey an exponential growth model

$$\frac{dx}{dt} = r\,x \tag{26}$$

when x values are near 0, since the term $1 - x/K \approx 1$ when $x \ll K$. Alternatively, one could arrive at the same conclusion using a Taylor expansion of the logistic equation about $x = 0$, which would yield the same linear approximation, but would also yield a remainder that quantifies the approximation error.

Approximation methods can also be used to study *transient dynamics*, although the approaches used vary by both the specifics of the model, and the question(s) being asked. For example, one might ask when an epidemic might peak in the SEIR model, and answering this question might require a different approach than is used to determine the conditions under which the Rosenzweig–MacArthur model exhibits damped oscillations that converge to a steady-state coexistence between predator and prey. In addition to computational investigations (see Sect. 2.6), the approximation approach described next in Sect. 2.5 is often used to disentangle some transient dynamics and long-term asymptotic dynamics in systems with multiple time scales, which commonly arise in biological applications.

2.5   Fast–Slow Analysis of Systems with Multiple Time Scales

In some systems of ODEs, sometimes called *stiff* systems, dynamics occur very rapidly in certain directions in state space, while other parts of the system change more slowly. These types of systems are said to have *multiple time scales* and these can often be analyzed by separately considering the dynamics on the different time scales using appropriate approximating models. In short, the slow time scale state variables can often be treated as fixed constants, yielding an approximate model that includes only the fast time scale variables. Once the dynamics of that fast subsystem are understood, that information can be used to create a complementary set of equations for the slow subsystem that approximates the long-term behavior of the model. Often, the fast variables converge quickly to a quasi-steady-state (i.e., an equilibrium point of the fast subsystem) it can typically be assumed that, over the slow time scale, the fast time scale variables closely track these quasi-equilibrium values. Thus, the ODEs governing the slow time scale variables can be approximated by replacing all occurrences of the fast time scale variables with these quasi-equilibrium expressions, thus reducing the system to a smaller number of ODEs.

---

[11]Here, integration by parts will yield an equation for the solution curves $x(t)$, but we proceed assuming no such curves are available as this is typically the case in practice.

To clarify, suppose an ODE model with n state variables can be separated into slow ($\mathbf{x}$) and fast ($\mathbf{y}$) time scale variables and written as

$$\frac{d\mathbf{x}}{dt} = \epsilon\, f_{\text{slow}}(\mathbf{x}, \mathbf{y}, \theta) \tag{27a}$$

$$\frac{d\mathbf{y}}{dt} = f_{\text{fast}}(\mathbf{x}, \mathbf{y}, \theta). \tag{27b}$$

Here $0 < \epsilon \ll 1$ and f_{slow} and f_{fast} are roughly the same order of magnitude, hence $d\mathbf{x}/dt$ will be very small and $\mathbf{x}$ will change much more slowly than $\mathbf{y}$. If the $\mathbf{x}$ values are treated as constants in Eq. (27b) (i.e., if we let $\epsilon \to 0$) and there are quasi-equilibrium expressions $\mathbf{y}_*(\mathbf{x}, \theta)$ that satisfy $f_{\text{fast}}(\mathbf{x}, \mathbf{y}_*(\mathbf{x}, \theta), \theta) = 0$ so that the slow time scale dynamics can be approximated by

$$\frac{d\mathbf{x}}{dt} = \epsilon\, f_{\text{slow}}(\mathbf{x}, \mathbf{y}_*(\mathbf{x}, \theta), \theta) \tag{28}$$

where we see that the right hand side is purely a function of $\mathbf{x}$ and θ. For examples of such fast–slow system analyses, see [17, 36, 37, 90, 91, 134, 142, 153] or see the more thorough treatment of the topic in [103].

To illustrate how a fast–slow analysis can yield a simplified model, consider the well-known Michaelis–Menten equation, which was originally used to model a biochemical reaction in which an enzyme catalyzes the hydrolysis of sucrose into simpler sugars [95]. In more general terms, the model is of a substrate S that binds to enzyme E to form the complex ES which either dissociates back into free enzyme and substrate, or undergoes a reaction that releases the enzyme and a reaction product P. This can be written in chemical reaction equation notation as

$$E + S \underset{k_u}{\overset{k_b}{\rightleftharpoons}} ES \overset{k_{cat}}{\to} E + P$$

where k_b is the binding rate, k_u is the unbinding rate, and k_{cat} is the catalytic rate. Let the concentrations of enzyme E, substrate S, complex ES, and product P at time t be denoted $z(t)$, $s(t)$, $x(t)$, and $p(t)$, respectively. A corresponding (mean field) ODE model of this reaction is

$$\frac{dz}{dt} = -k_b\, z\, s + k_u\, x + k_{cat}\, x \tag{29a}$$

$$\frac{ds}{dt} = -k_b\, z\, s + k_u\, x \tag{29b}$$

$$\frac{dx}{dt} = k_b\, z\, s - k_u\, x - k_{cat}\, x \tag{29c}$$

$$\frac{dp}{dt} = k_{cat}\, x \tag{29d}$$

Note that there are two quantities in this model that remain constant over time: If z_0 is the initial amount of enzyme, and s_0 is the initial amount of substrate, then $z + x = z_0$ and $s + x + p = s_0$. Thus, we can omit the first equation (29a) from the model since we can always infer z from x since $z = z_0 - x$. Likewise, we can omit Eq. (29c) since $x = s_0 - s - p$. This yields the equivalent, but simpler, model

$$\frac{ds}{dt} = -k_b \left(z_0 - (s_0 - s - p)\right) s + k_u \left(s_0 - s - p\right) \tag{30a}$$

$$\frac{dp}{dt} = k_{cat} \left(s_0 - s - p\right). \tag{30b}$$

Next, assume that the catalytic reaction is much faster than the other binding and unbinding reactions, so that Eqs. (30) become a slow-fast system where s very quickly reaches a quasi-equilibrium state that then slowly tracks the slower timescale changes in p. In short,[12] setting $ds/dt = 0$ and solving for that quasi-equilibrium relationship between s and p yields

$$p(s) = \frac{-k_b z_0 s}{k_u + k_b s} + s_0 - s \tag{31}$$

which, when substituted into Eq. (30b), yields

$$\frac{dp}{dt} = \frac{V_{max} s}{k_d + s} \tag{32}$$

where the maximum velocity (i.e., maximum rate of change of p) $V_{max} = k_{cat} z_0$ and the dissociation constant $k_d = k_u/k_b$. Other assumptions can lead to similar approximations for this same model [24, 95], and similar analyses for mathematically related interaction processes have been used to derive similar nonlinear response curves, e.g., compare the right hand side of Eq. (32) with the predation rate in Eqs. (23) (for more details, see [38, 125] and references therein).

2.6   Computing Numerical Solutions to ODEs

It can often be useful to supplement analytical approaches with numerical simulation studies, e.g., to assess how well an approximate model compares to the original model, or to investigate transient dynamics when analytical approaches are not practical. In typical introductory ODE class, students are taught methods for finding (or sometimes approximating) analytical solutions to a given set of ODEs. Unfortunately, analytical solutions rarely exist for the kinds of nonlinear

---

[12]Here we are glossing over the formal details from singular perturbation theory [103] for deriving equations that approximate the fast- and slow-timescale dynamics of this model.

ODE models encountered in applications. This section introduces how to compute numerical solutions to ODEs.

It is relatively straightforward to compute approximate numerical solutions to ODEs using standard computational methods available in nearly all mathematical software [138], or by coding up your own solver if the more advanced method that you need is otherwise unavailable. In R, the ode function in the deSolve package [156] implements various numerical methods that are each tailored to provide reliable numerical solutions by addressing various numerical challenges posed by different categories of ODEs. Similar tools exist in Matlab [115] and additional solvers can sometimes be found in other software, e.g., in [78]. Here we provide a simple introduction to computing numerical solutions in R, but leave it to the reader to further investigate the conditions under which different methods should, and should not, be applied. See the Appendices for resources to get started using R, and for more on numerical methods see [115, 138, 155, 160] and references therein.

2.6.1 Euler's Method

To understand how various methods for computing numerical solutions are implemented on a computer, it is instructive to see the very simple algorithm known as Euler's method. Like other methods, it approximates a continuous curve using a discrete-time approximation where information about the derivative of the curve (i.e., the right hand side of the ODE) is used to determine where the state variables move to on each time step. Consider the generic system of ODEs

$$\frac{d\mathbf{x}}{dt} = f(t, \mathbf{x}, \theta), \quad \mathbf{x}(0) = \mathbf{x_0} \tag{33}$$

and note that, if we back off of the limit implicit in the derivative on the left, we get that for small Δt,

$$\frac{\mathbf{x}(t + \Delta t) - \mathbf{x}(t)}{\Delta t} \approx f(t, \mathbf{x}(t), \theta) \tag{34}$$

which can be rearranged to yield

$$\mathbf{x}(t + \Delta t) \approx \mathbf{x}(t) + f(t, \mathbf{x}(t), \theta) \, \Delta t. \tag{35}$$

Euler's method approximates trajectories starting at $\mathbf{x}(0) = \mathbf{x_0}$ by iterating the above discrete-time map Eq. (34). This method has been improved upon in various ways, which range from adaptively picking smaller or larger time steps when appropriate, to incorporating information about the curvature of the trajectory to obtain a better increment than the linear approximation $f(t, \mathbf{x}(t), \theta)\Delta t$.

Consider Eq. (34) starting at time $t = 0$. The following information is needed to implement Euler's method and similar algorithms for computing numerical solutions to ODEs: a starting vector of *initial condition* values $\mathbf{x}(0) = \mathbf{x_0}$, parameter values θ, and a way of computing the derivative $f(\mathbf{t}, \mathbf{x}, \theta)$.

The R code below illustrates how this is implemented using the `ode` function in the `deSolve` package is illustrated below.

2.6.2   Numerical Solutions in R

The first step is to create a function that computes $f(t, \mathbf{x}, \theta)$ given a specific value of time t, the state of the system $\mathbf{x}$, and a set of parameter values θ. Consider the SIR model given by Eqs. (8). The right hand side of the model (the *derivative function f*) is a function of the state variables S, I, and R and parameter values β and γ. According to the R documentation[13] for the `ode` function we must create an R function that computes these derivatives based on three inputs—time, a vector of state variable values, and a parameter vector, in that order, i.e., $f(t, \mathbf{x}(t), \theta)$—and returns the vector of derivative values in an R object known as a `list`. This is implemented in the following R code.

```
# We need to load the ode() function in package deSolve
# install.packages("deSolve") Install once, if needed
library(deSolve); # load  deSolve  into the workspace
# Define an SIR function to use with ode()
SIR = function(tval, X, params) {
        S = X[["S"]] # one could also have used X[1]
        I = X[["I"]] # ... X[2]
        R = X[["R"]] # ... X[3]
        B = params[["beta"]]  # or  params[1]
        g = params[["gamma"]] # ... params[2]
        # Now compute the SIR model derivatives
        dS = -B*I*S
        dI = B*I*S - g*I
        dR = g*I
        # Return the derivatives in a list to use with ode()
        return(list(c(dS,dI,dR)))
}
```

Next, the `ode` function requires that we give it a set of initial conditions, a vector of (user specified) time values for which to return state variable values, and a set of parameter values. Thus, we must specify $S(0)$, $I(0)$, $R(0)$ and parameters β and γ. The `ode` function returns a matrix whose first column is the vector of time values, and the subsequent columns are the corresponding state vector values (in this case,

---

[13]To view the documentation, load the package `deSolve` with the command `library(deSolve)` then type `?ode` into the R console. See the Appendix for additional resources to get started using R.

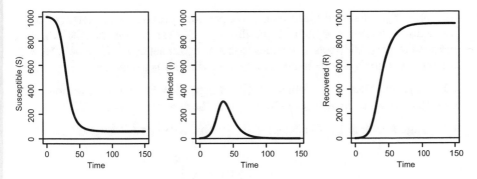

Fig. 2 Numerical solution to the SIR model, Eqs. (8). See the code in the main text for details

the values of S, I, and R). This is implemented in the following code, which also plots the three numerical solution curves as functions of time (Fig. 2).

```
X0 = c(S=998, I=2, R=0) # Initial state values
Pars = c(beta=0.0003, gamma=0.1)
Time = seq(0, 150, length=200) # time values for ode()
# Now we can compute a numerical solution
Xout - ode(X0, Time, SIR, Pars) # Default: method="lsoda"
head(Xout, 3) # Show the first three rows of output
```

	time	S	I	R
[1,]	0.0000000	998.0000	2.000000	0.0000000
[2,]	0.7537688	997.5131	2.324244	0.1626696
[3,]	1.5075377	996.9476	2.700737	0.3517011

```
# Plot the numerical solution curves ...
par(mfrow=c(1,3)) # ... in a figure with 1 row, 3 columns.
plot(Time, Xout[,2], ylab="Susceptible (S)", ylim=c(0,sum(X0)),
     type="l", lwd=2);  abline(h=0) # abline() draws the x-axis
plot(Time, Xout[,3], ylab="Infected (I)", ylim=c(0,sum(X0)),
     type="l", lwd=2);  abline(h=0)
plot(Time, Xout[,4], ylab="Recovered (R)", ylim=c(0,sum(X0)),
     type="l", lwd=2);  abline(h=0)
```

Exercise 7 Modify the example above so that the ODE solver only uses the equations for S and I, but not R. At the end of the code, after the output from ode has been obtained, add the R column to Xout using the relationship $R(t) = N(0) - S(t) - I(t)$ (to convince yourself $N = S + I + R$ is constant, note its derivative $dN/dt = 0$).

Challenge Problem 7 Modify the code above by adding an exposed class so that the code generates solutions of the SEIR model mentioned previously. Explore how solution change for different values of v.

Challenge Problem 8 Modify the code above to correspond to changing the quantity β to a positive-valued function of time, e.g., using a periodic function such as $\beta(t) = \beta_0 + A \sin(\omega t)$. This makes the previously autonomous system of ODEs into a non-autonomous system. To do this, modify the equations in SIR as well as the parameter vector (to assign values to the new parameters in the model). How do the solution curves change relative to the case where β is a constant?

Challenge Problem 9 Implement the Euler algorithm on your own, without using ode and the 4th order Runge–Kutta (RK4) algorithm (e.g., see page 33 in [160]).

Challenge Problem 10 Use integration by parts to show the θ-Logistic model

$$\frac{dx}{dt} = r x \left(1 - \left(\frac{x}{K} \right)^{\theta} \right) \tag{36}$$

has an analytical solution, then modify the code above to find a numerical solution using ode in R. Use initial conditions $x(0) = 10$, $r = 1$, and $K = 1000$, and explore θ values above, below, and at $\theta = 1$. Plot both curves together on the same plot to compare the exact and approximate solution curves.

2.6.3 Keeping Numerical Solutions Positive: The Log-Transform Trick

The above approach to computing numerical solutions to ODEs can sometimes be problematic for ODE models of biological systems. Often, in biological models (e.g., like the SIR model mentioned above) one or more equations are of the form

$$\frac{dx}{dt} = g(x) x. \tag{37}$$

Importantly, this implies that trajectories $x(t)$ slow as they asymptotically approach $x = 0$, and therefore can never pass through $x = 0$. However, numerical methods may take an approximated, discrete-time step that results in the simulated x trajectory erroneously crossing zero (which should be impossible!). This can lead to state variables running off to $\pm\infty$, and other undesirable behaviors. Fortunately, there is a clever trick that allows us to avoid these numerical errors: if we assume that $x(t) > 0$ over the time period for which we seek a numerical solution, then we can let $X = \log(x)$ and by the properties of the natural log function[14] it follows that

$$\frac{dX}{dt} = \frac{d}{dt} \log(x) = \frac{1}{x} \frac{dx}{dt}$$
$$= \frac{1}{x} g(x) x = g(\exp(X)) \tag{38}$$

---

[14] Here the notation log(x) is used for the natural log function, instead of ln(x), following the convention used in most modern scientific programming languages. Likewise, exp(x) $= e^x$.

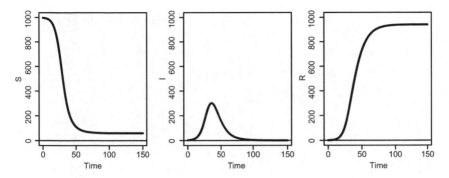

Fig. 3 A numerical solution to the log-transformed SIR model using the same parameters and initial conditions as in Fig. 2. See the code in the main text for details

That is, if we transform a strictly positive state variables x to a real-valued variable X using the natural log function, then numerical solutions of X can take on both positive and negative values that can later be transformed back to strictly positive x values, since $x = \exp(X)$. This transformation of variables is the default approach some modelers take to computing numerical solutions of models where trajectories should remain strictly positive over a finite time interval.

One caveat to this approach is that, since $X(t) \to -\infty$ as $x(t) \to 0$, it can lead to numerical errors due to finite limits on the size of floating point values on a computer. While it may seem contradictory to trade one numerical problem for another, it is usually far less common (and much more manageable) in practice to have a numerical solution diverging to $-\infty$ and hitting the computer's floating point limit. Thus, this approach should be reserved for situations where $x(t)$ remains bounded away from 0 or converges slowly to 0. A second caveat is that often times solutions that erroneously cross through 0 do so because of typos in the code that defines the derivative function. Thus, carefully checking that the equations, as implemented in the code, are free of errors is strongly advised before assuming that numerical methods are to blame for seemingly erroneous numerical solutions.

To illustrate this approach, here is the above SIR model code rewritten in terms of the log-transformed values $X = \log(S)$ and $Y = \log(I)$ where we have also omitted the R equation since the constant population size allows us to calculate $R(t)$ once the $S(t)$ and $I(t)$ curves are obtained using $R(t) = N(0) - S(t) - I(t)$ (Fig. 3).

```
# A function for ode() to compute numerical solutions
# of the log-transformed SIR model.
logSI = function(tval, Xs, params) {
    X = Xs[["X"]]            # X = log(S)
    Y = Xs[["Y"]]            # Y = log(I)
    B=params[["beta"]]
    g=params[["gamma"]]
    # Now compute the derivatives
    dX = -B*exp(Y)          # I = exp(Y)
    dY = B*exp(X) - g       # S = exp(X)
```

```
  # Return the derivatives in a list to use with ode()
  return(list(c(dX,dY)))
}

# Initial conditions, log-transformed
logX0 = c(X=log(X0[["S"]]), Y=log(X0[["I"]]))
logXout = ode(logX0, Time, logSI, Pars) # requires deSolve

# Convert back to S and I, and calculate R
Xout = cbind(Time,S=exp(logXout[,2]), I=exp(logXout[,3]),
             R=sum(X0) - exp(logXout[,2]) - exp(logXout[,3]))
head(Xout,3)
```

```
           Time        S        I             R
[1,]  0.0000000 998.0000 2.000000 -2.273737e-13
[2,]  0.7537688 997.5126 2.324242  1.631915e-01
[3,]  1.5075377 996.9464 2.700730  3.528366e-01
```

```
# Plot the numerical solution curves as before
par(mfrow=c(1,3)) # 3 panels in 1 row
plot(Time, Xout[,2], ylab="S", ylim=c(0,sum(X0)),
     type="l", lwd=2);   abline(h=0)
plot(Time, Xout[,3], ylab="I", ylim=c(0,sum(X0)),
     type="l", lwd=2);   abline(h=0)
plot(Time, Xout[,4], ylab="R", ylim=c(0,sum(X0)),
     type="l", lwd=2);   abline(h=0)
```

Exercise 8 Modify the code above to simulate the Rosenzweig–MacArthur model, Eqs. (23), using the parameter values from Fig. 1 and using both direct implementation and the log-transformation technique. Plot both results to compare approaches.

Challenge Problem 11 Look at the different methods available through the ode function in the deSolve package ([156]; type ?ode into R to view the documentation). When should you use one method over the other? Compare these to the methods available in Matlab, Python, or other computing platforms (see also [115, 138]).

Challenge Problem 12 It can be challenging to obtain numerical solutions to stiff systems. Find resources like [116, 138] to help you better understand which numerical methods work well for stiff (multiple time scale) systems in the R package deSolve as well as in Matlab or similar software, and look up some stiff models to use as examples. Use the different methods to find numerical solutions for the same parameter values and time values, and compare these to methods that should perform poorly (e.g., Euler). How different are the results from different methods? How do R's stiff solvers compare to those in Matlab or other software?

2.7 ODEs as Statistical Models

This final section in Sect. 2 introduces some useful statistical concepts and basic approaches to using an ODE model as the basis for a statistical model.[15] Increasingly, ODE models are being used as components of statistical models, e.g., for making inferences about underlying mechanisms or for forecasting. Applications include making inferences from model parameters estimated from time series data (i.e., a sequence of state variable measurements taken at multiple time points, that in some way corresponds to a sequence of state variable values), quantifying uncertainty in those parameter estimates, forecasting, and conducting statistical tests to determine whether a certain parameter is significantly different from zero or whether one model fits the data better than another model (i.e., *model selection*).

2.7.1 A Brief Overview of Key Statistical Concepts

Parameter estimation can be approached in various ways, but to start off it is helpful to first introduce some concepts from the statistical theory of *estimators* (e.g., see [85, 106]). You may recall that certain formulas exist for estimating parameters in specific contexts, e.g., for a normal (Gaussian) distribution with mean μ and variance σ^2, an estimate of μ can be obtained by calculating the *sample mean* $\bar{y} = \sum_{i=1}^{n} y_i / n$ of a *random sample*, i.e., n data points $y_1, \ldots, y_n$ where each y_i are independent draws from the given normal distribution. Formally, we refer to $\widehat{\mu} \equiv \bar{y}$ as an *estimator* because it defines a function of our data that yields a parameter estimate. The "hat" notation indicates that $\widehat{\mu}$ is an estimator of parameter μ. Note that we could have chosen other estimators to compute an estimate of parameter μ, e.g., (a) the median of the sample, (b) the geometric mean instead of the arithmetic mean, or (c) the mean of the extreme values of the sample (i.e., the mean of y_{min} and y_{max}).

Estimator theory in statistics deals in part with how to select an estimator with certain desired properties. For example, some estimators are more robust to outliers than others, and some may be more or less biased. A good place to begin discussing properties of estimators is to recognize that estimators are functions of data, and hence, we can think of them as *functions of random variables*. As such, estimators are themselves random variables and therefore follow some distribution, and it is that distribution that determines the properties of the estimator. For example, the sample mean $\bar{y}$ above is proportional to the sum of normal random variables, and is therefore itself normally distributed with a mean and variance that can be computed from the definitions of the y_i distribution [85, 106]. In a statistical context, a desirable property of an estimator is that it be *unbiased*, i.e., its expected value equals the true parameter value of interest (if not, we say the estimator is *biased*).

---

[15]For related resources in R, see the packages such as CollocInfer [87], deBInfer [19], or browse the relevant CRAN Task Views (https://cran.r-project.org/web/views/).

Estimators should also have as *small a variance as possible* so that estimates fall as close as possible to the mean (i.e., the true parameter value if the estimator is unbiased). Hence, we typically strive for *minimum variance, unbiased* estimators; however, in some circumstances we select estimators that may deviate from this ideal (e.g., because they may be more robust to outliers and thus in practice may perform better with "messy" data). In our simple normal distribution example, note that

$$E(\widehat{\mu}) = E\left(\sum y_i/n\right) = \sum E(y_i)/n = \sum \mu/n = \mu \qquad (39)$$

so this estimator is *unbiased*. But is it a *minimum variance estimator*? To answer that question requires taking a deeper look into the theory of estimators that is beyond the scope of this chapter. For a more in depth treatment of estimator theory, minimum variance estimators, and a very nice result known as the Cramér–Rao lower bound, see texts such as [28, 85, 106].

Alternatively, on a case-by-case basis, when trying to decide among different estimation procedures or assessing properties of a given estimator, one could simulate data repeatedly effectively sample from the estimator distribution(s), and thus make these assessments computationally. By repeatedly estimating parameters from simulated data, where the "true" parameter values are known, we can sample from an estimator's distribution a large number of times. That large sample of estimates can then be used to assess properties like the mean and variance of the estimator distribution. Comparing multiple estimators in this way can reveal differences in their variance, and any bias can be quantified by calculating how well the mean of the estimates compares to the parameter value used to generate samples from the estimator distribution. For more on computational approaches in statistics, see [70].

To illustrate, the following R code compares the three candidate estimators of μ in the above example by reconstructing each estimator distribution by random sampling: the standard arithmetic mean ($\bar{y}$), the geometric mean ($y_1 \cdot y_2 \cdots y_n)^{1/n}$, and the average of the two extreme values y_{min} and y_{max} (which we would expect to perform poorly given how much data it omits from the calculation). This is accomplished by repeatedly sampling from each estimator distribution by iteratively drawing a random sample from the given normal distribution and then calculating parameter estimates from that random sample.

```
# Begin by setting parameters for our simulations
mu = 25  # define our mean parameter for our normal distribution
sd =  2  # define our variance (variance = sd^2)
N  = 10  # to simulate data, use a sample size of N
mu.est1 = c() # empty list for our arithmetic mean estimates
mu.est2 = c() # empty list for our geometric mean estimates
mu.est3 = c() # empty list for our mean(min,max) estimates
# Sample our estimators 100000 times each...
set.seed(6.28) # set seed to get reproducible random numbers
for(i in 1:100000) { # simulate and estimate 100000 times
```

```
  y = rnorm(N,mu,sd) # Simulate data. Read ?rnorm for details
  mu.est1[i] = mean(y) # calculate the standard estimator
  mu.est2[i] = prod(y)^(1/N)      # the geometric mean
  mu.est3[i] = (min(y)+max(y))/2 # the mean of ymin, ymax
}
# Plot results
par(mfrow=c(1,3)) # plots in 1 row, 3 columns
# First panel: standard estimator (population mean)
plot(density(mu.est1), # empirical density function
    main=expression(hat(mu)==sum(y[i])/N),
    xlab=expression(hat(mu)), xlim=c(21.5,29), ylim=c(0,0.65))
hist(mu.est1, 30, freq=FALSE, add=TRUE,
    col=rgb(.6,.6,.6,.5), border=rgb(0,0,0,0.5))
text(27.5,0.6,paste("Mean =",signif(mean(mu.est1),4),  ))
text(27.5,0.57,paste("SE =",signif(sd(mu.est1),3),  ))
abline(h=0) # draw the horizontal axis
# Second panel: geometric mean
plot(density(mu.est2), # empirical density function
    main=expression(hat(mu)==(prod(y[i]))^{1/N}),
    xlab=expression(hat(mu)), xlim=c(21.5,29), ylim=c(0,0.65))
hist(mu.est2, 30, freq=FALSE, add=TRUE,
    col=rgb(.6,.6,.6,.5), border=rgb(0,0,0,0.5))
text(27.5,0.6,paste("Mean =",signif(mean(mu.est2),4),  ))
text(27.5,0.57,paste("SE =",signif(sd(mu.est2),3),  ))
abline(h=0) # draw the horizontal axis
# Third panel: mean(ymin,ymax)
plot(density(mu.est3), # empirical density function
    main=expression(hat(mu)==(y[min]+y[max])/2),
    xlab=expression(hat(mu)), xlim=c(21.5,29), ylim=c(0,0.65))
hist(mu.est3, 40, freq=FALSE, add=TRUE,
    col=rgb(.6,.6,.6,.5), border=rgb(0,0,0,0.5))
text(27.5,0.6,paste("Mean =",signif(mean(mu.est3),4),  ))
text(27.5,0.57,paste("SE =",signif(sd(mu.est3),3),  ))
abline(h=0) # draw the horizontal axis
```

The simulation output in Fig. 4 illustrates how estimators can differ in terms of bias and their variance. In practice, it is not always possible to obtain unbiased estimators that are also minimum variance estimators. However, choosing likelihood based estimation procedures is usually a good starting point because they often yield minimum variance estimates that are asymptotically unbiased (i.e., the bias vanishes as the sample size grows).

2.7.2   Likelihood Based Parameter Estimation

Maximum likelihood based parameter estimation is a widely used approach in statistical applications. Since the approach is not restricted to parameter estimation for ODEs, and can also be used for the other kinds of models mentioned in this chapter, it is worth reviewing some maximum likelihood basics before applying this approach using ODE models. For further details, see [21, 28, 85, 106].

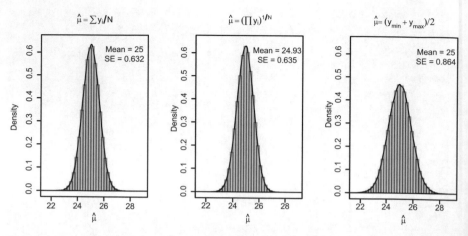

Fig. 4 A comparison to assess the bias and relative variability of three candidate estimators for the mean of a normal distribution: From left to right, the three panels show the result of simulating 100,000 samples (sample size of $n = 10$) and computing an estimate for the mean μ using either the standard arithmetic mean (left), the geometric mean (center), or the mean of the extreme values (right). The plots show a histogram with an empirical density function overlay. The empirical mean and standard deviation (standard error) of the estimator distribution are shown in the top right of each panel. The true value used to simulate the data was $\mu = 25$. These results suggest that the geometric mean may be slightly biased and have a slightly larger variance than the standard estimator, while the third option appears to be unbiased but has substantially higher variance. For more details, see the code in the main text

As mentioned above, statistical theory tells us that estimators derived by finding the parameter set that maximizes the *likelihood function* (defined next) are usually (but not always!) minimum variance estimators and often have little to no bias (again, we would like to check these properties hold to the extent it is possible to do so). We call these estimators *maximum likelihood estimators* (MLEs), and their desirable statistical properties, combined with the relative ease of computing MLEs, have made this a very popular approach to parameter estimation.

The likelihood function should be a familiar function to those who have taken a first course in probability: In short, the likelihood function is just the *joint probability density function* (for continuous data; or the *joint mass function* for discrete data), but where we have fixed the values of our random variables (i.e., our data) and instead treat the distribution parameters as the unknown independent variables. More specifically, for a set of independent observations $y = (y_1, \ldots, y_n)$ (i.e., our data) where the distribution of y_i is described by density function $f_i(y_i, \theta)$ with parameter vector θ, the likelihood of θ, given the data y, is given by

$$L(\theta|y) = \prod_{i=1}^{n} f_i(y_i, \theta). \tag{40}$$

That is, *both the likelihood and joint density functions have the same mathematical formulation*, but we think of them as distinct functions over two different domains: the familiar joint density function domain is the space of possible random variable values (e.g., $\mathbb{R}^n$ for a sample of size n from a normal distribution), while the domain of the likelihood function is the space of all possible parameter values θ (e.g., $\mu \in \mathbb{R}$ and $\sigma > 0$ in our normal distribution example).

In general, once we can write down a joint density function for our data, and thus can specify a likelihood function $L(\theta|y)$ for parameter vector θ given the data y, the MLE $\widehat{\theta}$ is given by the parameter set that maximizes the likelihood function:

$$\widehat{\theta} = \underset{\theta}{\text{argmax}} \ L(\theta|y). \tag{41}$$

In practice, it is often both analytically and computationally easier to minimize the *negative log likelihood* function $nLL(\theta|y) \equiv -\log(L(\theta|y))$ rather than maximize the likelihood function (the two yield equivalent estimators, due to the monotonicity of the log function). Thus, we also have the equivalent definition of the MLE

$$\widehat{\theta} = \underset{\theta}{\text{argmin}} \ -\log(L(\theta|y)). \tag{42}$$

In some cases (like the example below), formulas for the MLE can be found using the standard approach taught in calculus: take the partial derivatives of the negative log likelihood function nLL with respect to each unknown parameter, then set each partial derivative to zero and solve to find the parameter values (which will be functions of the data) that minimize the negative log likelihood (and thus maximize the likelihood). If finding the MLE analytically is not a viable option (e.g., see [101]), computational optimization approaches can be used to find parameter values that minimize nLL for a specific data set.

Continuing with our simple example from the previous section, suppose we would like to use MLEs to estimate the unknown mean μ_0 and standard deviation σ_0 for a normal distribution from a random sample of size n.[16] Let y_i denote the ith observation in our sample ($i = 1, \ldots, n$). The *likelihood function* in this case is given by the same formulation as the joint density function

$$L(\mu, \sigma|y) = \prod_{i=1}^{n} f(y_i, \mu, \sigma) \tag{43}$$

where function f is the normal density function evaluated at y_i using the given mean μ and standard deviation σ. Recall that our y_i values (our data) are now fixed constants, and the domain of this likelihood function is the upper half of $\mathbb{R}^2$ since our two independent variables (inputs) are $\mu \in \mathbb{R}$ and $\sigma > 0$. The maximum likelihood

---

[16]Here the subscript 0 is used to distinguish the true parameters values μ_0 and σ_0 from the candidate values μ and σ that one might plug in to the likelihood function.

principle states that the parameter values $\mu = \widehat{\mu}_{MLE}$ and $\sigma = \widehat{\sigma}_{MLE}$ that together maximize $L(\mu, \sigma | y)$ are the MLEs for unknown parameters μ_0 and σ_0^2. In this case, these can be found analytically, and are

$$\widehat{\mu}_{MLE} = \frac{1}{n} \sum_{i=1}^{n} y_i, \quad \text{and} \quad \widehat{\sigma^2}_{MLE} = \frac{1}{n} \sum_{i=1}^{n} (y_i - \widehat{\mu}_{MLE})^2. \tag{44}$$

That is, the MLE for the mean is the familiar *sample mean* (recall Fig. 4), while the MLE for the variance is the *unadjusted sample variance*. Importantly, while the MLE for the mean μ_0 is *unbiased*, since $E(\widehat{\mu}_{MLE}) = \mu_0$, the MLE for the variance σ_0^2 is biased, since $E(\widehat{\sigma^2}_{MLE}) = \frac{n}{n-1}\sigma_0^2$, but asymptotically unbiased since the bias vanishes as sample size $n \to \infty$. This example nicely illustrates how, in practice, minimum variance unbiased estimators do not always exist, and we often prefer to use unbiased estimators, even if they require a small increase in the variance of our estimates.

In general, it is worth noting that there are no guarantees that an MLE exists, or that when an MLE exists that it is unique.[17] When no MLE exists, other approaches to parameter estimation must be used or the model must be reformulated. In cases where multiple MLEs exist, special considerations must be made to identify the appropriate course of action (this is discussed below in the context of parameter identifiability).

Exercise 9 Consider the normal distribution example above. Suppose we would like to analyze a long-term series of water depth data (each depth measurement x_i has an associated time value t_i) in a coastal bay to quantify a rise in sea level (which we assume might be increasing linearly over the time period for which we have data). Assume that the data collection site in the bay experiences tidal fluctuations and that the expected value for a depth measurement at time t is assumed to follow the curve $d(t)$ given by

$$d(t) = d_0 + mt + A\sin(2\pi t / T) \tag{45}$$

where d_0 is a baseline depth, m is the slope of the long-term increase (or decrease) in sea level, and the last term describes the smaller timescale rise and fall of the tide (which here has been simplified to have a fixed period T and fixed amplitude A). Assume that wave activity and other factors introduce noise into these measurements that roughly follow a normal distribution so that a measurement x_i at time t_i is normally distributed with mean $d(t_i)$ and standard deviation σ. Write down the likelihood function for this data set.

---

[17] Sufficient criteria for the existence of one or more MLEs are that the parameter space is compact and the log likelihood surface is continuous. In practice, non-unique MLEs are more commonly the problem, especially when working with nonlinear ODE models.

Next, we look at an approach to using likelihood based parameter estimation for ODE models where, in contrast to the exercise above, the mean value for our likelihood calculation depends on a solution to an ODE. Since many ODEs do not have a closed form solution, the mean therefore must be calculated using a numerical solution to the ODE.

2.7.3 Likelihood Framework for ODEs

A popular framework for parameter estimation using ODE models is the following (e.g., see [32, 33, 46, 47, 123, 141] and references therein). First, in order to distinguish between the modeled "true" state of the system $\mathbf{x}$ (e.g., the number of rabbits in a forest) versus the data $\mathbf{y}$ (e.g., the count of rabbits along multiple transects), we define the *process model* Eq. (46a) (i.e., what we typically think of as our mechanistic mathematical model of our real-world study system) and a set of one (or more) variables that we refer to as the *observation model* Eq. (46b).

$$\frac{d\mathbf{x}}{dt} = f(\mathbf{x}, \theta), \quad \mathbf{x}(0) = \mathbf{x_0} \tag{46a}$$

$$\mathbf{y} = g(\mathbf{x}, \theta) \tag{46b}$$

Note that unknown parameters that one hopes to estimate may appear in both the process and observation models. Typically, the observations ($\mathbf{y}$ values) are assumed to be noisy, thus g above may represent a procedure for sampling $\mathbf{y}$ values from a probability distribution appropriately parameterized by the underlying state of the process model (i.e., its mean and/or variance, etc., might depend on $\mathbf{x}$).

To illustrate how this framework is used to estimate ODE model parameters, we use the following example of a logistic growth model where the observations y_i are assumed to be normally distributed with constant variance σ^2 and a mean that is proportional (by some factor c) to the state variable at time t_i, i.e., $E(y_i) = c\,x(t_i)$:

$$\frac{dx}{dt} = r\,x\left(1 - \frac{x}{K}\right), \quad x(0) = x_0, \tag{47a}$$

$$y_i \sim \text{Normal}(c\,x(t_i), \sigma^2). \tag{47b}$$

We may now consider estimating one or more of the parameters r, K, c, and x_0 within a *maximum likelihood* framework.[18] It is strongly advised that, when attempting a new-to-you parameter estimation procedure, you to implement it first

---

[18]It is worth pointing out that this model does not have a unique best-fit parameterization, unless certain parameters are held constant (i.e., are already known), even when estimating parameters from ideal data (e.g., a large sample with little or no noise)! This problem and the ways of resolving the issue are detailed in the latter half of this section.

34

P. J. Hurtado

on simulated data where the known "true" parameter values can be compared to estimates. Hence, for the above example, we will first simulate data by picking a series of time values $(t_1, t_2, \ldots, t_n)$ and generating corresponding $x(t_i)$ values by computing a numerical solution to the above logistic equation, Eq. (47a), then the corresponding y_i values can be sampled from the normal distributions given by Eq. (47b). This is implemented in the following R code, and provides a data set with known "true" parameters that can be used to evaluate and verify that the parameter estimation procedure works as intended.

```
set.seed(6.28) # for repeatable random number generation
library(deSolve) # load the ode() function into the workspace

# Function to simulate x(t) using ode() in the deSolve package
odefunc = function(t,x,ps) { # see ?ode in R for details
  with(as.list(ps),{  # Use named values in ps as variables
    dx = r*x*(1-x/K); #     (compare to SIR code above).
    return(list(dx)); } )
}

# Simulate data. First, obtain an ODE solution
Time = seq(0,7,length=30) # 30 evenly spaced values from 0 to 7
params = c(r=2, K=500, c=0.1, sigma=2) # Model parameters
x0 = c(x=50) # labels our state variable  x  in xout below
xout = ode(y=x0, times=Time, func=odefunc, parms=params)
head(xout,2) # Display the first two rows of xout
```

```
          time        x
[1,]  0.0000000 50.00000
[2,]  0.2413793 76.29257
```

```
# Second, sample y values from the given normal distributions
mydata = cbind(Time, y = rnorm(nrow(xout),
                mean = params["c"]*xout[,2], params["sigma"]))
head(mydata,2)
```

```
          Time        y
[1,]  0.0000000 5.539212
[2,]  0.2413793 6.369287
```

```
tail(mydata,2)
```

```
          Time        y
[29,]  6.758621 52.78624
[30,]  7.000000 48.87896
```

```
# Plot process model trajectory x(t) and data (t_i, y_i)
par(mfrow=c(1,2)) # Plot results in 1 row, 2 columns
plot(Time, xout[,2], type="l", ylim=c(0,550),
     main="Process Model", ylab="x")
plot(mydata, pch=19, col="gray50", ylim=c(0,55),
     main="Data (Simulated)")
```

The remainder of this section illustrates how to carry out the parameter estimation procedure, and how to conduct additional analyses to address a commonly encountered problem where there exits multiple "best-fit" parameter estimates. This is known as an *identifiability* problem (also called an *estimability* problem), and will be discussed in greater detail below.

2.7.4   Parameter Estimation as an Optimization Problem

Continuing with the example above, let us assume that it is known that $x_0 = 50$, and we seek to estimate the other parameter values from our (simulated) data. The likelihood of a given parameter set $\theta = (r, K, c, \sigma)$ (given those n simulated pairs of data $y_{\text{data}} = (t_i, y_i)$) can be computed using the *likelihood function* defined as

$$L(r, K, c, \sigma \,|\, y_{\text{data}}) = \prod_{i=1}^{n} f(y_i, \text{mean} = c\,x(t_i), \text{sd} = \sigma) \qquad (48)$$

where function f is the density function for a normal distribution[19] with the given mean and standard deviation, evaluated at y_i, and parameters x_0, r, and K are used to numerically find $x(t_i)$. Our goal is to computationally find the set of parameter values that maximizes L for a given data set. As stated above, it is often more practical to minimize the *negative log likelihood* function $nLL(\theta|y) \equiv -\log(L(\theta|y))$ rather than maximize the likelihood function. Thus, by properties of the natural log function,

$$nLL(\theta|y) = -\log\left(\prod_{i=1}^{n} f(y_i, \text{mean} = c\,x(t_i), \text{sd} = \sigma)\right)$$

$$= -\sum_{i=1}^{n} \log(f(y_i, \text{mean} = c\,x(t_i), \text{sd} = \sigma)).$$

We can now frame our parameter estimation problem as a computational optimization problem. Since solutions $x(t)$ typically have no closed form, we will use numerical solutions of our process model in order to compute likelihoods.

Optimization algorithms essentially take a real-valued *objective function* (in our case, a negative log likelihood function) which defines a surface over parameter space, and from an initial set of parameter values (i.e., a point in parameter space) moves through parameter space by moving "downhill" (or "uphill") on the objective function surface until the minimum (or maximum) objective function value is found. Thus, to use generic optimization tools in software like R, we must construct a

---

[19]See dnorm in R.

negative log likelihood function that conforms to the requirements of our chosen optimization routines.

The objective functions defined below are written specifically to be used with the generic optimization methods available under optim() and optimx() in R [129]. These functions require that the first argument to the objective function be a parameter vector, and any additional arguments are for passing in additional fixed values (in this case, the data). See the help documentation for optim in R for further details. The examples below use the default Nelder–Mead algorithm, which can be relatively slow,[20] and does not allow for the specification of explicit constraints (e.g., we would like to require the standard deviation parameter σ be positive), but often gives good results where gradient-based methods perform poorly (for additional optimization considerations and resources, see [128, 129, 161]).

In the context of our specific application, we additionally require that all parameters are positive-valued. This is a common constraint; however, generic *unconstrained* optimization methods like Nelder–Mead often work best when allowed to freely consider all real numbers as candidate values for each unknown parameter. To impose constraints on those parameter ranges requires us to either modify our objective function—to coerce the optimization method into avoiding negative parameter values—or to instead use a constrained optimization method. When using a box-constrained optimization method in R (e.g., method L-BFGS-B in optim) the limits on the range of possible parameter values can be specified explicitly as arguments to optim or optimx. Here we only illustrate two variations on the first approach, using unconstrained optimization with parameter constraints built into the objective functions.

The first objective function (nLL1) implements this constraint by returning a very large value if any parameter value is not strictly positive. This introduces a discontinuity into the likelihood surface, which can cause problems for many optimization methods, especially gradient-based methods (i.e., those methods that use derivatives to "move downhill" to find objective function minima).

```
nLL1 = function(theta, tydat) {
  # To disallow negative parameter values, nLL1 returns a large
  # number if the optimization algorithm inputs any negatives.
  # (See the text and nLL() below for better alternatives.)
  if(any(theta<=0)) { return(1*10^10) }

  # Otherwise, calculate our likelihood of theta given the data.
  Times = tydat[,1] # 1st column are the times
  Ys = tydat[,2]    # 2nd column are the y values

  # Simulate a trajectory x(t) under the given parameters (theta)
  x0 = c(x=50) # For now, we assume x0 is known.
  xout = ode(y=x0 ,times=Times, func=odefunc, parms=theta)
```

---

[20]The performance can be improved somewhat, e.g., by setting the flag kkt=FALSE to avoid unnecessary computations or using parscale when parameter values vary by multiple orders of magnitude. See the documentation for optim and optimx for details.

```
# Return the -log(Lik) value under a normal distribution. Here
# we use the built-in ability to calculate log-density values
# by setting the argument log=TRUE. See ?dnorm for details.
return(-sum(dnorm(Ys,theta["c"]*xout[,2],
            theta["sigma"],log=TRUE)))
}
```

The second (preferred) approach, implemented in nLL2 below, uses transformations (like those discussed in Sect. 2.6) to ensure parameter values remain strictly positive. Here, (positive-valued) parameters are transformed using the natural log and then unconstrained optimization is performed using these real-valued transformed quantities. The resulting log-transformed values can then be transformed back to positive parameter values using the exponential function $\exp(x) = e^x$. In the example above, if we want r to remain strictly positive, we can re-parameterize nLL1 using the transformed quantity $q = \log(r)$ (where $r = \exp(q)$ is positive for all $-\infty < q < \infty$). Likewise, defining $k = \log(K)$, $C = \log(c)$, and $S = \log(\sigma)$ so that $K = \exp(k)$, $c = \exp(C)$, and $\sigma = \exp(S)$, we can rewrite nLL1 above as a function of these new parameters $P = (q, k, C, S)$ where $\theta = \exp(P)$.

```
nLL2 = function(P, tydat) {
  Times = tydat[,1] # 1st column are the times
  Ys = tydat[,2]    # 2nd column are the y values
  # Next, obtain x(t) under the given parameters
  x0 = c(x=50) # As above, we assume x0 is known.
  xout = ode(x0, times=Times, func=odefunc,
             parms = c(r=exp(P[["q"]]), K=exp(P[["k"]])))
  # Return the log(Lik) value under a normal distribution
  return(-sum(dnorm(Ys,
         exp(P[["C"]])*xout[,2], exp(P[["S"]]), log=T)))
}
```

While, in theory, the optimal parameter values obtained from these two approaches should be identical, in practice round-off error and other factors can lead to different outcomes. Optimization methods tend to perform better using the log-transformation with unconstrained optimization, or using an explicit constrained optimization method instead of the approach implemented in nLL1 above.

To illustrate this, the following code obtains parameter estimates using both approaches described above. This requires that initial values are provided for the parameters we would like to estimate, as required by optim.

```
# Initial parameter values for optimization using nLL1
theta0 = c(r=1,  # Next, use the last y value for K, but
           # divide that y value by c to get an x value...
           K=as.numeric(mydata[nrow(mydata),2])/0.6,
           c=0.6, sigma=0.1)

# The log transformed initial values for nLL2
P0 = c(q = log(theta0[["r"]]), k = log(theta0[["K"]]),
       C = log(theta0[["c"]]), S = log(theta0[["sigma"]]) )
```

```
# The true values, for comparison (impossible with real data!)
Ptrue = c(q = log(params[["r"]]),  k = log(params[["K"]]),
          C = log(params[["c"]]),  S = log(params[["sigma"]]) )

# Fit the model both ways, then compare.
fit1 = optim(theta0, nLL1, tydat=mydata,
             control=list(maxit=1e9, reltol=1e-10))
fit2 = optim(P0, nLL2, tydat=mydata,
             control=list(maxit=1e9, reltol=1e-10))
```

The results from these two optimization runs are summarized below by showing the lowest negative log likelihood value found, the corresponding parameter values, the number of calls of the objective function (counts), and the convergence code[21] indicating that the optimization routine reported no errors.

```
# Combine rows into a single object to display the results
rbind(fit1 = c(fit1$par, nLL=fit1$value,
              counts=fit1$counts[[1]], conv.code=fit1$convergence),
      fit2 = c(exp(fit2$par), nLL=fit2$value,
              counts=fit2$counts[[1]], conv.code=fit2$convergence),
      true = c(params, nLL1(params,mydata), NA, NA) );
```

	r	K	c	sigma	nLL	counts	conv.code
fit1	1.8623	430.47	0.11738	2.2161	66.448	779	0
fit2	1.8619	430.29	0.11744	2.2171	66.448	295	0
true	2.0000	500.00	0.10000	2.0000	67.859	NA	NA

Observe that the estimates are effectively the same, but the number of function calls indicates that the optimization run on the log-transformed parameter values (fit2) converged much faster. Furthermore, notice that the best estimate is slightly off from the known ("true") values used to simulate these data, due to the added noise and finite sample size. For more on implementing and debugging this optimization approach to parameter estimation, see [128, 129, 161].

Exercise 10 In the example above, we may wish to further assume c is a proportion that cannot exceed 1. What other invertible functions might one use to map the open interval $(0, 1)$ to $\mathbb{R}$?

Challenge Problem 13 Read through R's help documentation for the optim function (and the optimx function in the optimx package; see also [161]), and identify one or more constrained optimization methods that can be used as an alternative to Nelder–Mead. Also, modify the code above to include a third objective function identical to nLL1 but with no checks on the sign of the parameter values. Apply one or more constrained optimization method(s), and provide a comparative summary of your results.

---

[21]Convergence code 0 indicates no errors, and convergence code 10 indicates degeneracy of the Nelder–Mead simplex. The code 10 often occurs when the model is not identifiable and reaches a "flat spot" in the objective surface.

2.7.5 Recognizing Identifiability (Estimability) Problems

In practice, before we attempt using optimization to find our best-fit parameter values, we must ask one important question that was neglected in the above example: *Does this model and estimation procedure have a unique best-fit parameter set, or might there be multiple sets of parameter values that give equally good fits?* That is, phrased in terms of the geometry of the negative log likelihood surface: Is it roughly bowl shaped (with a distinct minimum) or is it more like a valley with a connected line of minima (or worse!)? In the first case, we call the model *identifiable*—it has a unique likelihood-maximizing parameter set. In the second case, we say the model is *unidentifiable*. While such problems may not affect forecasting of your observable quantities, it is often the case that addressing any such identifiability problems is essential to properly performing and drawing inferences from the parameter estimates (e.g., see [11, 33, 46, 47, 57, 141]). Below, we introduce some ways of assessing whether identifiability problems exist, and some approaches to correcting the problem.

There are two predominant causes of identifiability problems that arise in practice. First, sometimes our data lack the information needed to estimate one or more parameters. For example, if our logistic model data were restricted to just the early exponential growth phase of the trajectory shown in Fig. 5, we might be able to estimate growth rate r, but our data would lack information about the carrying capacity K. This can make K very difficult to estimate, and may result in wildly different estimates of K depending on where the estimation procedure was initialized. This can sometimes (but not always) lead to numerical problems if the optimization methods used require the existence of a unique optimum. Perhaps a worse outcome is that no numerical errors are reported, and a seemingly good (but in reality, arbitrary) estimate of K is obtained. Without any further identifiability or uncertainty analysis, the conclusions drawn from that flawed estimate would be

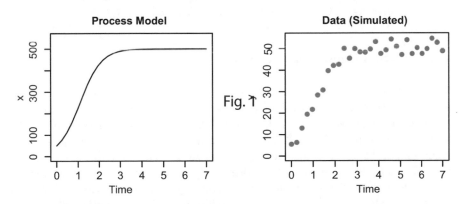

Fig. 5 A trajectory $x(t)$ of logistic model Eq. (47) (left) and a corresponding data set where the simulated data values y_i are normally distributed with a mean that is proportional to $x(t_i)$. For more details, see the corresponding code in the main text

spurious (e.g., if K being very large or very small were of great importance). This would be a *practical identifiability* problem, since it results from a shortcoming in the data, not our model.

The second cause for identifiability problems arises from an overparameterized model, e.g., where there are multiple parameter sets that yield the same outcomes for the deterministic part of our model used to define the mean of observation distributions. We call this a *structural identifiability* problem, and unlike practical identifiability problems, these persist even when estimating parameters from ideal data (e.g., full trajectories with no noise and arbitrarily large sample size).

For a simple example of a structurally unidentifiable model, consider the simple linear regression model where the slope parameter m is replaced with two a parameter expression $m = m_1 + m_2$, i.e., consider

$$E(y_i) = (m_1 + m_2) x_i + b. \tag{49}$$

It is no surprise that this model has infinitely many best-fit parameter sets! Specifically, if the standard regression model had a best-fit slope of say $\widehat{m} = 5$ and intercept of $\widehat{b} = 2$ then for our overparameterized model we would get equally good fits out of any set of parameters where $\widehat{m_1} + \widehat{m_2} = 5$. Notice that in this simple example, the problem has nothing to do with the amount of noise in the data, or the sample size, but it is a fundamental property of the deterministic part of our statistical model.

Since, in practice, both structural and practical identifiability problems manifest in the form of an objective function that does not have a unique and/or well defined optimum, a *practical identifiability analysis* is a good approach to assessing whether or not such problems exist. This can then be complemented by a *structural identifiability analysis* to look for identifiability problems that are caused by the deterministic part of our model, if needed.

Next, we look at some approaches to conducting these identifiability analyses to assess presence and extent of these problems, and see some ways to correct them.

2.7.6 Practical Identifiability Analysis

In practice, to assess whether or not an identifiability problem exists, the first step is often to perform a practical identifiability analysis. *Likelihood profiling* is a common way to perform a practical identifiability analysis when using a likelihood based estimation procedure [32, 46, 141], because this is often the method used for quantifying uncertainty in the parameter estimates (i.e., to approximate confidence intervals) and thus requires no additional analyses. A more simplistic alternative that can be used in other estimation frameworks (e.g., least-squares regression) is to run the estimation procedure using multiple different initial parameter values, then check to see if each converges to the same, or different, best-fit parameter estimates.

To illustrate this approach, we again consider estimating parameters for our logistic example above, but we now assume x_0 is unknown and thus a quantity

we would like to estimate. To implement this in R, the code below first defines
a negative log likelihood function nLL (an extension of nLL2) that takes in the
additional free parameter $X_0 = \log(x_0)$. Unlike the case above, where x_0 was
a known quantity, this additional free parameter makes the model structurally
unidentifiable.

Below, the estimation procedure described above was run for the full model (i.e.,
for nLL) using 1000 different initial parameter values. The pairwise relationships
among the best estimates out of those 1000 runs (based on nLL values) are plotted
in Fig. 6 to show their range of variation as well as any obvious patterns indicating
trade-offs between pairs of parameters. Here x_0 is initialized using the first value in
our data set, and the 1000 different initial parameter sets are obtained by random
sampling from a uniform distribution so that each parameter value is within ±30%
of the initial parameter values specified above.

```
# nLL for the full example model with log-transformed parameters
nLL = function(P, tydat) {
    Times = tydat[,1] # 1st column are the times
    Ys = tydat[,2]    # 2nd column are the y values
    xout = ode(c(x=exp(P[["X0"]])), times=Times, func=odefunc,
               parms=c(r=exp(P[["q"]]), K=exp(P[["k"]])))
    # Return the -log(Lik) value under a normal distribution
    return(-sum(dnorm(Ys, mean=exp(P[["C"]])*xout[,2],
                sd=exp(P[["S"]]), log=TRUE)))
}

# Use the first y value for a crude initial estimate of x0
P0 = c(X0 = log(as.numeric(mydata[1,2])/0.6),
       q = log(theta0[["r"]]), k = log(theta0[["K"]]),
       C = log(theta0[["c"]]), S = log(theta0[["sigma"]]) )

# initialize a data frame to store optimization output
fits=data.frame(x0=NA,r=NA,K=NA,c=NA,sigma=NA,nLL=NA,code=NA)
set.seed(123)  # for repeatable random number generation
for(i in 1:1000) {
    # Random parameter values are within +/- 30% of true values
    P0i = log( exp(P0) * runif(length(P0), 0.7, 1.3) )
    fiti = optim(P0i, nLL, tydat=mydata, control=list(maxit=1e5))
    fits[i,] = c(exp(fiti$par), nLL=fiti$value,
                 code=fiti$convergence)
}
# select the estimates within 0.01% of the best fit
fits = fits[order(fits$nLL,decreasing=FALSE), ]
fits2 = fits[(fits$code==0 | fits$code==10) &
             fits$nLL <= 1.0001 * min(fits$nLL, na.rm=T), ]
# Omit some extremes to clean up the graphical output below
fits2 = fits2[fits2$K != max(fits2$K), ]
fits2 = fits2[fits2$K != min(fits2$K), ]
# log-scale plots to show pairwise tradeoffs, parameter ranges
pairs(fits2[,1:5], log="xy",
      labels=c(expression(x[0]),"r","K","c",expression(sigma)))
```

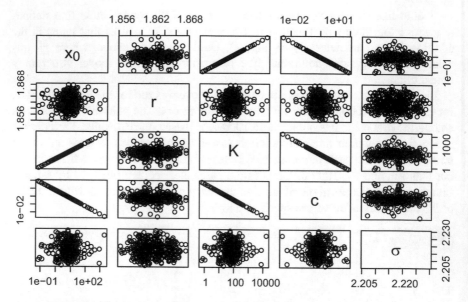

Fig. 6 Parameter estimates, on a log scale, obtained by running the optimization routine in our estimation procedure for 1000 different starting parameter sets, then taking the estimates within 0.01% of the best estimate's nLL value and plotting their pairwise relationships (note the best-fit negative log likelihood value of these 350 top fits is 66.4476 and the worst is 66.4488). These estimates (which exclude some extreme values to clarify patterns in the plots; see code for details) have the following coefficients of variation: $cv_{x_0} = 5$, $cv_r = 0.0013$, $cv_K = 5.1$, $cv_c = 3.2$, and $cv_\sigma = 0.0023$. The very close negative log likelihood values, combined with the large ranges of variation and correlations between some parameters indicate that there is an identifiability problem (i.e., there is apparently no unique best-fit parameter set, either due to there being more parameters than estimable quantities in the model, or the optimization method lacks the sensitivity to find an existing optimum due to a nearly flat objective function near that optimum). It seems only r and σ appear to be identifiable. Linear relationships on these log–log plots indicate linear, inverse, or power law relationships between best-fit parameters. These estimates are slightly better (lower negative log likelihood values) than the previous figure, where x_0 was held constant at a non-optimal value. See the code in the main text for further details

```
rownames(fits2)=c()  # remove row numbers for the output below
head(fits2[,1:6],5)  # Show the 5 best parameter estimates
```

```
         x0       r        K        c  sigma     nLL
1  19.35396  1.8624  166.5768  0.30336  2.2170  66.448
2  11.51700  1.8623   99.1655  0.50959  2.2196  66.448
3   5.25086  1.8615   45.1620  1.11897  2.2166  66.448
4   3.55392  1.8610   30.5587  1.65359  2.2161  66.448
5   0.79901  1.8616    6.8779  7.34744  2.2188  66.448
```

```
summary(fits2[,1:3])
```

```
      x0                    r                   K
 Min.   : 0.07      Min.    :1.86      Min.    :  0.6
 1st Qu.: 4.31      1st Qu.:1.86      1st Qu.: 37.2
```

Median :	7.25	Median :	1.86	Median :	62.6
Mean :	25.21	Mean :	1.86	Mean :	217.2
3rd Qu.:	13.79	3rd Qu.:	1.86	3rd Qu.:	119.1
Max. :	1934.69	Max. :	1.87	Max. :	16685.9

`summary(fits2[,4:6])`

c		sigma		nLL	
Min. :	0.003	Min. :	2.21	Min. :	66.4
1st Qu.:	0.424	1st Qu.:	2.21	1st Qu.:	66.4
Median :	0.807	Median :	2.22	Median :	66.4
Mean :	2.278	Mean :	2.22	Mean :	66.4
3rd Qu.:	1.359	3rd Qu.:	2.22	3rd Qu.:	66.4
Max. :	78.369	Max. :	2.23	Max. :	66.4

Note that, while the estimates for r and σ are tightly clustered, the others vary by multiple orders of magnitude! These wildly different parameter estimates, each fitting equally well to the data (i.e., they have nearly identical negative log likelihood values), are a strong indication of an identifiability problem.

Next, we conduct a structural identifiability analysis to assess whether or not these problems arise from a structural identifiability problem (and would therefore persist even for data with a very large sample size and little to no noise). Importantly, even though this does not always yield the desired results (see [46, 121] and references therein), it can often provide equations that exactly describe the parameter trade-offs evident in Fig. 6.

2.7.7 Structural Identifiability Analysis

Structural identifiability problems can often be identified (and sometimes, corrected for) by a structural identifiability analysis (e.g., see [11, 33, 46, 57, 122, 123, 141, 146] and references therein) that determines whether or not there is a one-to-one mapping between the model parameters and the quantities in the model that can be estimated (e.g., in our simple linear regression example above, Eq. (49), there is no one-to-one mapping between parameters m_1, m_2, and b and the estimable slope m and intercept b). These estimable quantities are also often called (identifiable) *parameter combinations*. A commonly used method of conducting a structural identifiability analysis is the *input–output equation approach* detailed in the references above, and the benefit of this approach is that it can give you generic analytical relationships (e.g., like the $m = m_1 + m_2$ relationship in Eq. (49)) that can be used to correct the problem, e.g., by fixing one of the problem parameters in order to make one or more others uniquely estimable.

Since the above practical identifiability analysis indicates an identifiability problem, we here apply the input–output equation approach to determine whether or not Eq. (46) is a structurally unidentifiable model, and if so, to discover the identifiable parameter combinations.

To do this, we first assume a zero-variance (deterministic) observation variable $y(t) = c\,x(t)$ and differentiate it with respect to t. Second, we substitute the process model Eq. (47a) and substitute $x = y/c$ to obtain a differential equation that is only in terms of the observation variable y. The resulting equation is referred to as the *input–output equation*, and for our logistic example it is given by

$$\frac{dy}{dt} - r\,y - \frac{r}{c\,K}y^2 = 0, \quad y(0) = c\,x_0. \tag{50}$$

Since this equation is a monic polynomial in terms of y and derivatives of y, it follows that are our identifiable parameter combinations (i.e., the quantities we can estimate from data) are the two coefficients in the input–output equation and the initial condition, i.e., one can only estimate values for

$$r, \quad \frac{r}{c\,K}, \quad \text{and} \quad c\,x_0. \tag{51}$$

What does this imply about the estimability of our model parameters? If there were a one-to-one mapping between these three identifiable parameter combinations and our four parameters, then we would expect there to be a unique best parameter estimate (given ideal data). But, note that we have three identifiable parameter combinations and four model parameters, hence we cannot have a one-to-one mapping between them. Thus, our example logistic model is structurally unidentifiable.

Obtaining these expressions for the trade-offs between parameters are a major benefit of conducting a structural identifiability analysis. From the above example, we have learned that we should expect a unique estimate for r, but any parameter values for c, x_0, and K that yield the same "best-fit" values of $c\,K$ and $c\,x_0$ will fit equally well. Thus, the pairwise plots in Fig. 6 should show an inverse relationship between c and x_0 and between c and K (or, a linear relationship with negative slope when viewed on a log–log plot), and a linear relationship with positive slope between x_0 and K. Also, by fixing one parameter—e.g., c or K or x_0 (but not r)— there would then be a one-to-one mapping between the three identifiable parameter combinations and the three unknown model parameters. Often, in applications, some parameters can be assigned values based on independent experiments, so holding constant the most well known of these parameter values might correct these identifiability problems and thus permit investigators to estimate the other unknown parameters from data.

Exercise 11 In the logistic example above, prove that fixing c, K, or x_0 corrects the identifiability problem, whereas fixing r does not. To do this, suppose there are two different sets of best-fit parameter values (r_1, K_1, x_{01}, c_1) and (r_2, K_2, x_{02}, c_2) that yield the same identifiable quantities (51), i.e., $c_1 x_{01} = c_2 x_{02}$, $r_1 = r_2$, and $\frac{r_1}{c_1 K_1} = \frac{r_2}{c_2 K_2}$. Fix each parameter in turn (e.g., fix $c_1 = c_2 = c$), and show whether or not the resulting system of equations permits two distinct parameter sets.

Exercise 12 Structural identifiability results can provide explicit functional forms describing trade-offs among unidentifiable parameters, like those shown in Fig. 6. These can be found by writing our identifiable quantities explicitly and solving for the implied relationships between parameters. For the example above, we have the identifiable quantities $\alpha_1 = c\,x_0$, $\alpha_2 = r$, and $\alpha_3 = r/(c\,K)$, which are uniquely estimable from an ideal data set. Use these expressions to show that the ratio K/x_0 is constant, and thus K and x_0 have a positive linear relationship consistent with Fig. 6. Verify the other pairwise relationships suggested by the log–log plots in Fig. 6.

Let us now use these results to correct our estimation procedure. Suppose that, in our full logistic growth example, we had good reason to believe that our sampling design captured roughly 10% of the population at any given time. That is, we can reasonably assume $c = 0.10$. In that case, our negative log likelihood function could be modified so that it takes in a vector of parameter values that does not include c (and instead holds it fixed at 0.10), and likewise c would be omitted from initial parameter value vector provided to `optim`.

These modifications are implemented in the practical identifiability analysis below, which replicates the parameter estimation procedure 500 times for different initial parameter sets. The results are shown in Fig. 7, where pairwise scatterplots of the best-fit parameter estimates can be used to assess whether or not replicates seem to be converging to a unique best-fit parameter set.

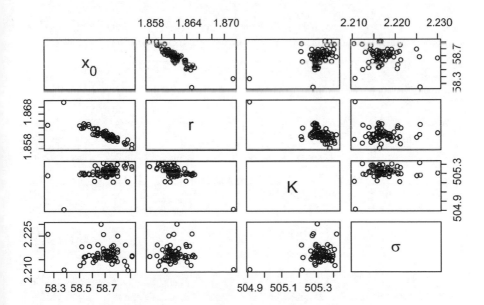

Fig. 7 Pairwise plots of parameter estimates for the logistic model Eq. (47) where $c = 0.1$ is fixed to correct the structural unidentifiability of the model (see Fig. 6). To illustrate the improved convergence of estimates, 500 different initial parameter sets were used, and for the 65 of those estimates with a negative log likelihood value within 0.01% of the best estimate, these had the coefficients of variation: $cv_{x_0} = 0.002$, $cv_r = 0.001$, $cv_K = 0.00013$, and $cv_\sigma = 0.0015$

```
nLLc = function(P, tydat) {
   Times = tydat[,1] # 1st column = times; 2nd = y values
   Ys = tydat[,2]
   x0 = c(x=exp(P[["X0"]]))
   xout = ode(x0, times=Times, func=odefunc,
              parms = c(r=exp(P[["q"]]),K=exp(P[["k"]])))
   # Return -log(Lik) value as in previous examples
   return(-sum(dnorm(Ys, 0.1*xout[,2], exp(P[["S"]]),log=T)))
}


# initial P values for all but C=log(c)
P00 = P0[-4] # exclude the 4th element, C=log(c), from P0

# Pairwise plots to verify the identifiability problem is fixed
fits = data.frame(x0=NA,r=NA,K=NA,sigma=NA,nLL=NA,code=NA)
for(i in 1:500) {
   # Random parameter values are within +/- 20% of true values
   P00i = log( exp(P00) * runif(length(P00),0.8, 1.2) )
   fiti = optim(P00i, nLLc, tydat=mydata,
               control=list(maxit=1e6, reltol=1e-10))
   fits[i,] = c(exp(fiti$par), nLL=fiti$value,
               code = fiti$convergence)
}
# Display the best-fit estimates and the pairwise scatterplots
fits  = fits[order(fits$nLL, decreasing=FALSE), ]
# Only include those with a nLL withing 0.01% of the best fit
fits2 = fits[(fits$code==0 | fits$code==10) &
            fits$nLL <= 1.0001 * min(fits$nLL, na.rm=T), 1:5]
summary(fits2[,-5]) # Omit nLL quartiles (all are very close)
```

x0	r	K	sigma
Min. :58.25	Min. :1.858	Min. :504.9	Min. :2.210
1st Qu.:58.64	1st Qu.:1.861	1st Qu.:505.3	1st Qu.:2.216
Median :58.72	Median :1.862	Median :505.3	Median :2.216
Mean :58.70	Mean :1.862	Mean :505.3	Mean :2.217
3rd Qu.:58.76	3rd Qu.:1.863	3rd Qu.:505.4	3rd Qu.:2.219
Max. :58.91	Max. :1.871	Max. :505.4	Max. :2.230

```
# Estimates should be close to params, used to simulate the data,
# but not exact due to noise and finite sample size.
c(x0=50, params)  # The values used to simulate the data.
```

x0	r	K	c	sigma
50.0	2.0	500.0	0.1	2.0

```
rownames(fits2)=c() # erase the old row numbers from `fits
head(fits2[,1:5], 4) # The best estimates obtained are...
```

	x0	r	K	sigma	nLL
1	58.68502	1.862500	505.3260	2.215778	66.44762
2	58.69906	1.862127	505.3115	2.218590	66.44763
3	58.73725	1.861839	505.3141	2.216148	66.44763
4	58.72716	1.861420	505.3199	2.215563	66.44764

```
pairs(fits2[,1:4], log="xy",
      labels=c(expression(x[0]), "r", "K", expression(sigma)))
```

From these pairwise plots of parameter estimates, we see there is still some small variation in our estimates, but the best estimates are now all clustered tightly around the same value, and close to the true values used to simulate the data (note the estimates would deviate more from the true values if we had fixed c at a value away from 0.1). Some deviation from the true values is always expected when the data set is noisy and has a small to moderate number of data points. Note that some parameters show correlations, e.g., r and x_0, indicating that there is some slight trade-off between them. This is often the case when the objective surface (i.e., the negative log likelihood surface) has a unique optimum but it lies along a nearly flat trough with only a slight amount of curvature near the optimum, which can cause the optimization routine to end prematurely. This suggests that our estimates might be refined, if needed, using a different optimization method or by modifying the control parameters so that the criteria for when to conclude that an optimum was found are more conservative.

Exercise 13 Suppose data for the θ-logistic model (see Eq. (36)) were simulated only for a short period of time starting with $x_0 \ll K$ so that only the exponential growth phase was reflected in the data. Would you expect to be able to estimate K or the exponent θ with much certainty? Likewise, for the logistic or θ-logistic models, would you expect to be able to estimate r with any certainty if the data were only sampled from a trajectory that started at carrying capacity $x_0 = K$? Discuss these in the context of *structural* versus *practical identifiability*.

Challenge Problem 14 There are still some correlations in Fig. 7 suggesting some "trade-off" between estimated parameter values. Explore different optimization routines and control settings. Are the correlations and variation in estimates improved by a more finely tuned optimization procedure? How does increasing or decreasing the noise, or the sample size, of our simulated data affect these estimates?

Challenge Problem 15 (Standardizing Parameters for Optimization) The Nelder–Mead algorithm used in the examples above (as well as many other optimization methods) may not perform well in situations where the parameter values in question span many orders of magnitude. A simple way to correct this problem is to standardize the optimization parameters. To do this, given p unknown parameter values, one can introduce a new set of p scalars (initially set at 1) that can be multiplied by a base parameter vector to create a new parameter vector. Optimization is then done on these multipliers using a new objective function that takes these new parameters as inputs, multiplies them by a base parameter set, and then computes a negative log likelihood value for the resulting set of model parameters using the original nLL function. To do this, begin with nLL above and a base parameter set params0 such that the usual optimization call would be optim(params0, nLL, ...). Next, write a new function nLLscaled that takes a vector of parameter values M (initially all set to 1) as an input, and also a second parameter vector basepars which can be passed in as an argument to optim using `optim(M, nLLscaled, basepars=params0, ...)`. The optimum parameter values returned by `optim` can be multiplied by `params0` to obtain the best-fit model parameters.

Implement this approach for the logistic example above, where $c = 0.1$ is assumed to be known, and compare its performance (e.g., in terms of rates of convergence and goodness of fit) to the unstandardized approach. Finally, note that `optim` and `optimx` have a built-in way of achieving this scaling using the `parscale` option (see the documentation for `optim` and `optimx`). Add this built-in option to your comparison of approaches.

2.7.8 Statistical Analyses Beyond Parameter Estimation

Once we are happy with our parameter estimation procedure, other statistical analyses can and should be conducted. For example, we typically should conduct some sort of *uncertainty quantification* to assess how much certainty we have in our parameter estimates. Uncertainty quantification is almost always necessary, as it provides an essential context for evaluating any inferences drawn from estimates obtained by fitting models to data. In the context presented above, uncertainty quantification can be done by approximating confidence intervals for the parameter estimates using *likelihood profiling* techniques, which, as mentioned above, can also be used for practical identifiability analysis [47]. Resampling methods like bootstrapping or other approaches to uncertainty quantification may also be used, e.g., see [1, 2, 19, 21, 32, 70, 79, 114] and references therein. Model comparisons can also be made, e.g., using likelihood ratio tests, relative AIC values, or similar criteria (e.g., see [21, 123] and references therein). While uncertainty quantification is not treated here in detail, readers fitting models to data are strongly encouraged to quantify the level of uncertainty in their parameter estimates, forecasts, etc. as part of any presentation or discussion of their final results.

2.7.9 Alternative Approaches to Parameter Estimation and Uncertainty Quantification

New and improved methods for parameter estimation and uncertainty quantification for ODE-based statistical models are still being actively developed, but in practice the above approach is a good place to start. As mentioned at the start of this section, there are other approaches than the one detailed above and some have even been implemented as R packages. For example, the `deBInfer` package can be used to implement Bayesian parameter estimation for ODE models [19]. A Bayesian framework offers many advantages, including a straightforward avenue to uncertainty quantification using the posterior parameter distributions. Sometimes likelihood surfaces that involve solutions to ODEs can deviate greatly from the ideal "hump shaped" likelihood surface with a distinct optimum, and thus make for a difficult computational optimization problem. Alternative methods have been developed that instead use a different measure of how well the model fits the data by comparing the right hand side of the ODE model (i.e., derivatives of the solution curves) to derivatives of a non-parametric smooth curve fit to time series data [139]. The package `CollocInfer` implements one of these *functional data analysis* methods in R [87, 88], and there is also Matlab software for use with the FDA package [86]. See also [14, 32] and the *probe matching* approach discussed in [50, 97, 151].

2.7.10 Closing Remarks on Fitting ODE Models to Data

In general, it is strongly advised that any parameter estimation procedure be assessed using data simulated from the exact statistical model being used, to verify that the implementation was done correctly and that the estimates look reasonable (additionally, this simulation-estimation check can be repeated for many simulated data sets to generate a "boot-strapped" empirical distribution for the estimates, which can then be used to assess estimator properties, e.g., to quantify bias). An identifiability analysis also further helps to ensure the parameter estimation procedure works as intended. If optimization challenges arise, other methods are available by selecting a different `method` option in `optim`, or multiple methods can be compared using `optimx` in the R package `optimx`. For even more optimization options in R, see the CRAN Task View page on Optimization [161] which lists additional methods available in various contributed R packages.

3 Identifying and Modifying Model Assumptions

The sections above introduce some useful methods for analyzing ODE models or otherwise using them in an applied scientific setting. They also introduce some of the simple ways in which new ODE models can be derived by altering the assumptions of an existing model. In the sections that follow, we take a closer look at the task critically evaluating ODE model assumptions, and formulating new models that in some cases lead to new non-ODE models.

To aid in identifying assumptions associated with a given ODE model that might be altered to yield a new ODE (or non-ODE) model, it helps to consider a few general features of ODE models. First, there is the continuous (vs. discrete) nature of ODE models, both in time and state space, as well as the deterministic (vs. stochastic) nature of ODE models. Modifying one or both of these assumptions may be both scientifically fruitful and mathematically interesting to investigate. Relaxing the memoryless (vs. history-dependent) nature of ODEs can often yield different results from a corresponding set of ODEs, as can altering the assumptions that a system is autonomous (vs. driven by external forcing terms), non-spatial (vs. being explicitly spatial), well-mixed, and/or composed of identical (vs. heterogeneous) individuals. Modifying model assumptions like those listed above will often lead to opportunities to formulate new non-ODE models, while altering other assumptions is perhaps more likely to yield new ODE models [4, 6, 45, 51, 100].

To illustrate, let us revisit the SIR model Eqs. (8) and its assumptions. First, because ODEs are inherently memoryless (i.e., the future state of the system depends only on the current state of the system, not the past), introducing any dependence on past states of the system may yield a non-ODE model, unless the necessary information about that history can be incorporated into the state variable values at the present time (e.g., the addition of an exposed class and the resulting SEIR model is a way of accounting for history by assuming an exponentially distributed lag that manifests as an intermediate exposed state). As stated above, the dwell times in states S and I follow Poisson process first event time distributions (e.g., the duration if infectiousness is exponentially distributed with rate γ), and altering those distributions (as shown below) can yield ODE and non-ODE models.

Second, the SIR model is seemingly agnostic of any spatial relationships between individuals, but (like many ODE models) it actually implicitly assumes that the system is well mixed as opposed to assuming a contact process between individuals that is, e.g., limited to nearest neighbors over some spatial domain. This implicit assumption also implies no other sources of spatial heterogeneity, e.g., areas where transmission rates are higher or lower than other areas.

Third, in the basic SIR model, individuals are assumed to be identical (or identically distributed) with respect to infectiousness and recovery times, as opposed to assuming multiple categories of individuals differing in their transmission risk (e.g., in modeling sexually transmitted infections, modelers typically acknowledge that sexual interactions are not random, so individuals are often partitioned into age classes or into groups that engage in low- versus high-risk behavior) or differ in disease progression (e.g., vaccinated versus unvaccinated individuals, or individuals that do or do not have a certain genotype that confers some protection from disease).

Fourth, the SIR model also assumes constant rates of per capita infection and recovery, versus rates that vary over time. However, the transmission rate for many childhood diseases is strongly driven by whether or not school is in session, as this increases the contact rate among infected and susceptible children.

Fifth, the SIR model is also continuous in both time and state space, which may not be desirable if one seeks to quantify the time it takes for an epidemic to end (i.e., for $I(t) = 0$, which cannot happen in finite time under the ODE model) or if one wants to use the SIR model as part of a statistical model for the analysis of weekly case count data, which might be more tractable with a stochastic discrete-time, discrete-state model.

Lastly, the SIR model makes very specific assumptions about the functional forms of the transmission and recovery rates. It is possible to instead define models that include more mathematically general terms (e.g., instead of defining the force of infection as $\lambda(I) \equiv \beta I$ one could more generally assume that the force of infection λ is instead some unknown strictly increasing function of I—and possibly a function of S and R too—where $\lambda(0) = 0$) which are quite amenable to mathematical analyses to explore what results can be obtained that apply across this broader class of models. This kind of model generalization based on more specific models can go a long way to draw more general conclusions from specific study systems, and can also provide important context for drawing attention to the full range of dynamics that one might observe in similar real systems (e.g., see [35]).

Modelers strive to find an appropriate balance between simplicity and reality. As you can see from the SIR model assumptions listed above, there are often myriad ways of altering the assumptions of an existing ODE model. Some of these may be important to investigate, e.g., to better understand how a certain simplifying assumption might impact a key results obtained from the model. On the other hand, incorporating too many complicating assumptions can quickly make models too complex to analyze, diminishing their utility for applications.

Students can often find an interesting research project by focusing on a single ODE model assumption (e.g., the presence or absence of an Exposed class in the SIR model) and that might be altered to yield a new set of ODEs that are still simple enough to analyze and compare to the original model. But, students are particularly encouraged to consider altering those simplifying assumptions that primarily serve to ensure the model is a system of ODEs, especially when such assumptions might seem to be an egregious oversimplification given the research question at hand, and given the specifics of the real-world system(s) being modeled.

In the next few sections, we briefly introduce some approaches to generalizing, or otherwise altering, these ODE model assumptions. To give a full treatment of the analytical and computational aspects of deriving and analyzing these models is beyond the scope of this chapter, so instead an effort has been made to provide references to resources that can be consulted to dig deeper into these topics.

3.1 Autonomous ODEs to Non-autonomous ODEs

Autonomous ODE models, which only depend on the current state of the system (i.e., that could be written with $f(\mathbf{x}, \theta)$ in the context of Eq. (1)), assume that the system dynamics are not governed by any other time-varying quantities. This is often a major simplifying assumption for biological systems where, for example, the true rates of growth, reproduction, and survival may vary greatly over time following changes in temperature, light levels, or evolved seasonal patterns of behavior. Accounting for this time dependence is often accomplished by incorporating time-varying terms into ODEs, especially when incorporating strong seasonal or other periodically varying quantities into the model (e.g., seasonal transmission patterns in infectious diseases; e.g., see [34, 71] and references therein). Aperiodic and/or stochastic environmental forcing terms can also often be represented by continuous functions of time, e.g., to include temperature dependencies in a population growth model using hourly temperature data interpolated to create a continuous function of time.

While the choice to use a non-autonomous ODE model may preclude something like an equilibrium stability analysis (since the notion of an equilibrium point is often irrelevant when the system is being forced), however such extensions can often make for a great follow-up study of a system previously modeled with autonomous ODEs [44, 143, 158]. Mathematical analysis can be done in some cases, e.g., where the forcing functions are bounded around some mean or are periodic (e.g., see [15, 108]), or where they can be approximated by step functions thus allowing one to represent the model with autonomous ODEs over specific time intervals [89].

For a simple example of a seasonally forced model, consider a logistic growth model where we have separated out the birth and death processes in the form

$$\frac{dx}{dt} = b\,x - (d + \mu\,x)\,x = (b - d)\,x \left(1 - \frac{x}{(b - d)/\mu}\right) \tag{52}$$

where the birth rate is modeled by an exponential-growth-like term with rate b, and the per capita mortality rate $d + \mu\,x$ is *density dependent* in that it increases from a baseline level d at rate $\mu > 0$ with increasing population size x. Rewritten in the standard logistic equation form (Eq. (21)), we see the model has an exponential growth rate $r = b - d$ (we may assume $b > d$ to ensure population growth) and a carrying capacity $K = (b - d)/\mu = r/\mu$.

The following exercises illustrate how one might modify such a model to include time-varying processes.

Exercise 14 Suppose Eq. (52) was to be used to model a population with density-dependent growth, where temperature was either to be held constant in one set of experiments or cycled between high and low temperatures to mimic a day–night temperature cycle. To make this model non-autonomous, one could make b a positive-valued function of time, and/or d a positive-valued function of time, and/or the quantity $b - d$ a function of time (e.g., by multiplying it by a function that oscillates between values a little above and a little below 1, and whose long-term average is 1), and/or μ a positive-valued function of time. For what scenarios might one of these be more appropriate than the other?

Challenge Problem 16 Numerically investigate two or more instances of the modified logistic models described in the exercise above.

Challenge Problem 17 Write down equations for a modified SIR model with a periodic transmission rate $\beta(t)$, e.g., using sine, cosine, or step functions (i.e., piecewise constant functions) that alternate periodically between two positive values. Modify the R code in previous sections to find numerical solutions to this seasonally forced SIR model, and compare model trajectories under different scenarios.

3.2   Deriving Deterministic Discrete-Time Models

For some applications, models that change according to discrete-time steps are sometimes desirable, thus one may want to modifying the assumptions of an existing ODE model to derive an analogous deterministic discrete-time model (or, as discussed in the next section, a stochastic discrete-time model). For example, one might want to take an existing model of a continuously breeding organism and modify it for use with an annual plant or an animal population that has a very distinct breeding period each year (e.g., salmon or various insects, like mayflies). Another example may be a desire to consider a discrete-time analog of an ODE model as part of a statistical model to use with, e.g., weekly or monthly data. This is often accomplished using an exact or approximate representation of how the (continuous-time) ODE model trajectories change over a fixed time interval. For example, in some cases a discrete-time map can be obtained when ODE model solutions can be found analytically as in [89]. Alternatively, over short time steps, Euler's method gives an approximate discrete-time map $x(t+\Delta t) = x(t) + f(x(t), \theta)\Delta t$. However, there is some merit in taking a different approach by going back to a set of system-specific assumptions and re-deriving an appropriate model from first principles.

For examples of biological applications of discrete-time models, readers are encouraged to explore the literature. Many discrete-time models exist for studying population dynamics, and various texts address formulating and analyzing discrete-time models, including [4, 45, 50, 100, 126].

Exercise 15 Take the standard logistic model (or the form above, written in terms of b, d, and μ) and use integration by parts to verify the analytical solution curve

$$x(t) = \frac{x_0 K}{x_0 + (K - x_0)\exp(-r t)}, \tag{53}$$

starting at $x(0) = x_0$. Next, use that solution to find α and β such that this system has a discrete-time map

$$x(k + \Delta t) = \frac{x(k)}{\alpha + \beta x(k)}. \tag{54}$$

Hint: Consider the solution curve from time $t = k$ to $t = k + \Delta t$.

Challenge Problem 18 Use a `for` loop to simulate a trajectory of the above discrete map analogue of the continuous-time logistic equation. Use code from previous sections to obtain a numerical solution to the continuous-time logistic model, and also implement Eq. (53) and overlay these three trajectories in one plot.

Challenge Problem 19 Construct a discrete-time model (using the results from the exercise above, and Eq. (52)) for the following scenario. Assume a migratory population spends a portion of the year ($0 < T_1 < 1$) on the breeding grounds, where the population grows according to a logistic equation, then a proportion ρ_1 survive a quick (assume it is instantaneous) migration to the wintering grounds. Once there, natural mortality leads to an exponential decline in population size over the remainder of the year. Then, of the remaining individuals, a fraction ρ_2 survive the return trip to the breeding grounds to start another annual cycle. Construct a map $x(t+1) = f(x(t))$ that reflects all of these processes described above. Use resources like one of the references given above to do an analytical (or simulation-based) equilibrium stability analysis of this map (see also [89, 150] for similar models).

3.3 Deriving Stochastic Models

Revising the assumptions of an ODE model to can sometimes result in the need to derive a stochastic model, e.g., to investigate the role of demographic stochasticity arising from individual births and deaths in a population, or to alter implicit assumptions like the probability distribution describing the duration of time individuals are infectious in an SIR-type disease model, or to use in deriving a new deterministic model. Stochastic models can take various forms, ranging from Markov chains to discrete maps to stochastic differential equations (SDEs), just to name a few. This section introduces some common types of the stochastic dynamic models found in applications. For further reading, see [3, 6, 21, 45, 131] and references therein.

3.3.1 Continuous-Time, Discrete-State Stochastic Models

A natural stochastic model analogue of a given ODE model (if it can be viewed as a mean field model of some unspecified stochastic process) is the particular continuous-time stochastic process based on Poisson processes mentioned in Sect. 2.1. This is often used to add demographic stochasticity to a model, or alter implicit individual-level assumptions to derive a new mean field model. This process can often be derived via the approaches outlined in texts like [3, 6] and references therein.

Another way to construct this model, especially for simulation purposes, is the Stochastic Simulation Algorithm (SSA; also known as the Gillespie algorithm) [68, 69, 133]. In short, the SSA is a straightforward algorithm for simulating the stochastic analog of a mean field ODE using the Poisson process intuition illustrated in the exponential growth example in Sect. 2.1. There are various extensions and refinements of the SSA that an interested reader may wish to pursue, e.g., for time-varying processes [162] or the simulation of large systems where computational efficiency is important [26, 27, 31, 124]. Here we introduce the basic SSA algorithm.

To implement the SSA we first must identify all state transition events that lead to a change in state variables, the corresponding state vector changes for each type of event, and the corresponding rates for each event type. For example, in our SIR example, the two basic event types are the transition from susceptible to infected and the transition from infected to recovered, as detailed in Table 1. This information is used to iterate the following two main steps of the SSA: first, simulate the time to the next event, then determine which event type occurs.

Table 1 Possible changes to the state vector $\mathbf{x} = (S, I, R)$ in a stochastic discrete-state discrete-time SIR model with time step Δt, and their approximate probabilities

Event	Update vector	Probability = Rate·Δt
Transmission	$(\Delta S, \Delta I, \Delta R) = (-1, 1, 0)$	$\beta\, S\, I\, \Delta t$
Recovery	$(\Delta S, \Delta I, \Delta R) = (0, -1, 1)$	$\gamma\, I\, \Delta t$

First, to simulate the time to the next event we need to specify its probability distribution. As mentioned in Sect. 2.1, the rate terms in our ODE that correspond to these different transition events can be interpreted as Poisson process rates. In an autonomous system, the time to the next event will be exponentially distributed with a rate given by the sum of each event rate, since these rates depend only on the state of the system, and so between events those rates are constant.[22] Once this overall event rate is calculated, the time to the next event is sampled from the given exponential distribution.

The second step in the SSA is to determine which event type occurred, so the state variables can be updated accordingly. To do this requires using the fact that the probability of the ith event type occurring, p_i, is given by the ratio of that event rate (i.e., the corresponding term from the ODE) divided by the total rate. To clarify this step, consider again the SIR model where our two events have corresponding rates $\beta S I$ (infection event) and γI (recovery event), i.e., the probabilities of a new infection or a recovery are approximately $\beta S I \Delta t$ and $\gamma I \Delta t$, respectively. Here we specify the overall rate of a transmission event occurring, as opposed to the per capita rate. Then the probability p that the next event is a transmission event is given by $p = \beta S I / (\beta S I + \gamma I)$. Thus, the event type is determined by sampling from this discrete probability distribution defined by these relative rates. This sampling is done using the inverse transform sampling method [40] (see example below), and then the state variables are updated accordingly. This process is then repeated to obtain the desired simulated trajectory of the stochastic model.[23]

To illustrate how to apply the Stochastic Simulation Algorithm, it is implemented below in R for the basic SIR model. The overall rate r that defines the time until the next event is $r = \beta S I + \gamma I$ as mentioned above, thus the probabilities of an event being a transmission or recovery event are $\beta S I / r$ and $\gamma I / r$, respectively. To sample which event occurs, we simply sample from a Uniform(0,1) distribution and check whether the sampled value falls below $\beta S I / r$ (indicating a transmission event) or above that value (indicating a recovery event).

These two steps are repeated until a specified time is reached, or until $I(t) = 0$ (at which point the state values become fixed for all future times).

---

[22] Since the time to the next event under multiple different Poisson processes—one for each event type—is the minimum of the corresponding exponentially distributed event times, it is itself exponentially distributed with a rate that is the sum of the individual rates.

[23] Additional intuition for this algorithm can be found in the section in [92] regarding "competing clocks" in a Poisson process framework.

```
# Using the same parameters and initial conditions above...
set.seed(8675309) # Repeatable random number generation
X = cbind(Time=c(0), S=c(998), I=c(2), R=c(0)) # Store output
B=0.0003 # parameter values
g=0.1

# Next a matrix of "update vectors" to add to the
# current state of the system, i.e., x(t), when either a
# transmission (row 1) or recovery (row 2) event occurs.
update = matrix(c(-1,1,0,  0,-1,1), nrow=2, ncol=3, byrow=TRUE)
# display the two row vectors
update
```

```
     [,1] [,2] [,3]
[1,]   -1    1    0
[2,]    0   -1    1
```

```
# Stochastic Simulation Algorithm iteration using a while loop
i = 1 # a counter for our while loop iterations
while(X[i,1] < 150 & X[i,3] > 0) { # while Time < 150 and I>0 ...
  rates = c(B*X[i,2]*X[i,3] , g*X[i,3]) # indiv. event rates
  r = sum(rates) # total rate of some event happening
  tstep = rexp(1,r) # time to the next event is Exponential(r)

  # Determine which event occurs
  p = rates/r  # Probability vector for each event type
  u = runif(1) # Sample a Uniform(0,1)
  j = which(u <= cumsum(p))[1] # Sample which event occurred
  # The cumulative sum defines intervals [0,p1], [p1,p1+p2] ...
  # with interval lengths p1, p2, ... The function which(...)
  # returns the index for each interval including and above u.
  # Index [1] returns the 1st of these -- the one containing u.
  # Thus, Prob(j=1)=p1, Prob(j=2)=p2, etc.

  # Store updated time and the state variables, in new row of X
  X = rbind(X, c(X[i,1] + tstep, X[i,2:4] + update[j,]))
  i = i+1 # increment our row index for X
} # end of while() loop
if(X[i,1] < 150) { # If I reached 0, extend simulation to t=150
  X = rbind(X, c(150, X[nrow(X),2:4]))
}
head(X,3) # Display first and last few simulated values
```

```
          Time   S I R
[1,] 0.000000 998 2 0
[2,] 2.080826 997 3 0
[3,] 2.523152 997 2 1
```

```
tail(X,3)
```

```
            Time  S I   R
[1884,] 119.9396 57 1 942
[1885,] 135.8091 57 0 943
[1886,] 150.0000 57 0 943
```

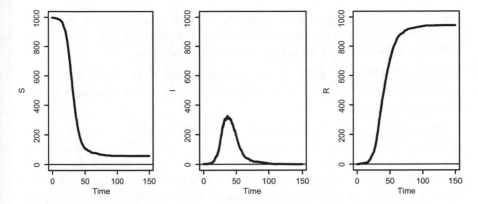

Fig. 8 One realization of the Stochastic Simulation Algorithm applied to the SIR model, which yields a simulated trajectory under the discrete-state, continuous-time analogue of the ODE-based SIR model defined by Eqs. (8). Compare to the ODE model trajectory in Figs. 2 and 3. See the code in the main text for further details

```
# Plot the stochastic trajectories as before
Time = X[,1]
par(mfrow = c(1,3)) # three panels in 1 row
plot(Time, X[,2], ylab="S", ylim=c(0,sum(X[1,2:4])),
     type="l", lwd=2);    abline(h=0) # Add a horizontal axis
plot(Time, X[,3], ylab="I", ylim=c(0,sum(X[1,2:4])),
     type="l", lwd=2);    abline(h=0)
plot(Time, X[,4], ylab="R", ylim=c(0,sum(X[1,2:4])),
     type="l", lwd=2);    abline(h=0)
```

These simulation results in Fig. 8 illustrate how this stochastic model captures the random fluctuations induced by the discrete nature of individuals in the system, and the probabilistic nature of the transmission and recovery processes. We call this kind of stochasticity *demographic stochasticity* to distinguish it from *environmental stochasticity* which might, for example, take the form of a non-constant transmission coefficient β fluctuating in time. This may be modeled using a pre-specified curve as discussed in the previous section, or by approximating the combined demographic and environmental noise within a stochastic differential equation (SDE) framework as discussed in the next section.

Exercise 16 Rerun the code above multiple times, but omit the set.seed() call so that a different stochastic simulation is generated each time. Do all simulations look the same, or not? How do these compare to Figs. 2 and 3?

Exercise 17 Combine the above SSA code with the code for plotting the numerical solution curve to the ODE version of the SIR model, and overlay the ODE output with the stochastic model output. How do the two trajectories compare?

Challenge Problem 20 Repeat the stochastic simulation above a few hundred times, store each of the simulations, and plot them together in one graph. How do they compare? Plot or otherwise summarize the distribution (across these different simulations) of any interesting quantities you can think of, e.g., peak number infected, total number infected, time to end of epidemic, etc.

Challenge Problem 21 Iterate these stochastic simulations many times and for each iteration store the first time at which $I(t) = 0$. Plot a histogram of these extinction times.

Challenge Problem 22 Implement the Stochastic Simulation Algorithm (SSA) for the Rosenzweig–MacArthur predator–prey model, Eqs. (23). Carefully consider the dependence or independence of different events (e.g., loss of prey via predation versus the birth of a new predator). As above, compare stochastic and ODE model output for the same parameter values and initial conditions. Increasing a through the Hopf bifurcation (see Fig. 1), noise may sometimes cause the predators or prey to go extinct, especially as the amplitude of the oscillations increases with increasing a. How does the average time to extinction vary with a above the threshold value that gives rise to predator–prey oscillations?

3.4 Stochastic Differential Equations (SDEs)

A common way to approximate a continuous-time discrete-state stochastic model is to derive a continuous-time, continuous-state model known as a *stochastic differential equation* (SDE) which retains the key average (mean) changes over time (like ODEs) as well as some of the covariance structure of the stochastic fluctuations about that average behavior. There are multiple ways to derive SDEs, but the derivation below follows Chapter 5 of [3].

First, consider the discrete-time approximation of the underlying continuous-time stochastic model for our SIR equation, and the possible update vector values corresponding to transmission and recovery events. These are summarized in Table 1 where as above we only focus on the S and I equations since R can be found later using $R(t) = N_0 - S(t) - I(t)$. Let the update vector $\Delta \mathbf{x}$ be a random vector that takes on these possible values according to their corresponding probabilities. The mean and second moment of the update vector $\Delta \mathbf{x}$ are, respectively,

$$E(\Delta \mathbf{x}) = \begin{bmatrix} -\beta S I \\ \beta S I - \gamma \end{bmatrix} \Delta t \tag{55}$$

$$E(\Delta \mathbf{x}(\Delta \mathbf{x})^{\mathsf{T}}) = \begin{bmatrix} \beta S I & -\beta S I \\ -\beta S I & \beta S I + \gamma I \end{bmatrix} \Delta t. \tag{56}$$

If we define $\boldsymbol{\mu} = E(\Delta \mathbf{x})/\Delta t$ and $\mathbf{B}$ such that $\mathbf{B}^2 = E(\Delta \mathbf{x}(\Delta \mathbf{x})^{\mathsf{T}})/\Delta t$ then the SDE for this scenario—using *drift coefficient* $\boldsymbol{\mu}$ and *diffusion coefficient* $\mathbf{B}$—is given by

$$dx = \mu\, dt + \mathbf{B}\, d\mathbf{W} \tag{57}$$

where $\mathbf{W}(t)$ is a vector of two independent Wiener processes (i.e., standard Brownian motion where for $t > s$ each $W_i(t) - W_i(s)$ is normally distributed with mean 0 and variance $t - s$). Since $\mathbf{W}(t)$ is not time-differentiable, the notation used in Eq. (57) implicitly assumes these quantities exist as the integrand of an integral over time [3, 6, 131]. Thus, an SDE approximation of our stochastic SIR model is

$$\begin{bmatrix} dS(t) \\ dI(t) \end{bmatrix} = \begin{bmatrix} -\lambda(t)\, S(t) \\ \lambda(t)\, S(t) - \gamma\, I(t) \end{bmatrix} dt + \mathbf{B}\, d\mathbf{W}. \tag{58}$$

Matrix $\mathbf{B}$ can be found analytically in this case, but in general may need to be computed numerically. Here, the covariance matrix $\mathbf{B}^2$ has multiple square roots, but only one is a symmetric positive definite matrix [159], thus we use

$$\mathbf{B} = \begin{bmatrix} \dfrac{\sqrt{BSI}\,\sqrt{gI}+BSI}{\sqrt{\left(\sqrt{gI}+\sqrt{BSI}\right)^2+BSI}} & -\dfrac{BSI}{\sqrt{\left(\sqrt{gI}+\sqrt{BSI}\right)^2+BSI}} \\ -\dfrac{BSI}{\sqrt{\left(\sqrt{gI}+\sqrt{BSI}\right)^2+BSI}} & \dfrac{gI+\sqrt{BSI}\,\sqrt{gI}+BSI}{\sqrt{\left(\sqrt{gI}+\sqrt{BSI}\right)^2+BSI}} \end{bmatrix}. \tag{59}$$

3.4.1 Numerical Solutions to SDEs

Numerical solutions to SDEs can be computed using algorithms similar to those discussed in the section above for simulating ODEs, e.g., see [3, 16, 76, 83, 138, 147]. Below, the stochastic fourth-order Runge–Kutta algorithm from [76] is used to numerically solve the above SDE version of the SIR model (Fig. 9). For additional details on deriving, analyzing, and generating numerical solutions of SDEs and related models (e.g., stochastic delay differential equations) see [3, 6, 93, 131].

```
###############################################################
# Stochastic Runge-Kutta-4 for Stratonovich calculus,
# as described in Hansen and Penland 2006.

# A function to calculate the deterministic part of our SDE,
dSI = function(X, ps) { # i.e., the drift coefficient mu.
    S = X[["S"]]        # Unpack our state variable values
    I = X[["I"]]
    b = ps[["beta"]]    # Unpack parameters
    g = ps[["gamma"]]
    return(matrix(c(-b*S*I, b*S*I - g*I), nrow=2, ncol=1))
}

sdeRK4 = function(X, dt, ps) {
    S = X[["S"]];       I = X[["I"]]; # As in dSI above.
    b = ps[["beta"]];   g = ps[["gamma"]];
```

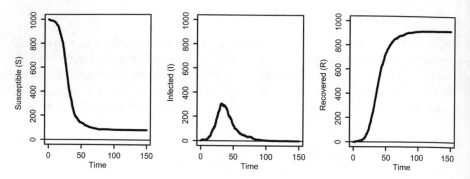

Fig. 9 One numerical solution to the SDE version of the simple SIR model defined by Eqs. (58) and (59), simulated using a stochastic fourth-order Runge–Kutta algorithm [76]. Compare to the ODE model trajectory in Fig. 2 and the Stochastic Simulation Algorithm (SSA) implementation of that model in Fig. 8. See the code in the main text for further details

```
# Standard normal values with variance dt
dW = matrix(rnorm(2,mean=0,sd=sqrt(dt)), nrow=2, ncol=1)

# Square root of the second moment matrix B^2
B = matrix(c(sqrt(b*S*I)*sqrt(g*I)+b*S*I, -b*S*I,
             -b*S*I, g*I + sqrt(b*S*I)*sqrt(g*I)+b*S*I)/
             sqrt((sqrt(g*I) + sqrt(b*S*I))^2+b*S*I),
             nrow=2, ncol=2)

# Define the terms of the stochastic RK4 scheme
# See Hansen and Penland 2006 for details.
k1 = dSI(X,ps)*dt + B %*% dW ## %*% is matrix multiplication
k2 = dSI(X + .5 * k1, ps)*dt + B %*% dW
k3 = dSI(X + .5 * k2, ps)*dt + B %*% dW
k4 = dSI(X + k3, ps)*dt      + B %*% dW
return(X + (1 / 6) * (k1 + 2*k2 + 2*k3 + k4))
}

# Parameters for our numerical solution
Pars = c(beta=0.0003, gamma=0.1)
N=1000
X = data.frame(S=998, I=2) # Assume that R(0)=0
Time = c(0)
tstep = 0.1

# Iterate sdeRK4() steps to simulate a full trajectory
set.seed(246) # For repeatable random number generation
i=2 # counter for our while loop below
while(Time[i-1] < 150 & X[i-1,2]>0) { # while t<150, I>0
   Time[i] = Time[i-1] + tstep
   X[i,] = sdeRK4(X[i-1,],tstep,Pars)
   i = i+1
}
```

```
# Combine columns into a single data frame
Xout = cbind(Time, X, R = N - rowSums(X))
tail(Xout,4)
```

	Time	S	I	R
1417	141.6	81.05222	0.02383346	918.9239
1418	141.7	81.05465	0.04914261	918.8962
1419	141.8	81.06478	0.02691306	918.9083
1420	141.9	81.07035	-0.01539963	918.9450

```
# If I(t) dropped below zero on the last iteration...
if(Xout[i-1,3] < 0) { # ... set I=0 and extend to t=150:
        Xout[i-1,3]  = 0  # First set negative value to 0, then
        Xout[i,]     = Xout[i-1,] # duplicate that last row and
        Xout[i,1]    = 150        # set the final time to 150.
}

# Plot the numerical solution curves as before
par(mfrow=c(1,3)) # Plot 3 panels in 1 row
plot(Xout[,1], Xout[,2], ylab="Susceptible (S)", ylim=c(0,N),
     xlab="Time", type="l", lwd=2); abline(h=0)
plot(Xout[,1], Xout[,3], ylab="Infected (I)", ylim=c(0,N),
     xlab="Time", type="l", lwd=2); abline(h=0)
plot(Xout[,1], Xout[,4], ylab="Recovered (R)", ylim=c(0,N),
     xlab="Time", type="l", lwd=2); abline(h=0)
```

Challenge Problem 23 Derive and implement SSA- and SDE-based stochastic simulations of the logistic equation with explicit birth and death rates given by Eq. (52). How do their assumptions differ? What similarities and differences can you identify between these two stochastic implementations of the ODE-based logistic growth model?

3.5 Distributed Delay Equations

Another widespread ODE model assumption that can be altered is the assumed exponentially distributed *dwell times*[24] that are so often encountered as simplifying assumptions in mean field ODEs. Choosing other dwell time assumptions can sometimes alter model behavior in ways that are important for a given application (e.g., [63, 64]). Typically, models that relax this assumption must be derived from stochastic model first principles, and often the end result is a set of *integral equations* or *integro-differential equations* rather than ODEs (e.g., [80]). Under certain distributional assumptions, these integral equations can also be realized as a

---

[24]The *dwell time*, or *passage time*, is the duration of time an individual spends in a given state. As discussed in Sect. 2.1, ODEs with linear loss rate terms implicitly assume that—in the context of Sect. 3.3.1—the times individuals spend in that state are exponentially distributed.

system of ODEs (e.g., see [64, 65, 92, 154]), and in some cases it is possible to write down those ODEs directly from the stochastic model assumptions (see Sect. 3.5.2 and references therein). The following sections introduce these *distributed delay* equation models, as well as the linear chain trick (LCT) and a generalized linear chain trick (GLCT) for approximating or exactly representing these with systems of ODEs. *Discrete delay* differential equations (DDEs), which can be thought of as the zero-variance limit of a distributed delay equation, are also introduced in Sect. 3.5.4.

3.5.1 Integral and Integro-Differential Equations

Integral and integro-differential equations arise as a more general way of writing mean field models for continuous-time stochastic processes, especially when the dwell time distributions and related assumptions do not permit an ODE representation of the mean field model (e.g., when the model is not memoryless [92]). Specific forms, known as *Volterra integral equations* and *Volterra integro-differential equations* (or sometimes just referred to as *Volterra equations*) are named after the prolific mathematician and physicist Vito Volterra (born 3 May 1860, died 11 October 1940) who, building upon his work on functionals in functional analysis, laid the groundwork to use integral and integro-differential equations to study biological problems.

One approach to derive a mean field distributed delay model in the form of an integral equation is to first derive a discrete-time mean field model that accounts for all the past movements into and out of the different states. This can be done for an arbitrary dwell time distribution, as shown in the following example.

For an example of such a derivation, consider again our SIR model (see also [92]). Suppose we seek to model $I(t)$—the number of infected individuals at time t—with an integral equation. To begin, the time interval $[0, t]$ is partitioned into a large number of N sub-intervals, each of duration Δt. We can refer to each sub-interval by indexing the start time of each interval as t_i, i.e., the ith sub-interval is $(t_i, \ t_i + \Delta t]$. Assume the number of new infections during the ith time interval is binomially distributed where there are $n = S(t_i)$ individuals who might become infected, each with probability $p = \beta I(t_i)\Delta t$, during that time period of duration Δt. Thus, by properties of binomial distributions, the expected number of new infections during the ith time interval is $n p = \beta S(t_i)I(t_i)\Delta t$. For each of these cohorts entering the infectious state during the same time interval, assume all follow the same dwell time distribution with domain $[0, \infty)$, cumulative distribution function (CDF) $F(t)$, and *survival function* $G(t) \equiv 1 - F(t) = P(X > t)$. It then follows that, for the cohort of individuals who entered the infectious state during the ith sub-interval of time, the expected proportion that remains in that state at time t (after $t - t_i$ time units) is $G(t - t_i)$.

With those details in hand, a discrete-time mean field model of the number of infected individuals at time t is obtained by summing over all N of the sub-intervals of time (where $t = N\Delta t$), accounting for (a) the expected number entering in each time interval, and (b) the proportion of each cohort[25] remaining in I at time t:

$$I(t) = \overbrace{I(0)\,G(t)}^{\text{initial cohort}} + \sum_{i=1}^{N} \overbrace{\beta S(t_i)I(t_i)\Delta t}^{\text{expected influx}} \underbrace{G(t - t_i)}_{\text{proportion remaining at time } t}. \tag{60}$$

Recognizing this as a Riemann sum, we can take the limit as $\Delta t \to 0$ $(N \to \infty)$ and replace the times t_i with the integration variable s to yield the continuous-time mean field model given by the (Volterra) integral equation

$$I(t) = I(0)\,G(t) + \int_0^t \beta S(s)I(s)G(t - s)\,ds. \tag{61}$$

Similarly, one can go on to derive the full generalization of the SIR model, which incorporates general distributions for the time spent in the susceptible state, and the time spent in the infectious state (i.e., the *duration of infectiousness* also called the *infectious period*) which each have the respective CDFs $F_S(t)$ and $F_I(t)$ (i.e., survival functions $G_i(t) = 1 - F_i(t)$). Written as above, the equations are

$$S(t) = S(0)G_S(t) \tag{62a}$$

$$I(t) = I(0)G_I(t) + \int_0^t \beta\,I(s)\,S(s)\,G_I(t - s)\,ds \tag{62b}$$

$$R(t) = N - S(t) - I(t) \tag{62c}$$

where $N = S(0) + I(0) + R(0)$. The ODE model given by Eqs. (8) is a special case of the above model. This can be shown by assuming the time spent in the susceptible state obeys the 1st event time under a Poisson process with rate $\lambda(t) = \beta\,I(t)$, and thus $G_S(t) = \exp\left(-\int_0^t \beta\,I(u)\,du\right)$, and further assuming $G_I(t) = \exp\left(-\gamma\,t\right)$. However, in this more general framework one could consider various other choices for the dwell time distribution that describes the duration of infectiousness. For more details and similar models see [29, 63, 92, 102, 113].

One can undertake analyses (analogous to those mentioned above for ODE models) to investigate integral equation dynamics; however, even basic equilibrium stability analyses and computing numerical solutions requires more careful approaches than for ODEs [12, 25, 80, 111]. In practice, a combination of analytical and numerical methods can be used to study these types of distributed delay equations. However, another alternative is to take advantage of the fact that, for some choices of the dwell time distribution, differentiating these integral equations can yield equivalent systems of ODEs, as detailed in the next section.

---

[25]For simplicity, we here assume the initial cohort enters state I at $t = 0$ and thus follows the same dwell time distribution G.

3.5.2   Linear Chain Trick

The *linear chain trick* (LCT; also known as the *gamma chain trick*) allows us to derive a new ODE model from an existing ODE model that assumes an exponential dwell time distribution in one or more states by replacing that assumption with an Erlang[26] distributed dwell time. It also allows for the derivation of a system of ODEs that approximates a distributed delay equation, or a delay differential equation (as illustrated in Sect. 3.5.4). Erlang distributions are the special set of gamma distributions that can be thought of as the sum of k independent exponential distributions, each with the same rate [85, 106, 145]. Importantly, the LCT allows modelers to write down this new ODE model directly, without going through the process of constructing a more general integral equation model (as in the above example), and/or without explicitly deriving ODEs from an integral equation (as discussed below).

The LCT works as follows. We seek to replace a target state and corresponding state variable x in an ODE—which has an implicit exponential dwell time distribution with a mean dwell time of $1/r$ (or equivalently, rate r, i.e., the loss rate in the dx/dt equation is $-r\,x$)—with a series of substates x_i that partition state x into k substates. Individuals enter the state into the first substate x_1, then transition to the second substate, and so on, in series. This ensures that the overall passage time through these substates is Erlang distributed, since the sum of k exponentially distributed random variables with rate rk yields an Erlang distribution with mean $k/(rk) = 1/r$, as desired.[27]

The LCT is applied to specific model equations as follows. Suppose a state x in the model has an input rate $\Lambda(t)$ and an exponential dwell time with rate μ (i.e., mean $1/\mu$), and thus the mean field ODE equation is of the form

$$\frac{dx}{dt} = \Lambda(t) - \mu\,x(t). \tag{63}$$

To instead assume an Erlang distributed duration of time with the same mean $(1/\mu)$, the equation for $x(t)$ can be replaced with equations for k substates $x_i(t)$, where $x(t) = \sum x_i(t)$, and the equation above is replaced by

$$\frac{dx_1(t)}{dt} = \Lambda(t) - \mu k\,x_1(t) \tag{64a}$$

$$\frac{dx_j(t)}{dt} = \mu k\,x_{j-1}(t) - \mu k\,x_j(t), \quad \text{for } j = 2, \dots, k. \tag{64b}$$

---

[26]Erlang distributions are gamma distributions with integer-valued shape parameters. Compared to exponential distributions, the Erlang density function is more hump shaped and the variance can be made arbitrarily small, as is sometimes desired in applications.

[27]Readers familiar with Poisson processes may recall that homogeneous Poisson processes have inter-event times that are exponentially distributed, and the time to the kth event under a homogeneous Poisson process with rate r is Erlang distributed with rate r and shape k.

Furthermore, one can choose a value of k in order to implement the desired variance (increasing k decreases the variance). Often, it is preferable to specify the desired *coefficient of variation*[28] instead, since the coefficient of variation (cv) for an Erlang distribution is determined solely by shape parameter k, where $cv = 1/\sqrt{k}$. Note cv can be specified to yield an integer-valued $k = 1/cv^2$.

Other model equations may also need to be modified when implementing the LCT, for example you may need to replace expressions involving state variable $x(t)$ with the sum of the new substate variables $x_i(t)$, and any input rates into other states that depend on the loss rate from x, formerly $\mu x(t)$, must instead depend on the new loss rate $\mu k x_k(t)$ (for an example, see below).

To illustrate, suppose that for our original SIR model Eqs. (8) we wanted to replace our exponentially distributed dwell time (rate γ; mean $1/\gamma$) in the infected state with an Erlang distribution with mean $1/\gamma$ and coefficient of variation of $1/3$ (i.e., $k = 9$). To do this using the LCT, we simply replace the ODE

$$\frac{dI}{dt} = \beta S I - \gamma I \tag{65}$$

as indicated above, and modify the transmission and recovery rate terms to yield

$$\frac{dS}{dt} = -\beta \left(\sum_{i=1}^{9} I_i(t) \right) S(t) \tag{66a}$$

$$\frac{dI_1}{dt} = \beta \left(\sum_{i=1}^{9} I_i(t) \right) S(t) - \gamma k I_1(t) \tag{66b}$$

$$\frac{dI_2}{dt} = \gamma k I_1(t) - \gamma k I_2(t) \tag{66c}$$

$$\vdots$$

$$\frac{dI_9}{dt} = \gamma k I_8(t) - \gamma k I_9(t) \tag{66d}$$

$$\frac{dR}{dt} = \gamma k I_9(t). \tag{66e}$$

In practice, the LCT may be more challenging to apply if there are additional model complexities like transitions to multiple states or substate transitions that should not "reset the clock" for the overall dwell time distribution in a state. For guidance on applying the LCT in those situations, see [92].

---

[28]The *coefficient of variation* is the standard deviation divided by the mean.

Challenge Problem 24 As illustrated above, the LCT can quickly increase the number of state variables. Explore the consequences of this increase in dimensionality either in terms of model dynamics, and/or the accuracy of numerical solutions.

3.5.3 Mathematical Foundations of the Linear Chain Trick

For readers interested in the mathematical machinery behind the LCT, the following is a brief overview of how to explicitly derive ODEs from integral equations.

The Erlang density and survival functions are given (respectively) by

$$g_r^k(t) = r \frac{(r\,t)^{k-1}}{(k-1)!} e^{-rt} \tag{67a}$$

$$G_r^k(t) = \sum_{j=1}^{k} \frac{1}{r} g_r^j(t). \tag{67b}$$

The following recursion relationship between Erlang density functions and their derivatives is the linchpin of the LCT, as summarized in Theorem 4 (adapted from [92], which follows Eqs. 7.11 in [154]); see [92] for extensions).

Theorem 4 *The Erlang density functions $g_r^j(t)$, with rate r and shape j, satisfy*

$$\frac{d}{dt} g_r^1(t) = -r g_r^1(t), \quad where \ \ g_r^1(0) = r, \tag{68a}$$

$$\frac{d}{dt} g_r^j(t) = r \left[g_r^{j-1}(t) - g_r^j(t) \right], \quad where \ \ g_r^j(0) = 0 \ for \ j \geq 2. \tag{68b}$$

This recursion relation allows one to differentiate integral equations like Eqs. (62) with assumed Erlang dwell times, and formally derive equivalent mean field ODEs like Eqs. (66). The linear chain trick can also be generalized to a much broader family of dwell time distributions that includes the *phase-type distributions* as detailed in [92]. More specifically, these phase-type distributions describe the distribution of times it takes to hit an absorbing state in a continuous-time Markov chain, and there is a straightforward formula for writing the ODEs for such cases based on the corresponding transition matrix and rate vector for the continuous-time Markov chain representation of the assumed phase-type distribution. Interested readers should consult [92, 154] and references therein for further details, as here we only introduce the standard linear chain trick.

Challenge Problem 25 Take the $I(t)$ equation from the Volterra integral form of the SIR model (Eqs. (62)) and assume an Erlang distribution survival function with rate r and shape k. Use the Leibniz rule for differentiating integrals to prove that the above set of equations for I_i are correct. *Hint: After differentiating the integral and applying Theorem 4, use the substitution*

$$I_j(t) = I_0 \frac{1}{r} g_r^j(t) + \int_0^t \beta S(s) I(s) \frac{1}{r} g_r^j(t - s) ds. \tag{69}$$

3.5.4 Delay Differential Equations

Another approach to modifying the exponential delay assumption implicit in many ODE models is to consider taking the limit of the dwell time distribution's variance to zero while the mean is fixed at some value τ. This yields equations of the form[29]

$$\frac{d\mathbf{x}}{dt} = f(t, \mathbf{x}(t), \mathbf{x}(t - \tau), \theta) \tag{70}$$

where $\tau > 0$ thus allowing the derivative of the current state (at time t) to depend on the past state of the system τ time units into the past. These are known as (discrete) *delay differential equations* (DDEs) [154], and while DDEs are slightly more difficult to analyze than ODEs (e.g., in terms of equilibrium stability analysis), they are easier to mathematically analyze and to compute numerical solutions for, relative to integral and integro-differential equations. It is also worth noting that DDEs often exhibit more complex dynamics than similar (non-delayed) ODE models (e.g., see the simple example in the Appendix of [67]).

There are many resources that address the analysis of DDEs (e.g., [8, 18, 154, 155]), and applications of DDEs are common in the scientific literature, so the various methods of mathematical analysis will not be reviewed here. Numerical solutions can be obtained using the dede function in the deSolve package in R [156], which is used in a similar fashion to the ode function but with a few caveats. First, unlike ODEs (which are memoryless and only depend on the current state) DDEs depend on the history of the system. Thus, to begin a numerical solution and compute forward in time starting at time $t = 0$ requires that $x(t)$ be defined over the time interval $t \in [-\tau, 0]$. That is, unlike the vector of initial state values needed to simulate trajectories for ODEs, DDEs require *initial functions* over $[-\tau, 0]$ to obtain numerical solutions (for example code in R, see the bottom of the help documentation for dede which illustrates how to specify these initial function values).

DDEs with a finite number of state variables are actually, in a sense, infinite dimensional in that all of the values of $x(t)$ between time $t - \tau$ and t can be thought of as state variables. To provide some intuition for why this is the case (and thus provide a means of approximating DDEs with ODEs, or vice versa) consider the

---

[29]More generally, if there are multiple lag values $\tau_i, i = 1, \ldots, k$, the right hand side is of the form $f(t, \mathbf{x}(t), \mathbf{x}(t - \tau_1), \ldots, \mathbf{x}(t - \tau_k), \theta)$.

following construction of the SIR model with a discrete delay based on Eqs. (66), the SIR model with an Erlang distributed duration of infection. Recall that, for a fixed mean duration of infectiousness $1/\gamma$, the shape parameter k represents the number of substates I_i necessary to include in the ODE when assuming an Erlang distributed delay between acquiring an infection, and recovery. Recalling that the coefficient of variation (and the variance) decreases as k increases, it follows that taking $k \to \infty$ should yield a model that has a discrete delay of exactly $\tau = 1/\gamma$. For all k, $\sum I_i(t) = I(t) = N_0 - S(t) - R(t)$, thus in the limit as $k \to \infty$ the number entering the recovered state during an infinitesimally small period of time approaches exactly the same number who became infected $\tau = 1/\gamma$ time units ago (i.e., the rate of recovery at t is $\beta I(t-\tau) S(t-\tau)$), one could write a DDE version of this model as

$$\frac{dS(t)}{dt} = -\beta (N_0 - S(t) - R(t)) S(t) \tag{71a}$$

$$\frac{dR(t)}{dt} = \beta (N_0 - S(t-\tau) - R(t-\tau)) S(t-\tau). \tag{71b}$$

Exercise 18 Differentiate $I(t) = N_0 - S(t) - R(t)$ and use Eqs. (71) to show that the full version of the model above can be written as

$$\frac{dS(t)}{dt} = -\beta I(t) S(t) \tag{72a}$$

$$\frac{dI(t)}{dt} = \beta I(t) S(t) - \beta I(t-\tau) S(t-\tau) \tag{72b}$$

$$\frac{dR(t)}{dt} = \beta I(t-\tau) S(t-\tau). \tag{72c}$$

Exercise 19 Consider repeating the above derivation of a DDE for the basic SEIR model, where we would like to change the assumption of an exponentially distributed latent period (with mean $\tau = 1/\nu$) to a discrete delay with the same mean, i.e., we would like to assume that the incubation period before an exposed individual becomes infectious is exactly τ units of time. Justify whether or not this model should be of the form

$$\frac{dS(t)}{dt} = -\beta I(t) S(t) \tag{73a}$$

$$\frac{dI(t)}{dt} = \beta I(t-\tau) S(t-\tau) - \gamma I(t) \tag{73b}$$

$$\frac{dR(t)}{dt} = \gamma I(t) \tag{73c}$$

by first generalizing the SEIR model using the linear chain trick, and then letting the variance of the latent period go to zero (i.e., letting $k \to 0$).

Lastly, the above procedure can also be reversed as a way of constructing a system of ODEs that approximates a DDE by relaxing the "zero-variance" assumption to instead assume an Erlang distributed delay with mean τ. This well-known approximation method is described at the end of Chapter 7 in [154].

For example, consider the following DDE version of the Rosenzweig–MacArthur model, where we assume that there is some period of time τ between predators consuming prey and the subsequent arrival of new predators in the system [167].

$$\frac{dN(t)}{dt} = r\,N(t)\left(1 - \frac{N(t)}{K}\right) - \frac{a\,P(t)}{k + N(t)}N(t) \tag{74a}$$

$$\frac{dP(t)}{dt} = \chi \frac{a\,N(t-\tau)}{k + N(t-\tau)}P(t-\tau) - \mu\,P(t) \tag{74b}$$

This may be reasonable if, for example, food resources are quickly turned into offspring and their overall maturation time has a small coefficient of variation (compare Eq. (74) to the model in §6.2 of [118]). Here we may treat the delayed rate of new predators entering the system (i.e., the *recruitment rate*)

$$\chi \frac{a\,N(t-\tau)}{k + N(t-\tau)}P(t-\tau) \tag{75}$$

as the input rate into this previously unwritten count of offspring that will eventually become adults. Approximating with the LCT introduces a series of intermediate state variables Y_i, $i = 1, \ldots, k$ reminiscent of the linear chain trick results above, where we assume the value of k such that it gives the desired coefficient of variation (recall $k = 1/c_v^2$) for an assumed Erlang distribution of incubation times. Relaxing the discrete delay assumption to include some non-zero variance in the incubation period therefore yields the following ODE model.

$$\frac{dN(t)}{dt} = r\,N(t)\left(1 - \frac{N(t)}{K}\right) - \frac{a\,P(t)}{k + N(t)}N(t) \tag{76a}$$

$$\frac{dY_1(t)}{dt} = \chi \frac{a\,N(t)}{k + N(t)}P(t) - \frac{k}{\tau}Y_1(t) \tag{76b}$$

$$\frac{dY_j(t)}{dt} = \frac{k}{\tau}Y_{j-1}(t) - \frac{k}{\tau}Y_j(t), \quad \text{for } j = 2, \ldots, k \tag{76c}$$

$$\frac{dP(t)}{dt} = \frac{k}{\tau}Y_k(t) - \mu\,P(t) \tag{76d}$$

It should be noted that these finite-state approximations of DDE models, especially for large k (i.e., small coefficient of variation), can have different dynamics relative to the original DDE, and they can also be numerically problematic when computing numerical solutions for large k values.

Challenge Problem 26 Use the dede and ode functions in deSolve to generate comparable numerical solutions to the two SIR models (and/or the two predator–prey models) above, and compare those solution curves for various values of k (i.e., for various values of the implicitly assumed variance about the mean delay value τ). Think carefully about the initial function values for the DDE model solutions (e.g., consider using ODE model solution curves for these values) so that the simulated results can be fairly compared.

3.6 Individual Heterogeneity

Different approaches exist for relaxing the ODE model assumption of homogeneity among individuals in a given state. This may be a reasonable assumption for the individual molecules in a solution of chemical reactants, but may be an oversimplification for a population of plants or animals. The most straightforward way to relax this assumption is to simply assume a finite number of distinct types of individuals. For example, this has been done with models of sexually transmitted infections to account for low- and high-risk groups [22, 81], and in multispecies interaction models where prey can occur in forms that are either undefended (a good food resource) or well-defended (a poor food resource) [36, 49, 168, 169].

Methods for modeling a continuum of individual heterogeneity also exist, e.g., to allow something like individual size to vary along a continuum as opposed to a finite set of possible size classes. For example, in a discrete-time case, this scenario can be modeled using *integral projection models* [53, 54, 120], which can offer benefits over discrete-state models when used for parameter estimation since smooth kernels with only a few parameters may have fewer parameters to estimate than a model with a larger number of discrete states, each with their own related parameters. In continuous-time, a PDE model can be used to describe the time-evolution of a continuous trait distribution across a population, and there are a variety of analytical and computational resources available for such models, e.g., [45, 52, 100, 127].

Another approach to constructing a model that incorporates individual heterogeneity is to implement an *agent based* (also known as *individual based*) *model* [72]. These computational models allow for the inclusion of more complex rules governing individual behavior than can typically be incorporated into a tractable mathematical model. These approaches have been used quite successfully to better understand phenomena like animal movements in groups or through heterogeneous environments (e.g., [50, 66, 72, 144]). While these typically do not allow much mathematical analysis, they can be useful for modeling specific systems that are data rich or are otherwise well understood, and can form the basis for subsequently deriving mathematical models that aim to address very specific phenomena observed in simulations.

3.7 Spatially Explicit Models

Spatial effects are perhaps one of the most important components of real-world systems that frequently get oversimplified by using ODE models, and some of the other types of models discussed above. ODE models typically assume that systems are well mixed and not constrained by spatial distances. However, incorporating these spatial dependencies can have a significant impact on how these models behave. There is a sizable body of literature on spatial models and methods for analyzing various types of spatially explicit models. Here we just scratch the surface of this topic and provide only a very brief introduction to a few of these modeling approaches, and leave it to the reader to seek out a more in depth treatment of spatially explicit models.

Readers in search of a research project that includes spatial effects are encouraged to derive a spatially explicit model (like one of the model types listed below) from an existing ODE model, or to slightly modify the assumptions of an existing spatially explicit model, for example a PDE model. One might also consider the more ambitious option of deriving a multipatch or network model based on an existing spatial (e.g., PDE) model, or vice versa.

One approach to deriving a spatially explicit model from an existing non-spatial model is to simply consider dividing the system up across multiple "patches." A well-known example of such models are the *metapopulation models* in population ecology [77, 109, 110, 132]. In these contexts, the dynamics within each patch can be governed by a simple model, such as an ODE model, and some explicit rules can be imposed that govern movement or other interactions between patches. Constructing these sorts of *multipatch models* is an easy way to explore how spatially dependent processes affect dynamics. These models can also be used to explore the effects of spatial heterogeneity by allowing model parameters to vary from patch to patch. In recent decades, such models have been revisited in the context of the burgeoning field of network science [130], allowing investigators to bring the tools and insights from the study of networks to bear on understanding how the patterns of patch connectedness (i.e., the network properties) shape system-wide dynamics.

Many examples of multipatch and network models exist in the literature, and some of the standard methods of ODE model analysis can be applied to these models. However, these methods must sometimes be adapted for the high dimensionality of these systems (e.g., see [30] and references therein; also compare methods in the literature for calculating epidemic thresholds on networks with the next generation operator approach to calculating the basic reproduction number $\mathscr{R}_0$ for small to medium sized model, as detailed in [43]). For an introduction to methods for network models, see [130, 135, 136] and also [58–62, 152].

Another approach to explicitly incorporating space into a dynamical systems model is to discretize a continuous 1-, 2-, or 3-dimensional spatial domain into a finite number of contiguous patches. This special case of the previously described patch models can then be analyzed using approximation methods such as *pair approximation* [48, 82, 110], or by taking the limit as the patch size shrinks to zero to derive a PDE model. As mentioned above, there are other resources available that cover methods for deriving, analyzing, and numerically simulating PDE models (including the methods available in the deSolve package in R). Interested readers may find more detailed treatments of these topics in [45, 50, 100, 127] and related publications (also, see the list of "books in related areas" in the preface of [45]).

4 Choosing a Research Project

I hope the preceding sections have inspired you to reevaluate a familiar ODE model, or to seek out an ODE model related to a topic of particular interest to you, and take a critical look at the explicit and implicit assumptions of that model. The sections above provide an overview of some commonly used methods for analyzing ODE models (or otherwise using them in a research setting), and an overview of some ways that ODE model assumptions can be altered to derive new models in pursuit of new avenues of research.

From the perspective of conducting research to advance our collective knowledge of how the world works, research projects that parallel an existing ODE model analysis (using a model derived by meaningfully altering the assumptions of that original model) can help incrementally build upon an existing body of work. These types of research projects can be great for student researchers, since they provide some guidance in terms of which questions are worth asking, and which types of analyses can be used to address those questions. Additionally, these alternative models can also provide opportunities to ask new questions about important real-world systems that may not be as practical (or possible) to answer using an existing ODE model (e.g., the distribution of extinction times is more appropriately addressed in a discrete-state stochastic model). For these reasons, the relationships between different models and their differing assumptions are a wonderful place to probe and challenge our understanding of the real-world systems we seek to understand through mathematical modeling.

4.1 Additional Project Topics

Scientifically, many open problems exist in the study of specific systems that are being investigated using ODE models. As stated in the introduction, studying multiple mathematical models of the same system, each with slight variations in assumptions, can yield a more clear understanding of which results are robust across models (e.g., see [1, 2, 99, 107]) and across similar real-world systems. These studies can also identify which results might be meaningfully influenced by simplifying assumptions. Below are a few additional project ideas to consider that have some potential to lead to yield meaningful contributions to the very broad field of mathematical biology.

Research Project 1 (Timing in Infectious Disease Models) Recently, increased attention has been given to the importance of properly modeling the presence or absence of latent periods in infectious disease models [164], and the importance of making proper assumptions about the distributions that describe the duration of time spent in different model states [64, 65]. Many models exist in the literature for various diseases in wildlife and humans (and for other types of multispecies interactions that do not involve infectious parasites), and improving a model in this manner can make a great project.

One approach to tackling this project is to do the following:

1. Search the literature for relatively recent publication using an SIR-type ODE model to study a specific infectious disease. Diseases in non-human hosts or diseases that are otherwise not heavily researched (e.g., not HIV or Malaria, Dengue, or Rabies) are perhaps more likely to be modeled using relatively simple or otherwise poorly refined models.
2. Identify key model assumptions and the research aims addressed in the paper, paying careful attention to timing and the presence or absence of latent periods. Which assumptions would be the most important to reconsider?
3. Identify a question you plan to answer, and deriving a new model based on modifying just one (or maybe two) relevant assumptions, e.g., adding a latent period.
4. If possible, consult with a public health or infectious disease expert who has some expert knowledge of the system(s) being modeled.
5. Focusing on one question asked in the original paper, recreate the published analyses to verify those results.
6. Conduct an analogous analysis of your new model to address the same question under your revised assumptions, and/or conduct additional analyses to address other important questions.

From a more mathematical or methodological perspective, there are also some open questions related to the analytical and computational tools that exist for the various model frameworks mentioned in the sections above. While these can be more challenging problems to tackle, and may require a preparation beyond the typical undergraduate curriculum, some of these questions are still accessible to undergraduate researchers and could make interesting and impactful research projects.

Research Project 2 (Numerical Solutions for Distributed Delay Equations) Methods to compute numerical solutions to integral and integro-differential equations (often called *Volterra* integral and integro-differential

(continued)

Research Project 2 (continued)

equations) are currently being developed and refined. Review the literature, and select two or more methods and compare their performance under different modeling scenarios (e.g., periodically forced systems or systems with multiple time scales). To provide a reference point for evaluating the accuracy of numerical solutions, the example models used are typically simple enough to have analytical solutions. A nice spin on this project would be to use distributional assumptions that allow integral or integro-differential equations to be fully reduced to an equivalent system of ODEs according to the linear chain trick (LCT, see Sect. 3.5.2) or the generalized linear chain trick (GLCT) [92]. How well do numerical solutions to those ODEs perform in these comparisons?

A project like this could be conducted as follows:

1. Identify some numerical methods for one or more types of distributed delay equations (e.g., just integral equations).
2. Do a literature search to find appropriate models to use as examples. Include some with analytical solutions, and some that could be written as ODEs via the LCT and therefore simulated using appropriate ODE solvers.
3. Establish a set of criteria by which to compare the different methods (e.g., computing time per numerical solution, implementation time or some other measure of the effort required leading up to computing numerical solutions, the accuracy relative to some baseline solution such as an analytical solution, and so forth).
4. Implement the planned comparisons.

Students interested in exploring statistical projects may choose to focus on ODE-based models or projects that involve other modeling frameworks.

As discussed in Sect. 2.7, investigating statistical properties of dynamic models can be done using simulated data so that the known "true" parameter values can be used to evaluating the parameter estimation procedure. Therefore one can take almost any dynamic model and assess the statistical properties of that model under certain observation process assumptions or using different estimation techniques. This is especially meaningful when done for models that were used in evaluating real data where it is unclear that models have identifiable parameters, and/or where additional analyses (e.g., a power analysis) might provide important context for interpreting the analysis of real data.

Research Project 3 (Estimator Properties) Investigate the statistical properties of a system-specific ODE model that has previously been (or might someday be) fit to empirical data.

This project could be approached as follows:

1. First, as above, find a peer-reviewed publication that includes an ODE model that was fit to data. Focus on searching in journals in application areas where such analyses are unlikely to be included in the publication.
2. Conduct a structural and/or practical identifiability analysis of the model.
3. Conduct a simulation-based study to characterize what if any bias exists for the estimation procedure used in the publication.
4. Assess what (if any) *uncertainty quantification* was done, and consider further analyses if warranted.

Research Project 4 (Fitting Models with Delays to Data) It is not well understood whether (or under which conditions) distributed delay equations can be fit to data from comparable discrete delay equations (and vice versa) and yield correct parameter inferences or predictions about future states of the system. While deriving general statements about fitting one model type to another may be challenging, these questions can be addressed in the context specific cases, perhaps laying the groundwork for identifying more general patterns.

For this project, consider doing the following:

1. Find a published distributed or discrete delay model, and derive the complementary model so that you have a discrete delay model paired with an analogous distributed delay model.
2. Use the linear chain trick to write a set of ODEs for the distributed delay model.
3. Design a simulation-based experimental comparison, thinking carefully about the differences in assumptions between models, then simulate data under each model. Fit each model to each data set.
4. How biased are the estimates in each case? Which quantities can or cannot be estimated well (e.g., other parameter values in the model unrelated to the form of the delays), and/or how well does each model forecast a continuation of those simulated data sets?

5   The Importance of Publishing

Finally, perhaps the single most important part of conducting research is to share that research with your colleagues, the broader scientific community, and the public. This might include preparing and presenting a poster at a local poster session or professional conference, or preparing a final manuscript for submission to a peer-reviewed journal, so that others may benefit from your work.

Once a project is underway, work with your research mentor to begin drafting your manuscript in LaTeX using templates and guidelines provided by journals for authors. Write up your results as if you were planning to submit the manuscript for publication. To select a journal, start with the one that published the paper you are basing your work on, and talk to your mentor about which journal might be most appropriate for you to submit your work to.

In addition to standard peer-reviewed journals, there are also a number of options available for publishing undergraduate research through the peer-review process. Examples include journals like *Spora: A Journal of Biomathematics*, *SIAM Undergraduate Research Online (SIURO)*, and regional or university-related journals like *The Minnesota Journal of Undergraduate Mathematics*, *The Journal of Undergraduate Research at Ohio State*, or the *Nevada State Undergraduate Research Journal*.

Acknowledgements My outlook on undergraduate research at the interface of biology, mathematics, and statistics has been shaped by many influential mentors and peers that I was lucky to have known as an undergraduate at the University of Southern Colorado (now Colorado State University-Pueblo), at the Mathematical and Theoretical Biology Summer Institute (MTBI), as a graduate student in the Center for Applied Mathematics at Cornell University, and as a postdoctoral researcher in the Aquatic Ecology Laboratory and the Mathematical Biosciences Institute at The Ohio State University. It has truly been a privilege to know and learn from such an outstanding collection of people. I thank the students in my courses for exposing me to a broad array of project topics that have further influenced my outlook on undergraduate research across the sciences. I thank my colleague and wife Dr. Deena Schmidt; my colleagues Michael Cortez, Marisa Eisenberg, Colin Grudzien, and Andrey Sarantsev; my students Amy Robards, Jace Gilbert, Adam Kirosingh, Narae Wadsworth, and Catalina Medina; and reviewers Andrew Brouwer and Kathryn Montovan for many helpful comments, criticisms, and suggestions that ultimately improved this chapter. Finally, I thank my children Alex (age 7), Ellie (age 3), and Nat (age 3) for their unyielding companionship and boundless energy, which were instrumental to my preparing this chapter over such an extended period of time.

Appendix

Getting Started Writing in LaTeX and Programming in R

Installing the free software LaTeX and R should be straightforward, but here are some installation tips for Microsoft Windows and Mac OS X users (Linux users can find similar installation instructions using the resources mentioned below). Readers are encouraged to ask other students and faculty at their institution about additional resources.

There are two pieces of software each, for LaTeX and R, that should be installed: the basic LaTeX and R software, and an enhanced user interface that facilitates learning for new users and helps established users with their day-today workflow (e.g., helpful menus, code autocompletion and highlighting, custom keyboard shortcuts, advanced document preparation capabilities, etc.). Educators will appreciate that both TeXstudio and R Studio have a consistent user interface across operating systems, making them ideal for group or classroom environments where students may be running a mix of operating systems.

If installing both LATEX and R (recommended), install the base software first (in either order) before installing TeXstudio and/or R Studio (in either order). More detailed instructions and resources are provided below.

Installing and Using LATEX

There are different implementations of LATEX available: *MiKTeX* is a popular Microsoft Windows option (http://miktex.org/), and *TeX Live* a popular Mac OS X option. *TeX Live* comes as part of a full 2 gigabyte installation called *MacTeX* (www.tug.org/mactex/; which includes the popular editors TeXstudio and TeXShop) or can be installed through a smaller 110 megabyte bundle *BasicTeX* (www.tug. org/mactex/morepackages.html). Configure LATEX to install packages "on the fly" without prompting you for permission. This can be done during (preferred) or after installation. Also download and install Ghostscript (www.tug.org/mactex/ morepackages.html).

Next, install the TeXstudio editor (www.texstudio.org/), preferably after R is installed. Various settings can be changed after installation, including color themes, and configuring TeXstudio to compile a type of LATEX document that includes blocks of R code known as a `Sweave` or `knitr` document (use `knitr`).

For additional LATEX resources, see the author's website (www.pauljhurtado.com/ R/), the LaTeX wikibook [166], the AMS Short Math Guide for LATEX [42], and references and resources listed therein.

Installing and Using R

Download R from www.r-project.org/ and use the default installation process. Once R is installed (and, preferably once LATEX is installed), install R Studio from www. rstudio.com. By installing R Studio after LATEX, you will be able to create multiple document types to generate PDFs, including *R Markdown* documents and `knitr` documents. Helpful online resources include the "cheat sheets" on the R Studio website, introductory courses by DataCamp (www.datacamp.com) and Software Carpentry (www.software-carpentry.org), [157] for a gentle introduction to R and some applications to population modeling, and R resources on the author's website (www.pauljhurtado.com/R/).

References

1. Adamson, M.W., Morozov, A.Y.: When can we trust our model predictions? Unearthing structural sensitivity in biological systems. Proceedings of the Royal Society A: Mathematical, Physical and Engineering Sciences **469**(2149) (2012). https://doi.org/10.1098/rspa.2012.0500
2. Adamson, M.W., Morozov, A.Y.: Defining and detecting structural sensitivity in biological models: Developing a new framework. Journal of Mathematical Biology **69**(6–7), 1815–1848 (2014). https://doi.org/10.1007/s00285-014-0753-3
3. Allen, E.: Modeling with Itô Stochastic Differential Equations, *Mathematical Modelling: Theory and Applications*, vol. 22. Springer Netherlands (2007). https://doi.org/10.1007/978-1-4020-5953-7
4. Allen, L.: An Introduction to Mathematical Biology. Pearson/Prentice Hall (2007)
5. Allen, L.J.S.: Mathematical Epidemiology, *Lecture Notes in Mathematics*, vol. 1945, chap. An Introduction to Stochastic Epidemic Models, pp. 81–130. Springer Berlin, Heidelberg (2008). https://doi.org/10.1007/978-3-540-78911-6_3
6. Allen, L.J.S.: An Introduction to Stochastic Processes with Applications to Biology. Chapman and Hall/CRC (2010)
7. Anderson, R.M., May, R.M.: Infectious Diseases of Humans: Dynamics and Control. Oxford University Press (1992)
8. Arino, O., Hbid, M., Dads, E.A. (eds.): Delay Differential Equations and Applications, *NATO Science Series*, vol. 205. Springer (2006). https://doi.org/10.1007/1-4020-3647-7
9. Armbruster, B., Beck, E.: Elementary proof of convergence to the mean-field model for the SIR process. Journal of Mathematical Biology **75**(2), 327–339 (2017). https://doi.org/10.1007/s00285-016-1086-1
10. Arrowsmith, D.K., Place, C.M.: An Introduction to Dynamical Systems. Cambridge University Press (1990)
11. Audoly, S., Bellu, G., D'Angio, L., Saccomani, M., Cobelli, C.: Global identifiability of nonlinear models of biological systems. IEEE Transactions on Biomedical Engineering **48**(1), 55–65 (2001). https://doi.org/10.1109/10.900248
12. Baker, C.T.: A perspective on the numerical treatment of Volterra equations. Journal of Computational and Applied Mathematics **125**(1–2), 217–249 (2000). https://doi.org/10.1016/S0377-0427(00)00470-2
13. Banks, H.T., Catenacci, J., Hu, S.: A Comparison of Stochastic Systems with Different Types of Delays. Stochastic Analysis and Applications **31**(6), 913–955 (2013). https://doi.org/10.1080/07362994.2013.806217
14. Banks, H.T., Cintrón-Arias, A., Kappel, F.: Mathematical Modeling and Validation in Physiology: Applications to the Cardiovascular and Respiratory Systems, chap. Parameter Selection Methods in Inverse Problem Formulation, pp. 43–73. Springer Berlin Heidelberg (2013). https://doi.org/10.1007/978-3-642-32882-4_3
15. Barrientos, P.G., Rodríguez, J.Á., Ruiz-Herrera, A.: Chaotic dynamics in the seasonally forced SIR epidemic model. Journal of Mathematical Biology **75**(6–7), 1655–1668 (2017). https://doi.org/10.1007/s00285-017-1130-9
16. Bayram, M., Partal, T., Buyukoz, G.O.: Numerical methods for simulation of stochastic differential equations. Advances in Difference Equations **2018**(1) (2018). https://doi.org/10.1186/s13662-018-1466-5
17. Bertram, R., Rubin, J.E.: Multi-timescale systems and fast-slow analysis. Mathematical Biosciences **287**, 105–121 (2017). https://doi.org/10.1016/j.mbs.2016.07.003. 50th Anniversary Issue
18. Beuter, A., Glass, L., Mackey, M.C., Titcombe, M.S. (eds.): Nonlinear Dynamics in Physiology and Medicine. Interdisciplinary Applied Mathematics (Book 25). Springer (2003)
19. Boersch-Supan, P.H., Ryan, S.J., Johnson, L.R.: deBInfer: Bayesian inference for dynamical models of biological systems in R. Methods in Ecology and Evolution **8**(4), 511–518 (2016). https://doi.org/10.1111/2041-210X.12679

20. Bolker, B.M.: Ecological Models and Data in R, chap. Dynamic Models (Ch. 11). Princeton University Press (2008). https://ms.mcmaster.ca/~bolker/emdbook/chap11A.pdf
21. Bolker, B.M.: Ecological Models and Data in R. Princeton University Press (2008)
22. Brauer, F., Castillo-Chavez, C.: Mathematical Models in Population Biology and Epidemiology, 2nd (2012) edn. Texts in Applied Mathematics (Book 40). Springer-Verlag (2011)
23. Brauer, F., van den Driessche, P., Wu, J. (eds.): Mathematical Epidemiology. Lecture Notes in Mathematics: Mathematical Biosciences Subseries. Springer-Verlag Berlin Heidelberg (2008). https://doi.org/10.1007/978-3-540-78911-6
24. Briggs, G.E., Haldane, J.B.S.: A note on the kinetics of enzyme action. Biochemical Journal 19(2), 338–339 (1925). https://doi.org/10.1042/bj0190338
25. Burton, T., Furumochi, T.: A stability theory for integral equations. Journal of Integral Equations and Applications 6(4), 445–477 (1994). https://doi.org/10.1216/jiea/1181075832
26. Cao, Y., Gillespie, D.T., Petzold, L.R.: Avoiding negative populations in explicit Poisson tau-leaping. The Journal of Chemical Physics 123(5), 054,104 (2005). https://doi.org/10.1063/1.1992473
27. Cao, Y., Gillespie, D.T., Petzold, L.R.: Efficient step size selection for the tau-leaping simulation method. The Journal of Chemical Physics 124(4), 044,109 (2006). https://doi.org/10.1063/1.2159468
28. Casella, G., Berger, R.: Statistical Inference, 2nd edn. Cengage Learning (2001)
29. Champredon, D., Dushoff, J., Earn, D.: Equivalence of the Erlang-Distributed SEIR Epidemic Model and the Renewal Equation. SIAM Journal on Applied Mathematics 78(6), 3258–3278 (2018). https://doi.org/10.1137/18M1186411
30. Chapman, A., Mesbahi, M.: Stability analysis of nonlinear networks via M-matrix theory: Beyond linear consensus. In: 2012 American Control Conference (ACC). IEEE (2012). https://doi.org/10.1109/ACC.2012.6315625
31. Chatterjee, A., Vlachos, D.G., Katsoulakis, M.A.: Binomial distribution based τ-leap accelerated stochastic simulation. The Journal of Chemical Physics 122(2), 024,112 (2005). https://doi.org/10.1063/1.1833357
32. Cintrón-Arias, A., Banks, H.T., Capaldi, A., Lloyd, A.L.: A sensitivity matrix based methodology for inverse problem formulation. Journal of Inverse and Ill-posed Problems pp. 545–564 (2009). https://doi.org/10.1515/JIIP.2009.034
33. Cobelli, C., DiStefano, J.J.: Parameter and structural identifiability concepts and ambiguities: A critical review and analysis. American Journal of Physiology-Regulatory, Integrative and Comparative Physiology 239(1), R7–R24 (1980). https://doi.org/10.1152/ajpregu.1980.239.1.R7
34. Conlan, A.J., Grenfell, B.T.: Seasonality and the persistence and invasion of measles. Proceedings of the Royal Society B: Biological Sciences 274(1614), 1133–1141 (2007). https://doi.org/10.1098/rspb.2006.0030
35. Cortez, M.H.: When does pathogen evolution maximize the basic reproductive number in well-mixed host–pathogen systems? Journal of Mathematical Biology 67(6), 1533–1585 (2013). https://doi.org/10.1007/s00285-012-0601-2
36. Cortez, M.H.: Coevolution-driven predator-prey cycles: Predicting the characteristics of eco-coevolutionary cycles using fast-slow dynamical systems theory. Theoretical Ecology 8(3), 369–382 (2015). https://doi.org/10.1007/s12080-015-0256-x
37. Cortez, M.H., Ellner, S.P.: Understanding rapid evolution in predator-prey interactions using the theory of fast-slow dynamical systems. The American Naturalist 176(5), E109–E127 (2010). https://doi.org/10.1086/656485
38. Dawes, J., Souza, M.: A derivation of Hollings type I, II and III functional responses in predator–prey systems. Journal of Theoretical Biology 327, 11–22 (2013). https://doi.org/10.1016/j.jtbi.2013.02.017
39. Dayan, P., Abbott, L.F.: Theoretical Neuroscience: Computational and Mathematical Modeling of Neural Systems. Computational Neuroscience. The MIT Press (2005)
40. Devroye, L.: Non-Uniform Random Variate Generation, chap. General Principles in Random Variate Generation: Inversion Method (§2.2). Springer-Verlag. (1986). http://luc.devroye.org/rnbookindex.html

41. Dhooge, A., Govaerts, W., Kuznetsov, Y.A.: MATCONT: A MATLAB package for numerical bifurcation analysis of ODEs. ACM Transactions on Mathematical Software **29**(2), 141–164 (2003). https://doi.org/10.1145/779359.779362

42. Downes, M., Beeton, B.: Short Math Guide for LaTeX. American Mathematical Society (2017). https://www.ams.org/tex/amslatex. (Accessed: 22 April 2019)

43. van den Driessche, P., Watmough, J.: Reproduction numbers and sub-threshold endemic equilibria for compartmental models of disease transmission. Mathematical Biosciences **180**(1–2), 29–48 (2002). https://doi.org/10.1016/S0025-5564(02)00108-6

44. Earn, D.J.: A simple model for complex dynamical transitions in epidemics. Science **287**(5453), 667–670 (2000). https://doi.org/10.1126/science.287.5453.667

45. Edelstein-Keshet, L.: Mathematical Models in Biology. Classics in Applied Mathematics (Book 46). Society for Industrial and Applied Mathematics (2005). https://doi.org/10.1137/1.9780898719147

46. Eisenberg, M.: Generalizing the differential algebra approach to input–output equations in structural identifiability. ArXiv e-prints (2013). https://arxiv.org/abs/1302.5484

47. Eisenberg, M.C., Hayashi, M.A.: Determining identifiable parameter combinations using subset profiling. Mathematical Biosciences **256**, 116–126 (2014). https://doi.org/10.1016/j.mbs.2014.08.008

48. Ellner, S.P.: Pair approximation for lattice models with multiple interaction scales. Journal of Theoretical Biology **210**(4), 435–447 (2001). https://doi.org/10.1006/jtbi.2001.2322

49. Ellner, S.P., Becks, L.: Rapid prey evolution and the dynamics of two-predator food webs. Theoretical Ecology **4**(2), 133–152 (2011). https://doi.org/10.1007/s12080-010-0096-7

50. Ellner, S.P., Guckenheimer, J.: Dynamic Models in Biology. Princeton University Press (2006)

51. Ellner, S.P., Guckenheimer, J.: Dynamic Models in Biology, chap. Building Dynamic Models (Ch. 9). Princeton University Press (2006). http://assets.press.princeton.edu/chapters/s9_8124.pdf

52. Ellner, S.P., Guckenheimer, J.: Dynamic Models in Biology, chap. Spatial Patterns in Biology (Ch. 7). Princeton University Press (2006). http://assets.press.princeton.edu/chapters/s7_8124.pdf

53. Ellner, S.P., Rees, M.: Integral projection models for species with complex demography. The American Naturalist **167**(3), 410–428 (2006). https://doi.org/10.1086/499438

54. Ellner, S.P., Rees, M.: Stochastic stable population growth in integral projection models: theory and application. Journal of Mathematical Biology **54**(2), 227–256 (2006). https://doi.org/10.1007/s00285-006-0044-8

55. Ermentrout, B.: XPP/XPPAUT Homepage. http://www.math.pitt.edu/~bard/xpp/xpp.html. (Accessed: April 2019)

56. Ermentrout, B.: Simulating, Analyzing, and Animating Dynamical Systems: A Guide to XPPAUT for Researchers and Students. Software, Environments and Tools (Book 14). Society for Industrial and Applied Mathematics (1987)

57. Evans, N.D., Chappell, M.J.: Extensions to a procedure for generating locally identifiable reparameterisations of unidentifiable systems. Mathematical Biosciences **168**(2), 137–159 (2000). https://doi.org/10.1016/S0025-5564(00)00047-X

58. Feinberg, M.: Complex balancing in general kinetic systems. Archive for Rational Mechanics and Analysis **49**(3), 187–194 (1972). https://doi.org/10.1007/BF00255665

59. Feinberg, M.: On chemical kinetics of a certain class. Archive for Rational Mechanics and Analysis **46**(1), 1–41 (1972). https://doi.org/10.1007/BF00251866

60. Feinberg, M.: Chemical reaction network structure and the stability of complex isothermal reactors—i. The deficiency zero and deficiency one theorems. Chemical Engineering Science **42**(10), 2229–2268 (1987). https://doi.org/10.1016/0009-2509(87)80099-4

61. Feinberg, M.: Foundations of Chemical Reaction Network Theory. Springer International Publishing (2019). https://doi.org/10.1007/978-3-030-03858-8

62. Feinberg, M., Ellison, P., Ji, H., Knight, D.: Chemical reaction network toolbox. https://crnt.osu.edu/CRNTWin. (Accessed: April 2019)

63. Feng, Z., Thieme, H.: Endemic models with arbitrarily distributed periods of infection I: Fundamental properties of the model. SIAM Journal on Applied Mathematics **61**(3), 803–833 (2000). https://doi.org/10.1137/S0036139998347834

64. Feng, Z., Xu, D., Zhao, H.: Epidemiological models with non-exponentially distributed disease stages and applications to disease control. Bulletin of Mathematical Biology **69**(5), 1511–1536 (2007). https://doi.org/10.1007/s11538-006-9174-9

65. Feng, Z., Zheng, Y., Hernandez-Ceron, N., Zhao, H., Glasser, J.W., Hill, A.N.: Mathematical models of Ebola—Consequences of underlying assumptions. Mathematical biosciences **277**, 89–107 (2016)

66. Fiechter, J., Rose, K.A., Curchitser, E.N., Hedstrom, K.S.: The role of environmental controls in determining sardine and anchovy population cycles in the California Current: Analysis of an end-to-end model. Progress in Oceanography **138**, 381–398 (2015). https://doi.org/10.1016/j.pocean.2014.11.013

67. Ghil, M., Zaliapin, I., Thompson, S.: A delay differential model of ENSO variability: Parametric instability and the distribution of extremes. Nonlinear Processes in Geophysics **15**(3), 417–433 (2008). https://doi.org/10.5194/npg-15-417-2008

68. Gillespie, D.T.: A general method for numerically simulating the stochastic time evolution of coupled chemical reactions. Journal of Computational Physics **22**(4), 403–434 (1976). https://doi.org/10.1016/0021-9991(76)90041-3

69. Gillespie, D.T.: Exact stochastic simulation of coupled chemical reactions. The Journal of Physical Chemistry **81**(25), 2340–2361 (1977). https://doi.org/10.1021/j100540a008

70. Givens, G.H., Hoeting, J.A.: Computational Statistics, 2nd edn. Computational Statistics. John Wiley & Sons (2012)

71. Grassly, N.C., Fraser, C.. Seasonal infectious disease epidemiology. Proceedings of the Royal Society B: Biological Sciences **273**(1600), 2541–2550 (2006). https://doi.org/10.1098/rspb.2006.3604

72. Grimm, V., Railsback, S.F.: Individual-based Modeling and Ecology. Princeton University Press (2005)

73. Guckenheimer, J., Holmes, P.: Nonlinear Oscillations, Dynamical Systems, and Bifurcations of Vector Fields, corr. 6th printing, 6th edn. Applied Mathematical Sciences. Springer (2002)

74. Guckenheimer, J., Myers, M.: Computing Hopf Bifurcations. II: Three Examples From Neurophysiology. SIAM Journal on Scientific Computing **17**(6), 1275–1301 (1996). https://doi.org/10.1137/S1064827593253495

75. Guckenheimer, J., Myers, M., Sturmfels, B.: Computing Hopf Bifurcations I. SIAM Journal on Numerical Analysis **34**(1), 1–21 (1997). https://doi.org/10.1137/S0036142993253461

76. Hansen, J.A., Penland, C.: Efficient approximate techniques for integrating stochastic differential equations. Monthly Weather Review **134**(10), 3006–3014 (2006). https://doi.org/10.1175/MWR3192.1

77. Hanski, I.A.: Metapopulation Ecology. Oxford Series in Ecology and Evolution. Oxford University Press (1999)

78. Heitmann, S.: Brain Dynamics Toolbox (2018). http://bdtoolbox.org. (Accessed: 20 June 2019) Alt. URL https://github.com/breakspear/bdtoolkit/

79. Helton, J., Davis, F.: Latin hypercube sampling and the propagation of uncertainty in analyses of complex systems. Reliability Engineering & System Safety **81**(1), 23–69 (2003). https://doi.org/10.1016/S0951-8320(03)00058-9

80. Hethcote, H.W., Tudor, D.W.: Integral equation models for endemic infectious diseases. Journal of Mathematical Biology **9**(1), 37–47 (1980). https://doi.org/10.1007/BF00276034

81. Hethcote, H.W., Yorke, J.A.: Gonorrhea Transmission Dynamics and Control. Lecture Notes in Biomathematics. Springer Berlin Heidelberg (1984). https://doi.org/10.1007/978-3-662-07544-9

82. Hiebeler, D.E., Millett, N.E.: Pair and triplet approximation of a spatial lattice population model with multiscale dispersal using Markov chains for estimating spatial autocorrelation. Journal of Theoretical Biology **279**(1), 74–82 (2011). https://doi.org/10.1016/j.jtbi.2011.03.027

83. Higham, D.J.: An algorithmic introduction to numerical simulation of stochastic differential equations.   SIAM Review **43**(3), 525–546 (2001).   https://doi.org/10.1137/S0036144500378302

84. Hirsch, M.W., Smale, S., Devaney, R.L.: Differential Equations, Dynamical Systems, and an Introduction to Chaos, 3rd edn. Elsevier (2013). https://doi.org/10.1016/C2009-0-61160-0

85. Hogg, R.V., McKean, J.W., Craig, A.T.: Introduction to Mathematical Statistics, 7th edn. Pearson (2012)

86. Hooker, G.: MATLAB Functions for Profiled Estimation of Differential Equations (2010). http://faculty.bscb.cornell.edu/~hooker/profile_webpages/

87. Hooker, G., Ramsay, J.O., Xiao, L.: CollocInfer: Collocation inference in differential equation models. Journal of Statistical Software **75**(2), 1–52 (2016). https://doi.org/10.18637/jss.v075.i02

88. Hooker, G., Xiao, L., Ramsay, J.: CollocInfer: An R Library for Collocation Inference for Continuous- and Discrete-Time Dynamic Systems (2010). http://faculty.bscb.cornell.edu/~hooker/profile_webpages/

89. Hurtado, P.J.: The potential impact of disease on the migratory structure of a partially migratory passerine population. Bulletin of Mathematical Biology **70**(8), 2264 (2008). https://doi.org/10.1007/s11538-008-9345-y

90. Hurtado, P.J.: Within-host dynamics of mycoplasma infections: Conjunctivitis in wild passerine birds. Journal of Theoretical Biology **306**, 73–92 (2012). https://doi.org/10.1016/j.jtbi.2012.04.018

91. Hurtado, P.J., Hall, S.R., Ellner, S.P.: Infectious disease in consumer populations: Dynamic consequences of resource-mediated transmission and infectiousness.   Theoretical Ecology **7**(2), 163–179 (2014). https://doi.org/10.1007/s12080-013-0208-2

92. Hurtado, P.J., Kirosingh, A.S.: Generalizations of the 'Linear Chain Trick': Incorporating more flexible dwell time distributions into mean field ODE models. Journal of Mathematical Biology **79**, 1831–1883 (2019). https://doi.org/10.1007/s00285-019-01412-w

93. Iacus, S.M.: Simulation and Inference for Stochastic Differential Equations. Springer New York (2008)

94. Izhikevich, E.M.: Dynamical Systems in Neuroscience: The Geometry of Excitability and Bursting. Computational Neuroscience. MIT Press (2010)

95. Keener, J., Sneyd, J.: Mathematical Physiology I: Cellular Physiology, 2nd edn.   Springer (2008)

96. Keener, J., Sneyd, J.: Mathematical Physiology II: Systems Physiology, 2nd edn.   Springer (2008)

97. Kendall, B.E., Briggs, C.J., Murdoch, W.W., Turchin, P., Ellner, S.P., McCauley, E., Nisbet, R.M., Wood, S.N.: Why do populations cycle? A synthesis of statistical and mechanistic modeling approaches.   Ecology **80**(6), 1789–1805 (1999).   https://doi.org/10.1890/0012-9658(1999)080[1789:WDPCAS]2.0.CO;2

98. Kermack, W.O., McKendrick, A.G.: A contribution to the mathematical theory of epidemics. Proceedings of the Royal Society of London. Series A, Containing Papers of a Mathematical and Physical Character **115**(772), 700–721 (1927)

99. Koopman, J.: Modeling infection transmission.   Annual Review of Public Health **25**(1), 303–326 (2004). https://doi.org/10.1146/annurev.publhealth.25.102802.124353

100. Kot, M.: Elements of Mathematical Ecology. Cambridge University Press (2014)

101. Kozubowski, T.J., Panorska, A.K., Forister, M.L.: A discrete truncated Pareto distribution. Statistical Methodology **26**, 135–150 (2015). https://doi.org/10.1016/j.stamet.2015.04.002

102. Krylova, O., Earn, D.J.D.: Effects of the infectious period distribution on predicted transitions in childhood disease dynamics. Journal of The Royal Society Interface **10**(84) (2013). https://doi.org/10.1098/rsif.2013.0098

103. Kuehn, C.: Multiple Time Scale Dynamics, *Applied Mathematical Sciences*, vol. 191, 1 edn. Springer International Publishing (2015). https://doi.org/10.1007/978-3-319-12316-5

104. Kuznetsov, Y.: Elements of Applied Bifurcation Theory, 3 edn.   Applied Mathematical Sciences. Springer New York (2004)

105. Lande, R., Engen, S., Saether, B.E.: Stochastic Population Dynamics in Ecology and Conservation. Oxford University Press (2003). https://doi.org/10.1093/acprof:oso/9780198525257.001.0001
106. Larsen, R.J., Marx, M.L.: Introduction to Mathematical Statistics and Its Applications, 5th edn. Pearson (2011)
107. Lee, E.C., Kelly, M.R., Ochocki, B.M., Akinwumi, S.M., Hamre, K.E., Tien, J.H., Eisenberg, M.C.: Model distinguishability and inference robustness in mechanisms of cholera transmission and loss of immunity. Journal of Theoretical Biology 420, 68–81 (2017). https://doi.org/10.1016/j.jtbi.2017.01.032
108. Lee, S., Chowell, G.: Exploring optimal control strategies in seasonally varying flu-like epidemics. Journal of Theoretical Biology 412, 36–47 (2017). https://doi.org/10.1016/j.jtbi.2016.09.023
109. Levin, S.A., Powell, T.M., Steele, J.W. (eds.): Patch Dynamics. Springer Berlin Heidelberg (1993). https://doi.org/10.1007/978-3-642-50155-5
110. Liao, J., Li, Z., Hiebeler, D.E., Iwasa, Y., Bogaert, J., Nijs, I.: Species persistence in landscapes with spatial variation in habitat quality: A pair approximation model. Journal of Theoretical Biology 335, 22–30 (2013). https://doi.org/10.1016/j.jtbi.2013.06.015
111. Linz, P.: Analytical and Numerical Methods for Volterra Equations. Society for Industrial and Applied Mathematics (1985). https://doi.org/10.1137/1.9781611970852
112. Liu, Y., Khim, J.: Taylor's theorem (with Lagrange remainder). https://brilliant.org/wiki/taylors-theorem-with-lagrange-remainder/. (Accessed 15 April 2019)
113. Ma, J., Earn, D.J.D.: Generality of the final size formula for an epidemic of a newly invading infectious disease. Bulletin of Mathematical Biology 68(3), 679–702 (2006). https://doi.org/10.1007/s11538-005-9047-7
114. Marino, S., Hogue, I.B., Ray, C.J., Kirschner, D.E.: A methodology for performing global uncertainty and sensitivity analysis in systems biology. Journal of Theoretical Biology 254(1), 178–196 (2008). https://doi.org/10.1016/j.jtbi.2008.04.011
115. Mathworks: MATLAB Documentation: Choose an ODE solver. https://www.mathworks.com/help/matlab/math/choose-an-ode-solver.html. (Accessed: April 2019)
116. Mathworks: MATLAB Documentation: Solve Stiff ODEs. https://www.mathworks.com/help/matlab/math/solve-stiff-odes.html. (Accessed: April 2019)
117. May, R.: Stability and Complexity in Model Ecosystems. Landmarks in Biology Series. Princeton University Press (2001)
118. McCann, K.S.: Food Webs. Monographs in Population Biology (Book 57). Princeton University Press (2011)
119. Meiss, J.D.: Differential Dynamical Systems, Revised Edition. Society for Industrial and Applied Mathematics, Philadelphia, PA (2017). https://doi.org/10.1137/1.9781611974645
120. Merow, C., Dahlgren, J.P., Metcalf, C.J.E., Childs, D.Z., Evans, M.E., Jongejans, E., Record, S., Rees, M., Salguero-Gómez, R., McMahon, S.M.: Advancing population ecology with integral projection models: a practical guide. Methods in Ecology and Evolution 5(2), 99–110 (2014). https://doi.org/10.1111/2041-210X.12146
121. Meshkat, N., Eisenberg, M., DiStefano, J.J.: An algorithm for finding globally identifiable parameter combinations of nonlinear ODE models using Gröbner bases. Mathematical Biosciences 222(2), 61–72 (2009). https://doi.org/10.1016/j.mbs.2009.08.010
122. Meshkat, N., zhen Kuo, C.E., Joseph DiStefano, I.: On finding and using identifiable parameter combinations in nonlinear dynamic systems biology models and COMBOS: A novel web implementation. PLoS ONE 9(10), e110,261 (2014). https://doi.org/10.1371/journal.pone.0110261
123. Miao, H., Dykes, C., Demeter, L.M., Wu, H.: Differential equation modeling of HIV viral fitness experiments: Model identification, model selection, and multimodel inference. Biometrics 65(1), 292–300 (2008). https://doi.org/10.1111/j.1541-0420.2008.01059.x
124. Moraes, A., Tempone, R., Vilanova, P.: Hybrid Chernoff Tau-Leap. Multiscale Modeling & Simulation 12(2), 581–615 (2014). https://doi.org/10.1137/130925657

125. Murdoch, W.W., Briggs, C.J., Nisbet, R.M.: Consumer–Resource Dynamics, *Monographs in Population Biology*, vol. 36. Princeton University Press, Princeton, USA (2003)
126. Murray, J.D.: Mathematical Biology: I. An Introduction. Interdisciplinary Applied Mathematics (Book 17). Springer (2007)
127. Murray, J.D.: Mathematical Biology II: Spatial Models and Biomedical Applications. Interdisciplinary Applied Mathematics (Book 18). Springer (2011)
128. Nash, J.C.: On best practice optimization methods in R. Journal of Statistical Software **60**(2), 1–14 (2014). http://www.jstatsoft.org/v60/i02/
129. Nash, J.C., Varadhan, R.: Unifying optimization algorithms to aid software system users: optimx for R. Journal of Statistical Software **43**(9), 1–14 (2011). http://www.jstatsoft.org/v43/i09/
130. Newman, M.: Networks, 2nd edn. Oxford University Press (2018)
131. Øksendal, B.: Stochastic Differential Equations: An Introduction with Applications, 6th ed. (6th corrected printing 2013) edn. Springer-Verlag Berlin Heidelberg (2003). https://doi.org/10.1007/978-3-642-14394-6
132. Ovaskainen, O., Saastamoinen, M.: Frontiers in Metapopulation Biology: The Legacy of Ilkka Hanski. Annual Review of Ecology, Evolution, and Systematics **49**(1), 231–252 (2018). https://doi.org/10.1146/annurev-ecolsys-110617-062519
133. Pineda-Krch, M.: GillespieSSA: Gillespie's Stochastic Simulation Algorithm (SSA) (2010). https://CRAN.R-project.org/package=GillespieSSA. R package version 0.5-4
134. Poggiale, J.C., Aldebert, C., Girardot, B., Kooi, B.W.: Analysis of a predator–prey model with specific time scales: A geometrical approach proving the occurrence of canard solutions. Journal of Mathematical Biology (2019). https://doi.org/10.1007/s00285-019-01337-4
135. Porter, M.A., Gleeson, J.P.: Dynamics on Networks: A Tutorial. ArXiv e-prints (2015). http://arxiv.org/abs/1403.7663v2
136. Porter, M.A., Gleeson, J.P.: Dynamical Systems on Networks: A Tutorial, *Frontiers in Applied Dynamical Systems: Reviews and Tutorials*, vol. 4. Springer (2016). https://doi.org/10.1007/978-3-319-26641-1
137. R Core Team: R: A Language and Environment for Statistical Computing. R Foundation for Statistical Computing, Vienna, Austria (2019). https://www.R-project.org/
138. Rackauckas, C.: A Comparison Between Differential Equation Solver Suites In MATLAB, R, Julia, Python, C, Mathematica, Maple, and Fortran. The Winnower (2018). https://doi.org/10.15200/winn.153459.98975
139. Ramsay, J.O., Hooker, G., Campbell, D., Cao, J.: Parameter estimation for differential equations: A generalized smoothing approach. Journal of the Royal Statistical Society: Series B (Statistical Methodology) **69**(5), 741–796 (2007). https://doi.org/10.1111/j.1467-9868.2007.00610.x
140. Rand, R.H.: Lecture notes on nonlinear vibrations. Cornell eCommons (2012). https://hdl.handle.net/1813/28989
141. Raue, A., Kreutz, C., Bachmann, J., Timmer, J., Schilling, M., Maiwald, T., Klingmüller, U.: Structural and practical identifiability analysis of partially observed dynamical models by exploiting the profile likelihood. Bioinformatics **25**(15), 1923–1929 (2009). https://doi.org/10.1093/bioinformatics/btp358
142. Reynolds, A., Rubin, J., Clermont, G., Day, J., Vodovotz, Y., Ermentrout, G.B.: A reduced mathematical model of the acute inflammatory response: I. Derivation of model and analysis of anti-inflammation. Journal of Theoretical Biology **242**(1), 220–236 (2006). https://doi.org/10.1016/j.jtbi.2006.02.016
143. Rinaldi, S., Muratori, S., Kuznetsov, Y.: Multiple attractors, catastrophes and chaos in seasonally perturbed predator-prey communities. Bulletin of Mathematical Biology **55**(1), 15–35 (1993). https://doi.org/10.1007/BF02460293
144. Rose, K.A., Fiechter, J., Curchitser, E.N., Hedstrom, K., Bernal, M., Creekmore, S., Haynie, A., ichi Ito, S., Lluch-Cota, S., Megrey, B.A., Edwards, C.A., Checkley, D., Koslow, T., McClatchie, S., Werner, F., MacCall, A., Agostini, V.: Demonstration of a fully-coupled end-to-end model for small pelagic fish using sardine and anchovy in the California Current.

Progress in Oceanography **138**, 348–380 (2015). https://doi.org/10.1016/j.pocean.2015.01.012

145. Ross, S.: Introduction to Probability Models, 11th edn. Elsevier (2014). https://doi.org/10.1016/C2012-0-03564-8

146. Saccomani, M.P., Audoly, S., DAngiò, L.: Parameter identifiability of nonlinear systems: The role of initial conditions. Automatica **39**(4), 619–632 (2003). https://doi.org/10.1016/S0005-1098(02)00302-3

147. Sauer, T.: Computational solution of stochastic differential equations. Wiley Interdisciplinary Reviews: Computational Statistics **5**(5), 362–371 (2013). https://doi.org/10.1002/wics.1272

148. Schelter, W.F.: Maxima (2000). http://maxima.sourceforge.net/. (Accessed: April 2019)

149. Segel, L., Edelstein-Keshet, L.: A Primer on Mathematical Models in Biology. Society for Industrial and Applied Mathematics, Philadelphia, PA (2013). https://doi.org/10.1137/1.9781611972504

150. Shaw, A.K., Binning, S.A.: Migratory recovery from infection as a selective pressure for the evolution of migration. The American Naturalist **187**(4), 491–501 (2016). https://doi.org/10.1086/685386

151. Shertzer, K.W., Ellner, S.P., Fussmann, G.F., Hairston, N.G.: Predator–prey cycles in an aquatic microcosm: Testing hypotheses of mechanism. Journal of Animal Ecology **71**(5), 802–815 (2002). https://doi.org/10.1046/j.1365-2656.2002.00645.x

152. Shinar, G., Alon, U., Feinberg, M.: Sensitivity and Robustness in Chemical Reaction Networks. SIAM Journal on Applied Mathematics **69**(4), 977–998 (2009). https://doi.org/10.1137/080719820

153. Shoffner, S., Schnell, S.: Approaches for the estimation of timescales in nonlinear dynamical systems: Timescale separation in enzyme kinetics as a case study. Mathematical Biosciences **287**, 122–129 (2017). https://doi.org/10.1016/j.mbs.2016.09.001

154. Smith, H.: An introduction to delay differential equations with applications to the life sciences, vol. 57. Springer (2010)

155. Society for Industrial and Applied Mathematics: DSWeb Dynamical Systems Software. https://dsweb.siam.org/Software. (Accessed: 1 April 2019)

156. Soetaert, K., Petzoldt, T., Setzer, R.W.: Solving differential equations in R: Package deSolve. Journal of Statistical Software **33** (2010). https://doi.org/10.18637/jss.v033.i09

157. Stieha, C., Montovan, K., Castillo-Guajardo, D.: A field guide to programming: A tutorial for learning programming and population models. CODEE Journal **10**, Article 2 (2014). https://doi.org/10.5642/codee.201410.01.02. https://scholarship.claremont.edu/codee/vol10/iss1/2/

158. Stone, L., Olinky, R., Huppert, A.: Seasonal dynamics of recurrent epidemics. Nature **446**(7135), 533–536 (2007). https://doi.org/10.1038/nature05638

159. Strang, G.: Introduction to Linear Algebra, 5th edn. Wellesley – Cambridge Press (2016). https://math.mit.edu/~gs/linearalgebra/

160. Strogatz, S.H.: Nonlinear Dynamics and Chaos: With Applications to Physics, Biology, Chemistry, and Engineering, 2nd edn. Studies in Nonlinearity. Westview Press (2014)

161. Theussl, S., Schwendinger, F., Borchers, H.W.: CRAN Task View: Optimization and mathematical programming (2019). https://CRAN.R-project.org/view=Optimization. (Accessed: April 2019)

162. Vestergaard, C.L., Génois, M.: Temporal Gillespie Algorithm: Fast simulation of contagion processes on time-varying networks. PLOS Computational Biology **11**(10), e1004, 579 (2015). https://doi.org/10.1371/journal.pcbi.1004579

163. Vodopivec, A.: wxMaxima: A GUI for the computer algebra system maxima (2018). https://github.com/wxMaxima-developers/wxmaxima. (Accessed: April 2019)

164. Wearing, H.J., Rohani, P., Keeling, M.J.: Appropriate models for the management of infectious diseases. PLOS Medicine **2**(7) (2005). https://doi.org/10.1371/journal.pmed.0020174

165. Wiggins, S.: Introduction to Applied Nonlinear Dynamical Systems and Chaos, *Texts in Applied Mathematics*, vol. 2, 2nd edn. Springer-Verlag New York (2003). https://doi.org/10.1007/b97481

166. Wikibooks: LaTeX — Wikibooks, The Free Textbook Project (2019). https://en.wikibooks. org/w/index.php?title=LaTeX&oldid=3527944. (Accessed: 22 April 2019)
167. Xia, J., Liu, Z., Yuan, R., Ruan, S.: The effects of harvesting and time delay on predator–prey systems with Holling type II functional response. SIAM Journal on Applied Mathematics **70**(4), 1178–1200 (2009). https://doi.org/10.1137/080728512
168. Yoshida, T., Hairston, N.G., Ellner, S.P.: Evolutionary trade–off between defence against grazing and competitive ability in a simple unicellular alga, *Chlorella vulgaris*. Proceedings of the Royal Society of London. Series B: Biological Sciences **271**(1551), 1947–1953 (2004). https://doi.org/10.1098/rspb.2004.2818
169. Yoshida, T., Jones, L.E., Ellner, S.P., Fussmann, G.F., Hairston, N.G.: Rapid evolution drives ecological dynamics in a predator–prey system. Nature **424**(6946), 303–306 (2003). https:// doi.org/10.1038/nature01767

A Tour of the Basic Reproductive Number and the Next Generation of Researchers

Carlos W. Castillo-Garsow and Carlos Castillo-Chavez

Abstract The Mathematical and Theoretical Biology Institute (MTBI) is a national award winning Research Experience for Undergraduates (REU) that has been running every summer since 1996. Since 1997, students have developed and proposed their own research questions and derived their research projects from them as the keystone of the program. Because MTBI's mentors have no control over what students are interested in, we need to introduce a suite of flexible techniques that can be applied to a broad variety of interests. In this paper, we walk through examples of some of the most popular techniques at MTBI: epidemiological or contagion modeling and reproductive number analysis. We include an overview of the next generation matrix method of finding the basic reproductive number, sensitivity analysis as a technique for investigating the effect of parameters on the reproductive number, and recommendations for interpreting the results. Lastly, we provide some advice to mentors who are looking to advise student-led research projects. All examples are taken from actual student projects that are generally available through the MTBI website.

Suggested Prerequisites *Ordinary differential equations, mass action, SIR model, flow diagrams, equilibria, stability, Jacobian matrix, and eigenvalue.*

C. W. Castillo-Garsow
Mathematics Department, Eastern Washington University, Cheney, WA, USA
e-mail: ccastillogarsow@ewu.edu

C. Castillo-Chavez (✉)
School of Human Evolution and Social Change, Simon A. Levin Mathematical Computational Modeling Science Center, Arizona State University, Tempe, AZ, USA

Division of Applied Mathematics, Brown University, Providence, RI, USA
e-mail: ccchavez@asu.edu

© Springer Nature Switzerland AG 2020
H. Callender Highlander et al. (eds.), *An Introduction to Undergraduate Research in Computational and Mathematical Biology*, Foundations for Undergraduate Research in Mathematics, https://doi.org/10.1007/978-3-030-33645-5_2

87

1 Introduction

The Mathematical and Theoretical Biology Institute (MTBI) at Arizona State University is a summer Research Experience for Undergraduates (REU) that has been running every summer since 1996. From 1996 through its 2018 summer program, MTBI has recruited and enrolled a total of 507 regular first-time undergraduate students and 78 advanced (returning) students. To date MTBI students have completed 211 technical reports, many of which have been converted into publications (including, but not limited to, projects used as examples in this paper [4, 30, 38, 41, 57]).

The program has received external recognition in the form of multiple national awards. The Director of MTBI was awarded a Presidential Award for Excellence in Science, Mathematics and Engineering Mentoring (PAESMEM) in 1997 and the American Association for the Advancement of Science Mentor Award in 2007. Also in 2007, the AMS recognized MTBI as a Mathematics Program that Makes a Difference, and MTBI was awarded a Presidential Award for Excellence in Science, Mathematics and Engineering Mentoring (PAESMEM) in 2011.

A key feature of MTBI is that students start with their own research topics and associated questions [21, 23]. As a result, students often know more about the topic being investigated than the mentors. This means that the mathematical techniques taught at MTBI need to be extremely flexible and well suited to a large population of research problems. In this paper, we give an overview of some of the mathematical techniques that are most commonly used at MTBI, and how they can be put together to form a complete research project.

1.1 Overview

The MTBI summer program runs for 8 weeks. In the first 3 weeks, students attend lectures in theoretical ecology and various epidemiological-contact-contagion modeling techniques, as used in broadly understood dynamical systems: nonlinear systems of difference and differential equations, discrete and continuous time Markov chains, partial differential equations and agent based modeling. In the final 5 weeks of the program, students form self-selected groups of three to five undergraduates, and investigate a problem of their own choosing. They research the background of the problem, identify a question, construct a model to address the question, analyze the results, and write a technical report describing their project. While the faculty mentors' experience is primarily in mathematical biology, students' interests can be quite diverse [21]. Past projects have included such diverse topics as epidemiology [1, 4, 5, 12, 27, 38, 39, 47, 48], eating disorders [40, 41], party politics [13, 57], prison education [2, 49], immigration [25], the menstrual cycle [37], education [29, 30, 32, 42], and ecology [28, 45, 55, 56]. Because of the broad variety of student selected topics, students frequently know a lot more about the modeling application than the mentors do. Students take the lead on the project and provide subject matter knowledge, while mentors provide general expertise in mathematical modeling techniques that can be applied to a broad variety of topics.

The most common modeling method in MTBI is systems of ordinary differential equations, with a focus on equilibrium analysis and the basic reproductive number (R_0). By far the most popular techniques for reproductive analysis at MTBI have been the next generation matrix (NGM) and sensitivity analysis. These are extremely powerful, accessible, and flexible techniques that can be applied to a broad variety of situations. Since the founding of the REU in 1996 to the present year of 2018, students have written 211 technical reports. Of these reports, 34.6% have used sensitivity analysis, 22.7% have used the NGM, and 12.3% have used both. Since the techniques became popular with MTBI students in 2003, there have been 140 technical reports: 50.7% have used sensitivity analysis, 31.4% have used the NGM, and 19.3% have used both [51].

In this article, we provide an introduction to the next generation method of finding the basic reproductive number (R_0), details of how the reproductive number might be checked of accuracy and meaningfully interpreted, and an introduction to sensitivity analysis of R_0 as a way of evaluating the impact of possible interventions. This introduction is followed by examples of some complexities that can arise in using these techniques and how to handle them. All examples come from actual student projects, the full texts of which can be found on the MTBI website [51]. Lastly, we will provide some notes for mentors with an interest in student-led projects.

All together, this paper provides a nearly complete template for project in epidemiological modeling. We do not discuss how to formulate a model here, except for some notes to the mentors, so that part of the template is not complete. Also there are many other techniques that can be included in an epidemiological/contagion research paper that can add further insights into the model. A research project is never really done. However, this paper will provide a guide for turning the model that a student comes up with into a story that includes some results, conclusions, and some recommendations. So that while this research effort might just be the first chapter of the story that you tell about your research interests, it nevertheless has a satisfying ending.

1.2 What Is the Basic Reproductive Number and Why Is It Important?

Epidemiologists study the spread and control of diseases,[1] and so the first question they are usually interested in is whether or not it is possible to eradicate/eliminate

---

[1] Also other contagions. Epidemiological modeling it is quite common as it accounts for the interactions of individuals and the possibility that contacts may lead to transitions in the state of the individuals involved (contagion) and the possibility that cumulative transitions may or may not be effective in spreading the contagion. The contagion is usually a disease, but it does not have to be. A rumor, a meme, or an addiction, for example, would make a good contagion.

the disease or whether or not it is possible to prevent the disease from invading a new population. The basic reproductive number is related to responses to both questions.

In most, but not all epidemic models, an equilibrium state exists where the disease does not exist. This is a state where the population is made up entirely of uninfected people, that is, the infected populations are zero. It is typically called the *disease-free equilibrium.*

1.2.1   Definitions of R_0

Biologically, the *basic reproductive number* (R_0) is often interpreted as "the expected[2] number of new infections created by a single (typical) infectious individual in a population that is otherwise completely uninfected." The typical units structure for R_0 looks something like:

$$R_0 = \frac{\text{infections}}{\text{time} \cdot \text{infected}} (\text{duration of infection}). \tag{1}$$

We can interpret the left-hand factor $\frac{\text{infections}}{\text{time} \cdot \text{infected}}$ as the expected number of people that one person will infect per unit time. When we multiply by (duration of infection), we get the expected number of people that one person will infect over the entire time that they have the disease. Now imagine that a disease is invading for the first time. There is one infected person. If R_0 number is greater than 1, then—on average—infected people more than replace themselves (increase the infected population by more than 1) before they recover, (reducing the infected population by 1) and the disease grows. If this number is less than 1, then—on average—infected people do not replace themselves before they recover and the disease dies out.

However, as we will see in our examples, both this structure and its biological interpretation can be subject to considerable variation. In order to really explore R_0, we need to establish what it means independent of any particular model. This means we need a mathematical definition.

Practically, R_0 is a number calculated from the parameters of the model that can be used to answer fundamental questions about disease invasion, persistence, and control. When $R_0 < 1$, values close to the disease-free equilibrium tend toward the disease-free equilibrium making it impossible for a new disease to invade. When $R_0 > 1$, values close to the disease-free equilibrium tend away from the disease-free equilibrium making it impossible for a disease to completely die out as long as there is sufficient supply of susceptibles. Examining the conditions under which R_0 changes from greater than one to less than one can suggest possible interventions to control a disease. This gives us enough information to define *basic reproductive number* mathematically.

---

[2]Alternatively "typical" or "average."

Definition 1 We formally define R_0 a function of the parameters of the model such that $R_0 < 1$ implies that the disease-free equilibrium is locally asymptotically stable,[3] and $R_0 > 1$ implying that the disease-free equilibrium is unstable.[4]

1.2.2 Methods of Finding R_0

Historically, R_0 was found by directly investigating the stability of the disease-free equilibrium, and this is an approach that is still used today. But for models with a lot of compartments or a lot of non-linearities, a direct analysis of the disease-free equilibrium can be extremely algebraically intensive. In 1990, Diekmann et al. [33] developed a new approach to finding the basic reproductive number that uses a reduced form of the model. This reduces the computational complexity of the problem of stability and makes complex models more accessible. It is a method that has become very popular with our own students, but tutorials can be difficult to find, so we decided to share this method of finding R_0 here, along with some techniques for interpreting the results.

1.3 Additional Reading

For an introduction to the techniques used in MTBI in general, we recommend the MTBI course book [9]. Alternative books with similar content include [35] or [54]. For more information on formulating a research question, we recommend [58]. For more details on how to derive a model, we recommend [54]. For more detailed treatment of the next generation method, see [33, 34]. For a deeper dive into sensitivity analysis, see [3]. For a full collection of MTBI tech reports, see [51]. For further reading into the design of MTBI's mentorship model, we recommend reading [14, 18, 23, 24].

2 An Introductory Example: Student–Teacher Ratio

Katie Diaz, Cassie Fett, Grizelle Torres-Garcia, and Nicholas M. Crisosto were the authors of the project we will use in our first example [32]. While most students at MTBI take on ecology or biology projects, these students were particularly interested in problems related to the quality of education in the USA. They worried about students feeling discouraged by working in stressful school conditions with little teacher support. As budget cuts started to affect class sizes, they were

---

[3]Populations close to the equilibrium tend toward the equilibrium.

[4]Populations close to the equilibrium tend away from the equilibrium

concerned that problems with student attitudes and dropping out would increase. So they designed a model to investigate the problem of how student–teacher ratio in classes affects class sizes. In this model, Diaz, Feet, Torres-Garcia, and Crisosto modeled a discouraged attitude as something that both students and teachers could have. By social interaction, teachers and students could both pass on (metaphorically infect others with) this negative attitude.

This student project example [32] is 2003, when the NGM and sensitivity where first becoming popular with MTBI students and mentors. It received a poster award at the AMS/MAA Joint Meetings in 2004. The project uses a relatively simple model to study the development of positive or discouraged attitudes in the interactions between students and teachers, and tells the full story of using the NGM to find R_0, interpreting the expression of R_0, and using sensitivity analysis of R_0 to explore the effect of model parameters on the behavior of the system. Because the model is relatively simple, this project serves as an excellent introductory example. We will explore possible complications that you might encounter in your own project with later examples. Like all student project examples used in this paper, the full text of the students' technical report can be found on the MTBI website [51], and we strongly encourage students to review the original reports and related materials.

2.1   The Model

The researchers constructed a model with four classes: Teachers with positive attitudes P_1, discouraged teachers D_1, students with positive attitudes P_2, and discouraged students D_2. They assumed that each group influenced every other group through a random mixing contagion process, resulting in the following system of four differential equations[5]:

$$\dot{P}_1 = \mu_1 q_1 + \lambda_1 P_1 D_1 + \lambda_2 P_2 D_1 - \beta_1 D_1 P_1 - \beta_2 D_2 P_1 - \mu_1 P_1$$

$$\dot{D}_1 = \mu_1(1 - q_1) - \lambda_1 P_1 D_1 - \lambda_2 P_2 D_1 + \beta_1 D_1 P_1 + \beta_2 D_2 P_1 - \mu_1 D_1$$

$$\dot{P}_2 = \mu_2 q_2 + \lambda_1 P_1 D_2 + \lambda_2 P_2 D_2 - \beta_1 D_1 P_2 - \beta_2 D_2 P_2 - \mu_2 P_2 \tag{2}$$

$$\dot{D}_1 = \mu_2(1 - q_2) - \lambda_1 P_1 D_2 - \lambda_2 P_2 D_2 + \beta_1 D_1 P_2 + \beta_2 D_2 P_2 - \mu_2 D_2$$

where μ_i represents the rate of flow of group i into and out of the school, q_i is the proportion of new recruits into group i with positive attitudes, λ_j is the rate at which a discouraged person is encouraged by an interaction with a positive person from group j, and β_j is the rate at which a positive person is discouraged by an interaction

---

[5] $\dot{P}$ is notation used in the sciences for $\frac{dP}{dt}$. Historically, dot notation comes from Newton, while fraction notation is from Leibniz. Newton had a strong influence on the sciences, particularly physics, and many scientific fields working with calculus continue to use his notation.

with someone with a discouraged person from group j. Because the total population of each group is constant ($\dot{P_i} + \dot{D_i} = 0; i = 1, 2$), the researchers chose to define the classes as proportions rather than populations, and set $P_i + D_i = 1; i = 1, 2$.

2.2 Finding R_0 with the Next Generation Matrix

The next generation matrix method for finding R_0 is a relatively recent and useful reformulation for determining the stability of a disease-free or contagion-free equilibrium, that was introduced by Diekmann et al. [33]. First we will explain the general strategy that inspired this approach to finding R_0, and then we will describe precisely how to carry out that strategy mathematically, using the student–teacher model as an example.

2.2.1 The Intuition Behind the NGM Strategy

Intuitively, the next generation method is based on the common structure of R_0 described above (Eq. (1)). To aid in the discussion, we have reproduced that structure here.

$$R_0 = \frac{\text{infections}}{\text{time} \cdot \text{infected}}(\text{duration of infection}). \tag{3}$$

In most models, the duration of the infection (units of time) is represented as a recovery rate (units of $\frac{1}{time}$). For example, in the simple differential equation $(\dot{I}) = -\mu I$, the expected time that an individual would stay in I is $\frac{1}{\mu}$. If we also observe that the left factor of (3) represents the rate at which individuals infect others, then we can rewrite Eq. (3) as

$$R_0 = \frac{\text{infections}}{\text{time} \cdot \text{infected individual}} \frac{1}{\text{recovery rate}} \tag{4}$$

$$R_0 = (\text{individual infection rate})(\text{individual recovery rate})^{-1}.$$

The next generation method extends this idea to matrices. In situations when there are multiple types of infection to keep track of, the NGM separates the infected system into two matrices of rates. Traditionally, these matrices of rates are called F and V:

$$R = [\text{infection rates}][\text{recovery rates}]^{-1}$$

$$R = FV^{-1}. \tag{5}$$

However, the R in this equation is a matrix, not a number. So it cannot represent the basic reproductive number. Instead it represents a sort of measure of how each class changes multiplicatively to the next generation. If we imagine a vector of infected classes Y, then near the equilibrium

$$Y(t + 1) \approx RY(t)$$
$$Y(t) \approx R^t Y(0). \tag{6}$$

In order to explore how the *size* of Y changes, we use the eigenvector equation.[6] If Y is an eigenvector then:

$$RY = \lambda Y$$
$$Y(t) \approx \lambda^t Y(0). \tag{7}$$

We see that when $|\lambda| < 1$, then Y decreases. Since Y represents infected classes, this means the infection dies out. When $|\lambda| > 1$, then Y increases and the disease grows.

In order to be sure the disease dies out, we need all eigenvalues $|\lambda| < 1$. If any eigenvalue is greater than 1 in absolute value, then there is a path for the disease to grow. This means we only need to check the eigenvalue that is largest in absolute value (called the dominant eigenvalue). If this dominant eigenvalue is in $(-1, 1)$, all eigenvalues are within $(-1, 1)$ and the disease dies out. If the dominant eigenvalue is greater than 1 in absolute value, then the disease can grow. So the dominant eigenvalue has the properties we are looking for in a basic reproductive number. $R_0 = \lambda = \rho(FV^{-1})$, the dominant eigenvalue of FV^{-1}.

2.2.2 The Mathematical Approach

Formally, the next generation method begins by separating the classes of the model into two vectors: a vector X of "uninfected" classes and a vector Y of "infected" classes. In this model, the infection process is bi-directional. From one point of view, positive people are "infecting" discouraged people with positivity. From another perspective, discouraged people are "infecting" positive people with discouragement. Given the choice, the researchers chose to think of discouragement as the infection. In Exercises 1 and 2 you will have the opportunity to show what happens if you make the opposite choice.

The researchers chose discouragement as the infection, so $X = [P_1, P_2]^T$, and $Y = [D_1, D_2]^T$. Now that we have defined the infection, we proceed to find the infection-free equilibrium. In this case, a discouragement-free equilibrium does

---

[6]For a more thorough treatment of the biological interpretation of eigenvalues and eigenvectors, see the chapters on discrete time modeling from Bodine et al. [7].

not exist unless $q_1 = q_2 = 1$, so this becomes a condition of the model. In the scenario where all people arrive with positive attitudes, then the discouragement-free equilibrium is: $X^* = [1, 1]^T$, and $Y^* = [0, 0]^T$. We will also define a point for the discouragement-free equilibrium that includes all four classes $W^* = (P_1^*, D_1^*, P_2^*, D_2^*) = (1, 0, 1, 0)$.

The main advantage of the NGM method is that it allows the researcher to ignore any uninfected classes and focus only on the infected classes. This reduces the complexity of calculations in the model. In this model, X is discarded and the new equation for the model is

$$\dot{Y} = \begin{bmatrix} \dot{D}_1 \\ \dot{D}_2 \end{bmatrix} = \begin{bmatrix} \mu_1(1 - q_1) - \lambda_1 P_1 D_1 - \lambda_2 P_2 D_1 + \beta_1 D_1 P_1 + \beta_2 D_2 P_1 - \mu_1 D_1 \\ \mu_2(1 - q_2) - \lambda_1 P_1 D_2 - \lambda_2 P_2 D_2 + \beta_1 D_1 P_2 + \beta_2 D_2 P_2 - \mu_2 D_2 \end{bmatrix}.$$
(8)

The next step is to separate the Y equation for discouraged classes into two separate rate vectors: $\mathscr{F}$ represents the rates of all flows from X to Y, and $\mathscr{V}$ represents rates of all other flows. We also adjust signs so that

$$\dot{Y} = \mathscr{F} - \mathscr{V}.$$
(9)

In the student–teacher model, the separation is

$$\dot{Y} = \begin{bmatrix} \beta_1 D_1 P_1 + \beta_2 D_2 P_1 \\ \beta_1 D_1 P_2 + \beta_2 D_2 P_2 \end{bmatrix} - \begin{bmatrix} -\mu_1(1 - q_1) + \lambda_1 P_1 D_1 + \lambda_2 P_2 D_1 + \mu_1 D_1 \\ -\mu_2(1 - q_2) + \lambda_1 P_1 D_2 + \lambda_2 P_2 D_2 + \mu_2 D_2 \end{bmatrix}$$
(10)

so

$$\mathscr{F} = \begin{bmatrix} \beta_1 D_1 P_1 + \beta_2 D_2 P_1 \\ \beta_1 D_1 P_2 + \beta_2 D_2 P_2 \end{bmatrix}$$
(11)

$$\mathscr{V} = \begin{bmatrix} -\mu_1(1 - q_1) + \lambda_1 P_1 D_1 + \lambda_2 P_2 D_1 + \mu_1 D_1 \\ -\mu_2(1 - q_2) + \lambda_1 P_1 D_2 + \lambda_2 P_2 D_2 + \mu_2 D_2 \end{bmatrix}.$$
(12)

Note that although the terms $\mu_1(1 - q_1)$ and $\mu_2(1 - q_2)$ represent rates of new infections, they are not rates flows from X to Y, but instead rates for flows from outside the system into Y. Because of this, these terms are included in $\mathscr{V}$ and not $\mathscr{F}$. However, these terms are also problematic because they prevent a discouragement-free equilibrium from existing. Because $q_1 = q_2 = 1$ is a necessary condition for the discouragement-free equilibrium to exist, we set $q_1 = q_2 = 1$ and $\mathscr{V}$ becomes

$$\mathscr{V} = \begin{bmatrix} \lambda_1 P_1 D_1 + \lambda_2 P_2 D_1 + \mu_1 D_1 \\ \lambda_1 P_1 D_2 + \lambda_2 P_2 D_2 + \mu_2 D_2 \end{bmatrix}.$$
(13)

Next we define F and V as the Jacobian matrices of $\mathscr{F}$ and $\mathscr{V}$ evaluated at the discouragement-free equilibrium $W^* = (P_1^*, D_1^*, P_2^*, D_2^*) = (1, 0, 1, 0)$

$$F = \left(\frac{\partial \mathscr{F}}{\partial Y}\right)\Bigg|_{W^*}, \quad V = \left(\frac{\partial \mathscr{V}}{\partial Y}\right)\Bigg|_{W^*}. \tag{14}$$

So for the student–teacher discouragement model

$$F = \begin{bmatrix} \beta_1 P_1 & \beta_2 P_1 \\ \beta_1 P_2 & \beta_2 P_2 \end{bmatrix}\Bigg|_{(1,0,1,0)} = \begin{bmatrix} \beta_1 & \beta_2 \\ \beta_1 & \beta_2 \end{bmatrix} \tag{15}$$

$$V = \begin{bmatrix} \lambda_1 P_1 + \lambda_2 P_2 + \mu_1 & 0 \\ 0 & \lambda_1 P_1 + \lambda_2 P_2 + \mu_1 \end{bmatrix}\Bigg|_{(1,0,1,0)}$$

$$= \begin{bmatrix} \lambda_1 + \lambda_2 + \mu_1 & 0 \\ 0 & \lambda_1 + \lambda_2 + \mu_2 \end{bmatrix}. \tag{16}$$

This allows us to calculate our next generation matrix and the basic reproductive number.

Definition 2 The next generation matrix is FV^{-1}, where F is the Jacobian of the rates of flows from uninfected to infected classes evaluated at the disease-free equilibrium, and V is the Jacobian of the rates of all other flows to and from infected classes evaluated at the disease-free equilibrium (Eqs. (9) and (14)).

Definition 3 The basic reproductive number (R_0) is the spectral radius (largest eigenvalue) of the next generation matrix: $R_0 = \rho(FV^{-1})$ [33, 34].

So in the student–teacher model, the next generation matrix is

$$FV^{-1} = \begin{bmatrix} \beta_1 & \beta_2 \\ \beta_1 & \beta_2 \end{bmatrix} \begin{bmatrix} \dfrac{1}{\lambda_1 + \lambda_2 + \mu_1} & 0 \\ 0 & \dfrac{1}{\lambda_1 + \lambda_2 + \mu_2} \end{bmatrix}$$

$$= \begin{bmatrix} \dfrac{\beta_1}{\lambda_1 + \lambda_2 + \mu_1} & \dfrac{\beta_2}{\lambda_1 + \lambda_2 + \mu_2} \\ \dfrac{\beta_1}{\lambda_1 + \lambda_2 + \mu_1} & \dfrac{\beta_2}{\lambda_1 + \lambda_2 + \mu_2} \end{bmatrix} \tag{17}$$

and the basic reproductive number is

$$R_0 = \rho(FV^{-1}) = \frac{\beta_1}{\lambda_1 + \lambda_2 + \mu_1} + \frac{\beta_2}{\lambda_1 + \lambda_2 + \mu_2}. \tag{18}$$

Exercise 1 If you interpret a positive attitude as the infection instead of discouragement, how does the form of the resulting R_0 change?

2.3   Interpreting R_0

Traditionally, the basic reproductive number (R_0) is interpreted as "the average[7] number of new infections created by a single (typical) infectious individual in a population that is otherwise completely uninfected." If $R_0 < 1$, then an infection does not replace itself, and the disease dies out. If $R_0 > 1$ then the infection more than replaces itself and the disease spreads.

An important feature of this interpretation is that by describing R_0 as people per person or infections per infection, R_0 is a *dimensionless* quantity. We can use this feature as a way to check our result in Eq. (18).

Since P_1, P_2, D_1, and D_2 are proportions, they are also dimensionless. Examining Eq. (2), we see that in order for units to match on both sides of the equation, this means that λ_i, β_i, and μ_i must all have units of $\frac{1}{\text{time}}$. Then in Eq. (18), we see that R_0 has units of $\frac{1/\text{time}}{1/\text{time}}$ and so is dimensionless. This does not mean that we necessarily calculated R_0 correctly, but it is a good safety check.

The basic reproductive number is interpreted in a population that is mostly susceptible, meaning that it must be interpreted in situations very close to the disease-free equilibrium. When we compute the Jacobian around the disease-free equilibrium, we are approximating the original model with an easier to interpret linear model. In short the analysis is local and mathematically speaking we are talking about local asymptotic stability.

$$\dot{Y} = \mathcal{F} - \mathcal{V} \approx FY - VY. \tag{19}$$

In the case of the student–teacher discouragement model, this gives us

$$\begin{bmatrix} \dot{D_1} \\ \dot{D_2} \end{bmatrix} \approx \begin{bmatrix} \beta_1 D_1 + \beta_2 D_2 \\ \beta_1 D_1 + \beta_2 D_2 \end{bmatrix} - \begin{bmatrix} (\lambda_1 + \lambda_2 + \mu_1)D_1 \\ (\lambda_1 + \lambda_2 + \mu_2)D_2 \end{bmatrix}. \tag{20}$$

Linear models are relatively simple to interpret in the context of populations. In the contexts of proportions the interpretation is a little bit more complex. So for now let us imagine that the classes D_1 and D_2 represent populations, and we will correct for proportions afterwards.

In population models, $\lambda_1 + \lambda_2 + \mu_1$ can be interpreted as a general "death rate" for the D_1 (discouraged teacher) class. Since average lifespan is the inverse of the exit rate in a linear model, $\frac{1}{\lambda_1 + \lambda_2 + \mu_1}$ is the average length of a single episode of teacher discouragement. Similarly $\frac{1}{\lambda_1 + \lambda_2 + \mu_2}$ is the average length of a single episode of student discouragement. Similarly β_1 and β_2 can be interpreted as birth rates. But β_1 and β_2 are both used as inputs into both equations. So $1/\beta_1$ is the average length

---

[7] Alternatively "typical" or "expected".

of time it takes for a discouraged teacher to create two new discouraged people: a new discouraged student and a new discouraged teacher, and $1/\beta_2$ is the average length of time for a discouraged student to create a new discouraged person and a new discouraged teacher.[8]

Putting these interpretations together, we have that $\frac{1/(\lambda_1+\lambda_2+\mu_1)}{1/\beta_1}$ is the average time that it takes a teacher to get discouraged divided by the average time it takes a teacher to discourage a new student–teacher pair. This then calculates the average number of student–teacher pairs that a single teacher discourages. Similarly $\frac{\beta_2}{\lambda_1+\lambda_2+\mu_2}$ is the average number of student–teacher pairs that a single student discourages. Put together we have that R_0 is the average number of new discouraged student–teacher pairs created by a discouraged teacher and a discouraged student.

There are two problems with this interpretation, and both problems are tied to the idea that D_1 and D_2 are *proportions* not *populations*. Our interpretation of R_0 is based on the idea of a single discouraged teacher and a single discouraged student. But $D_1 = D_2 = 1$ is not the case of a single teacher and a single student. Instead these are the cases of the entire populations being discouraged. So there are two problems with this. The first problem is that $D_1 = D_2 = 1$ represent different numbers of teachers and students, because the total populations are different. The second problem is that in order for us to interpret R_0, we need the population to be mostly uninfected, which means that D_1 and D_2 need to be small. So we need an interpretation that is meaningful for D_1 and D_2 as proportions, with $D_1 = D_2 \ll 1$.

So an overall better interpretation of this basic reproductive number is that it measures the combined percentage of discouraged teachers and percentage of discouraged students produced by a small percentage of discouraged teachers and equivalent small percentage of discouraged students throughout the duration of time that those small percentages are discouraged. For example, if R_0 is 2, then we could interpret this as saying that when the percentage of discouraged people is small, the percentage of the next generation of discouraged people will be approximately twice as big.

This structure of calculating the number of new infections per infection as $\frac{\text{infection rate}}{\text{output rates}}$ is a very common structure that you will see in a lot of basic reproductive numbers. So it is important to remember both in interpreting your R_0 and in using that interpretation to check your calculations.

Exercise 2 Continuing from Exercise 1: If you interpret a positive attitude as the infection instead of discouragement, how does the interpretation of the resulting R_0 change?

---

[8]Remember that these approximations are only valid in a population that is almost entirely uninfected, so these interpretations cannot really be used for parameter estimation of λ_i or β_i.

2.4 Sensitivity Analysis

Since the basic reproductive number models whether an initial infection spreads or dies out, researchers typically want to know the effect of the parameters of the model on R_0. The problem is that the model typically has a significant number of parameters. This makes it difficult to do a full exploration of R_0. For example, we typically cannot plot R_0 as a function of every parameter in N-dimensional space. There are a few solutions to the problem. If there are only one or two parameters that can be controlled by policy, and the other parameters have known values, then a typical approach is to create a graph of R_0 as a function of those controllable parameters, and look at the contour where $R_0 = 1$. Sensitivity analysis is a different approach that can be used to efficiently examine many parameters at once by focusing on local changes in the parameter values [3]. When parameter values are unknown, sensitivity analysis can sometimes provide some useful general results. When parameter values can be estimated, sensitivity analysis can be used to suggest an intervention.

The key idea of sensitivity analysis it to look at how a small percentage change in one parameter affects the corresponding percentage change in a quantity of interest (in this case, R_0). So for example, a sensitivity index of 2 would mean that for a very small increase in the parameter, R_0 increases by twice that percentage. A sensitivity index of -0.5 would mean that for a very small increase in the parameter, R_0 decreases by half that percentage.

The reason why sensitivity analysis uses percentages rather than bare changes is so that different parameters with different units can be compared on an even playing field. If a parameter is measured in miles, a change of 0.0001 is much more dramatic a difference than if the parameter is measured in feet. But a change of 0.01% would be the same regardless of unit.

To calculate the percentage change in the parameter, we calculate the percentage change in the parameter ξ as $\frac{\Delta\xi}{\xi}$. Similarly, the percentage change in R_0 would be $\frac{\Delta R_0}{R_0}$. The sensitivity index of R_0 with respect to the parameter ξ is then just the quotient of these two percent change, as long as the percentage change is sufficiently small. The way we make sure that the change in percentage is small enough is by taking a limit, so that $\Delta\xi$ becomes $\partial\xi$ and ΔR_0 becomes ∂R_0.

Definition 4 The sensitivity index S_ξ of R_0 with respect to parameter ξ is given by Eq. (21):

$$S_\xi = \frac{\xi}{R_0}\frac{\partial R_0}{\partial\xi}. \tag{21}$$

In the student–teacher discouragement model, the researchers calculated the sensitivity of R_0 with respect to each parameter.

$$S_{\beta_1} = \frac{\beta_1}{R_0(\lambda_1 + \lambda_2 + \mu_1)} \qquad\qquad S_{\mu_1} = -\mu_1 \frac{S_{\beta_1}}{(\lambda_1 + \lambda_2 + \mu_1)}$$

$$S_{\beta_2} = \frac{\beta_2}{R_0(\lambda_1 + \lambda_2 + \mu_2)} \qquad\qquad S_{\mu_2} = -\mu_2 \frac{S_{\beta_2}}{(\lambda_1 + \lambda_2 + \mu_2)}$$

$$S_{\lambda_1} = -\lambda_1 \left(\frac{S_{\beta_1}}{(\lambda_1 + \lambda_2 + \mu_1)} + \frac{S_{\beta_2}}{(\lambda_1 + \lambda_2 + \mu_2)} \right)$$

$$S_{\lambda_2} = -\lambda_2 \left(\frac{S_{\beta_1}}{(\lambda_1 + \lambda_2 + \mu_1)} + \frac{S_{\beta_2}}{(\lambda_1 + \lambda_2 + \mu_2)} \right). \tag{22}$$

Because all parameter values are positive and R_0 is positive, the sensitivity indices S_{β_1} and S_{β_1} are positive, while the indices S_{λ_1}, S_{λ_2}, S_{μ_1}, and S_{μ_2} are negative. This gives us an expected result. Increasing the rates at which discouraged teachers and teachers successfully convert other students and teachers (β_i) increases to growth in discouragement R_0. Increasing the rate at which positive teachers and students convert other students and teachers (λ_i) reduces R_0. What might be less expected is that increasing the rate of student and teacher turnover (μ_i) also decreases R_0. This is because of the assumption that $q_1 = q_2 = 1$ that was needed to find an R_0. Because all new people entering the school have positive attitudes, turnover removes discouraged people and replaces them with positive people.

Next we can look at which sensitivity indices have the largest impact on the system. Because $\mu_i < \lambda_1 + \lambda_2 + \mu_i$ we have that $|S_{\beta_i}| > |S_{\mu_i}|$. So changing the rate that discouraged teachers or students recruit always has a stronger impact than the same percentage change in the corresponding turnover rate. Because $\lambda_1 + \lambda_2 < \lambda_1 + \lambda_2 + \mu_i$ we have that $|S_{\beta_1} + S_{\beta_2}| > |S_{\lambda_1} + S_{\lambda_2}|$, which suggests that targeting the spread of negative attitudes overall is more effective than targeting positive attitudes overall.

Exercise 3 Continuing from Exercises 1 and 2: Use sensitivity analysis to investigate the effect of the model parameters on the spread of positive attitudes and make a recommendation.

The researchers also used parameter estimates to calculate the sensitivity indices numerically, although the exact values they used are not included in the original paper. Using these parameter estimates they found that in order from largest to smallest absolute value $S_{\beta_2} = 0.5959$, $S_{\lambda_2} = -0.5224$, $S_{\beta_1} = 0.4041$, $S_{\lambda_1} = -0.435$, $S_{\mu_1} = -0.04004$, $S_{\mu_2} = -0.002159$. This means that in a school that would be modeled by these parameter estimates, the effects that students have on the attitudes of teachers and students (β_2 and λ_2) have the strongest impact on the spread of discouragement, but the impact of teachers (β_1 and λ_1) is not that much less. Turnover (μ_1 and μ_2) has little impact in this scenario. Which interventions the researchers recommend, however, would depend on other factors, such as how difficult or costly it could be to alter a particular parameter.

2.5 Other Full Examples

The full process of analysis in this example was: (1) Use the next generation matrix to find the basic reproductive number, (2) interpret the basic reproductive number as a way of verifying the calculation, and (3) use sensitivity analysis to investigate the effect of parameters on the model and make recommendations for intervention. The student–teacher model [32] is one full example of this process with a small model. Because details necessarily had to be omitted, we highly recommend that you read the original 2003 paper at https://mtbi.asu.edu/tech-report.

Other MTBI student papers that give detailed examples of the full process are:

- A 2012 study on HIV and malaria co-infection [4]
- A 2012 study on prison reform [2]
- A 2007 study on HIV and Tuberculosis [27]
- A 2006 study on the economics of sex work [31]

As you are reading some of the cited papers and technical reports, you will notice that they do not always use notation in the same way. While the calculations for F and V are always the same, notation for intermediate steps can vary. Some researchers define X as the infected class, and Y as the uninfected class, so that $X = \mathscr{F} - \mathscr{V}$. Some researchers do not define their uninfected X and infected Y classes explicitly, and you have to identify how they separated their classes from the $\mathscr{F}$ and $\mathscr{V}$ expressions. Other researchers will define $\mathscr{F}$ and $\mathscr{V}$ for all the classes in the model (in the student–teacher model $\mathscr{F}$ and $\mathscr{V}$ would be 4×4 instead of 2×2) but still only calculate the Jacobian for the infected classes, so the resulting F and V are still the same (2×2) matrices as above. Regardless of intermediate steps and choice of notation, F and V are used consistently as in Eq. (14) and Definition 2.

Exercise 4 No paper ever includes all the details of the mathematics. Choose one of the papers above and reconstruct the missing steps in the NGM method. Define X, Y, $\mathscr{F}$, and $\mathscr{V}$. Then take the Jacobian to find F and V, calculate the next generation matrix FV^{-1}, and find the spectral radius R_0. Verify your results with the results from the paper.

Exercise 5 No paper ever includes all the details of the mathematics. Choose one of the papers above and find their reproductive number, then calculate the sensitivity indices for each parameter included in the reproductive number. If the paper includes detailed parameter estimates, then use these estimates to also calculate the sensitivity indices numerically. Use the parameter definitions given in the paper to interpret the sensitivity indices and suggest a possible intervention. Verify your results with the results from the paper.

3   Challenges in the of the Next Generation Matrix

Because we introduced these techniques with a simple model, we have really only spent time with a "best case" scenario. Since your own project will likely involve a more complicated model, this also means that it is likely that a situation could arise where you are not quite sure what the best decision is. In the next few sections, we will discuss some complications that can arise when using these techniques to study a model of your own creation. We will discuss each of these scenarios briefly using examples from actual student projects, with a focus on the decisions that needed to be made. As always, we encourage you to read the original papers [51] as a way of filling in some of the missing details. And you should always consult your research mentor about complications in your own model.

In this first section, we will discuss some complications that can arise when finding and using a next generation matrix. Most of these complications revolve around decisions related to how to define the uninfected (X) and infected (Y) classes.

3.1   Defining the Infected Class: Infected but Not Infectious

This complication arose in research project from 2018 on the spread of herpes [1]. Herpes (HSV-2) is a disease that "hides" in the nervous system of the host, and only periodically emerges to cause symptoms on the skin. Current treatments can force the disease into the asymptomatic stage, but cannot eliminate the disease from the body entirely, so it always returns. When the disease emerges, initial symptoms resemble a lot of other diseases, while later symptoms are distinctive. Current protocols only administer treatment when the disease has progressed to the later stage, and the patient has already been infections for some time. The question the Luis Almonte-Vega, Monica Colón-Vargas, Ligia Luna-Jarrń, Joel Martinez, and Jordy Rodriguez-Rincón, wanted to study was the cost effectiveness of treating during the early stages when the disease is less infectious, but money would be wasted on treating false positives. Could spending extra on treating people who only might have herpes save money in the long run by reducing the prevalence of the disease?

The model the students constructed had seven population compartments: S for susceptible, I_1 for mildly symptomatic individuals, I_2 for strongly symptomatic individuals, L for asymptomatic individuals, T_1 for individuals with mild symptoms undergoing treatment, T_2 for strongly symptomatic individuals undergoing treatment, and X for false positive undergoing treatment.

During the project, the question arose of how to separate the infected and uninfected classes for the next generation matrix. To illustrate, let us consider a simple SIR model with partial immunity. Susceptible people become infected, infected people recover, and recovered people have a chance of being infected again.

So the flows would be $S \to I \leftrightarrow R$. In this scenario, the only population that has the disease is I, so the separation of classes for the NGM would be $X = [S, R]^T$ uninfected and $Y = [I]$ infected.

But there could be another interpretation of this diagram where the disease is never fully eradicated. In the case of herpes, the asymptomatic stage resembles recovery, but it is really the same infection. So the structure of the model $S \to I \leftrightarrow L$. In this scenario, the both I and L have the disease, so the separation of classes for the NGM would be $X = [S]$ uninfected and $Y = [I, L]^T$ infected. The key distinction here is that the flow from $L \to I$ in the SIL model is considered a new stage in the same infection, while the flow from $R \to I$ in the SIR model is a completely new infection.

So in the HSV-2 model, the students separated the classes into $[S, X]^T$ uninfected, and $[I_1, I_2, L, T_1, T_2]$ infected. L, T_1, and T_2 were included because even though the people in these compartments had no symptoms and were not contagious, they still carried the disease in their bodies, so they were still infected.

We will discuss how to interpret the reproductive number from this model in Sect. 4.4.

3.2 Defining the Infected Class: Multiple Levels of Infection

This complication arose in a research project from 2001 on collaborative learning [29, 30], and another research project from 2005 on the success of political third parties [13]. The 2005 paper on third parties is more detailed, so it is the one we will discuss here, and also the one we recommend that you read first.

Karl Calderon, Clara Orbe, Azra Panjwani, and Daniel Romero were inspired by the 2000 election, when election analysts attributed the defeat of Al Gore to splitting the base. Ralph Nader, a Green Party candidate who won 2% of the vote, which many believed could otherwise have gone to Gore and turned the election. Using the growth of the Green Party as a source of data, the researchers wanted to study the growth of grassroots political movements, and explore voter recruitment strategies.

The third party model had a relatively simple structure, where the researchers considered three levels of engagement. A population S of people who were susceptible to the messages of the political party, a population V of people who voted for the third party, and a population M of active members of the party. Both voters and members could convince susceptible to become voters, although members were more effective than voters. Also members could recruit voters to become members. Lastly, voters could change their mind. So the structure of the model was $S \leftrightarrow V \to M$.

This model was interesting because it had two processes that could be considered an infection: the recruitment of voters from $S \to V$ by voters and members, or the recruitment of members from $V \to M$ by members. In the first case, where voting is the infection, the separation of classes would be $X = [S]$ uninfected, and

$Y = [V, M]^T$ infected. In the second case, where membership is the infection, the separation of classes would be $X = [S, V]^T$ uninfected and $Y = [M]$ infected. Unlike the case of HSV-2 above, where the researchers needed to make a decision to identify the correct separation, here there is not correct separation. The researchers needed both perspectives to understand the behavior of the model. So this model did the NGM process twice (once with each separation) and had two reproductive numbers.

We will discuss how to interpret these reproductive numbers in Sect. 4.3.

Exercise 6 The collaborative learning paper [29, 30] has a very similar model to the third parties paper [13], but their NGM section is extremely light on details. Use the model from the collaborative learning paper and the NGM method to find both reproductive numbers for this model. Verify your results with the results from the paper.

3.3 Unclear Which Eigenvalue Is Largest: Co-infection

Another complication that can arise is with the definition of R_0 as the spectral radius of the NGM. The spectral radius is the largest eigenvalue, but sometimes it is not clear which eigenvalue of the NGM is largest. This is a common problem when studying the interaction between two diseases that can co-infect. Two projects that encountered this problem were a 2012 paper on HIV and malaria co-infection [4, 5], and a 2007 project on HIV and tuberculosis co-infection [27]. Both papers are very similar. We will use the malaria model for this example.

Kamal Barley, Sharquetta Tatum, and David Murillo were inspired by the high prevalence of both HIV and malaria in the Republic of Malawi. Both diseases increase the impact on the other. HIV weakens the immune system, which makes it easier for malaria to infect people, and malaria increases the viral load of HIV. In particular the researchers were interested in studying how the interactions between these two diseases might increase mortality.

In the malaria-HIV model, there were six classes: S susceptible humans, I_M humans infected with malaria, I_H humans infected with HIV, I_{HM} humans infected with both diseases, V susceptible mosquitos, and I_V mosquitos carrying malaria. So in this model, there were four infected classes $Y = [I_M, I_H, I_{HM}, I_V]^T$ and the researchers calculated the NGM to be

$$
\begin{bmatrix}
0 & 0 & 0 & \dfrac{\beta_{VM}\Lambda}{\alpha N \mu_V} \\[2mm]
0 & \dfrac{\beta_H}{\mu_H + \alpha} & \dfrac{\beta_H k\gamma}{(\mu_H + \alpha)(\mu_{HM} + k\gamma + \alpha)} + \dfrac{\rho_2 \beta_H}{\mu_{HM} + k\gamma + \alpha} & 0 \\[2mm]
0 & 0 & 0 & 0 \\[2mm]
\dfrac{\beta_{MV} N\alpha}{\Lambda(\mu_M + \gamma + \alpha)} & 0 & \dfrac{\rho_4 \beta_{MV} N\alpha}{\Lambda(\mu_{HM} + k\gamma + \alpha)} & 0
\end{bmatrix}
\tag{23}
$$

In this matrix, subscript H is for HIV, M for malaria, and HM for both. Λ is recruitment rate for humans, N is the total population of humans, the βs are infection rates, μs and α are mortality rates, γ is the recovery rate from malaria, and k is the reduction in this recovery rate due to HIV infection.

This model has two eigenvalues that could both be largest, depending on parameter values:

$$R_{0H} = \frac{\beta_H}{\mu_H + \alpha}$$

$$R_{0M} = \sqrt{\frac{\beta_{MV}\beta_{VM}}{\mu_V(\mu_M + \gamma + \alpha)}}.$$

(24)

So the basic reproductive number is whichever of these two eigenvalues happens to be the largest.

$$R_0 = \max\{R_{0H}, R_{0M}\}.$$

(25)

This is a very common structure for R_0 in the case of co-infection. We will discuss how to interpret this result in Sects. 4.2 and 4.5.

Exercise 7 The malaria paper [4] also has a reduced model with four compartments. Read the paper to identify the reduced model, and find R_0 for the reduced model by using the NGM method. Verify your results with the results from the paper.

4 Challenges in Interpreting R_0

We have already discussed one scenario that complicated the interpretation of R_0: the situation where we are adding proportions instead of populations. However, the overall the structure of the basic reproductive number in the student–teacher model followed a relatively simple $\frac{\text{infection rate}}{\text{output rates}}$ structure that is common to many basic reproductive numbers. In this section, we will discuss other forms that R_0 can take that commonly appear in student projects, and how to interpret those structures. In order to keep these sections short, we will primarily be using flow diagrams to discuss the models. As always we strongly encourage you read the original papers for more details on the models and analysis. Furthermore, every paper cited in this section also used a next generation approach to finding the basic reproductive number, so these are all good sources if you want additional examples of that approach.

4.1   Multiple Infection Pathways: The Case of SARS

This example project is from 2004 on the spread of SARS [38, 39]. Julijana Gjorgjieva, Kelly Smith, and Jessica Snyder were inspired by the prominent news coverage of the 2002 SARS epidemic. In particular, they were interested in the control plan for SARS. Because the disease was new and there was no treatment, control focused on isolating people who showed symptoms, and tracing their contacts to identify other people who might have the disease. Since no treatment had been developed, the researchers saw an opportunity to compare the current tracing-and-isolation control plan to a possible future vaccination strategy.

The control plan for SARS involved tracing an infected individual's contacts in order to identify who else was at risk of contracting SARS, but this tracing process was not always successful. Treatment for SARS generally involved isolation to prevent further spread. So the model used classes for S susceptible, E_i traced latent, E_n untraced latent, I infectious, W isolated, R recovered, and D dead. This resulted in a model where an individual could take many different paths. For example, an individual could never be identified by doctors and pass through $S \rightarrow E_n \rightarrow I \rightarrow D$ (Fig. 1), or an individual could be caught in the infectious stage and pass through isolation and treatment and survive, passing through $S \rightarrow E_n \rightarrow I \rightarrow W \rightarrow R$ (Fig. 1).

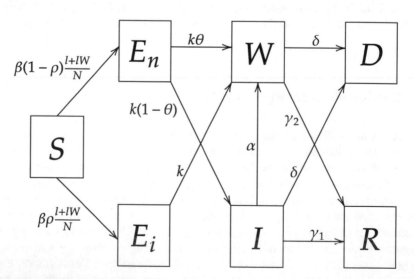

Fig. 1 Flow diagram for the SARS model with contact tracing: S susceptible, E_n untraced latent, E_i traced latent, I untraced infectious, W diagnosed and isolated infectious, D dead, and R recovered [39]. β is the infection rate, k is the rate of developing symptoms, θ is the probability that the patient is diagnosed while latent, α is the rate that infected individuals become isolated, δ is the death rate due to infection, and γs are recovery rates

Using the next generation approach, the researchers found the following basic reproductive number[9]:

$$R_0 = \beta \left(\frac{(1-\rho)l\theta}{\delta + \gamma_2} + \frac{(1-\rho)l(1-\theta)\alpha}{(\delta + \gamma_2)(\alpha + \delta + \gamma_1)} + \frac{(1-\rho)(1-\theta)}{\alpha + \delta + \gamma_1} + \frac{\rho l}{\delta + \gamma_2} \right).$$

(26)

This R_0 follows the standard $\frac{\text{infection rate}}{\text{output rates}}$ pattern, but because there are multiple pathways to get to I or W, these rates need to be averaged together by the proportion of individuals who follow each path. Examining the flow diagram (Fig. 1), the only path to I is $S \to E_n \to I$. Individuals are infected at a rate β. Of these individuals, a proportion $(1 - \rho)$ enter E_n, and of these individuals in E_n, a proportion $(1 - \theta)$ enters I. Looking at the outflows of I: average time in I is $\frac{1}{\alpha + \delta + \gamma_1}$, so the $\frac{\text{infection rate}}{\text{output rates}}$ for I would be $\frac{\beta}{\alpha + \delta + \gamma_1}$, weighted by the proportion of infected individuals who enter I: $(1 - \rho)(1 - \theta)$, which is what we see in Eq. (26).

W is less infectious than I, so the rate of those infected by W is $\beta l < \beta$. All the terms with l in the numerator then involve pathways to W. Let us examine the second term, $\frac{\beta l(1-\rho)(1-\theta)\alpha}{(\delta + \gamma_2)(\alpha + \delta + \gamma_1)}$. In this term, $\frac{\beta l}{(\delta + \gamma_2)}$ is $\frac{\text{infection rate}}{\text{output rates}}$ for W. $(1 - \rho)$ is the proportion moving from $S \to E_n$. $(1 - \theta)$ is the proportion moving from $E_n \to I$. Lastly, $\frac{\alpha}{(\alpha + \delta + \gamma_1)}$ is the rate from $I \to W$ divided by the total rate out of I, so it is the proportion of individuals moving from I to W. So this term represents the infection contribution from the $S \to E_n \to I \to W$ path.

Similarly the first term $\frac{\beta l(1-\rho)\theta}{\delta + \gamma_2}$ represents the infection contribution from the $S \to E_n \to W$ path, and the final term $\frac{\beta l \rho}{\delta + \gamma_2}$ represents the infection contribution from the $S \to E_i \to W$ path.

Exercise 8 Imagine that each of the proportions in this model is instead a probability. Combine each of these four paths together to create a probability tree diagram. Interpret R_0 as the expected value of $\frac{\text{infection rate}}{\text{output rates}}$ for a single individual passing through the system at random.

4.2 The Maximum of Two Reproductive Numbers: Co-infection

This example project is from 2007 on the spread of HIV and tuberculosis [27]. Like the malaria project above, Diego Chowell-Puente, Brenda Jiménez-González, and Adrian Smith were interested in studying how two deadly diseases interacted in South Africa, where both are common. TB is often carried latently, and HIV

---

[9]In the original model, the researchers also included a vaccination parameter in R_0, but we have omitted it from this example.

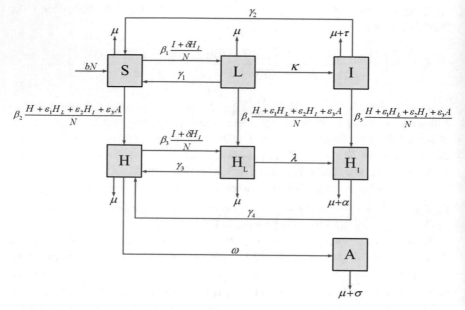

Fig. 2 Flow diagram for the HIV-TB co-infection model: S susceptible, L infected with latent TB, I infectious stage TB, H infected with HIV, H_L infected with HIV and latent TB, H_I infected with HIV and infectious stage TB, and A developed AIDS [27]

increases the chance of TB developing symptoms. So the researchers wanted to construct a model that would suggest strategies to control both epidemics.

Tuberculosis has a latent stage, during which a person is infected but uninfectious. HIV has an asymptomatic infectious stage, and the symptomatic stage, when the infected person has developed AIDS. In order to track the behavior of a population, the researchers needed to develop a model that captures every possible combination of stages, as seen in Fig. 2. Note that the infection terms need to be summed over a large number of compartments, because a large number of different compartments all carry the same disease. For example, H, H_L, H_I, and A all carry HIV and can infect a susceptible with HIV, so the HIV infection rate must include all of these compartments.

Like most co-infection models, the basic reproductive number for this model included a maximum.

$$R_0 = \max\{R_0^{TB}, R_0^{HIV}\}$$

$$R_0^{TB} = \frac{\beta_1}{\gamma_2 + \mu + \tau} \cdot \frac{k}{\gamma_1 + k + \mu} \qquad (27)$$

$$R_0^{HIV} = \frac{\beta_2}{\omega + \mu} + \frac{\beta_2 \epsilon_3}{\mu + \sigma} \cdot \frac{\omega}{\omega + \mu}.$$

In order to interpret this reproductive number, let us examine each component using the tools from Sect. 4.1. Beginning with R_0^{TB}, we see that the first factor $\frac{\beta_1}{\gamma_2+\mu+\tau}$ is the infection rate for TB divided by the sum of the rates out of I. So this factor takes the form $\frac{\text{infection rate}}{\text{output rates}}$ for TB by itself. The second factor $\frac{k}{\gamma_1+k+\mu}$ takes the form of rate from L to I divided by total rate out of L. So this is the proportion of people in L who enter I. Taken together, this is the basic reproductive number for the $S \to L \to I$ pathway by itself. This would be the path that TB takes if there were no HIV.

Similarly, R_0^{HIV} is the reproductive number for HIV by itself. $\frac{\beta_2}{\omega+\mu}$ is $\frac{\text{infection rate}}{\text{output rates}}$ for the H class, $\frac{\beta_2\epsilon_3}{\mu+\sigma}$ is the $\frac{\text{infection rate}}{\text{output rates}}$ for the A class, and $\frac{\omega}{\omega+\mu}$ is the proportion of people moving from $H \to A$.

The basic reproductive number for the whole system is the maximum of these two reproductive numbers because R_0 only tracks infected or uninfected. Once a person in infected with either of the two diseases, they are counted as infected. Being infected with both involves first being infected with one, and then the other. Because R_0 is only concerned with the disease-free state, that second co-infection is not relevant. Only the first infection affects the disease-free state, so in examining the disease-free state, we are only concerned with whichever infection is strongest.

For another example of this type of co-infection R_0, see the HIV-malaria model [5] discussed in Sects. 3.3 and 4.5.

4.3 Two Thresholds and Backwards Bifurcations

In this section we will return to the 2005 project on third parties from Sect. 3.2. In this project [13], there were three classes of population: S susceptible, V voters, and M members. Because either voting or membership could be considered the infection, there were also two reproductive numbers:

$$R_1 = \frac{\beta - \phi}{\mu + \epsilon}$$

$$R_2 = \frac{\gamma}{\mu}\left(1 - \frac{1}{R_1}\right). \tag{28}$$

In this model (Fig. 3), when a susceptible encounters a voter, there are two possible outcomes. Either the voter can influence the susceptible to become a voter ($S \to V$), or the susceptible can convince the voter not to vote ($S \leftarrow V$). The former happens at rate β and the latter occurs at rate ϕ, so the net rate of people being influenced to vote is $\beta - \phi$, and this becomes the infection rate, while $\mu + \epsilon$ is time in V. So R_1 follows our standard pattern of $\frac{\text{infection rate}}{\text{output rates}}$.

For R_2, γ is the infection rate, and μ is the exit rate. Following our pattern from Sect. 4.1, the remaining factor $\left(1 - \frac{1}{R_1}\right)$ should be some sort of proportion

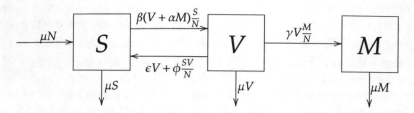

Fig. 3 Flow diagram for the third party model: S susceptible, V voters, and M members [13]

of the population. To figure this out, let us go back to an earlier interpretation of R_0 from Sect. 2.3. We think of R_0 as $\frac{\text{infection rate}}{\text{output rates}}$, but because our rates typically have units of 1/time, we can also think of R_0 as $\frac{\text{total time infectious}}{\text{time it takes to infect someone}}$. So then $\frac{1}{R_1}$ would be $\frac{\text{time it takes to convince someone to vote}}{\text{total time as a voter}}$. In other words, it is the proportion of voters who are currently convincing a susceptible to vote. Then $\left(1 - \frac{1}{R_1}\right)$ is the proportion of voters who are not currently occupied convincing someone else to vote, which means they have free time to interact with a member and be susceptible to membership.

An interesting feature of this two threshold model is that R_1 only measures the impact of voters on susceptibles, but members can also convince susceptibles to vote. This means it is possible for members to exist as a population when $R_1 < 1$, as long as enough members exist in the population. This is known as a backward bifurcation. Also note that when $R_1 < 1$, $R_2 < 0$. This negative value for R_2 is a little difficult to interpret, but it is occurring because members are recruiting susceptibles to become voters and then almost immediately recruiting those voters to become members, so voters have no time to recruit susceptibles as voters. This maintains the member population without maintaining the voter population.

4.4 Geometric Series: Returning to the Infectious Class

In this section, we continue the discussion of the 2018 herpes model [1]. A key feature of herpes discussed in Sect. 3.1 is that a single infection goes through multiple cycles of infectiousness. This leads to a potentially infinite number of paths to I: $S \rightarrow I$, $S \rightarrow I \rightarrow L \rightarrow I$, $S \rightarrow I \rightarrow L \rightarrow I \rightarrow L \rightarrow I$, etc. Each of these paths is still a single infection, so each path needs to be considered in the basic reproductive number. In order to study this phenomenon, the researchers started with a simplified model before moving on to their full model. The simplified model only has compartments for S susceptible, I infected, and L latent (Fig. 4). The basic reproductive number for this simplified model was:

$$R_0 = \frac{\beta}{\mu + \gamma} \cdot \frac{1}{1 - \frac{\gamma}{\mu+\gamma} \cdot \frac{r}{\mu+r}}. \tag{29}$$

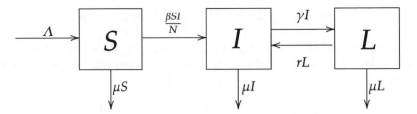

Fig. 4 Flow diagram for the simplified herpes model: S susceptible, I infectious, and L latent [1]

In this basic reproductive number, β is the infection rate, divided by $\mu + \gamma$ is the time in I, so this forms our standard basic reproductive number. Our previous examples have suggested that the remaining factor should be a proportion. $\frac{\gamma}{\mu+\gamma}$ is the proportion of people who leave I to enter L, and $\frac{r}{\mu+r}$ is the proportion of people who leave L to return to I. So the product $p = \frac{\gamma}{\mu+\gamma} \cdot \frac{r}{\mu+r}$ is the proportion of people in I who make a single round trip back to I ($I \to L \to I$).

Because people can make this round trip any number of times, we need to account for all possible pathways. Infection for people who make zero round trips would be $\frac{\beta}{\mu+\gamma}$, from one round trip would be $\frac{\beta}{\mu+\gamma}p$, two round trips: $\frac{\beta}{\mu+\gamma}p^2$, three round trips: $\frac{\beta}{\mu+\gamma}p^3$, etc.

$$R_0 = \frac{\beta}{\mu+\gamma} \sum_{k=0}^{\infty} p^k. \tag{30}$$

Because the limit of a geometric series with common ratio p is $\frac{1}{1-p}$, we have the form of R_0 given in Eq. (29).

Exercise 9 Other student projects with examples of this kind of looping R_0 include a 2018 study of prison recidivism [49] and a 2010 study of immigration [25]. Choose one of these two projects, describe a model, use unit analysis to identify the units of each parameter, verify that the reproductive number is unitless, and interpret the reproductive number using the techniques you have learned so far.

4.5 Geometric Mean: Indirect Transmission

The project is from 2005 on the spread of HIV between two sexes: male truck drivers and female sex workers [47]. Titus Kassem wanted to focus on the interaction between two populations that were both at high risk of HIV in Nigeria. In particular, truck drivers travel frequently and for great distances and interact with sex workers in many different places. High numbers of sexual partners and travel lead to a disproportionally large influence on the geographic spread of HIV. So Kassem wanted to study these two at-risk groups, and how their different ideas about condom usage affected the spread of HIV.

The model used six classes, S_m susceptible males, I_m HIV infected males, A_m males with AIDS, and similarly for females. The paper did not provide a flow diagram, so instead, we duplicate the equations here.

$$\dot{S}_m = \mu_m N_m + \psi_m A_m - \mu S_m - \beta_1 S_m \frac{I_f}{K_f}$$

$$\dot{I}_m = \beta_1 S_m \frac{I_f}{K_f} - (\gamma_m + \mu_m) I_m$$

$$\dot{A}_m = \gamma_m I_m - (\psi_m + \mu_m) A_m$$

$$\dot{S}_f = \mu_f N_f + \psi_f A_f - \mu S_f - \beta_1 S_f \frac{I_m}{K_m}$$ (31)

$$\dot{I}_f = \beta_2 S_f \frac{I_m}{K_m} - (\gamma_f + \mu_f) I_f$$

$$\dot{A}_f = \gamma_f I_f - (\psi_f + \mu_f) A_f.$$

Exercise 10 Construct a flow diagram for this HIV model. Label each arrow with the corresponding term from Eq. (31).

The researchers found that the basic reproductive number was:

$$R_0 = \sqrt{\frac{\beta_1 \beta_2}{(\gamma_m + \mu_m)(\gamma_f + \mu_f)}}.$$ (32)

Exercise 11 The paper has omitted the details of these steps. Use a NGM to verify the basic reproductive number found in Eq. (32).

The basic reproductive number has an interesting form of the geometric mean between two values: $\frac{\beta_1}{(\gamma_m + \mu_m)}$ is the $\frac{\text{infection rate}}{\text{output rates}}$ for males, and $\frac{\beta_2}{(\gamma_f + \mu_f)}$ is the $\frac{\text{infection rate}}{\text{output rates}}$ for females.

The reason for the geometric mean is because the disease has to make a round trip. In order for a male to infect another male, the disease must first infect a female, and vice-versa. So the number of new infected males produced by a single infected male is the number of infected females produced by that infected male, times the number of infected males produced by each new infected female. For example, if a male infects 8 females, and each female infects 2 males, then the number of newly infected males produced by that original male would be $8 \cdot 2 = 16$. But this is a two-step process. In order to scale the process back to a single infection step, we need to average the two values. Because the process is multiplicative, we average with a geometric mean instead of an arithmetic mean. Note that $\sqrt{8 \cdot 2} = 4$ and $4 \cdot 4 = 16$, while an arithmetic mean does not quite work: $\frac{8+2}{2} = 5$ and $5 \cdot 5 = 25$.

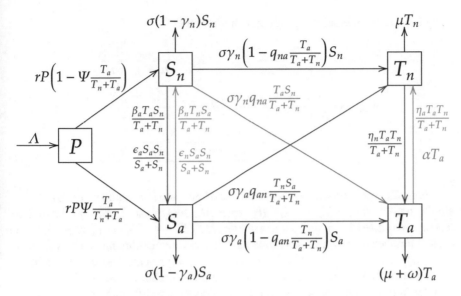

Fig. 5 Flow diagram for the math anxiety model: P for primary student, S_n for non-anxious secondary student, S_a for anxious secondary student, T_n for non-anxious teacher, and T_a for anxious teacher [42]

Exercise 12 Return to the HIV-malaria model [5] discussed in Sects. 3.3 and 4.2. Draw the flow diagram, explain the model, conduct a unit analysis, and interpret the basic reproductive number using the techniques you have learned so far.

4.6 Not Quite a Geometric Mean: Generational Models

This example comes from a 2017 study on the spread of math anxiety in a school [42]. Arie Gurin, Guillaume Jeanneret, Meaghan Pearson, and Melissa Pulley were inspired by their own past struggles with mathematics, and concern for the impact that math anxiety has on the recruitment and retention of women and minorities into STEM. Because many women with math anxiety become elementary school teachers who teach mathematics, the researchers wanted to study how generations of teachers might influence students attitudes toward mathematics as those students become teachers themselves. The researchers were looking to identify a best point of intervention to break the cycle of teachers with math anxiety creating new generations of teachers with math anxiety.

The model had five classes: P for primary student, S_n for non-anxious secondary student, S_a for anxious secondary student, T_n for non-anxious teacher, and T_a for anxious teacher (Fig. 5).

In this model, the basic reproductive number was

$$R_S = \frac{\epsilon_a}{\sigma + \epsilon_n + \beta_n}, \qquad R_T = \frac{q_{na}\mu + \eta_a}{\mu + \eta_n + \omega + \alpha}$$

$$R_{CTS} = \frac{\sigma \gamma_a (1 - q_{an})}{\mu + \eta_n + \omega + \alpha}, \qquad R_{CST} = \frac{\frac{\Psi \mu}{\gamma_n} + \frac{\beta_a \mu}{\sigma \gamma_n}}{\sigma + \epsilon_n + \beta_n} \qquad (33)$$

$$R_0 = \frac{R_S + R_T + \sqrt{(R_S - R_T)^2 + 4 R_{CTS} R_{CST}}}{2} \qquad (34)$$

where R_S represents the typical number of new anxious students recruited by a single anxious student ($S_n \rightarrow S_a$), R_T represents the typical number of new anxious teachers recruited by a single anxious teacher ($T_n \rightarrow T_a$), R_{CTS} represents the typical number of new anxious students mentored into becoming anxious teachers by a single anxious teacher ($S_n \rightarrow T_a$), and R_{CST} represents the typical number of new anxious students recruited by a single anxious teacher (P, $S_n \rightarrow S_a$).

Interpreting the reproductive number requires two useful mathematical facts:

$$\sqrt{A + B} \leq \sqrt{A} + \sqrt{B} \qquad (35)$$

$$\max\{A, B\} = \frac{A + B + |A - B|}{2}. \qquad (36)$$

Applying these to the form of R_0 we have

$$\max\{R_S, R_T\} \leq R_0 \leq \max\{R_S, R_T\} + \sqrt{R_{CTS} R_{CST}}. \qquad (37)$$

R_S and R_T are within generation effects, and the form of R_0 tells us that either of these is strong enough by itself to support an epidemic of math anxiety. If enough students recruit their peers, or enough teachers recruit their peers, then anxiety is sustained.

The geometric mean term represents a generational effect that takes place in two paths: in the first path, an anxious teacher mentors a non-anxious student into becoming an anxious teacher ($S_n \rightarrow T_a$). In the second path, an anxious teacher mentors a non-anxious student into becoming an anxious student (P, $S_n \rightarrow S_a$). These are the two ways that an anxious teacher can pass on their anxiety to the next generation.

However, this is not a true geometric mean. Unlike the HIV model, the two paths do no share endpoints, so this is not an average of two events happening sequentially. Instead what is happening is that neither of these paths makes a full cycle. One path is student to teacher, and the other path is teacher to student. So each represents only a half-step of infection and not a full teacher to teacher or student to student cycle, so each needs to be square-rooted.

Lastly, note that if there are no peer-pressure effects ($R_S = R_T = 0$), then the generational effects by themselves may not be enough to sustain the epidemic.

However, generational effects can create an epidemic where peer-pressure by itself might not be enough ($R_S = R_T < 1$). This is not a straight sum because the two methods of spread (peer-pressure and generational) are competing with each other. For example, when teachers recruit more primary students to become anxious secondary students, it creates fewer non-anxious secondary students to be recruited by their peers so the peer effect is weaker. Another example is when anxious students recruit more non-anxious students through peer-effects, it leaves fewer non-anxious students to be recruited though mentoring effects.

Exercise 13 Other student projects with similar forms for R_0 include a 2001 study on bulimia [40, 41] and 2018 study on hospital screening for MRSA [12]. Choose one of these papers, draw the flow diagram, explain the model, conduct a unit analysis, and interpret the basic reproductive number using the techniques you have learned so far.

5 Ways to Complicate Your Sensitivity Analysis

Sensitivity analysis is an extremely flexible technique and there is really no end to the number of ways that you can complicate it. For example, in a simple variation, you could study the sensitivity of the equilibria to the parameter values, or the sensitivity of the equilibria to R_0. You can also use more complex methods and numerical simulations to look at sensitivity of a compartment to a particular parameter as a function of time. Examples of student projects using this type of forward sensitive analysis include a 2018 study on the menstrual cycle [37] and a 2018 study of MRSA [12]. Using this approach, George et al. [37] was able to identify specific days in the menstrual cycle that intervention would be most effective.

All of these approaches go beyond studying the basic reproductive number and are outside of the scope of this paper. For details on these and other ways to use sensitivity analysis, we really recommend that you read an in-depth paper on the subject [3] and explore some of the many student projects that use sensitivity analysis in ways not described here [51].

6 Conclusions for Students

In this paper, we have shown you an overview of the most popular techniques used at MTBI, as well as examples of variety of topics that you can investigate with these techniques. All together, this article forms a sort of "research project in a box." It is up to you to unpack the contents of the box, consult your research mentor, and add your own personal interests to generate a unique research project.

This approach does some with some warnings: While reproductive number analysis is the most commonly used approach at MTBI, it is not suitable for every research project. This is an extremely flexible approach than can be used to investigate a lot of projects, not every problem is well suited to an ODE contagion model. Sometimes a different modeling approach, such as difference equations [15], Markov chains [53], cellular automata [43], agent based modeling [28], a probability model [50], or a statistical approach [46] is better suited to a particular research project.

Even within the realm of differential equations, not every question is a question that can be answered by reproductive analysis. For example: sometimes a problem is an equilibrium problem [49], or a cost problem [1], or an optimal control problem [11]. The reproductive number can be useful in investigating these types of problems, but it does not provide the whole story.

What is most important is that you choose a research project that you are passionate about. Our best projects come from students who have chosen a topic personally meaningful to them. Many of our student projects are driven by students who have suffered from the very diseases or social problems they choose to study. This passion gives students the drive to really dig deep into the topics that they choose to study. It improves their understanding of the problem, the accuracy of their model, the effort they put into their mathematics, and the quality of their interpretations and recommendations.

So our advice for building a project is as follows: (1) Keep in mind what kinds of techniques are possible, both techniques described in this paper and other techniques you might have learned about. (2) Think about what kinds of questions can be answered with the different techniques are your disposal. (3) Be aware that there are more techniques that your mentor can help you with that you might not think of on your own. (4) Find a topic that you are passionate about. (5) Define a research question that is interesting to you. (6) Let the research question choose your techniques, do not let your techniques choose your research question.

7 Generating Possible Research Questions

In developing a project, we prefer to that you choose a topic that has a personal meaning to you, as passion tends to make the best projects. But a research problem should also be of interest to others. Generating research questions/projects often emerges from reading what was said or left untouched in articles at the interface of the biological, computational, mathematical, and social sciences. Here are some directions and possibilities:

Research Project 1 The official confirmation of an outbreak of Ebola hemorrhagic fever in West Africa took place in 2014. Efforts to assess its impact and what could be learned from it were launched [19]. Models made use of earlier estimates of the basic reproductive number [26] with modeling results providing increased understanding of Ebola dynamics and helping assess the impact of various control efforts [17]. Soon after an effective vaccine was discovered and tested. Yet, as the 2018 Ebola epidemic in the Republic of Congo evolved, we began to experience one of the worst outbreaks. In short, fighting Ebola, vaccine in hand, did not prevent its devastating impact on the affected populations, Why?

Research Project 2 What is the impact of cross-immunity on influenza strain dynamics and how does its role compare to the effect of partially effective vaccines? What if we had a partially effective universal flu vaccine? What percentage of the population should be vaccinated to ameliorate the impact of an outbreak?

One of the most challenging family of viruses is that associated with influenza A. Currently, there are three subtypes of influenza A, with variants of each subtype, called strains, being continuously generated across the world and transported primarily by the mobility patterns of billions of individuals. Influenza involves what it is known as seasonal and pandemic variants. The kind spread and its impact depends on the variation between strains. The 1918 pandemic was devastating and it has been linked to at least 50 million deaths. Vaccines, typically effective against seasonal influenza, are in general ineffective against novel strains, the most likely of drivers of pandemic influenza. As an outbreak sweeps a region, it alters the average immunological profile of a population. By how much? Well it depends on levels of protection, cross-immunity, that may protect against future strains. Some relevant references include [8, 10, 52].

Research Project 3 The HIV epidemic, the re-emerging tuberculosis (TB) epidemic and their synergistic impact on each other, including the growth of antibiotic resistant TB as well as the need to reduce the evolution of resistance on HIV treatment are ongoing challenges. Studying the modeling history of these diseases brings tremendous insights and would help those willing to go through the mathematics, and offers the opportunity to learn new methods as well. We recommend the following references [16, 20, 22].

Research Project 4 A most important challenge facing the study of epidemics come from key questions at the intersection of three fields, demography, epidemiology, and genetics. These problems are particularly challenging because they involve multiple temporal scales: the epidemiological, the demographic, and the evolutionary time scales. What models and approaches have we developed to address the joint dynamics of these three processes? A good place to start may be [36].

Research Project 5 Moving from the field of epidemiology but still looking at problems that involve some form of contagion can also be addressed in frameworks built to address particular data patterns. We have work, for example, on the spread of scientific ideas and on whether or not there is a copycat effect on that temporal patterns generated by school mass shootings [6, 59]. In general, social science provides many opportunities to innovate with epidemiological techniques. Many of our students have had success with projects in this area.

8 Notes for Mentors

Guiding a student-led research project is a very different experience from assigning a research project. Assigning a research project allows the mentor to anticipate and plan for any difficulties that might arise, but it comes at the cost of student passion. A student-led research project is a passionate research project, but it will often be in an area outside your area of expertise. The mentor of a student-led research project is always playing catch-up. She (or he) relies on her experience and expertise to think faster than the students and anticipate problems that might arise while the project is being developed.

The primary role of the mentor of a student-led project is to serve as an academic advisor in almost the same way that you would mentor a graduate thesis. The mentor pushes to the student farther, provides instruction when students suffer from a skill gap, and generally encourage students to work to find their own answers instead of providing answers.

Part your role as a mentor will be to push the students to really understand the context. Make sure students do research into the subject and look at previous models. Use your knowledge of the literature and your academic connections to suggest places that students might research.

One of the most difficult things for students to do is to develop a research question. Help the students make sure their research question is clearly defined, academically interesting, and small enough that students can get some results within the time limit of their project. In general, the best research questions tend to be about choosing between mechanisms or strategies. Make sure that students avoid yes/no questions or questions where the answer is obvious.

For example, the Herpes group in 2018 [1] went through a number of research questions including "What is the effect of early treatment on HSV-2?" This is not a good research question because the answer "HSV-2 infections decrease" is obvious. The group further refined their research question to being about the cost of treatment, then about the relative cost of two different treatment strategies, and finally about finding the most cost-effective combination of both strategies.

In developing a research question and a corresponding model, it is important that mentors help students maintain a delicate applied mathematics balance. Students are often driven by a mathematical question or a biological question and tend to forget the other half of the model. When a model or question that is too mathematical, students tend to only pay lip-service to the biology. The biology becomes inaccurate, or the interpretation of the results becomes uninteresting. When a model or question is too biological, typically the students want to include every possible detail, the problem becomes too complex, and mathematical results become impossible within the time limit of the project. For more information on the delicate balance between biology and mathematics, the different ways in which researchers from different fields value models, and how these differences affect students selecting a research question, we highly recommend reading Smith et al. [58].

Similarly, do not let students get away with stating bare mathematical results. Students have tendency to assume that bare mathematical results speak for themselves. But students are supposed to provide topic level expertise. Force discussion, interpretation, and recommendations. There is a reason why we have included such a long section on the interpretation of R_0. Students should not anticipate that the readers of their papers will do the work of converting results into academic significance.

A major part of the role of a mentor is to make sure that mathematical models do not get too big. Use your modeling experience to anticipate the complexity of the model students propose. Count compartments and non-linearities to anticipate the complexity of the algebra. In many ways, Sect. 4 on interpreting the basic reproductive number is more useful for mentors than students. Mentors can use this approach backwards to anticipate the form of R_0 and how complex it will be. For example, if students want to construct a model has a generational effect, a loop-back latency period, and co-infection, squash that immediately. The resulting R_0 would be a total mess.

Much of the role of mentor is to provide mathematical expertise. Students will often come up with interesting research questions or models where the necessary analysis does not match the mathematical techniques they have learned in class. Much of mentoring is providing just-in-time tutoring in mathematical techniques that students either do not know about or have not learned very well. Occasionally, you will be missing tools as well, and a problem will require an approach or a technique that you are not familiar with. When you encounter this scenario of missing tools, learn the tools. One of the benefits of expertise and experience is that you learn mathematics much faster and much better than a student. Use this ability to learn to tutor your students in the missing tools that you both need.

Lastly, the most important role of a mentor is to inflame the passions that students have for a topic. Students work much harder when they are passionate about a topic. They innovate in ways we would not think to and study ideas that we would not normally study. They push us to innovate and improve our own practice of research. This is the primary reason why MTBI has used student-led projects since 1997.

Motivating students can be a tricky and often frustrating business. We have found that having a social agenda greatly improves the interest of both students and funding agencies in our program [23]. MTBI is a program driven by social impact. The entire program is intentionally designed to improve the representation of minorities in the sciences by creating a pipeline to graduate school. This intentional design is not hidden from students. We share our social and political passions with students, and this in turn encourages them to explore social and biological problems that they are passionate about. They see that mathematics can have a direct impact on the world and are fueled by their own passions to make the world a better place.

In the past, we have met with some resistance to this idea. We have encountered faculty who have strongly believed that student researchers and funding agencies should be motivated primarily by the intellectual merit of a research project without emphasizing broader impacts. We understand and share the frustrations of many of these researchers. It is a difficult thing to feel as if the research we are passionate about is not valued by others on its own merits. But the reality is the broader impacts will always be an important motivator for students and a determinant in attracting diverse populations to the mathematical sciences. And while we can say that intellectual merit is definitely a necessary motivator of students [44], the appeal of making a change in society does not take away from that motivation, it only adds to it.

Acknowledgements Special thanks to MTBI mentors Leon Arriola, Christopher Kribs, Anuj Mubayi, Karen Rios-Soto, and Baojun Song, whose contributed expertise in mentoring students formed the foundation for this paper. Student research that formed the basis for this paper was supported by NSF grants DMS-0502349, DMS-9977919, DMS-1757968, DMPS-0838705, MPS-DMS-1263374, HRD 9724850, and DEB 925370; NSA grants H98230-06-1-0097, H98230-11-1-0211, H98230-09-1-0104, H98230-J8-1-0005, MDA 904-00-1-0006, MDA 904-97-1-0074, and MDA 904-96-1-0032; The Sloan Foundation; T-Division of Los Alamos National Lab; The Data Science Initiative at Brown University, The Office of the Provost of Cornell University; and the Office of the Provost of Arizona State University. The views and opinions expressed in this paper and cited research are solely those of the authors and do not necessarily reflect the ideas and/or opinions of the funding agencies or organizations.

References

1. Almonte-Vega, L., Colón-Vargas, M., Luna-Jarrń, L., Martinez, J., Rodriguez-Rincón, J., Patil, R., Espinoza, B., Murillo, A., Arriola, L., Viswanathan, A., Mubayi, A.: A Cost-Effective Analysis of Treatment Strategies for the Control of HSV-2 Infection in the U.S.: A Mathematical Modeling – Based Case Study. Tech. rep., Arizona State University (2018)

2. Alvarez, W.E., Fene, I., Gutstein, K., Martinez, B., Chowell, D., Mubayi, A., Melara, L.: Prisoner reform programs, and their impact on recidivism. Tech. Rep. MTBI-09-06M, Arizona State University (2012)
3. Arriola, L., Hyman, J.M.: Sensitivity analysis for uncertainty quantification in mathematical models. In: Mathematical and Statistical Estimation Approaches in Epidemiology, pp. 195–247. Springer (2009)
4. Barley, K., Murillo, D., Roudenko, S., Tameru, A., Tatum, S., et al.: A mathematical model of HIV and malaria co-infection in sub-Saharan Africa. Journal of AIDS and Clinical Research 3(7), 1–7 (2012)
5. Barley, K., Tatum, S., Murillo, D., Roudenko, S., Tameru, A.M.: A mathematical model of HIV and malaria co-infection in sub-Saharan Africa. Tech. Rep. MTBI-04-06M, Arizona State University (2007)
6. Bettencourt, L.M., Cintrón-Arias, A., Kaiser, D.I., Castillo-Chávez, C.: The power of a good idea: Quantitative modeling of the spread of ideas from epidemiological models. Physica A: Statistical Mechanics and its Applications 364, 513–536 (2006)
7. Bodine, E.N., Lenhart, S., Gross, L.J.: Mathematics for the life sciences. Princeton University Press (2014)
8. Brauer, F., Castillo-Chavez, C.: Mathematical models for communicable diseases, vol. 84. SIAM (2012)
9. Brauer, F., Castillo-Chavez, C.: Mathematical Models in Population Biology and Epidemiology, second edn. Springer, New York (2012)
10. Brauer, F., Castillo-Chavez, C., Feng, Z.: Mathematical Models in Epidemiology. Springer (2018)
11. Burkow, D., Duron, C., Heal, K., Vargas, A., Melara, L.: A mathematical model of the emission and optimal control of photochemical smog. Tech. Rep. MTBI-08-07M, Arizona State University (2011)
12. Butler, C., Cheng, J., Correa, L., Preciado, M.R., Rios, A., Espinoza, B., Montalvo, C., Moreno, V., Kribs, C.: Comparison of screening for methicillin-resistant Staphylococcus aureus (MRSA) at hospital admission and discharge. Tech. rep., Arizona State University (2018)
13. Calderon, K., an Azra Panjwani, C.O., Romero, D.M., Kribs-Zaleta, C., Rios-Soto, K.: An epidemiological approach to the spread of political third parties. Tech. Rep. MTBI-02-02M, Arizona State University (2005)
14. Camacho, E.T., Kribs-Zaleta, C.M., Wirkus, S.: The mathematical and theoretical biology institute - a model of mentorship through research. Mathematical Biosciences and Engineering (MBE) 10(5/6), 1351 – 1363 (2013)
15. Carrera-Pineyro, D., Hanes, H., Litzler, A., McCormack, A., Velazquez-Molina, J., Kribs, C., Mubayi, A., Rios-Soto, K.: Modeling the dynamics and control of Lyme disease in a tick-mouse system subject to vaccination of mice populations. Tech. rep., Arizona State University (2018)
16. Castillo-Chavez, C.: Mathematical and statistical approaches to AIDS epidemiology, vol. 83. Springer Science & Business Media (2013)
17. Castillo-Chavez, C., Barley, K., Bichara, D., Chowell, D., Herrera, E.D., Espinoza, B., Moreno, V., Towers, S., Yong, K.: Modeling ebola at the mathematical and theoretical biology institute (MTBI). Notices of the AMS 63(4) (2016)
18. Castillo-Chavez, C., Castillo-Garsow, C.W., Chowell, G., Murillo, D., Pshaenich, M.: Promoting research and minority participation via undergraduate research in the mathematical sciences. MTBI/SUMS-Arizona State University. In: J. Gallian (ed.) Proceedings of the Conference on Promoting Undergraduate Research in Mathematics. American Mathematical Society (2007)
19. Castillo-Chavez, C., Curtiss, R., Daszak, P., Levin, S.A., Patterson-Lomba, O., Perrings, C., Poste, G., Towers, S.: Beyond ebola: Lessons to mitigate future pandemics. The Lancet Global Health 3(7), e354–e355 (2015)

20. Castillo-Chavez, C., Feng, Z., Huang, W.: On the computation of Ro and its role on. Mathematical approaches for emerging and reemerging infectious diseases: an introduction **1**, 229 (2002)
21. Castillo-Chavez, C., Kribs, C., Morin, B.: Student-driven research at the mathematical and theoretical biology institute. The American Mathematical Monthly **124**(9), 876–892 (2017)
22. Castillo-Chavez, C., Song, B.: Dynamical models of tuberculosis and their applications. Mathematical biosciences and engineering **1**(2), 361–404 (2004)
23. Castillo-Garsow, C.W., Castillo-Chavez, C.: Why REUs matter. In: M.A. Peterson, A. Rubinstein Yanir (eds.) Directions for Mathematics Research Experience for Undergraduates. World Scientific (2015)
24. Castillo-Garsow, C.W., Castillo-Chavez, C., Woodley, S.: A preliminary theoretical analysis of an REU's community model. PRIMUS: Problems, Resources, and Issues in Mathematics Undergraduate Studies **23**(9), 860 – 880 (2013)
25. Catron, L., La Forgia, A., Padilla, D., Castro, R., Rios-Soto, K., Song, B.: Immigration laws and immigrant health: Modeling the spread of tuberculosis in Arizona. Tech. Rep. MTBI-07-06M, Arizona State University, http://mtbi.asu.edu/research/archive/paper/immigration-laws-and-immigrant-health-modeling-spread-tuberculosis-arizona (2010)
26. Chowell, G., Hengartner, N.W., Castillo-Chavez, C., Fenimore, P.W., Hyman, J.M.: The basic reproductive number of ebola and the effects of public health measures: the cases of Congo and Uganda. Journal of theoretical biology **229**(1), 119–126 (2004)
27. Chowell-Puente, D., Jimenez-Gonzalez, B., Smith, A.N., Rios-Soto, K., Song, B.: The cursed duet: Dynamics of HIV-TB co-infection in South Africa. Tech. Rep. MTBI-04-08M, Arizona State University (2007)
28. Costo, D., Denogean, L., Dolphyne, A., Mello, C., Castillo-Garsow, C.W.: The recovery and ecological succession of the tropical Montserrat flora from periodic volcanic eruptions. Tech. Rep. BU-1513-M, Cornell University (1998)
29. Crisosto, N.M., Castillo-Chavez, C., Kribs-Zaleta, C., Wirkus, S.: Who says we r0 ready for change? Tech. Rep. BU-1586-M, Cornell University (2001)
30. Crisosto, N.M., Kribs-Zaleta, C.M., Castillo-Chavez, C., Wirkus, S.: Community resilience in collaborative learning. Discrete and Continuous Dynamical Systems Series B **14**(1), 17–40 (2010)
31. Davidoff, L., Sutton, K., Toutain, G.Y., Sánchez, F., Kribs-Zaleta, C., Castillo-Chavez, C.: Mathematical modeling of the sex worker industry as a supply and demand system. Tech. Rep. MTBI-03-06M, Arizona State University (2006)
32. Diaz, K., Fett, C., Torres-Garcia, G., Crisosto, N.M.: The effects of student-teacher ratio and interactions on student/teacher performance in high school scenarios. Tech. Rep. BU-1645-M, Cornell University, http://mtbi.asu.edu/research/archive/paper/effects-student-teacher-ratio-and-interactions-studentteacher-performance-hig (2003)
33. Diekmann, O., Heesterbeek, J.A.P., Metz, J.A.J.: On the definition and the computation of the basic reproduction ratio R0 in models for infectious diseases in heterogeneous populations. Journal of Mathematical Biology **28**(4), 365–382 (1990)
34. Van den Driessche, P., Watmough, J.: Reproduction numbers and sub-threshold endemic equilibria for compartmental models of disease transmission. Mathematical biosciences **180**(1-2), 29–48 (2002)
35. Edelstein-Keshet, L.: Mathematical Models in Biology, first edn. McGraw-Hill, New York (1988)
36. Feng, Z., Castillo-Chavez, C.: The influence of infectious diseases on population genetics. Mathematical Biosciences and Engineering 3(3), 467 (2006)
37. George, S.S., Mercado, L.O.M., Oroz, C.Y., Tallana-Chimarro, D.X., Melendez-Alvarez, J.R., Murrillo, A.L., Castillo-Garsow, C.W., Rios-Soto, K.R.: The effect of gonadotropin-releasing hormone (GnRH) on the regulation of hormones in the menstrual cycle: a mathematical model. Tech. rep., Arizona State University (2018)

38. Gjorgjicva, J., Smith, K., Chowell, G., Sanchez, F., Snyder, J., Castillo-Chavez, C.: The role of vaccination in the control of SARS. Journal of Mathematical Biosciences and Engineering **2**(4), 753 – 769 (2005)
39. Gjorgjieva, J., Smith, K., Snyder, J., Chowell, G., Sánchez, F.: The role of vaccination in the control of SARS. Tech. Rep. MTBI-01-5M, Arizona State University (2004)
40. González, B., Huerta-Sánchez, E., Ortiz-Nieves, A., Vázquez-Alvarez, T., Kribs-Zaleta, C.: Am i too fat? bulimia as an epidemic. Tech. Rep. BU-1578-M, Cornell University (2001)
41. González, B., Huerta-Sánchez, E., Ortiz-Nieves, A., Vázquez-Alvarez, T., Kribs-Zaleta, C.: Am i too fat? bulimia as an epidemic. Journal of Mathematical Psychology **1**(47), 515 – 526 (2003)
42. Gurin, A., Jeanneret, G., Pearson, M., Pulley, M., Salinas, A., Castillo-Garsow, C.W.: The dynamics of math anxiety as it is transferred through peer and teacher interactions. Tech. Rep. MTBI-14-05M, Arizona State University (2017)
43. Gurin, A., Manosalvas, P., Perez, L., Secaira, H., Ignace, C., Arunachalam, V., Castillo-Garsow, C.W., Smith, A.: Spatial dynamics of myeloid-tumor cell interactions during early non-small adenocarcinoma development. Tech. rep., Arizona State University (2018)
44. Harel, G.: Intellectual need. In: K.R. Leatham (ed.) Vital Directions for Mathematics Education Research, pp. 119–151. Springer New York, New York, NY (2013)
45. Huynh, M., Leung, M.R., Marchand, M., Stykel, S., Arriola, L., Flores, J.: The Effect of Localized Oil Spills on the Atlantic Loggerhead Turtle Population Dynamics. Tech. rep., Arizona State University (2010)
46. Izquierdo-Sabido, A., Lasky, J., Muktoyuk, M., Sabillon, S.A.: Mean time to extinction of source-sink metapopulation for different spatial considerations. Tech. Rep. BU-1421-M, Cornell University (1997)
47. Kasseem, G.T., Roudenko, S., Tennenbaum, S., Castillo-Chavez, C.: The role of transactional sex in spreading HIV/Aids in Nigeria: A modeling perspective. Tech. Rep. MTBI-02-13M, Arizona State University (2005)
48. Kasseem, G.T., Roudenko, S., Tennenbaum, S., Castillo-Chavez, C.: The role of transactional sex in spreading HIV/Aids in Nigeria: A modeling perspective. In: A. Gumel, C. Castillo-Chavez, D. Clemence, R. Mickens (eds.) Mathematical Studies on Human Disease Dynamics: Emerging Paradigms and Challenges, vol. 410, pp. 367 – 389. American Mathematical Society (2006)
49. Lopez, A., Moreira, N., Rivera, A., Amdouni, B., Espinoza, B., Kribs, C.M.: Economics of Prison: Modeling the Dynamics of Recidivism. Tech. rep., Arizona State University (2018)
50. Mohanakumar, C., Offer, A.E., Rodriguez, J., Espinoza, B., Moreno, V., Nazari, F., Castillo-Garsow, C.W., Bichara, D.: Mathematical model for time to neuronal apoptosis due to accrual of DNA DSBs. Tech. Rep. MTBI-12-05, Arizona State University (2015)
51. MTBI: Technical reports. https://mtbi.asu.edu/tech-report (2018)
52. Nuno, M., Chowell, G., Wang, X., Castillo-Chavez, C.: On the role of cross-immunity and vaccines on the survival of less fit flu-strains. Theoretical Population Biology **71**(1), 20–29 (2007)
53. Ortiz, J., Rivera, M.A., Rubin, D., Ruiz, I., Hernandez, C.M., Castillo-Chavez, C.: Critical response models for foot-and-mouth disease epidemics. Tech. Rep. BU-1620-M, Cornell University (2002)
54. Otto, S., Day, T.: A Biologist's Guide to Mathematical Modeling in Ecology and Evolution. Princeton University Press (2011). URL https://books.google.com/books?id=vnNhcE9UwYcC
55. Quintero, S., Machuca, V., Cotto, H., Bradley, M., Rios-Soto, K.: A mathematical model of coral reef response to destructive fishing practices with predator-prey interactions. Tech. rep., Arizona State University (2016)
56. Rodriguez-Rodriguez, L., Stafford, E., Williams, A., Wright, B., Kribs, C., Rios-Soto, K.: A Stage Structured Model of the Impact of Buffelgrass on Saguaro Cacti and their Nurse Trees. Tech. rep., Arizona State University (2017)

57. Romero, D.M., Kribs-Zaleta, C.M., Mubayi, A., Orbe, C.: An epidemiological approach to the spread of political third parties. arXiv (arXiv:0911.2388) (2009)
58. Smith, E., Haarer, S., Confrey, J.: Seeking diversity in mathematics education: mathematical modeling in the practice of biologists and mathematicians. Science and Education **6**, 441–472 (1997)
59. Towers, S., Gomez-Lievano, A., Khan, M., Mubayi, A., Castillo-Chavez, C.: Contagion in mass killings and school shootings. PLOS One **10**(7), e0117,259 (2015)

The Effect of External Perturbations on Ecological Oscillators

Eli E. Goldwyn

Abstract Population interactions cause oscillations across a variety of ecological systems. These ecological population oscillators are subject to external forces including climatic effects and migration. One way of understanding the impact of these perturbations is to study their effect on the phase of the oscillations. We introduce tools commonly used to describe the phase of oscillating phenomena in cardiac dynamics, circadian rhythms, firing neurons, and more recently, coupled predator–prey dynamics. We walk the reader through the applications of these methods to several common ecological models and present research opportunities to extend these techniques to other models to address specific ecological issues and phenomena.

Suggested Prerequisites *A first course in differential equations is essential, including the study of equilibrium and stability. An acquaintance with mathematical modeling and simulation is recommended.*

1 Introduction

Oscillatory behavior is ubiquitous in the physical and natural world. Prominent examples include the number of daylight hours over the year, the position of a mass on a frictionless spring, seasonal outbreaks of childhood diseases in the pre-vaccination era, action potentials in a neuron, and predator–prey populations. Mathematical modeling of oscillatory behavior has widespread applications including improving the control and management of disease epidemics, treating pathological conditions such as Parkinson's disease and epilepsy and better understanding and protecting ecological populations. More specifically, perturbations to ecological

E. E. Goldwyn (✉)
University of Portland, Portland, OR, USA
e-mail: goldwyn@up.edu

© Springer Nature Switzerland AG 2020
H. Callender Highlander et al. (eds.), *An Introduction to Undergraduate Research in Computational and Mathematical Biology*, Foundations for Undergraduate Research in Mathematics, https://doi.org/10.1007/978-3-030-33645-5_3

oscillators can cause nearby populations to synchronize or desynchronize with each other, which has important implications for species persistence and extinction, especially in the face of global climate change.

The goal of this chapter is to give students sufficient background on oscillators and their properties to work on the research projects in population ecology described in Sect. 4. Each of the projects involve oscillatory systems in ecology and most of them involve finding, using, and/or interpreting the phase response curve (PRC), a function that measures the effect of an external perturbation on an oscillator, to better understand the underlying properties of the oscillator. The majority of the projects require only mathematics up to a first course in undergraduate differential equations. Familiarity with simulating differential equations is not required (it is recommended). Although we ask the student to interpret models and results in their ecological context, no background knowledge of ecology is necessary (interest in ecology is recommended). To limit the amount of coding required for the project, we provide some MATLAB code and XPP code (see Appendix 1 and 2).

This chapter is organized so that Sect. 2 introduces oscillatory systems, provides a few ecological examples, and introduces several standard differential equation tools to understand these systems. Much of this material may be familiar from an introductory differential equations class. Section 3 introduces the more advanced topic of describing oscillating systems by their phase of the oscillation and includes several examples. Section 4 provides several in-depth research projects. The Appendix 1, 2, 3 describes and provides links to several relevant programs in MATLAB and XPP and adjustable graphs using Desmos. It also describes additional chemical oscillating system that can be used in one of the projects and contains an optional section on non-dimensionalization and relaxation oscillators. Exercises are included throughout the chapter and are written to reinforce the important mathematical skills necessary for the research projects and to provide opportunities to reflect on their ecological importance. The individual exercises vary greatly in difficulty and likely time necessary for completion.

1.1   Further Reading

The online peer-reviewed encyclopedia Scholarpedia (http://www.scholarpedia.org/article/Main_Page) is an excellent resource that has in-depth pages on most of the major topics discussed in the chapter, think Wikipedia but authored by experts in the field and peer-reviewed. Scholarpedia has pages on many of the models discussed here, including: predator–prey models including the Lotka–Volterra model [15], the FitzHugh–Nagumo model [16]. There are also pages on phase models [17], isochrons [18], and phase response curves [19], as well as a page on XPP [20].

Many introductory differential equations texts will cover the eigenvalue and eigenvector method of finding and classifying stability as well as nullcline analysis for understanding the qualitative behavior of a system of differential equations. For a particularly accessible treatment of these topics, try the Adler calculus book [10]. For a more ecologically focused text, [9] also covers these topics while providing more ecological modeling context. It also covers non-dimensionalizing systems of differential equations. For a more advanced treatment, [7] also covers these topics and is often used as a textbook for a second course in differential equations. Many examples of biological oscillators can be found in [12], while the book [8] provides an in-depth but not too technical look at many examples of synchrony in biology (an important phenomenon involving multiple interacting oscillators).

The topics in Sect. 3 are often covered in more mathematically advanced texts that may be very difficult for undergraduate mathematicians. Chapter 10 of [5] (available for free at www.izhikevich.org) does an excellent job of describing phase models, isochrons, the PRC, and the three different methods of deriving the infinitesimal PRC, in a way that is more approachable than the original texts on these topics. Though a challenging read for undergraduates, Arthur Winfree's classic book [11], is a deep look at oscillatory systems and even describes his invention of isochrons.

2   Oscillators

In this section we define several terms describing oscillations, give examples and exercises involving differential equations that produce oscillatory behavior, and describe several techniques to find if and when a system produces oscillations. The goal of this section is to familiarizing the reader with oscillatory systems and techniques used to understand them. Some or all of the definitions and techniques may be review for more advanced students.

Definition 1 A function F is periodic if $F(t + P) = F(t)$, for all t. The smallest such $P > 0$ is the period of F.

We define oscillatory behavior as equivalent to a periodic function and we use the term periodic orbit to refer to a periodic solution to a differential equation. Furthermore, we define any ecological phenomenon described by a periodic function as an ecological oscillator. Perhaps the most common physical example of oscillatory behavior is the frictionless spring-mass system, i.e., a simple harmonic oscillator. This system is described in many introductory differential equation courses.

Example 1 Let $S(t)$ describe the number of hours of sunlight in a day. The number of sunlight hours in a day attains its maximum at the summer solstice, its minimum at the winter solstice, and has a period of 1 year. This function has discrete domain (the integers 0–365 ignoring leap years), however, we approximate it using a continuous function for the exercise below.

Exercise 1 Let the function $S(t) = A + B \sin(\omega t + \phi)$ describe the number of sunlight hours in a day (with t in days) in your hometown. Look up the date and number of sunlight hours for the summer and winter solstices and the spring and fall equinox of your hometown this year. Use this information to find A, B, ω, ϕ. Find the amplitude and period of the oscillator.

Example 2 Consider an object with mass m on a spring with spring constant k. Let the system start with no velocity and stretch x_0 units from its resting state. Assume the system has no friction and define $x(t)$ as the distance between the mass and its resting position at time t. The distance the mass is from its resting state can be described by the following second order differential equation with initial values:

$$m\frac{d^2x}{dt^2} = -kx$$

$$x(0) = x_0$$

$$x'(0) = 0 \tag{1}$$

The only functions that have second derivatives that are a negative number times the original function are the sine and cosine functions. After applying the initial conditions, we find the solution is $x(t) = x_0 \sin(\omega t)$, with $\omega = \sqrt{\frac{k}{m}}$. This spring-mass system is an oscillator with period $\frac{2\pi}{\omega}$. That is, the mass will continue to oscillate between x_0 and $-x_0$ with the above period indefinitely. We start with an introduction of the frictionless spring-mass system because it is a relative straightforward example of an oscillatory physical system and we can analytically find its solution, period, and amplitude.

Definition 2 An equilibrium of a system of differential equations exists when all variables are constant over time, i.e., when every differential equation in the system equals 0.

Exercise 2 Define $y = x'$ and rewrite Eq. (1) as a system of two first-order differential equations y' and x' (hint: if $y = x'$, then $y' = x''$). Show that the origin is an equilibrium by plugging into the system and showing that both differential equations equal zero. Classify this equilibrium as a linear center by showing that the eigenvalues of the coefficient matrix are purely imaginary.

Example 3 Consider a model of the dynamics of love affairs described in Chapter 5 of [7]. The variable $R(t)$ represents Romeo's love ($R(t) > 0$) or hate ($R(t) < 0$) for Juliet and vice versa for $J(t)$. A linear model describing the change in their feelings for each other can be written as:

$$\frac{dR}{dt} = aR + bJ$$

$$\frac{dJ}{dt} = cR + kJ \tag{2}$$

Exercise 3 Interpret each of the four parameters in (2). Specifically, describe what positive values or negative values indicate. What values for a, b, c, k lead to oscillatory behavior? Would that make for a "healthy" or "stable" relationship? Does the initial state of the relationship matter?

There are several different methods that can be used to classify the stability of the equilibrium in Exercise 3:

1. Use simulation software such as MATLAB or XPP and vary the parameter values until oscillatory behavior occurs.
2. Calculate the nullclines and perform phase-plane analysis (see below).
3. Calculate the eigenvalues of the Jacobian matrix evaluated at the equilibrium (see below).

Since the Romeo–Juliet model is linear, method (3) will work. Method (2) often leaves it unclear whether the system has a closed loop and therefore oscillates, or spirals inwards to a stable equilibrium, or outwards. For that reason, it is often helpful to combine methods (2) and (3). Unfortunately, for non-linear systems, even combining a sketch of the phase plane with nullclines and arrows is not enough to demonstrate oscillations. Hence the importance of performing simulations (or using more sophisticated techniques).

Definition 3 The phase plane for a system of two differential equations is a graph with one variable (population) on the horizontal axis and the other variable (population) on the vertical axis (time is not explicitly plotted on the phase plane).

While a more complete description of phase-plane analysis can be found in many differential equations texts and in [7] and [9], here is a quick synopsis for the Romeo–Juliet model. First draw a graph with vertical axis $J(t)$ and horizontal axis $R(t)$. This is the phase plane. The R- and J-nullclines correspond to points (R, J) for which $\frac{dR}{dt} = 0$ and $\frac{dJ}{dt} = 0$, respectively. Notice that equilibrium points correspond to intersections of R- and J-nullclines. Solving for $\frac{dR}{dt} = 0$, we find only one R-nullcline, the line $J = -\frac{a}{b}R$. Next calculate the J-nullcline and sketch both nullclines on the phase plane. Note that the solution can only cross the R-nullcline vertically and the J-nullcline horizontally. Unless the parameters are such that the two nullclines are the same line, this will divide the phase plane into 4 regions. Since the differential equations are continuous, the sign of $\frac{dR}{dt}$ and $\frac{dJ}{dt}$ can only change when its corresponding nullcline is crossed and cannot change inside a region. Therefore the direction of the solution on the phase plane (up/down and right/left) is the same in each region. By plugging values into our differential equations, we can find the cardinal direction of the solution in each region of the phase plane separated by the nullclines, see Fig. 1. We can then sketch an arrow pointing in those cardinal directions on the phase plane.

Definition 4 A solution to a system of differential equations is a function that satisfies the initial conditions and each individual differential equation.

We can roughly sketch solutions to the differential equations by following the arrows and the rules about which direction a nullcline can be crossed. By examining the possible arrow configurations and nullclines, we can often figure out the stability

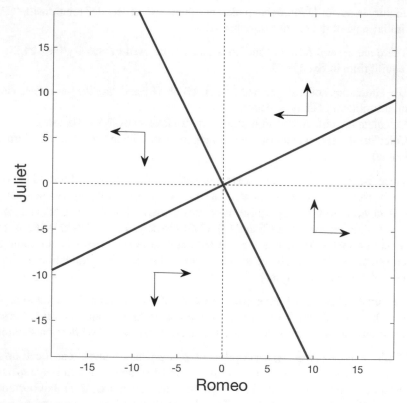

Fig. 1 The R-nullclines is plotted in blue and the R-nullcline is plotted in red for the Romeo–Juliet model, with parameter values $a = 1, b = -2, c = 2, k = 1$ and initial conditions $R(0) = J(0) = 1$. Black arrows indicate the cardinal direction of the solution

of an equilibrium and whether oscillations occur. As described above, this nullcline analysis may not be sufficient to determine oscillations. See [23] for nullclines of this system with adjustable parameter values plotted using Desmos.

Definition 5 The Jacobian matrix is the matrix of all first-order partial derivatives of the system of differential equations.

Consider a system of two differential equations as in (2). Let $F(R, J) = \frac{dR}{dT}$ and $G(R, J) = \frac{dJ}{dT}$.

$$Jacobian = \begin{bmatrix} \dfrac{dF}{dR} & \dfrac{dF}{dJ} \\[2mm] \dfrac{dG}{dR} & \dfrac{dG}{dJ} \end{bmatrix} \tag{3}$$

The local stability of any equilibrium can be determined by finding the eigenvalues of the Jacobian matrix evaluated at the equilibrium. The dynamics of even a seemingly simple system of differential equations can be quite complicated and oftentimes it is necessary to perform both a phase-plane analysis and to classify the stability of equilibria by calculating the eigenvalues of the Jacobian matrix in order to determine if a system has a periodic orbit. Since the Romeo-Juliet model is linear, oscillatory behavior is possible only if the two eigenvalues are purely imaginary. As we will see in subsequent sections, for non-linear systems of differential equations, this condition of purely imaginary eigenvalues is not necessary for oscillations (nor is it sufficient).

Example 4 Calculate the Jacobian matrix for the Romeo–Juliet model, find the eigenvalues, and classify the stability for various parameter values.

The partial derivative of the function $\frac{dR}{dt}$ with respect to R is a and its partial derivative with respect to J is b. Likewise, the partial derivative of the function $\frac{dJ}{dt}$ with respect to R is c and its partial derivative with respect to J is k. This yields a Jacobian matrix:

$$Jacobian = \begin{bmatrix} a & b \\ c & d \end{bmatrix}$$

The eigenvalues, λ, of the Romeo–Juliet model (or any system of differential equations) satisfy:

$$\det(Jacobian - \lambda I) = 0$$

For a system of two differential equations, the above equation, with $\tau = trace(Jacobian)$ and $\Delta = \det(Jacobian)$, can be written as the characteristic polynomial:

$$\lambda^2 - \tau\lambda + \Delta = 0$$

2.1 Predator–Prey Models

The interaction between predator and prey species can cause large scale oscillation. Perhaps the most famous data set in population ecology is that of the snowshoe hare and its predator, the Canadian lynx. Thanks to the Hudson's Bay Company, there are nearly 100 years of data showing fairly regular oscillations of these species with a 10-year period [2]. In this section, we introduce several different two-variable differential equation predator–prey models. Depending on the structure of the model and the parameter values, these systems can exhibit oscillatory behavior similar to that of the hare-lynx and other populations. The general form of this model is:

$$\frac{dx}{dt} = f(x) - g(x, y)$$

$$\frac{dy}{dt} = cg(x, y) - h(y) \tag{4}$$

The populations (usually described in terms of biomass as opposed to population number) at time t of the prey and predator are $x(t)$ and $y(t)$, respectively. The function $f(x)$ describes the growth rate of the prey in the absence of predation and depends only on the prey biomass. The predator death rate, or loss of biomass, depends only on the predator population and is modeled with the function $h(y)$. The predation rate depends on the interaction between the two populations and therefore depends on both of their biomasses. Here, $g(x, y)$ is the prey loss due to predation and the parameter $0 < c < 1$ is the conversion rate from prey biomass to predator biomass. While the variables x and y can instead measure population size, using biomass allows for a more natural interpretation of c and allows for the populations to be modeled by continuous variables.

2.1.1   Lotka–Volterra Predator–Prey Model

The Lotka–Volterra (LV) predator–prey model was derived independently by Alfred Lotka and Vito Volterra in the 1920s and is a staple of most introductory differential equations courses. The LV model follows the formulation of Eq. (4) with exponential growth of the prey in the absence of predators, exponential decay of the predator in the absence of prey, and a linear per-predator predation rate (sometimes called Holling Type I functional response):

$$f(x) = rx$$

$$g(x, y) = axy$$

$$h(y) = my \tag{5}$$

Each of the parameters r, a, and m in Eq. (5) is positive, r is the growth rate of the prey, m is the death rate of the predator, and a is the per capita predation rate. The left column of Fig. 2 graphs the solution for LV model with the top panel showing a time-series plot (the populations on the vertical axis and time on the horizontal axis) and the bottom row showing the phase planes. We can see from this figure that the solution to the LV model is oscillatory.

Exercise 4 Derive the analytical solution to the LV model in implicit form. First find $\frac{dy}{dx}$ by dividing $\frac{dy}{dt}$ by $\frac{dx}{dt}$. Separate and integrate this differential equation. Unfortunately, we cannot solve this expression explicitly for y in terms of x, however, we can find an implicit expression relating x and y. Pick your own initial conditions and parameter values and use a graphing program (the browser-based graphing calculator at desmos.com works well) to see the shape of the oscillation

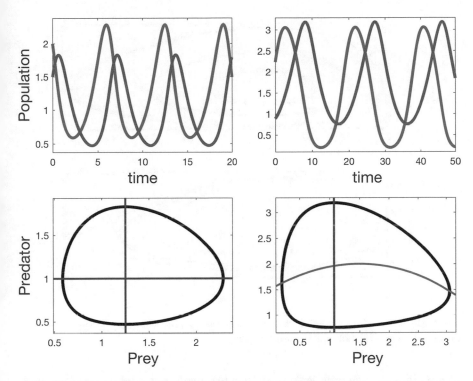

Fig. 2 The left column shows simulations of the Lotka–Volterra oscillator (Eq. (5)) with parameter values $a = 1, r = 1, m = 1, c = 0.8$, and initial conditions $x(0) = 2, y(0) = 1.5$, and the Rosenzweig–MacArthur oscillator (Eq. (6)) with parameter values $a = 1, r = 1, m = 1/3, c = 0.8, b = 1.5, K = 4.5$, and initial conditions $x(0) = 2.23, y(0) = 0.89$. Top row are the time-series plots with prey in blue and predator in red. The bottom row is the phase plane in black with the prey and predator nullclines also plotted in blue and red, respectively

on the x–y plane. Change the values of your initial conditions and parameters one at a time and observe their effect on the shape and amplitude of the oscillation. Does that fit your intuition as to what would happen?

Unlike the more complicated systems that we will explore later, the LV model can be solved (see Exercise 4). To understand the qualitative behavior of a system of differential equations, we first find and then classify the stability of all the equilibria. The LV model has two equilibria, the trivial equilibrium at extinction $(x^*, y^*) = (0, 0)$, and a coexistence equilibrium with $(x^*, y^*) = (\frac{m}{ca}, \frac{r}{a})$. The trivial equilibrium has one positive and one negative equilibrium $(\lambda_{1,2} = r, -m)$ making it a saddle. The eigenvalues of the coexistence equilibria are purely imaginary, $\lambda_{1,2} = \pm i\sqrt{mr}$. Purely imaginary eigenvalues suggest that an equilibrium may be a center but are inconclusive and can exhibit a variety of behaviors depending on higher order terms. In LV model, and in the frictionless spring-mass system, every initial condition (with the exception of the coexistence equilibrium and one species

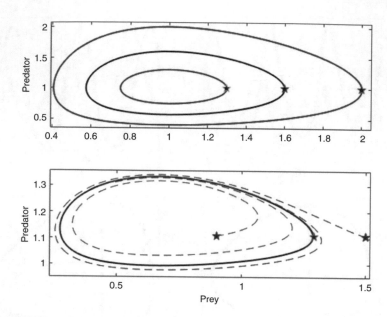

Fig. 3 Simulation of the Lotka-Volterra oscillator (top) and the Rosenzweig-MacArthur model (bottom) for three different initial conditions (stars)

being extinct) lies on a periodic orbit. In other words, every initial condition leads to oscillations as long as each population is greater than zero and the system is not at the coexistence equilibrium. The implicit relationship between x and y that you found in Exercise 4 and the top panel of Fig. 3 demonstrates that there are infinitely many periodic orbits. MATLAB and XPP code simulating the LV model can be found in the Appendix 1 and 2.

Exercise 5 Calculate the Jacobian matrix for the LV model and evaluate it at the two equilibria described above. Find the eigenvalues at each equilibrium. Check to make sure you get the same values as above.

Exercise 6 Calculate the prey and predator nullclines and graph them on the phase plane. Find the direction in each of the regions and sketch some trajectories for different initial conditions. Can we determine if the system is oscillatory from the phase plane alone or do we also need information from the Jacobian?

2.1.2 Rosenzweig–MacArthur Model

The Rosenzweig–MacArthur Model (RM) is formulated similarly to the LV model (Eqs. (4) and (5)), but with different functional forms for prey-growth rate $f(x)$ and the predation $g(x, y)$. While the LV model assumes unbounded exponential growth of the prey in the absence of the predator and unbounded per-predator predation as the prey population gets infinitely large, the RM model assumes that each of these terms has natural maxima. The RM assumes logistic growth of the prey in the absence of predation, with carrying capacity K and a per-predator predation rate that includes predator satiation (Holling Type II functional response) with a maximum rate of a and a half-saturation constant of b. Predator death has the same functional form as in the LV model. The individual terms in the RM model using the formulation of Eq. (4) are now

$$f(x) = rx(1 - x/K)$$
$$g(x, y) = \frac{axy}{b + x}$$
$$h(y) = my \tag{6}$$

While the RM model is more complex than the LV model as it has two additional parameters, it is generally considered to be a better description of many predator–prey populations.

Exercise 7 Sketch the per capita growth rate of the prey in the absence of predation for both the LV and RM model, $\frac{f(x)}{x}$ in (5) and (6). Why is the logistic growth rate in the RM model generally considered to be more realistic? Are there any circumstances where using the exponential growth rate from the LV model may be reasonable?

Exercise 8 Let us compare the difference in the per-predator predation rates, $p(x) = \frac{g(x,y)}{y}$, between the LV and RM models. For the RM model, calculate $p(0)$, $p(b)$, $\lim_{x \to \infty} p(x)$, and notice that $p'(x)$ is a decreasing function. Put that information together to sketch $p(x)$ for the RM model. Sketch $p(x)$ for the LV model. Examine how the parameters a and b affect predation. Describe why the predator satiation model is generally accepted as a more appropriate model of predator behavior than the linear model.

While the LV model produces oscillations for any non-zero parameter values and non-zero initial populations, the RM model only produces oscillations under a specific range of parameter values. For other parameter values the system will go to a stable coexistence equilibrium. Additionally, while the LV model produces oscillations that return to the initial conditions, the RM model has a unique stable limit cycle.

Definition 6 An asymptotically stable limit cycle is a periodic orbit upon which nearby trajectories converge.

This means that if the RM system starts with initial condition near this limit cycle, then the system will always approach this limit cycle. The top panel on Fig. 3 shows that different initial conditions for the LV model can lead to different periodic orbits, while solutions to the RM model approach a unique asymptotically stable limit cycle (bottom panel). Initial conditions outside the limit cycle spiral inwards towards it, while initial conditions inside the limit cycle spiral outwards towards it.

Exercise 9 Perform phase-plane analysis for the RM model (sketch the nullclines for the RM and find the cardinal directions of the solution in each region). Note that the position of the nullclines depends on the relative values of certain parameter combinations (assume $K > b$). What inequalities involving the parameter values must be true to produce oscillations? Note that the coexistence equilibrium is an unstable spiral and the system has an asymptotically stable limit cycle if the vertical nullcline is to the left of the vertex of the parabolic nullcline (see right column of Fig. 2).

Consider the RM model with parameters that produce oscillations as in the right column of Fig. 2. Ecological oscillators with low population levels in the trough of the oscillation can be at risk of extinction in the event of a negative event such as a drought or if there are so few individuals that it becomes difficult to find a mate. Intuitively, it seems that one way to prevent low prey populations at the trough of the oscillation would be to improve the habitat for the prey (increase the carrying capacity K). Counter-intuitively, increasing K leads to larger amplitude oscillations, lower troughs, and therefore a greater chance of prey extinction. In fact, decreasing K decreases the amplitude of oscillations and eventually changes the equilibrium from an unstable spiral to a stable spiral which will stabilize the population. This phenomenon of an improving prey habitat increasing the likelihood of prey extinction was originally described by Michael Rosenzweig as the "Paradox of Enrichment."

Exercise 10 Starting with the parameter values in the right column of Fig. 2, increase K and simulate the RM model. Notice that the amplitude of the oscillation increases and that the minimum prey population decreases. Next decrease K until the system is no longer oscillatory. Observe this with simulations and by noticing that the vertical nullcline is now to the right of the vertex of the parabolic nullcline. Determine the stability of the coexistence equilibrium by calculating the eigenvalues with the original parameters and the new lower K.

The phenomenon described in the above exercise where an equilibrium that is an unstable spiral which produces a stable limit cycle becomes a stable spiral as a parameter is varied is one type of Hopf-bifurcation [22]. In addition to predator–prey systems, Hopf-bifurcations appear in a variety of other physical systems [7], including certain chemical oscillators described in the project section and the Appendix 4.

Exercise 11 Consider a model with predation following a Holling Type III function response $g(x, y) = \frac{ax^2y}{b^2+x^2}$. Repeat your analysis from Exercise 8, but notice $p'(x)$ is no longer strictly decreasing and that $p'(0) = 0$.

Exercise 12 Examine the per predator predation for each of the Holling Types I–III functional responses (Exercises 8 and 11). What can we say about the predator's behavior that each functional response describes?

Predators are often described as specialists (only consume one prey species) or generalists (consuming a variety of prey species). Think about the above exercise in terms of both of these types of predators.

Exercise 13 Find parameter values that lead to limit cycle behavior of the system with predation following a Type III function response. This can be done either by guessing and check using XPP or MATLAB simulations, or by nullcline analysis (it may be helpful to first non-dimensionalize this model). For additional help see the Desmos graphs with nullclines plotted with sliders to manually change the parameter values [23].

3 The Phase of an Oscillator

Imagine a runner doing counter-clockwise laps at a constant speed around a circular track of radius r. We can think of the runner's position on the phase plane as a unique pair of points (x, y). Switching to polar coordinates, we can label each point by its angle and distance from the origin using (r, θ) and $\theta \in [0, 2\pi)$. Each point is mapped using the trigonometric relationship $x = r \cos \theta$ and $y = r \sin \theta$. Since the track is a circle, r is constant and therefore the position of the runner can be described by a single variable called the phase. The phase, $\theta(t)$, measures how far around the track the runner is at time t with $\theta(2\pi) = \theta(0)$ as a runner who is just starting is in the same position as a runner who has completed exactly one lap. The phase of the oscillator therefore describes the proportion (if we divide it by 2π) of the way around the track the runner is in terms of time.

Example 5 The Stuart–Landau (SL) oscillator can be written as a system of two differential equations using polar coordinates. As written below, the SL oscillator has a stable limit cycle flowing counter-clockwise on the unit circle with period 2π. This is a mathematical description of the example of the runner described above with added information describing what happens when the runner is inside or outside of the track.

$$\frac{dr}{dt} = r - r^3$$

$$\frac{d\theta}{dt} = 1. \tag{7}$$

Exercise 14 Show that the SL oscillator has an unstable equilibrium at the origin and that the unit circle is asymptotically stable limit cycle (find and classify all equilibria of the differential equation $\frac{dr}{dt}$).

Note that for now we are only concerned with behavior on the limit cycle, we save discussion for behavior off the limit cycle for the next two sections. For the SL oscillator, the convention is to define the initial phase $\theta = 0$ as the point on

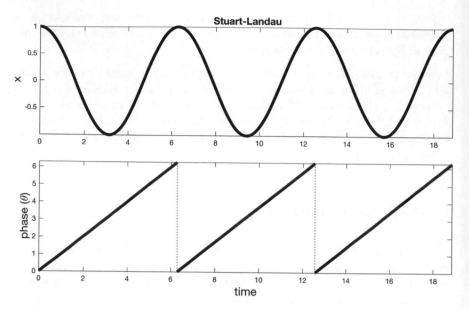

Fig. 4 Top panel is a time-series plot of x-component of the Stuart–Landau oscillator. The bottom panel is a time-series plot of the phase, θ

the plane where $x = 1$ and $y = 0$, though that is arbitrary. Since the change in phase is constant (the runner is going a constant speed when on the circular track in the SL oscillator), phase can be written as $\theta(t) = t$ (mod 2π). This can be seen in Fig. 4 where the phase increases linearly, resetting every 2π time units, while the runner's position on the x-axis is the cosine function. Returning to our track-running metaphor, the SL oscillator is the example where a runner is running at a constant speed around the unit circle (left column of Fig. 5).

We can describe all of the predator–prey oscillators previously discussed by just the phase (so long as the system is on a periodic orbit) as opposed to two variables describing the populations of the prey and the predator. Just like the SL oscillator has the property that $\theta(0) = \theta(2\pi)$, predator–prey oscillators with a period of T necessarily have the property that $x(0) = x(T)$ and $y(0) = y(T)$. Unlike the SL oscillator, ecological and other biological oscillators do not move around the phase plane at constant speed. Figure 5 shows phase-plane solutions to the SL, LV, and RM models. Squares are plotted every 1/5 of the period to demonstrate that the speed around the periodic orbit is constant for the SL model, but not for the LV or RM models. Returning to our running metaphor, the LV and RM models have irregularly shaped tracks and a runner whose speed changes depending on where they are in the track. Notice that for the parameter values in Fig. 5, the LV model is slower when both populations are low and that the prey population changes very little between boxes 3 and 4.

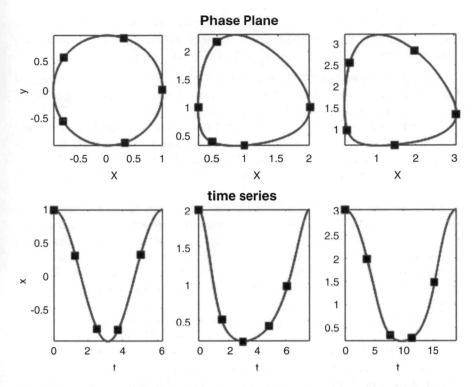

Fig. 5 The phase plane and time-series plots for the Stuart–Landau oscillator (left), Lotka–Volterra model (middle), and Rosenzweig–MacArthur model (right), with parameter values from Fig. 2. We choose initial conditions as the maximum value of the x-component for each model. The squares are plotted every 1/5 increment of the period

3.1 Isochrons

The Greek word isochron is translated as "equal time" and refers to the set of initial conditions of an oscillator that result in the same phase of an oscillator after an infinite amount of time. The notion of isochrons allows us to extend our notion of phase to points off of the limit cycle. Returning again to our running metaphor, imagine two runners whose speeds are dictated by a system of differential equations such as the SL or RM models. If the two runners start in different positions (with at least one of them not initially on the limit cycle), but eventually end up in the same position, then they started on the same isochron. An isochron therefore consists of all starting running positions where the runners eventually end up at the same place. More plainly, the points (x_1, y_1) and (x_2, y_2) are on the same isochron of an oscillator if they approach each other after a long time.

Definition 7 Consider a point $(x_1(0), y_1(0))$ on an asymptotically stable limit cycle and another point $(x_2(0), y_2(0))$ not on the limit cycle. These points are on the same isochron if $(x_2(t), y_2(t)) \to (x_1(t), y_1(t))$ as $t \to \infty$.

Every point on an individual isochron has the same phase, therefore two points that are on the limit cycle cannot be on the same isochron as they have different phases. The notion of isochrons cannot be applied to systems like the LV model and the frictionless spring-mass system (as opposed to the SL and RM oscillators) because while they are oscillators, they do not have an asymptotically stable limit cycle. Consider a runner obeying the LV model. If the runner is pushed off the track, she will continue to run in a loop, but would never return to the track (top panel in Fig. 3).

Unfortunately, analytically calculating isochrons is very difficult for even seemingly simple systems. The SL oscillator is an exception in that it has radial isochrons since the rate of change of the phase is always constant, see Fig. 6. We provide a MATLAB program that numerically computes the isochrons of an oscillator [23] (see Appendix 1 for a description on how this program works). The phase plane for any system with a stable limit cycle is dense with isochrons. In Fig. 6, we plot ten equally spaced in time isochrons for the SL and RM model.

Figure 7 shows the response of the SL oscillator to both horizontal and vertical perturbations. In the left column, shifting the system horizontally while starting at the point $(1, 0)$, leaves the system on the same isochron and therefore they approach each other as time increases. In the right column, shifting the system vertically advances (red) or delays (blue) the phase. This is illustrated in the bottom right corner of Fig. 7 by a horizontal shift in the phase of the oscillator, called a phase shift.

With ten equally spaced (in terms of time) isochrons plotted, the space between any two isochrons is one-tenth of the period. The SL oscillator has equally spaced isochrons which indicates that the system is moving around the oscillator at a constant speed. More tightly packed isochrons indicate that the system is moving slower in that region, while more spaced out isochrons indicate a faster moving section. The isochrons for the RM model in Fig. 6 indicate a system that is moving faster in the top right corner when the prey and predator populations are relatively large and slower on the left side when the prey population is small. Isochron figures for RM model can be found in [4] for additional parameter values.

Exercise 15 What important feature do we notice is different between isochrons coming from a system that produces rapid increases and decreases in prey population versus those that come from a system with smoother oscillations (see Exercise 10)? For the RM model, either numerically compute the isochrons yourself or refer to [4].

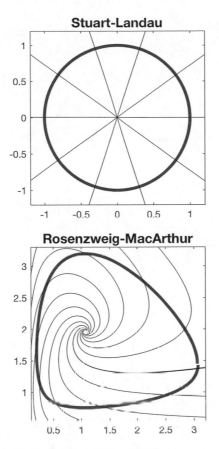

Fig. 6 Ten isochrons plotted in black along with the stable limit cycle in red for the Stuart–Landau oscillator and the Rosenzweig–MacArthur model with parameters as in Fig. 2

3.2 The Effects of Perturbations on Phase

Return to our runner and imagine she is bumped (perturbed) so that she is now a few feet inside the track on the grass, like the inner curve of the top-left panel of Fig. 7. Instead of immediately returning to the track, she slowly winds back out to the track (see $\frac{dr}{dt}$ in Eq. (7)). The phase response curve (PRC) measures how far she is ahead or behind of where she would have been had she not been bumped. Note that the PRC may depend on her phase when she was bumped and on how far off of track she was bumped.

Definition 8 The phase response curve measures the transient change in the period of an oscillator induced by a perturbation as a function of the phase at which it is received and of the magnitude of the perturbation.

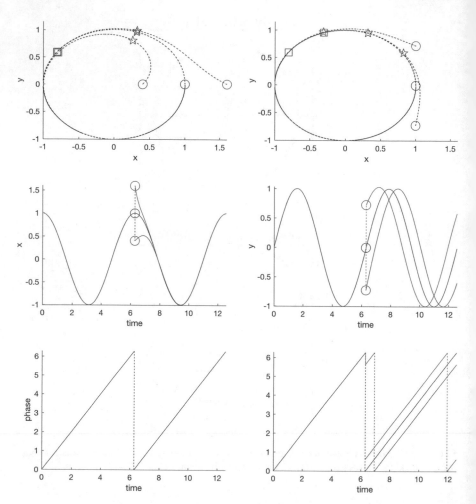

Fig. 7 Simulations of the Stuart–Landau oscillator starting with initial condition $(1, 0)$ and then perturbed after the completion of one cycle. The open circles refers to the position after the perturbation, the star $1/5$ of an oscillation later, and the square after $2/5$ of an oscillation later. The left column describes perturbations of x leading to positions $(0.4, 0)$ blue, $(1, 0)$, black, and $(1.6, 0)$ red, while the right column describes perturbations of y such that the positions are $(1, -\tan(\pi/5))$, blue, $(1, 0)$, black, and $(1, -\tan(\pi/5))$, red. The top row is the phase plane with the values for x and y plotted, the middle row shows the time-series plot of the perturbed variable, and the bottom row is the phase plotted as a function of time. The horizontal perturbations kept the system on the same isochron, and therefore the three points converged and the phase was unchanged. The vertical perturbation of $\pm \tan(\pi/5)$ advance or delays the phase by one isochron (6), or exactly $1/10$ of a cycle

In this chapter we only consider instantaneous perturbations, i.e., the runner experiences a one-time shove, as opposed to a longer lasting shouldering out of the way. As defined above, the PRC is a function of two variables and is defined only on the limit cycle. Let the vector $\vec{\epsilon}$ be the magnitude and direction of the perturbation:

$$PRC(\theta; \vec{\epsilon}) = \theta_{new} - \theta. \tag{8}$$

Similarly to isochrons, finding the PRC analytically is usually not possible, however, as with isochrons, we can calculate the PRC for the SL oscillator.

Exercise 16 Use trigonometry to find an implicit expression for the new phase, θ_{new}, of the SL oscillator if it is perturbed horizontally with magnitude b at phase θ.

The solution to the above exercise is worked out in [12] (they refer to it as the Poincaré oscillator):

$$\cos(2\pi\theta_{new}) = \frac{b + \cos 2\pi\theta}{\sqrt{1 + 2b\cos 2\pi\theta + b^2}}.$$

To numerically approximate the PRC, we perturb the system, run time forward, and then compare the phase of the perturbed system (θ_{new}) to the unperturbed system (θ). See the Appendix 1 for more details on the numerical simulations. The PRC is positive at a certain phase if a perturbation at that phase leads to a phase advance (red curves in the right column of Fig. 7), the PRC is negative if the perturbation leads to a phase delay (blue curves in the right column of Fig. 7), and it is zero if the perturbation does not affect the phase. The horizontal perturbations of the SL model when $\theta = 0$ shown in the left column of Fig. 7 did not change the phase of the runner, i.e., the runner ends up in the same position whether she was perturbed to the right 0.6 units, to the left 0.4 units, or not at all. Another way to determine the PRC is to measure how many isochrons (if you plot a finite number of isochrons) the system crosses due to the perturbation. A horizontal perturbations of the SL when $\theta = 0$ (or $\theta = \pi$) leaves the solution on the same isochron, as can be seen in the top-left panel of Fig. 6. Therefore the $PRC = 0$ for those perturbations, with the exception that if the perturbation is large enough, the system jumps past the unstable equilibria at the origin and speeds up an entire half cycle (Fig. 8).

Consider the perturbation in the vertical direction by $\pm\tan(\pi/5)$ in the right column of Fig. 7. In the red curve, the oscillator is perturbed exactly one isochron ahead of where it was before the perturbation (see the isochrons of the SL model in the top panel of Fig. 6). Since there are ten isochrons this is a phase advance of 10% of the phase (top panel of Figs. 6 and 8). Since the period of the oscillator is equal to 2π, this corresponds to a PRC value of $2\pi/10$. In general, when the PRC has a large positive value perturbations will increase the phase of the oscillator substantially, when the PRC has a large negative value a perturbation will delay the phase, and when the PRC has a value close to zero, a perturbation will leave the phase nearly unchanged.

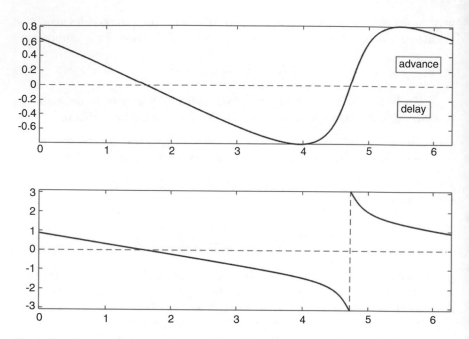

Fig. 8 The phase response curve for the Stuart–Landau oscillator for vertical perturbations. The top figure has a perturbation magnitude of $\tan(\pi/5)$ and the bottom figure has a perturbation of 1.2 in the vertical direction. When the phase of the oscillator is $3\pi/2$ the large vertical perturbation goes from slowing down the phase by half the period to speeding it up by half the period

Exercise 17 Calculate by hand the magnitude of perturbation necessary to advance the phase of the SL oscillator by 10% if the perturbation increases x and y the same amount (i.e., the angle of the perturbation is $45°$ in the phase plane) and it occurs at the point $(1, 0)$.

Exercise 18 Calculate the PRC for small positive prey perturbation of the RM model for parameter values leading to smooth oscillations and again for parameters leading to rapid increases and decreases of prey populations (see Exercise 10) Describe the differences in the PRCs that come from prey perturbations of these two different types of behavior. What can we conclude about the differing effects on the populations of this perturbation?

The infinitesimal PRC (iPRC) is the limit of the PRC as the magnitude of the perturbation goes to zero. In theory, the iPRC can be found numerically using the same method as the regular PRC, however, numerical errors mean that great care must be used for these calculations. Fortunately the iPRC can be computed by numerically solving the so-called adjoint equation [5]. Unfortunately, the mathematics behind these calculations is beyond the scope of this chapter and is difficult for most undergraduate students (see chapter 10 of [5] for three different derivations of the iPRC). We include this discussion of the iPRC because the program XPP has a built in routine that approximates the iPRC, which may be more approachable then the MATLAB code for those who want to limit coding and computational time. Detailed step-by-step instructions on using XPP can be found in the Appendix 2 of this chapter and in [6]. The iPRC (as opposed to the PRC) is useful for understanding the effect of frequent and very small perturbations and leads to the beautiful study of "weakly coupled oscillators" [14]. In predator–prey models, small perturbations can represent small weather effects or the effect of small amounts of migration in spatially separated populations [3]. We can also use results from studies of small perturbations to approximate the effect of somewhat larger perturbations.

4 Suggested Research Projects

As discussed throughout this chapter, finding the isochrons and the PRC of an oscillator is the key to understanding the effect of external perturbations on these systems. Ecological populations are regularly hit with both small and large perturbations in the form of climatic events, migration, and other external shocks. To that end, the projects all involve calculating isochrons and/or PRCs. Unfortunately analytically finding the PRC or isochrons of even relatively simple systems is often very difficult. Because of this, each of the projects includes at least some computational work. We provide several MATLAB routines to help the student get started [23] and give instructions on using the program XPP as it requires minimal actual programming. This section is structured so that the projects are written in increasing order of difficulty.

> **Research Project 1** Numerically calculate the PRC of the Rosenzweig–MacArthur model for different magnitudes and types of perturbations.

The PRC for small and large prey-only perturbations in the RM model was explored in [3] and [4], respectively. PRCs for predator-only perturbations or perturbations for both populations have not been explored. Many realistic perturbations in ecology would likely include effects on both populations as it may be due to severe weather, for example. Therefore exploring how sensitive the PRC is to perturbations of both populations may shed more insight than the existing work. Try varying the magnitude of the perturbations on each population. Think of a specific predator–prey system, what type of perturbation would affect the prey population more than the predator population, or vice versa? Also, compare the difference between perturbations that are proportional to the size of the populations and perturbations that are of fixed size.

Research Project 2 Numerically calculate the PRC and the isochrons for various parameter sets of a predator–prey model with logistic growth of the prey, linear death of the predator, and predation following a Type III functional response.

As we saw in Exercise 11, different ecological populations are better described by different functional responses, both for prey-growth and predation. The PRC in the RM model for a prey perturbation has been described in [3, 4]. However, we are not aware of any work on this for an ecological oscillator where predation follows a Type III functional response. As many predators have predation functions that follow this functional response, calculating the PRC of these systems will better describe the effect of a perturbation on these systems. Before attempting to calculate the PRC and isochrons, be sure that the parameter values that you choose lead to a stable limit cycle.

Research Project 3 Find your own oscillator and calculate the PRC and isochrons.

There are many oscillators in biology, chemistry, and physics that were not described in this chapter. Pick your own system (just make sure it has an asymptotically stable limit cycle and that the oscillation is not forced), and numerically calculate the PRC and isochrons. What do these results tell us about the effect of relevant external perturbations to the system? Interesting examples include the Selkov Model which describes glycolysis (see Appendix 4), as well as The Brusselator (see Appendix 4) and The Oregonator (a system of three differential equations) [21], which model autocatalytic chemical reactions.

Research Project 4 Explore the effect of periodic forcing on an ecological oscillator.

The perturbations explored so far in this chapter were all one-time events. However, many ecological systems are subject to regular repeated perturbations, notably seasonal effects. Examine the effect that periodic forcing has on the RM model and how a different frequency of forcing can have different effects on the oscillator. Start by having an external perturbation hit with the same frequency of the oscillator. What is the effect of these perturbations on the period and amplitude of oscillations? How does varying the timing within the cycle and the magnitude of the perturbation change these results. Does knowing the PRC of this oscillator help predict the effect of this periodic forcing? Next try varying the frequency of the perturbation so that it occurs just a little more (or less) frequently then oscillator completes a cycle. How does these perturbations effect the period and amplitude of oscillation? What if the periodic forcing occurs every other oscillation, or twice per oscillation? More generally, explore the effect that these perturbations have on the period and the amplitude of the oscillator as you vary their frequency and magnitude.

Research Project 5 Use the PRC to inform the step-size of an adaptive time step numerical differential equation solver. Compare the computational efficiency and accuracy of this method to other popular methods such as Runge–Kutta 4 and MATLAB's ode45.

Euler's method and Runge–Kutta 4 are two of the most commonly used explicit methods for approximating solutions to differential equations such as the one described in this chapter. Explicit methods such as these employ a fixed time step for calculations. Depending on the nature of the differential equation, it can be more computationally efficient to use a method that has an adaptive time step, like the Dormand–Prince method that MATLAB uses in its ode45 method. Practically, use a very small step-size or error threshold to accurately approximate the solution to one of the above oscillators with a stable limit cycle. Calculate the PRC for a small perturbation of the oscillator and make the step-size of your differential equation solver a function of the PRC. Compare the accuracy and computation time of your solver to other existing solvers. Probably the most straightforward comparison would be how accurate the method approximates the period. If the system is a relaxation oscillator, does that affect your results? This project is more computationally intensive than the other and probably requires some previous programming experience as well as experience with differential equation solvers or a numerics course.

Acknowledgements Thanks to Joshua Goldwyn for helping to brainstorm potential project ideas involving the phase response curve and for helpful comments on earlier drafts. Additional thanks to Timothy Lewis, Meredith Greer, and Per Sebastian Skardal for many helpful ideas for improvement from previous drafts.

Appendix 1: MATLAB Code

Some MATLAB code is provided to assist the reader in getting started on the exercises and the projects. See webpage [23] to download the MATLAB files that:

- Calculate isochrons of an oscillator—adapted from [5].
- Calculate the PRC of an oscillator—adapted from [5].

- Simulate the SL, LV, and RM models.

Summary of isochron.m

- You must provide the period of the oscillator and an initial condition on the asymptotically stable limit cycle
- Input the number of isochrons desired in the output figure
- Choose the time step of integration and the spatial resolution of the figure
- The program approximates the rest of the limit cycle
- The program then picks points near the center of the limit cycle and integrates backwards to find other points on the isochron
- The program plots the limit cycle and the isochrons

Summary of prc.m

- You must provide the period of the oscillator and an initial condition on the asymptotically stable limit cycle
- Input the number of points on the phase that the program will calculate the prc for
- Choose the magnitude and direction of the perturbation
- Choose the number of cycles necessary for the system to return sufficiently close to the limit cycle after the perturbation
- The program approximates the rest of the limit cycle and finds the number of time steps to a chosen point on the limit cycle (usually easiest to pick a point where the prey population is maximized)
- The program then runs the solution for several cycles starting from the perturbed location
- The program then measure how far this solution is from where it would have been without a perturbation
- The program then plots the PRC

Appendix 2: XPP Code

For those not familiar with or wanting to learn MATLAB, this appendix provides instructions for using free program XPP. This program was developed by Bard Ermentrout and can be used to approximate solutions to differential equations, graph these solutions, and calculate the iPRC. (XPP has many other tools and applications for dynamical systems not related to this chapter). Instructions for downloading and using XPP can be found at: http://www.math.pitt.edu/~bard/xpp/xpp.html. Two important pluses for XPP are that it is free to download and requires little to know actual programming. It is essentially a java applet. The major downside to XPP is that it was originally developed for MSDOS, so the interface can be a bit cumbersome at first. For more in-depth details and tutorial on XPP, including all the uses of XPP not relevant to this chapter, see the instruction book [6].

When opening XPP, you need to open it with a system of differential equations already chosen. These equations must be in the form of "*.ode." Figure 9 shows the

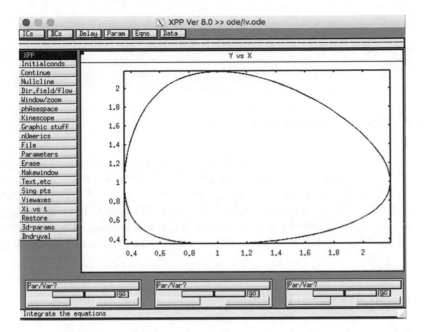

Fig. 9 Screen shot of the phase-plane solution to the Lotka-Volterra model as calculated by XPP

phase-plane solution to the LV model plotted. The clickable top row can be used to edit the initial conditions, "*ICs*," the parameter values, "*Param*," or to look at the approximate solution in numeric form. "*Data*." You can also check but not edit the system of differential equations by clicking on "*Eqns*." In order to actually compute anything you need to navigate through the left menus. These menus can be accessed either by clicking on them or by using hot keys with are just the capital letter. The below list includes instructions to the most relevant commands for this chapter.

- To simulate a model: "*Initialconds*," "*Go*"
- To graph the phase plane: "*Viewaxes*," "*2D*," and then adjust the axis as you prefer.
- To output the results to a .csv or other file: click on "*Data*" on the top and then "*Write*."
- To calculate the iPRC

    - First limit the time to exactly one period: "*nUmerics*," "*Total*," then enter the length of one period.
    - Return to the main screen then: "*nUmerics*," "*Averaging*," "*Adjoint*"

- XPP is not set up to calculate the PRC with a non-infinitesimal perturbation. This can be done manually by perturbing the system off its limit cycle, simulating the system, then seeing how far ahead (or behind) the perturbed system is from the unperturbed system.

In addition, extra XPP files are provided at [23] for the following models:

- LV model
- RM model

Appendix 3: Desmos Graphs

To help understand the structure of the nullclines and when certain predator–prey models will oscillate, we provide links to two different Desmos pages on the webpage [23]. One Desmos page has a graph of the nullclines of the Romeo–Juliet model and the other has graphs for the RM model, and the RM but with predation following a Type III functional response. Each graph has sliders so that the viewer can see the effect of each parameter value on the shape and location of the nullclines.

Appendix 4: Other Oscillators

The Selkov model for glycolysis can be written as:

$$\frac{dx}{dt} = --x + ay + x^2y \tag{9}$$

$$\frac{dy}{dt} = b - ay - x^2y. \tag{10}$$

Like the RM model, this system undergoes a Hopf-bifurcation. The parameter regime where oscillations occur can be found using nullcline analysis and simulations, as described previously in this chapter.

The Brusselator is a model for certain autocatalytic reactions including the Belousov–Zhabotinsky reaction. It can be written as:

$$\frac{dX}{dt} = A + X^2Y - BX - X \tag{11}$$

$$\frac{dY}{dt} = BX - X^2Y. \tag{12}$$

Let $A, B > 0$. This system also experiences a Hopf-bifurcation. The bifurcation occurs when $B = 1 + A^2$. This system often has relaxation-oscillator type behavior, so great care needs to be made to avoid numerical error if you use prc.m or isochron.m.

Appendix 5: Non-dimensionalization and Relaxation Oscillators*

This is an optional section that describes relaxation oscillators and the process of transforming a dimensional system like the LV and RM models described above into dimensionless systems. Non-dimensionalizing a system of differential equations reduces the number of parameters in a model to make techniques like finding the eigenvalues of a Jacobian and phase-plane analysis easier to compute and interpret. It is also useful in helping to determine the relationship between parameter values that yield different qualitative dynamics. A quick overview of non-dimensionalization can be found in [7] and [9]. For a more in-depth explanation, try [13] or appeal to the Buckingham π theorem.

The LV model has four parameters, r, a, c, m. Non-dimensionalization is an important technique that reduces the number of parameters in a model and therefore makes the above calculation easier. Rewriting Eqs. (4) and (5) with the following substitutions: $X = \frac{ca}{r}x$, $Y = \frac{a}{r}y$, $\tau = rt$, and $\mu = \frac{m}{r}$ yields

$$\frac{dX}{d\tau} = x - xy$$

$$\frac{dY}{d\tau} = xy - \mu y. \tag{13}$$

Exercise 19 Find all the equilibria in the system of differential equations (13). Classify the stability of each equilibrium by evaluating the Jacobian at the equilibrium point. You should get analogous results to what we found in the original dimensional model. Find the implicit relationship between Y and X and graph it.

Exercise 20 Find a different non-dimensional substitution of the LV model that also reduces the number of parameters in the system to one.

The FitzHugh–Nagumo model (FN) is a classic example of a relaxation oscillator (if $a = b = 0$ then Eq. (14) is the Van der Pol oscillator, described in detail Chapter 10 of [5]). The FN model is a simplified description of a spiking neuron with voltage described by $v(t)$ and the recovery variable described by $w(t)$:

$$\frac{dv}{dt} = v - \frac{v^3}{3} - w + I_{ext}$$

$$\frac{dw}{dt} = \frac{1}{\tau}(v + a - bw). \tag{14}$$

From the differential equations, we can see that a large τ indicates $v(t)$ changing much faster than $w(t)$. This difference in characteristic time scales between how quickly the two variables change describes a relaxation oscillator. Compare the top plots from Figs. 2 and 10. The time-series solutions for the LV model are more sinusoidal while the solution for $v(t)$ in the FitzHugh–Nagumo model is more jagged with periods of very fast increase or decrease representing the separation

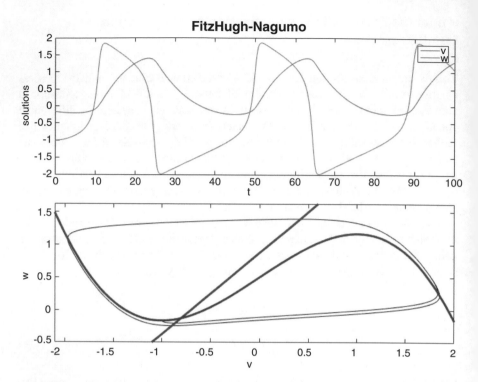

Fig. 10 Solutions for the FitzHugh-Nagumo model with $a = 0.7$, $b = 0.8$, $I_{ext} = 0.5$ and initial conditions $v(0) = -1$ and $w(0) = -1/6$. The top figure shows the time-series solutions. The bottom figure shows the phase plane with the solution in black. The w-nullcline is the blue line and the v-nullcline is the red cubic

in time scales. To further illustrate this, consider the initial conditions chosen at the local minimum of the cubic $(-1, -1/6)$ as in the bottom panel of Fig. 10. As time increases, $v(t)$ will rapidly increase while $w(t)$ will change very little. This is illustrated in Fig. 10: the top panel is the time-series and the blue curve, $v(t)$, increases very rapidly near $t = 10, 50, 90$, and in the bottom panel which is the phase plane and the black curve moves nearly straight to the right very quickly. After moving to the right, the system will slowly work its way up the right side of the red cubic until it hits the local maximum $(1, 7/6)$, this can be seen in the slower increase in the $w(t)$ and decrease in $v(t)$ in the top figure. After the system reaches the local max of the cubic, it will quickly shoot across to the left side of the cubic, this corresponds to the very rapid decrease in $v(t)$ in the top figure (blue). We will see in subsequent sections that this difference in time scales has important ramifications on the phase response curve.

Exercise 21 Complete the nullcline analysis by finding the direction of the solution in each section between the nullclines. Given the separation of time scales, what has to be true about the intersection of the nullclines for the system to have a stable limit cycle (as opposed to a stable equilibrium at the intersection)?

As with the LV model, we can reduce the number of parameters in the RM model by non-dimensionalizing [3]:

$$\frac{dX}{d\tau} = \frac{1}{\epsilon}\left(X(1 - \alpha X) - \frac{XY}{1 + X}\right)$$

$$\frac{dY}{d\tau} = \frac{XY}{1 + X} - \mu Y. \tag{15}$$

Much like the parameter τ in Eq. (14), ϵ represents the difference in time scales between the prey and the predator, with $\epsilon < 1$ indicating a prey population which responds faster than the predator population. By graphing the nullclines and performing phase-plane analysis, we find that conditions $\alpha < 1$ and $\mu < \frac{1-\alpha}{1+\alpha}$ are required for a stable limit. If both inequalities hold, the system has three equilibria, the trivial equilibrium at extinction, and the prey only equilibria with $X^* = 1/\alpha$ and predator extinct, both of which are saddles, and the coexistence equilibrium which is an unstable spiral.

Consider Eq. (15) with a very small ϵ. The prey population would change much faster than the predator population and a time-series graph of the prey population would look less like a smooth sine curve that we see with the hare and lynx and would have very rapid increases followed by slower changes as the predator reacts slowly to the new prey population. This difference in time scales is found in a variety of insect outbreak populations, a famous such example is the that of the spruce-budworm and the balsam fir and is studied in [1].

Exercise 22 Plot the solutions for $X(\tau)$ and $Y(\tau)$ for $\epsilon = 0.1$, $\alpha = 0.4$, $\mu = 0.4$ as in the bottom panel of Fig. 3. Decrease ϵ and observe the prey population exhibiting sharper "boom-bust" type dynamics. Can you get this same type of behavior by varying α and/or μ? Interpret why (Hint: think about the nullclines).

References

1. Ludwig, D. and Jones, D.D. and Holling, C.S., Qualitative analysis of insect outbreak systems: the spruce budworm and forest. J. Animal Ecol. **47**, 315–332 (1978)
2. Moran, PAP. The statistical analysis of the Canadian lynx cycle. II. Synchronization and meteorology. Aust. J. Zool. **1**, 291–298 (1953)
3. Goldwyn, EE., Hastings A. When can dispersal synchronize populations? TPB. **73**,395–402 (2008)
4. Goldwyn, EE., Hastings A. The roles of the Moran effect and dispersal in synchronizing oscillating populations. JTB. **289**,237–246 (2011)
5. Izhikevich, E., Dynamical Systems in Neuroscience: The Geometry of Excitability and Bursting. The MIT Press (2007)
6. Ermentrout G.B., Simulating, Analyzing, and Animating Dynamical Systems. SIAM (2002)

7. Strogatz, S.H., Nonlinear Dynamics and Chaos, 2nd edition. CRC Press (2015)
8. Strogatz, S.H., Sync: the emerging science of spontaneous order. Hyperion (2003)
9. Hastings, A., Population Biology: Concepts and Models. Springer-Verlag (1996)
10. Adler, F.R., Modeling the Dynamics of Life, 3rd edition. Brooks/Cole (2013)
11. Winfree, A.T., The Geometry of Biological Time 2nd edition. Springer-Verlag (2001)
12. Glass, L. , Mackey, M., From Clocks to Chaos. Princeton University Press (1988)
13. Holmes, M.H., Introduction to the Foundations of Applied Mathematics. Springer (2009)
14. Schwemmer, M.A., Lewis, T.J. The Theory of Weakly Coupled Oscillators. Phase Response Curve in Neuroscience. Springer (2012)
15. Hoppensteadt, F.: Predator-Prey Model http://www.scholarpedia.org/article/Predator-prey_model. Cited 26 Oct 2018
16. Izhikevich, E.M., FitzHugh, R.: FitzHugh-Nagumo Model http://www.scholarpedia.org/article/FitzHugh-Nagumo. Cited 26 Oct 2018
17. Izhikevich, E.M.,Ermentrout, G.B. Phase Model http://www.scholarpedia.org/article/Phase_model. Cited 26 Oct 2018
18. Josic, K., Shea-Brown, E.T., Moehlis, J: Isochron http://www.scholarpedia.org/article/Isochrons. Cited 26 Oct 2018
19. Canavier, C.C..: Phase Response Curve http://www.scholarpedia.org/article/Phase_response_curve. Cited 29 Oct 2018
20. Ermentrout, G.B.: XPPAUT http://www.scholarpedia.org/article/XPPAUT. Cited 26 Oct 2018
21. Field, R.J.: Oregonator http://www.scholarpedia.org/article/Oregonator. Cited 12 June 2019
22. Kuznetsov, Y.A.: Andronov-Hopf Bifurcation http://www.scholarpedia.org/article/Andronov-Hopf_bifurcation. Cited 21 June 2019
23. Goldwyn, E.E.: MATLAB and XPP code for Phase Sensitivity of Ecological Oscillators https://sites.up.edu/goldwyn/phase-sensitivity-of-ecological-oscillators/. Cited 26 Oct 2018

Exploring Modeling by Programming: Insights from Numerical Experimentation

Brittany E. Bannish and Sean M. Laverty

Abstract This chapter aims to provide students the background necessary to use computational methods and numerical experimentation to pursue mathematical research. We provide a brief refresher on differential equations, model building, and programming, followed by in-depth discussions of how numerical experimentation was critical to three undergraduate research projects involving parasite transmission in feral cats, the role of protein regulation in establishing circadian rhythms, and treatment costs in an influenza outbreak. Code (written in R) is provided for most of the models and nearly all of the figures presented in the chapter, so that the reader may learn the programming as they go along. We also propose ideas for extending the research projects presented in the text. Both deterministic and stochastic models are presented and the critical role of computational methods in obtaining solutions is illustrated.

Suggested Prerequisites *Exposure to mathematical modeling is encouraged and an interest in biology or the life sciences is beneficial. Basic knowledge of differential equations is required. Some exposure to computer programming is helpful but not required.*

1 Introduction

This chapter aims to provide students the background necessary to use computational methods and numerical experimentation to inform their mathematical

---

Electronic supplementary material The online version of this chapter (https://doi.org/10.1007/978-3-030-33645-5_4) contains supplementary material, which is available to authorized users.

B. E. Bannish · S. M. Laverty (✉)
University of Central Oklahoma, Edmond, OK, USA
e-mail: bbannish@uco.edu; slaverty@uco.edu

research. We also hope this chapter provides a convincing argument for the role of programming in mathematical modeling. Mathematical models of life science systems are presented. Our focus is less on building the models themselves (we assume students have some background knowledge on model building), and more on how computing can be used to generate results and inform the direction of research. Most models of real-world phenomena are too complicated (or impossible) to solve by hand, but all hope is not lost; computers can be a researcher's best tool. Numerical experimentation can provide "quick" insight about the model being studied, and can confirm (or deny) hypotheses about the basic mechanisms underlying the system of interest. The results of numerical experimentation can suggest directions to pursue for more rigorous mathematical analysis, and for new model modifications.

In this chapter we give a brief refresher of mathematical modeling (Sect. 1.2) and programming (Sect. 1.3), then focus on the myriad ways numerical experimentation can improve model analysis and solving. We introduce basic programming and analysis of math models with the simple predator–prey model (Sect. 2). From there we scale up to more complicated models that our own students have studied: parasite transmission in wild cats (Sect. 3), the role of protein regulation in establishing circadian rhythms (Sect. 4), and treatment costs in an influenza outbreak (Sect. 5). We conclude with an overview of simple stochastic models and stochastic extensions to the predator–prey and influenza models presented earlier (Sect. 6). We include several open-ended projects that students should be able to make some headway on through numerical experimentation.

With few exceptions, complete, annotated code is given in the text that can be used to reproduce the contents of many figures. In a few instances where the code is bit too long or complicated to include in the text itself, we provide information in the Appendix. We do encourage typing at least some of the code from the printed page to your computer (rather than copy and pasting); this will allow you to make errors and practice debugging and editing, which are useful skills for this type of mathematical research. Additionally, all code is available at the chapter's GitHub page (https://github.com/seanteachesmath/FURM), where it is organized in folders by project.

1.1 Differential Equations Refresher

We start with a brief review of differential equations. For a more thorough treatment, we direct the reader to several excellent books on the subject [5, 7]. A *differential equation* is an equation that contains derivatives of an unknown function. For example,

$$\frac{dP}{dt} = \lambda P \tag{1}$$

is a differential equation that says the derivative of $P(t)$ is equal to constant λ times $P(t)$. The *solution* of the differential equation is the function $P(t)$ that satisfies Eq. (1). You can verify that $P(t) = Ce^{\lambda t}$, for arbitrary constant C, is a solution. The particular value of C can be determined by an *initial condition*, $P(t_0) = P_0$. An initial condition gives the value of the unknown function at a single time point. For example, if we know that $P(0) = 15$, then the solution to the *initial value problem* (differential equation along with an initial condition) is $P(t) = 15e^{\lambda t}$. For a quantity that changes in time, the differential equation describing that change is often called a *dynamical system*. For example, Eq. (1) is a dynamical system for P.

Exercise 1 Verify that $P(t) = 15e^{\lambda t}$ is a solution to the differential equation given in Eq. (1), with initial condition $P(0) = 15$.

A differential equation is *linear* if all terms are linear in the unknown function and its derivatives. For example, Eq. (1) is linear, but $\dfrac{dP}{dt} = rP\left(1 - \dfrac{P}{K}\right)$ is *nonlinear* because of the P^2 term on the right-hand side. In general, it is much easier to solve linear differential equations than nonlinear differential equations; nonlinear differential equations are often "solved" numerically on a computer.

Exercise 2 Identify which differential equations are linear and which are nonlinear. In all examples, P and t are variables; all other letters represent constants.

1. $\dfrac{dP}{dt} = -r(P - K)$

2. $\dfrac{dP}{dt} = rP$

3. $\dfrac{dP}{dt} = rP\left(\dfrac{P}{A} - 1\right)\left(1 - \dfrac{P}{K}\right)$

4. $\dfrac{dP}{dt} = rPe^{-mP}$

5. $\dfrac{dP}{dt} = a + rP$

A *system of differential equations* is two or more differential equations in two or more unknown variables. For example,

$$\frac{dP}{dt} = r_1 P\left(1 - \frac{P}{K_1}\right) - \delta_1 PQ \tag{2}$$

$$\frac{dQ}{dt} = r_2 Q\left(1 - \frac{Q}{K_2}\right) - \delta_2 PQ \tag{3}$$

with $r_1, r_2, \delta_1, \delta_2, K_1, K_2$ constant, is a nonlinear, two-dimensional (2D) system of differential equations. It is nonlinear because of the P^2, Q^2, and PQ terms, and it is 2D because there are two equations. Additionally, this system of equations is said to be *coupled* since P shows up in the $\frac{dQ}{dt}$ equation and Q shows up in the $\frac{dP}{dt}$ equation. In other words, you cannot solve one of the differential equations in isolation; you must know both P and Q simultaneously. Systems of equations can be even more difficult than single differential equations to solve analytically, so we rely heavily on graphical and computational techniques, which are introduced in Sect. 1.3.

1.2 Mathematical Modeling Refresher

Differential equations are often used to model biological or physical systems. A *mathematical model* is an equation or system of equations that describes a real-world process. For example, Eq. (1) is a model of a population experiencing exponential growth (if $\lambda > 0$) or decay (if $\lambda < 0$). A number or constant in a model, like λ, is called a *parameter*. If we let the *variable* (or state variable) $P(t)$ describe the population at time t, then the model equation says that the change in population ($\frac{dP}{dt}$) is proportional to the current population (λP). Similarly, the system given by Eqs. (2) and (3) is a mathematical model of two species that compete for a resource. Let $P(t)$ be the population of species 1, and $Q(t)$ be the population of species 2. Then the model says each species grows logistically in the absence of the other species (the $r_1 P(1 - \frac{P}{K_1})$ and $r_2 Q(1 - \frac{Q}{K_2})$ terms), but that interactions between the species negatively affect both species (the $-\delta_i PQ$ terms). *Logistic growth* takes into account the fact that most populations have limited resources (food, water, space, etc.), and so are unable to grow exponentially forever. In Eq. (2), for example, K_1 represents the *carrying capacity*—the maximum population the environment can support. When P is small relative to K_1, the term $(1 - \frac{P}{K_1})$ is close to 1, and the population grows approximately exponentially. However, as P increases and gets closer to K_1, the term $(1 - \frac{P}{K_1})$ is close to 0, and the population growth rate slows to about 0. The last term in each equation is obtained using the *law of mass action* which says that the rate at which the species interact is directly proportional to their populations [25]. For example, if there are a lot of individuals of species 1, and only a handful of species 2, then the rate at which the two species interact will be smaller than if there were also a lot of species 2. For much more detail about model building and classical techniques of analysis, see Refs. [13, 24, 29, 30, 34, 35, 44] which are excellent resources for beginners and beyond.

When we set out to build a mathematical model, we start by learning as much as we can about the real-world system we are trying to explain. It is helpful to make note about what quantities are changing (these will be the variables in the model), and how the various components interact with each other. It is often critical to make some simplifying assumptions in order to keep the model small enough to be tractable. However, it is also important to put in enough detail that the model is a decent representation of reality. (Modelers have adapted the quotation attributed to Albert Einstein: "A model should be as simple as possible, but no simpler.")

Assuming we are studying a system that only changes in time (not also in space, for example), then we can model the change using the types of differential equations discussed above. We put the derivative on the left-hand side of the equation, and then on the right-hand side use the law of mass action, Michaelis–Menten kinetics [29], or any other appropriate argument to mathematically write all the ways the variable can change. For example, consider the spread of an infectious disease through a population. The population can be divided into two categories: susceptible people (S), who do not have the disease now but could contract it in the future; and infectious people (I), who have the disease and are able to spread it to others. Once a person recovers from the disease, they are immediately susceptible again; there is no period of immunity. Ignore all births and deaths, and assume that susceptible people can become infectious after having contact with an infectious individual. Then we can write the "SIS model" [32]

$$\frac{dS}{dt} = -\beta SI + \gamma I \tag{4}$$

$$\frac{dI}{dt} = \beta SI - \gamma I. \tag{5}$$

Equation (4) says that the number of susceptible people decreases (change in susceptible people, $\frac{dS}{dt}$, is negative) when susceptible people come into contact with infectious people and become infectious themselves ($-\beta SI$). The number of susceptible people increases (change in susceptible people, $\frac{dS}{dt}$, is positive) when infectious people recover and become susceptible again ($+\gamma I$). Similarly, in Eq. (5), there is an increase in infectious people when susceptible people become infectious ($+\beta SI$), and a decrease in infectious people when they recover ($-\gamma I$). The epidemiological and modeling literature is rich with variations of this type of infectious disease model. For further reading, see [2, 8, 26, 37].

The *SIS model* is an example of a *compartmental model*, in which we divide the total population into different "compartments" (in this case, a "susceptible" compartment and an "infectious" compartment), and draw arrows to indicate how individuals in the population move between compartments. All individuals in a given compartment are assumed to behave uniformly and have the same characteristics; there are no differences between individuals in a particular compartment. Figure 1 shows the compartmental SIS model for the parameters described in the previous paragraph. Most of the examples presented in this chapter can be described as compartmental models.

Exercise 3 Consider an infectious disease moving through a population. Individuals in the population can be susceptible (disease-free, but able to get infected), infectious (currently infected and able to spread the disease), or recovered (no longer infected and unable to be reinfected). Susceptible people can become infectious through contact with infectious individuals. Infectious people can recover at some fixed rate. Ignore all births and deaths. Write a model based on the given word story.

Exercise 4 Consider the infectious disease model below, where R represents recovered people. Explain in words what each term of each equation means biologically. Can you think of a disease that might be modeled this way?

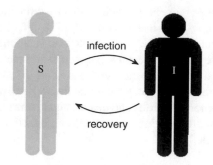

Fig. 1 Compartmental SIS model corresponding to Eqs. (4) and (5). Susceptible individuals move into the infectious compartment, at a rate βSI, proportional to the product of the numbers of susceptible and infectious individuals. Infectious individuals recover at a constant rate γ and the flow back to the susceptible compartment is given by the term γI

$$\frac{dS}{dt} = -\beta SI + \omega R \tag{6}$$

$$\frac{dI}{dt} = \beta SI - \gamma I \tag{7}$$

$$\frac{dR}{dt} = \gamma I - \omega R \tag{8}$$

We have, to this point, ignored a difficult part of modeling: determining realistic parameter values. To gain insight into the behaviors of the model using numerical approaches, we need to specify numerical values for the rates and interaction parameters (e.g., β and γ) in the model equations. This can often be a major hurdle to overcome at any stage of research, and depending on the level of experience of the researcher, can be an intimidating, or even off-putting, task. Often specifying parameters can be done by reviewing related models where similar parameters have been estimated or reported, which could require detailed reading of sometimes highly specialized literature. Care must be taken to ensure that parameter values from published sources apply to your model. Does that value reported in a paper about mouse experiments make sense in your model of humans, for example? These are important decisions to make. Reading the subject matter literature can often be inspiring or enlightening, but it can also be easy to fall down a rabbit hole and get lost in a trail of scientific papers. Strive for a good balance of reading enough of the literature to get ballpark values for some parameters, but not getting so bogged down in the literature that you do not have time for anything else.

Luckily, some parameters can be readily approximated. For example, taking into account the appropriate unit of measure, mortality rates of simple models are often approximated by taking the reciprocal of the lifespan of the organism in question. Otherwise, given no better starting point, we like to take the approach of setting all parameters to a value of one in order to ensure that the model is coded correctly (you can read much more about coding in Sect. 1.3 and throughout the rest of this chapter). This allows the researcher to execute code, or debug to isolate and correct errors, before refining and experimenting with parameters. It can be illuminating to increase or decrease parameters, one at a time, by powers of ten to see the effects on dynamics. From a purely mathematical and computational perspective, this could lead the researcher towards interesting behaviors of the dynamical system. From the mathematical biology perspective, this could help constrain parameters based on what model behaviors are realistic or not. To some degree, the uncertainty in some parameters is what provides an opportunity for experimentation when working with mathematical models, and is a way that mathematical modeling can benefit the life science communities.

1.3 Programming Refresher

As models are often systems of nonlinear differential equations that are hard or impossible to solve analytically, we rely heavily on the freely available, open-source program (and programming language) R for generating numerical and graphical results of our mathematical models [42]. We provide R code for the models discussed in this chapter, but the reader can adapt the code to work in their favorite software. For first-time coders, we recommend downloading R and using the provided code as a way to get familiar with mathematical software. For the projects highlighted in this chapter, we use built-in methods in the package deSolve [43]. Our emphasis is not on developing numerical methods, but instead on using existing, well-established methods to study our own mathematical models. Note that throughout this chapter we provide code that replicates the data in the figures, but not necessarily the aesthetics of the figures. We attempt to provide the minimal code necessary to get started on research projects, and leave it to the reader to learn more about plotting options. We speak briefly about these options in the Appendix.

Considering our simplest population model from Eq. (1), we will briefly illustrate the idea behind numerical algorithms to solve differential equations, before showing how the model can be solved with built-in methods. The most basic method for numerically solving a first-order differential equation is *Euler's method*. A detailed explanation of Euler's method can be found in any differential equations textbook [5, 7]; here we give the basic idea behind the algorithm, using Eq. (1) as our model. Euler's method uses the fact that over a short interval the tangent line to a function is a good approximation to the actual function. If we can link up tangent lines in a systematic way, we can produce a piecewise linear function that approximates the true solution.

Rewrite Eq. (1) as $\frac{dP}{dt} = f(t, P)$ where $f(t, P) = \lambda P$, and assume initial condition $P(t_0) = P_0$. Since the derivative at a point gives the slope of the tangent line at that point, we interpret $f(t_0, P_0)$ as the slope of the tangent line to P at point (t_0, P_0). We will discretize time into equal intervals of length h by computing $t_i = t_0 + ih$, $i = 0, 1, \ldots, n$. Then, we use the point-slope equation for a line to find the equation of the tangent line:

$$P = f(t_0, P_0)(t - t_0) + P_0. \tag{9}$$

If our second time point, t_1, is close enough to t_0, we can approximate the solution $P(t_1)$ using Eq. (9). Let P_1 denote the approximation to true solution $P(t_1)$, so

$$P_1 = f(t_0, P_0)(t_1 - t_0) + P_0 \tag{10}$$

$$P_1 = hf(t_0, P_0) + P_0, \tag{11}$$

where we used the fact that the length of the time interval, $t_1 - t_0$, is h. This process can be iterated as many times as desired. To find the approximate value of P at time t_2, call it P_2, we use

$$P_2 = hf(t_1, P_1) + P_1. \tag{12}$$

In general, to find the approximate value of $P(t_{i+1})$, compute

$$P_{i+1} = hf(t_i, P_i) + P_i. \tag{13}$$

Equation (13) is the main equation in Euler's method.

For a concrete example of Euler's method, consider

$$\frac{dP}{dt} = f(t, P) = 0.3P, \tag{14}$$

with initial condition $P(0) = 8.0$, and discrete time steps of size $h = 0.1$. Say we want to approximate the solution P over the time interval $[0, 10]$. Then

$$
\begin{aligned}
P_1 &= hf(t_0, P_0) + P_0 \\
&= 0.1f(0, 8.0) + 8.0 \\
&= 0.1\Big((0.3)(8.0)\Big) + 8.0 = 8.2400, \tag{15} \\
P_2 &= hf(t_1, P_1) + P_1 \\
&= 0.1f(0.1, 8.2400) + 8.2400 \\
&= 0.1\Big((0.3)(8.2400)\Big) + 8.2400 = 8.4872, \tag{16} \\
P_3 &= hf(t_2, P_2) + P_2 \\
&= 0.1f(0.2, 8.4872) + 8.4872 \\
&= 0.1\Big((0.3)(8.4872)\Big) + 8.4872 = 8.7418. \tag{17}
\end{aligned}
$$

We could continue in this manner 97 more times until we reach

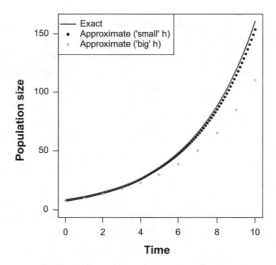

Fig. 2 Exact and approximate solutions to Eq. (1) with $\lambda = 0.3$ and initial condition $P(0) = 8$. The exact solution was found by integrating and the approximate solutions were calculated using Euler's method with step sizes of $h = 0.1$ ("small" h, black dots) and $h = 1$ ("big" h, gray dots)

$$P_{100} = 0.1 f(9.9, P_{99}) + P_{99}, \tag{18}$$

which is the approximation of $P(t)$ at time point $t = t_{100} = 10$. This would be incredibly tedious to do by hand. A computer, however, is very efficient at repeated calculations such as these. We can program Euler's method and quickly get the entire approximation. Figure 2 shows the plot of the true solution, $P(t) = 8e^{0.3t}$ and two different Euler approximations: one obtained with a step size of $h = 0.1$, and one with a step size of $h = 1$. Notice that the approximation is much better with the smaller step size. This is because the tangent line is a good approximation to the function over a short distance, so the shorter the time step we choose, the better the approximation.

The code parallels the work described in the calculations with one exception—in the code we set the model parameter value, find the total length of time, and plot the true solution. This might seem strange—to plot an exact solution, then to work to generate approximations to this solution—but this approach allows us to better understand how the numerical solutions compare to the exact solution.

The code provided throughout this chapter uses a tab to indicate where the longer line (that could not fit on one line of text) continues onto the next line. If you are copying and pasting code from an electronic source, you may have to delete these tabs in order for the code to run correctly. Additionally, you may have to "find and replace" the single quote ('), as printed on the page, with the standard single quote from your keyboard. More thorough details about the code syntax are described in Sect. 2.2, but you should not be afraid to search online for documentation, examples, and further explanation. Useful commands are summarized in the Appendix, specifically in Tables 4 and 5.

The first line of the block of code below is just a comment indicating that we are defining the parameter, the initial condition, and the maximum time. R does not execute lines of code that begin with the number sign (#), so we can type comments to ourselves in the code by using #'s. The next three lines of code define the values for the parameter, initial condition, and maximum time, using the assignment operator $<-$ in place of where you might have expected an equals sign. After the next comment indicating we are about to look at the true solution, we plot the true solution. The `plot()` command takes the true solution as input (`function(t)` tells R that what follows is a function of t, and the true solution, $P(t) = P_0 e^{\lambda t}$, is written in a form R understands, using $*$ for multiplication), and then specifies various plotting options. `lwd=2` sets the line width of the graph to be 2-times the default width. `col=2` sets the line color to be red, the second color in R's default list of colors. `xlim=c(0,tmax)` sets the plot range for the x-axis to be from 0 to `tmax` (which earlier we set to be 10). The command `c(...)` makes a vector whose entries are specified by the values inside the parentheses. Finally, `xlab='time'`, `ylab='p'` tells R what we want the x-axis and y-axis to be labeled (in this case "time" and "p").

```
## parameter, initial, and max time values
    lambda <- 0.3
    p0 <- 8.0
    tmax <- 10
## true solution
    plot(function(t) p0*exp(lambda*t), lwd=2, col=2,
    xlim=c(0, tmax), xlab='time', ylab='p')
```

Since we have already plotted the exact solution, we can use the next three chunks of code to specify the step size h, calculate the number of steps, and generate and plot the corresponding numerical approximation. Repeat the next three chunks of code with another value of h to generate additional approximate solutions. The first chunk of code sets the step size and initializes the P and t values, the second performs the calculation, and the third handles the graphics.

```
## length of step
    h <- 0.1
## number of steps
    n <- ceiling(tmax/h)
## initialize
    ts <- 0.0
    ps <- p0
```

From here the method itself is quite simple, as given below. We use a common programming technique called a *for loop*, in which a chunk of code is executed repeatedly. `for (i in 2:n) {...}` tells R that we want to start with i=2, execute the code contained in the curly brackets, and then update i to equal 3, execute the code contained in the curly brackets, and then update i to equal 4, etc., all the way until i=n (which we previously calculated, and is 100 for this example). The code inside the curly brackets is where Euler's method is actually being computed. We save the P estimates, ps, as a set. To access the first number in the set, which we defined to be p0, we would type ps [1]. To access the tenth number in the set, we would type ps [10], etc. So the code within the curly braces below calculates the next P estimate, ps [i] by taking the previous estimate, ps [i-1], plus the time step times λ times the previous estimate, h* (lambda*ps [i-1]), for i from 2 to n. The next time point, ts [i], is calculated by adding h to the previous time point, ts [i-1].

```
## method
for (i in 2:n) {
    ## approximations
    ps[i] <- ps[i-1] + h*(lambda*ps[i-1])
    ts[i] <- ts[i-1] + h
}
```

Finally, we plot each of the approximate solutions on the same graph we already plotted the exact solution. The command lines (ts, ps, . . .) takes coordinates given by ts (the t-values) and ps (the approximated p-values) and connects the corresponding points with line segments. The optional arguments lty, lwd, pch set the line style, line width, and point symbol, respectively.

```
## plotting approximates
    lines(ts, ps, type='b', lty=3, lwd=2, pch=19)
## adding a legend
    legend(`bottomright', c('exact', 'approx'),
    lwd=2, col=c(2, 1), lty=c(1, 3), bg='white')
```

The exact solution and the results of Euler's method with the two step sizes are given in Table 1.

Built-in differential equations solvers in mathematical software (like R) use more sophisticated methods than Euler's method, but the idea is the same: find a way to approximate the continuous solution at discrete points, in a way that minimizes the error between the true solution and the approximation. Common "higher order methods" (so-called because they approximate solutions to a higher order of accuracy) are *Heun's method* and the *Runge–Kutta methods* [9]. The interested reader can learn much more about higher order methods by reading the classic textbook [9], but rather than code up these methods ourselves, we will use a built-in solver in R. To solve the same initial value problem

$$\frac{dP}{dt} = 0.3P, \qquad P(0) = 8 \tag{19}$$

Table 1 Exact solution and Euler's method numerical approximations with two different time step values to the initial value problem $\frac{dP}{dt} = 0.3P$, $P(0) = 8.0$. The column labeled "Time" represents the actual t value, not the number of time steps. For instance, "Time" 3 corresponds to the third time step (t_3) for $h = 1$, but to the 30th time step (t_{30}) for $h = 0.1$

Time	Exact	'Small' step ($h = 0.1$)	'Big' step ($h = 1$)
0	8.00	8.00	8.00
1	10.8	10.75	10.40
2	14.58	14.45	13.52
3	19.68	19.42	17.58
4	26.56	26.10	22.85
5	35.85	35.07	29.70
6	48.40	47.13	38.61
7	65.33	63.34	50.20
8	88.19	85.13	65.26
9	119.04	114.40	84.84
10	160.68	153.75	110.29

using deSolve in R, we would use the code given below. The note in the code (which refers back to the text) refers to the step of loading the deSolve package. This only has to be done once per session, but the package may need to be installed the *very first time* that it is to be used. This can be done relatively simply by navigating the user menus or with the command install.packages(''deSolve'') prior to loading the package.

```
## load package
library(deSolve) ## ** see text **

## give model
POPmod <- function(t, x, parms)  {
   with(as.list(c(parms, x)), {
   dP <- lambda*P
   list(c(dP))
})}
```

Now we can specify parameters, the times for which we want solution values, and the initial conditions. With those specified, we can run the numerical solver. In general we prefer to combine the last two lines applying as.data.frame() directly to the lsoda() command, but as presented below it fits better on the printed page. lsoda is a slightly more sophisticated (than Euler's method) numerical solver, which tends to make for a good, fast, reliable default. Other numerical solvers available in deSolve are discussed in Appendix Table 5.

```
## list parameter(s)
   parms   <- c(lambda = 0.3)
## specify times solution should be returned
   times <- seq(0, 10, by=0.01)
## specify initial condition(s)
   xstart <- c(P = 8.0)
## apply solver
   out <-   lsoda(xstart, times, POPmod, parms)
   out <-   as.data.frame(out)
```

We could use `lines(out$time, out$P)` or `points(out$time, out$P)` with optional line or point style arguments to add this numerical solution to an existing plot. However, as we have done earlier, we can plot the solution values in a new window.

```
## plot solution
   plot(out$time, out$P, type='l', col='red',
   lwd=2, ylim=c(0, max(out$P)), las=1)
```

The $ symbol accesses a named column from a named data frame. For example, `lsoda` creates an output (which we called `out`) whose first column is named `time` and whose second column is named whatever the dependent variable is (in our case, P). So, `out$time` accesses the column called `time` in the `out` data frame.

The results from the `lsoda(...)` solver from `deSolve` (not shown) are closer to the true solution than either of the Euler's method approximations. In fact, the solutions from the built-in solver match the true solution to at least 6 decimal places over the interval $0 \le t \le 10$. This may seem like a large amount of work to solve an equation whose solution can be found by integrating, but the approach here scales relatively easily to more complex models. All that must be done to apply this approach to more complex models, as is done in later examples in the text (e.g., Sect. 2.2), is to give additional equations, parameters, and initial conditions.

Exercise 5 Modify the given Euler's method code to solve the initial value problem

$$\frac{dP}{dt} = -0.1P, \quad P(2) = 14$$

for $2 \le t \le 5$. Experiment with several different h values. Then, modify the given code to solve the initial value problem using the built-in `deSolve` command.

Exercise 6 Modify the given Euler's method code to solve the initial value problem

$$\frac{dP}{dt} = -3P + e^{-t}, \quad P(0) = 0$$

for $0 \le t \le 1$. Note that the right-hand side explicitly contains t, so the t_i values should appear explicitly in the code. In particular, instead of

```
ps[i] <- ps[i-1] + h*(lambda*ps[i-1])
```

we will now need

```
ps[i] <- ps[i-1] + h*(lambda*ps[i-1]+exp(ts[i-1]))
```

in the main Euler's method step. Plot the numerical solutions for several different h values, along with the true solution, $P(t) = \frac{1}{2}e^{-3t}\left(e^{2t} - 1\right)$.

Exercise 7 This exercise is identical to Exercise 6, except the `lambda` value is an order of magnitude bigger. Modify the given Euler's method code to solve the initial value problem

$$\frac{dP}{dt} = -30P + e^{-t}, \quad P(0) = 0$$

for $0 \le t \le 1$. Plot the numerical solutions for h values 0.1, 0.05, 0.01, and 0.001, along with the true solution, $P(t) = \frac{1}{29}e^{-30t}\left(e^{29t} - 1\right)$. Note that Euler's method did a poor job approximating the solution until the smallest step size. Using a more sophisticated method like `lsoda` or Runge–Kutta 4 (`rk4`) in R produces a much better approximation by $h = 0.01$. Verify this by plotting the true solution and the numerical solution obtained using the built-in `deSolve` command.

2 Predator–Prey

To practice what we reviewed in Sect. 1, we will consider a specific example involving animal populations. We will use differential equations to build a model, and then use computational techniques to solve the model and make predictions about the populations.

Consider a population of prey (say hares, $H(t)$) and of predators (lynx, $L(t)$) where H, L are measured in number of animals per area (also called "density"). Assume that in the absence of lynx, the hare population increases exponentially at constant rate α, measured in units $\frac{1}{s}$. In the absence of hares, the lynx population exponentially decays at a constant rate of δ $\left(\frac{1}{s}\right)$, meaning that the lynx cannot survive on any food source other than hares. The interaction of both lynx and hares negatively affects the hares (whose population decreases at a constant rate of β $\left(\frac{1}{density \cdot s}\right)$), and positively affects the lynx (whose population increases at a constant rate of γ $\left(\frac{1}{density \cdot s}\right)$), proportionally to how many animals are present. The increase in lynx population upon interaction with hares is due to increased reproduction resulting from feeding; eating hares keeps the lynx healthy enough to successfully reproduce. Given these assumptions, a basic "*predator–prey*" *model* (also known as the "Lotka-Volterra" model [7]) can be written:

$$\frac{dH}{dt} = \alpha H - \beta HL \tag{20}$$

$$\frac{dL}{dt} = \gamma HL - \delta L. \tag{21}$$

This is a two-dimensional, nonlinear model (because of the 2 state variables, H and L, and the $-\beta HL$ and γHL terms). For a thorough derivation and analysis of this predator–prey model, see [5, 7, 13]. We assume that $\gamma < \beta$, to reflect the fact that there is not a one-to-one relationship between the harm to prey, and the benefit to predators, caused by predation. In other words, when a hare gets eaten, that is *really* bad for the prey (it died), but it may only be slightly beneficial for the predator; the predator might need several meals like that to reproduce and produce new lynxes.

2.1 Analytical Results

Due to the nonlinearity of model (20) and (21), pencil-and-paper analysis is difficult. While we cannot derive explicit solutions $H(t)$ and $L(t)$, we *can* find a parametric relationship between the variables using a method called *separation of variables* [7]. Notice that dividing $\frac{dL}{dt}$ by $\frac{dH}{dt}$ gives

$$\frac{dL}{dH} = \frac{L(\gamma H - \delta)}{H(\alpha - \beta L)}, \tag{22}$$

or, upon separating variables,

$$\frac{\alpha - \beta L}{L} dL = \frac{\gamma H - \delta}{H} dH. \tag{23}$$

Integrating each side with respect to its indicated variable yields

$$\alpha \log(L) - \beta L = \gamma H - \delta \log(H) + c, \tag{24}$$

where c is a constant that depends on the initial conditions $H(0)$ and $L(0)$. Equation (24) can be written in standard parametric form as

$$\alpha \log(L) - \beta L - \gamma H + \delta \log(H) = c. \tag{25}$$

Simply deriving Eq. (25) does not provide much insight into what happens to the populations of hares and lynx. Here we see another instance of the benefit of computing. Using technology to plot the parametric curve for $c \approx 3.343$ (obtained from initial conditions $H(0) = 50$, $L(0) = 8$), we obtain a closed curve (Fig. 3). A plot in which one dependent variable is plotted as a function of the other dependent variable is often called a *phase plane*. Tracing the curve in the phase

Fig. 3 Solutions to the predator–prey model in the phase plane. The solutions begin from initial conditions $H(0) = 50, L(0) = 8$ (indicated by the solid dot), and quantities move along the curve in the direction of the arrow as time progresses

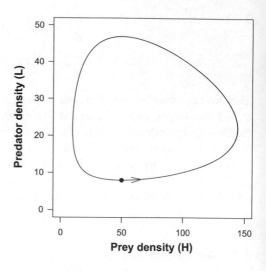

plane counterclockwise shows the population levels of the 2 species as time goes on. We see that the populations oscillate. For example, starting from initial condition $(H(0), L(0)) = (50, 8)$, the population of hares is moderate, while the population of lynx is fairly low. As time goes on, the hare population increases quite a bit (almost to 150) because there are not that many lynxes to eat them, while the lynx population only increases a modest amount. However, after some time, the lynx population starts to increase by a large amount (there are lots of hares to eat, so the lynxes are able to reproduce at a higher rate), while the hare population actually decreases (they are being eaten by all the new lynxes). After the hare population gets to about ten individuals, the lynx population decreases precipitously (because there are no longer enough hares to eat), while the hare population only shows a modest decline. The hare population then rebounds and starts to increase (because there are fewer lynxes to eat them), while the lynx population barely increases, until we reach the original starting point, and the whole process repeats.

Another analytical technique is to find the *steady states* (or *equilibria* or *fixed points*) of the model. These are the values at which the populations are unchanging (i.e., where $\frac{dH}{dt} = 0$ and $\frac{dL}{dt} = 0$). Setting the derivatives equal to 0 and using algebra to solve the resulting system of equations yields two steady states: $H = 0, L = 0$ and $H = \delta/\gamma, L = \alpha/\beta$. See [5, 7] for details on how to do stability analysis to classify the steady states. For our purposes, we prefer to numerically solve model (20)–(21), which allows us to clearly see the dynamics.

Exercise 8 Apply separation of variables to find a closed form solution to the SIS model given in Eqs. (4) and (5). Is there anything surprising about the result? It may be helpful to consider the equation that describes the rate of change of the quantity $N(t) = S(t) + I(t)$.

2.2   Numerical Results

To explicitly see how the two populations change in time, we can numerically solve the nonlinear system of equations. Assuming that we have already installed and loaded the packages (see Sect. 1.3 otherwise), we can define the model by creating the function PPmod.

```
## a Predator--Prey model
PPmod <- function(t, x, parms) {
   with(as.list(c(parms, x)), {
   dH <- alpha*H - beta*H*L
   dL <- gamma*H*L - delta*L
   list(c(dH, dL))
})}
```

If we change the model (as we may do in the exercises that follow), it would make sense to change the text after the comment, as well as the name of the function. The first line of the block of code is just a comment indicating the name of this particular model. The second line of the block defines the function as one that accepts the arguments t (for time), x (for the states), and parms (for the parameters of the model). The next two lines define the differential equations of our model, with the assignment operator <- in place of where you might have expected an equals sign. The shorthand dH is used to indicate the left-hand side of the differential equation, $\frac{dH}{dt}$, and on the right-hand side * is used for multiplication. The first line after the model has been specified lists the values returned by the function PPmod, and the final line closes up parentheses and brackets opened earlier. One minor detail has slipped past until now: the third line with(as.list(c(parms, x)), { serves the purpose of making our parameter and variable values available during the execution of the function.

Next we parameterize and initialize our model, as well as provide a list of times for which we want solution values. Here c(...) makes a vector whose entries can be named and seq(...) automates making a vector of a given length or spacing.

```
## parameters
   (parms <- c(alpha=1.1, beta=0.05, gamma=0.01,
   delta=0.5))
## vector of time steps
   (times <- seq(0, 500, by=0.1))
## initial conditions
   (xstart <- c(H = 50, L = 8))
```

An alternative approach to the above section omits the outer set of parentheses on each line. What we have shown in the text is a shortcut that allows for simultaneous assignment and printing of an object. Without these parentheses each object is "silently" assigned without being printed to the screen. In many cases we like the approach above as it provides another opportunity to troubleshoot and error check.

The initial conditions, time values, model, and parameter values are fed to a built-in solver, here `lsoda()`, and the output is processed into a "data frame" by `as.data.frame(...)`. Treating the output as a data frame means that we can conveniently access columns by name for graphing or calculations. As earlier, we print this on separate lines for readability, but the two lines can be combined.

```
out <- lsoda(xstart, times, PPmod, parms)
out <- as.data.frame(out)
```

It would be worth taking a few moments to explore the output by using a few built-in commands including `head(...)` which gives the first 6 rows of output (e.g., `head(out)` gives the first 6 rows of the data frame out), `tail(...)` which gives the last 6 rows, and `summary(...)` which robustly gives column by column summaries of its argument. Go ahead, try. We will wait. Next, it is possible to access specific columns by their name, such as `out$time`, or their position in the output, such as `out[ , 1]` for the first column. To look at the first row of output, perhaps to verify our initial conditions were applied correctly, we would use `out[1, ]`. Leaving the row or column blank gives all entries of the requested column or row. For example `out[10, 2]` gives the tenth value of the second column (in this case `out[10, 'H']` is equivalent). Depending on your purpose it may be easier (or safer) to refer to a quantity by name or number.

We can then plot results of the model by making two separate plots (as shown) or by specifying the figure layout (such as `par(mfrow=c(1,2))`). The command `matplot(...)` allows us to plot two (or more) quantities simultaneously.

```
## plot 1
   matplot(out$time, out[, c('H', 'L')], type='l',
   lwd=2, col=c('blue', 'red'), lty=1,
   ylim=c(0, 150), xlab='Time', ylab='Population
   densities (number/area)', las=1)
   legend('topleft', c('Prey (H)', 'Predator (L)'),
   col=c('blue', 'red'), lty=1, bg='white')

## plot 2
   plot(out$H, out$L, type='l', xlim=c(0, 150),
   ylim=c(0, 50), xlab=`Prey (H)',
   ylab='Predator (L)', las=1)
```

Figure 4 shows the solutions as a function of time (i.e., the "dynamics"), as well as the solution curve in the phase plane. In the dynamics plot, we see that as time goes on, the populations oscillate, with the peaks in predator population (red curve) lagging behind the peaks in prey population (blue curve). In fact, this time course data is information that we were unable to get from analysis alone, as we were unable to explicitly solve for $H(t)$ and $L(t)$. In the next section, we will apply the techniques we learned from the predator–prey system analysis to a more complicated model of parasite transmission in cats.

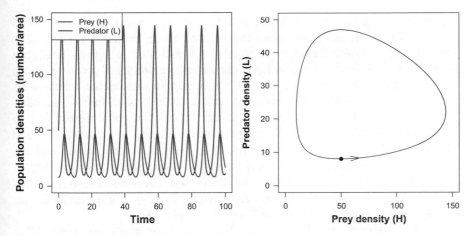

Fig. 4 Solutions to the predator–prey model obtained using our PPmod code. Parameter values $\alpha = 1.1, \beta = 0.05, \gamma = 0.01, \delta = 0.5$, and initial conditions $H(0) = 50, L(0) = 8$ were used. (Left) Solution densities plotted against time. (Right) Solutions in the phase plane

Exercise 9 Modify the provided code to solve the model where instead of alpha*H, the hare population grows at a constant rate alpha. In this case, the constant rate alpha models immigration—hares entering the population at a rate that does not depend on the current population density of hares.

Exercise 10 Modify the predator–prey model and code to study a system with a second predator species. Assume both predators eat the same prey species, but do not directly interact with each other. What happens if the second predator species has a higher benefit rate (the γ term) than the first species, but also a higher death rate (δ)?

Exercise 11 Modify the predator–prey model and code to study a system with a second prey species. Assume both prey species are eaten by the predator, but do not directly interact with each other. What happens if the second prey species has a higher growth rate (α) than the first species, but is also preferred by the predator species (β is higher)?

3 *Toxoplasma gondii* Transmission in Cats

Armed with the confidence we accrued in Sects. 1 and 2 to build and numerically solve mathematical models, we will begin applying these techniques to open-ended research projects. The models presented in Sects. 3–5 all started as undergraduate research projects, and they are good starting points for independent research.

Toxoplasma gondii (*T. gondii*) is one of the most successful parasites on Earth, infecting cats, rats, seals, and more than a third of the human population [27]. Animals contract the parasite by consuming infected food or coming in contact with infected cat feces. Toxoplasmosis, the disease caused by *T. gondii*, presents flu-like symptoms in humans, but can be deadly to endangered monk seals. Pregnant women with toxoplasmosis can have miscarriages, stillbirths, and other birth anomalies [40]. Cats are the only animal in which *T. gondii* reproduces, so it is critical to understand how the parasite moves through a cat population, and then potentially spills into other species. In a year-long undergraduate research project, we built and analyzed a between-host differential equations model to study transmission of *T. gondii*; specifically, we studied how pathogen infectiousness and shedding rate into the environment affect the proliferation of infection in the cat population. For additional models of *T. gondii*, see [17, 23].

A schematic of the model is shown in Fig. 5. The basic idea of our model is that infectious cats shed *T. gondii* oocysts (a specific life-stage of a parasite) into the environment. Susceptible cats that are exposed to these oocysts in the environment can become infected with the parasite. The model equations are

$$\text{Susceptible Cats} \qquad \frac{dS}{dt} = b - \omega E S - mS \qquad (26)$$

$$\text{Acutely Infectious Cats} \qquad \frac{dI}{dt} = \omega E S - \gamma I - \delta_I mI \qquad (27)$$

$$\text{Chronically Infectious Cats} \qquad \frac{dC}{dt} = \gamma I - \delta_C mC \qquad (28)$$

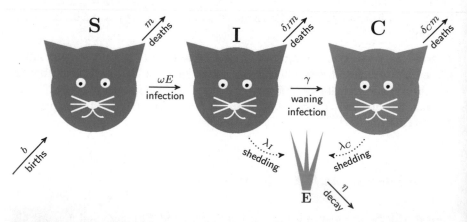

Fig. 5 Schematic model of *T. gondii* transmission in cats. *S* is the number of susceptible cats, *I* is the number of acutely infectious cats, *C* is the number of chronically infectious cats, and *E* is the number of oocysts in the environment. This figure was modified from [31], and is reproduced here with permission. Copyright ©2018 Society for Industrial and Applied Mathematics. Reprinted with permission. All rights reserved

$$\text{Oocysts in Environment} \qquad \frac{dE}{dt} = \lambda_I I + \lambda_C C - \eta E \qquad (29)$$

where b is the birth rate of cats $\left(\frac{\#\,\text{cats}}{\text{week}}\right)$, m is the death rate of cats $\left(\frac{1}{\text{week}}\right)$, γ is the loss of acute infection in cats $\left(\frac{1}{\text{week}}\right)$, δ_I is the infection-associated mortality in acutely infectious cats (unitless), δ_C is the infection-associated mortality in chronically infectious cats (unitless), η is the oocyst degradation rate in the environment $\left(\frac{1}{\text{week}}\right)$, ω is the rate of infection in cats $\left(\frac{1}{\#\,\text{oocysts}\times\text{week}}\right)$, λ_I is the oocyst shedding rate per acutely infectious cat $\left(\frac{\#\,\text{oocysts}}{\text{cat}\times\text{week}}\right)$, and λ_C is the oocyst shedding rate per chronically infectious cat $\left(\frac{\#\,\text{oocysts}}{\text{cat}\times\text{week}}\right)$, where oocysts are measured in billions.

Many parameter values can be found in, or estimated from, the literature (see citations in [31]). Other parameter values are harder to estimate. In the following section we discuss how numerical experimentation can help overcome this difficulty.

3.1   Role of Technology

In this project, we focus specifically on how the *rate of infection* (ω) affects steady-state population sizes. First, we use numerical solutions to generate intuition about the model behavior and biological system. Then we use programming to compute substantial calculations (solving the model dozens of times for dozens of different parameter values) to generate patterns. Then, insight gleaned from those patterns guides our use of analytical techniques to derive concrete evidence for the predictions obtained via numerical experimentation.

The model in Eqs. (26)–(29) can be translated to the code below. The student began with computer code for numerically solving the two-dimensional predator–prey model (see Sect. 2, specifically Sect. 2.2 for examples of the preliminary code) and updated the code to reflect the new variables and equations. In the code we use a "dot" where we might otherwise use an underscore to denote a subscript.

```
TOXOmod <- function(t, x, parms)   {
    with(as.list(c(parms, x)), {
## susceptible cats
    dCs <- B - Omega*E*Cs - M*Cs
## infected cats
    dCi <- Omega*E*Cs - Gamma*Ci - Delta.i*M*Ci
## chronically infected cats
    dCc <- Gamma*Ci - Delta.c*M*Cc
## environment
    dE <- (Lambda.i*Ci + Lambda.c*Cc) - eta*E
    list(c(dCs, dCi, dCc, dE))
})}
```

The model then needs parameters, a list of times for which to calculate the solution, and initial conditions. It is very important that the initial conditions match the order of the output listed in the model, and that the order of the output listed in the model matches the order of the listed equations.

```
## The parameters
   parms <- c(
      B = 60*(1/52), ## birth rate of kittens per week
      Omega = 0.046, ## rate of infection
      M = 0.2*(1/52), ## mortality of cats
      Gamma = 11/15,  ## loss of acute infection
      Delta.i = 1.05, ## added mortality from acute
      Delta.c = 1.0005, ## added mortality from chronic
      eta = 4*(1/52), ## oocyst decay rate
      Lambda.i = (0.07/11), ## acute shedding rate
      Lambda.c = 0) ## chronic shedding rate
## vector of timesteps
   times <- seq(0, 50*52, by=1)
## initial conditions
   xstart <- c(Cs = 495, Ci = 5, Cc = 0, E = 0)
```

Finally, we solve the system of differential equations. Again, we use the built-in solver lsoda, which is a reliable default solver.

```
## run model
   out <- lsoda(xstart, times, TOXOmod, parms)
   out <- as.data.frame(out)
```

In Sects. 3.2 and 3.3 we discuss various ways to use technology to visualize the model results and run more in-depth numerical experiments.

3.2 Dynamics

To illustrate the model dynamics we plot the cat solutions with an inset for the relatively small acutely infectious cat population. The fig arguments of the second instance of par() set the internal plot window for the inset. Here the student chose a somewhat sophisticated color scheme for the plotting to match the colors of the diagram in Fig. 5. With this choice of colors, the use of line styles makes the plot more readable in black and white or for the vision-impaired.

```
## plot susceptibles
plot(out$time, out$Cs, type='l', col='#B276B2', lty=1,
   ylim=c(0, 500), las=1, lwd=3,
   xlab='Time (in weeks)', ylab='Cats')
## add acute then chronic
```

```
lines(out$time, out$Ci, col='#F17CB0', lty=3, lwd=3)
lines(out$time, out$Cc, col='#FAA43A', lty=2, lwd=3)
## legend
legend(`topright', c('S', 'I', 'C'), lty=c(1, 3, 2),
    lwd=2, col=c('#B276B2', '#F17CB0', '#FAA43A'),
        bg='white')
## acute infection closeup
par(fig = c(0.25, 0.80, 0.5, 0.95), new = T)
plot(out$time, out$Ci, type='l', col='#F17CB0', lty=3,
    lwd=3, xlim=c(0, 600), ylim=c(0, 35),
    xlab='Time (in weeks)', ylab=`Infectious', las=1)
```

We also plot the environmental parasite burden in a separate window since its scaling is quite different from that of the cat population.

```
## plot oocysts
plot(out$time, out$E, type='l', col='#60BD68', las=1,
    lwd=3, ylim=c(0, 2), xlab='Time (in weeks)',
    ylab='Oocysts in Environment (billions)')
## legend
legend('topright', 'E', lwd=2, col='#60BD68',
    bg='white')
```

Figure 6 shows the dynamics of the system for $\omega = 0.046 \frac{1}{\text{\# oocysts} \times \text{week}}$, with other parameter values as given in the caption. Note that rather than trying

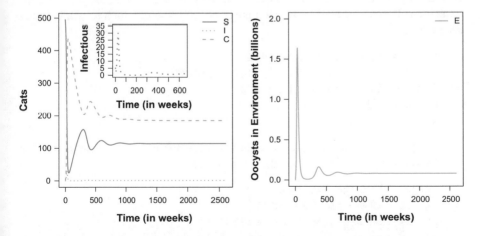

Fig. 6 Numerical solution of model equations. Inlaid are transient dynamics for acutely infectious cats on a shorter time scale. Color and line styles are *Susceptible* (*S*): purple, solid; *Acutely Infectious* (*I*): pink, dotted; *Chronically Infectious* (*C*): orange, dashed; and *Oocysts in Environment* (*E*): green, solid. Parameter values used for these results are as follows: $b = \frac{60}{52}, m = \frac{2}{520}, \gamma = \frac{11}{15}, \delta_I = 1.05, \delta_C = 1.0005, \eta = \frac{4}{52}, \omega = 0.046, \lambda_I = \frac{7}{1100}, \lambda_C = 0$. Initial conditions were $S(0) = 495, I(0) = 5, C(0) = 0$, and $E(0) = 0$

to solve the nonlinear system of 4 coupled differential equations analytically, we relatively quickly and easily solved the system numerically. We see that for the given parameters, each cat subpopulation displays decaying oscillations and eventually levels out at a fixed value (i.e., reaches a steady state). The number of oocysts in the environment also oscillates until reaching a small, but nonzero, steady state. In the long run, there will be very few acutely infectious cats (the pink dotted line is almost 0), and more chronically infectious than susceptible cats (the dashed orange line is above the solid purple line).

3.3   Stability Analysis

For the results above, we picked $0.046 \frac{1}{\text{\# oocysts} \times \text{week}}$ as the value of the unknown rate of infection parameter, ω. But should we expect similar results if we pick a different value for ω? We do not want to overgeneralize our results and assume that because the chronically infectious subpopulation of cats reaches a nonzero steady state, that it will also reach a nonzero steady state for any set of parameter values. What if the true value of ω is $0.01 \frac{1}{\text{\# oocysts} \times \text{week}}$, for instance, and results are totally different? Then by only looking at the case $\omega = 0.046 \frac{1}{\text{\# oocysts} \times \text{week}}$, we will be making claims about cat subpopulations and suggesting methods for controlling *T. gondii* transmission that are not based on the true system. Enter numerical experimentation. We can quickly test a range of parameter values and see how the model behaves.

Given the results of well-planned numerical experimentation, we determined that it would be worth studying the effect of the rate of infection parameter ω on the dynamics. To investigate this we set a range of values with which to parameterize the model.

```
## sets range of parameters
   omegas <- seq(0, 0.15, length=101)
## initialize 'vals' for storage
   vals <- NULL
   sols <- NULL
```

From here we calculated and stored the complete solution corresponding to each value and also recorded the final value of the solution.

```
## steps through omegas
for(i in 1:length(omegas)){
   parms['Omega'] <- omegas[i]

## solve and store solution
   out <- lsoda(xstart, times, TOXOmod, parms)
   sols[[i]] <- as.data.frame(out)

## saves last row, 2nd through last column of solution
```

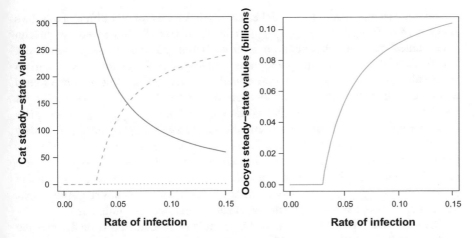

Fig. 7 Plot of stable steady-state values of the cat subpopulations for varying rates of infection, ω. Color and line styles are *Susceptible* (S): purple, solid; *Acutely Infectious* (I): pink, dotted; *Chronically Infectious* (C): orange, dashed. Parameter values used for these results are as follows: $b = \frac{60}{52}, m = \frac{2}{520}, \gamma = \frac{11}{15}, \delta_I = 1.05, \delta_C = 1.0005, \eta = \frac{4}{52}, \lambda_I = \frac{7}{1100}, \lambda_C = 0$, and ω was varied from 0 to 0.15 by steps of 0.0015

```
    vals <- rbind(vals, out[nrow(out), 2:ncol(out)])
}
```

Now we can plot the suspected equilibrium values against the corresponding parameter values. At the moment we say "suspected" because it is possible that the solution has not reached equilibrium by the time we stop our code. A fast, but unrigorous, way to test this is to inspect the plot in Fig. 7 for any jumps or bumps that might indicate that the solution has yet to reach equilibrium at the time that the numerical solution is stopped. If we see jumps or bumps, we inspect the dynamics of the corresponding stored solution and, if necessary, calculate the solution for a longer time until it reaches equilibrium. Additionally (or alternatively), we could numerically check for equilibria by looking at the variable values at the last several time points; if the values are the same up to some tolerance that makes sense for the given problem (say, 6 decimal places), then we consider equilibrium to be reached.

```
## panel 1
    matplot(omegas, vals[,1:3], type='l', lty=c(1,3,2),
        lwd=2, col=c('#B276B2', '#F17CB0', '#FAA43A'),
        xlab='Rate of infection',
        ylab='Cat steady-state values', las=1)

## panel 2
    plot(omegas, vals[,4], type='l', lty=1, lwd=2,
        col='#60BD68', las=1, xlab='Rate of infection',
        ylab='Oocyst steady-state values (billions)')
```

We numerically solved the model equations 101 times for 101 different ω values ranging from 0 to 0.15 $\frac{1}{\text{\# oocysts}\times\text{week}}$ by steps of 0.0015, and saved the steady-state values of each subpopulation from each run (as seen in the code above). Figure 7, where we plot the steady-state values as functions of ω, summarizes the results. Note that the results are actually discrete points, but we have joined them with a line. We learn quite a lot about the system from this graph. Firstly, we see that for "small enough" ω values (below about 0.03 $\frac{1}{\text{\# oocysts}\times\text{week}}$), $T.$ gondii is completely eradicated from the system; the steady-state subpopulations of acutely and chronically infectious cats are 0, and we only end up with susceptible cats in the population. For rates of infection above about 0.03 $\frac{1}{\text{\# oocysts}\times\text{week}}$, $T.$ gondii remains, mostly in the form of chronically infectious cats. Notice also that at $\omega \approx 0.06$ $\frac{1}{\text{\# oocysts}\times\text{week}}$, we switch from a population in which susceptible cats dominate, to a population in which chronically infectious cats dominate.

Here we should pause to note that it *is* possible to generate the same figure using analysis, rather than numerics. Much literature is dedicated to the subject of the "basic reproductive number," R_0, or the number of new infections caused by a single acutely infectious individual in a fully susceptible population. We direct the reader to Refs. [11, 12] for a general discussion, and to Ref. [31] for the R_0 calculation for model Eqs. (26)–(29). Using this analysis one can show that $T.$ gondii will be eradicated from the population if

$$\omega < m^2 \eta \delta_C (\gamma + \delta_I m) \left(\frac{\delta_C(\gamma + \delta_I m) + \gamma}{\gamma b(\gamma \lambda_C + \delta_C \lambda_I m)} \right). \tag{30}$$

For the parameter values used in this study, that means $\omega < 0.0596$ $\frac{1}{\text{\# oocysts}\times\text{week}}$. So while it is possible to do some analysis on this system, it is faster and more illuminating, in our opinion, to do the numerical study first. After getting a feel for the problem, and recognizing that there is a switch in behavior, we can do the pen-and-paper analysis.

3.4   Implications

The actual rate of infection is unknown, but if our other parameter values are realistic, then Fig. 7 can help us determine a range of possible ω values for a given cat population. For instance, if we know that a particular population of cats has a steady population of chronically infectious cats, then we know that ω must be larger than 0.03 $\frac{1}{\text{\# oocysts}\times\text{week}}$ (because below that value the model predicts there are no chronically infectious cats).

Rate of infection depends on two separate things: the probability of infection when a susceptible cat encounters the parasite, and the rate at which a cat encounters the parasite. It is unlikely that we can change the probability of infection, but perhaps it is possible to change the rate at which a cat encounters the parasite. Or, we can aim to reduce the total number of oocysts in the environment, by treating the environment with an agent that kills *T. gondii* oocysts.

The modeling and numerical experimentation helped elucidate potential targets for reducing *T. gondii* infection in cats. At this stage in mathematical biology research, it is useful to discuss the model and/or results with a biologist familiar with the system you are studying. The biologist can give you a feel for how feasible the model results and implications are, and suggest further avenues of study. Also, your model results may cause the biologist to think differently about the system, which could lead to exciting new experiments the biologist might not have thought of otherwise.

In the projects below, we pose a lot of questions to get the reader/researcher started. We do not know the answers to these questions, or even if the answers will be interesting. Do not constrain yourself to our questions. If you see some interesting results, dig into them. Ask your own questions.

Research Project 1 (Additional Parameter Studies) In the model described in Sect. 3, we assumed the rate at which chronically infected cats shed oocysts into the environment, λ_C, is 0 $\frac{\text{\# oocysts}}{\text{cat} \times \text{week}}$. How do results change if λ_C is small but nonzero? Which situation seems most realistic? There is a range of values we have estimated for most of the parameters in the model [31], but for the results shown above, we usually used the midpoint of the range (with the exception of the δ parameters). What happens if you use the upper range of all of the parameters? What about the lower range? When you change so many values at once, it is hard to tell which one(s) is/are responsible for the differences in results. Can you systematically change parameters in such a way to identify which are having the biggest effect on results? (For example, making λ_I its maximum value seems to drastically change the quantitative, but not qualitative results.) Are there parameters that change the qualitative results, too? Is there ever a situation in which acutely infectious cats have a large steady-state population? Is there ever a situation in which the populations oscillate forever, without reaching a steady state?

Research Project 2 (Birth Function) In the model described in Sect. 3, we assumed that cats were born at a constant rate, free of the *T. gondii* parasite, irrespective of the number of cats present. Can you think of other reasonable assumptions for cat reproduction? (For example, should the rate at which cats are born depend on the number of cats in the population? If so, should it depend on the numbers in all subpopulations, or only some subpopulations? Does a logistic growth term make more sense? etc.) How would you

(continued)

Research Project 2 (continued)
implement these different rates into the model? Once implemented, can you repeat the types of numerical studies described above? What conclusions do you draw about how *T. gondii* spreads through a cat population? How could humans attempt to reduce the spread of *T. gondii*? Note: using different birth functions could fundamentally change model results. You may find other, more interesting avenues to pursue, rather than simply looking at how steady-state values change with varying ω. Be curious as you analyze the dynamics of your model, and ask (then try to answer) new questions.

Research Project 3 (Cat Ecology) Consider what changes to the model would be necessary to account for a prey species upon which cat reproduction depended (see the predator–prey model as a start) or if we considered predators of the cats themselves. Make an assumption about cats as predators or prey and add another species to the model. Does this second species have a direct role in the transmission of infection or does it mostly act to manipulate the cat population?

Research Project 4 (New Model System) The model described in Sect. 3 was for *T. gondii* transmission in a population of cats. Read about a different parasite or disease, and build a new model of its transmission through its host population. Now use numerical techniques to study your model and make predictions about how to best maintain a healthy population.

4   Circadian Rhythms and Alcohol Dependence

In Sect. 3 we studied a problem on the large scale of environments and populations. Similar techniques can be used to study smaller-scale phenomena, like activity within a cell.

The mammalian *circadian oscillator* is the body's internal, 24-h clock that controls brain wave activity, energy production, and other biological activities (and, incidentally, is responsible for the jet lag experienced when changing time zones). Acute and chronic alcohol consumption disrupts a regular circadian rhythm, which subsequently affects mood regulation, sleep cycles, blood pressure, and other biological rhythms [28]. For the following project, we started with a model from the literature [4], tweaked it to include an effect of alcohol, and then analyzed the model results numerically to gain biological insight. Further modeling ideas are listed as Projects 5 through 7.

Deoxyribonucleic acid (DNA) contains the genes that regulate almost everything in a living organism. During transcription, *messenger ribonucleic acid (mRNA)* is made from DNA. Then the mRNA is decoded to produce active proteins in the process of translation. Proteins are critical to organismic life; proteins are necessary for the structure, regulation, and function of tissues and organs. In this project, we consider four proteins that are involved in regulating circadian rhythms.

For a more thorough explanation of the biology and the model, see Becker-Weimann et al. [4]. In the brief explanation that follows, mRNA is represented by *italics*, and protein is represented by ALL CAPS. Figure 8 shows a schematic of the simplified biology used to build the mathematical model. Within the nucleus of a neuron, the transcriptionally active BMAL1/CLOCK (B_a) heterodimer (macromolecule formed from two distinct proteins) activates transcription of *Per2* and *Cry* genes, resulting in more *Per2/Cry* mRNA (P_m), and consequently more PER2/CRY protein (P_c) in the cytosol. The PER2/CRY complex is transported to the nucleus (P_n), where it inhibits *Per2/Cry* transcription and activates *Bmal1* transcription. This increases the amount of *Bmal1* mRNA (B_m) and BMAL1 protein (B_c) in the cytosol. The BMAL1 protein can be transported to the nucleus (B_n), where upon activation to the transcriptionally active form, BMAL1* (B_u), it is able to activate transcription of *Per2* and *Cry*, starting the cycle all over again. See Becker-Weimann et al. for an explanation of model assumptions [4].

The model described by Fig. 8, as presented in Becker-Weimann et al., is

$$\text{\textit{Per2/Cry} mRNA} \qquad \frac{dP_m}{dt} = \frac{v_P(B_a + c)}{k_{MP}(1 + (P_n/k_i)^s) + (B_a + c)} - d_{Pm}P_m \quad (31)$$

$$\text{PER2/CRY (cyto.)} \qquad \frac{dP_c}{dt} = k_b P_m{}^q - d_{Pc}P_c - k_P^{in}P_c + k_P^{out}P_n \quad (32)$$

$$\text{PER2/CRY (nuc.)} \qquad \frac{dP_n}{dt} = k_P^{in}P_c - k_P^{out}P_n - d_{Pn}P_n \quad (33)$$

$$\text{\textit{Bmal1} mRNA} \qquad \frac{dB_m}{dt} = \frac{v_B(P_n)^r}{k_{MB}^r + (P_n)^r} - d_{Bm}B_m \quad (34)$$

$$\text{BMAL1 prot. (cyto.)} \qquad \frac{dB_c}{dt} = k_t B_m - d_{Bc}B_c - k_B^{in}B_c + k_B^{out}B_n \quad (35)$$

$$\text{BMAL1 prot. (nuc.)} \qquad \frac{dB_n}{dt} = k_B^{in}B_c - k_B^{out}B_n - d_{Bn}B_n + k_d B_a - d_{Ba}B_n \quad (36)$$

$$\text{BMAL1 (trans. active)} \qquad \frac{dB_a}{dt} = k_a B_n - k_d B_a - d_{Ba}B_a \quad (37)$$

where "cyto." refers to "in the cytoplasm," "nuc." refers to "in the nucleus," "prot." refers to "protein," and "trans. active" refers to "transcriptionally-active form."

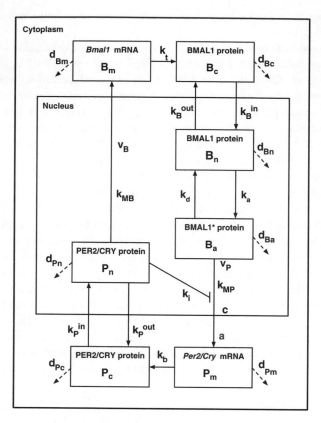

Fig. 8 Model of mammalian circadian oscillator. (Adapted from [4].) The asterisk on BMAL1* protein B_a indicates that this BMAL1 protein is transcriptionally active, and is able to start transcription of *Per2/Cry* genes. Dashed arrows represent degradation of mRNA and protein, solid arrows represent a positive effect, and the solid line with a vertical bar at the end represents a negative effect of P_n on activation of P_m. The red parameter, a, is our novel modification to the model

As a brief modeling aside, the function $f(P_n) = \frac{(P_n)^r}{k_{MB}^r + (P_n)^r}$ on the right-hand side of Eq. (34) is called a *Hill function* [29]. Hill functions, which have a sigmoidal shape, are commonly used in modeling because they describe cooperativity of smaller molecules (called ligands) binding to macromolecules (like enzymes or receptors). In the variables of our model, P_n is the concentration of unbound ligand and k_{MB} is the ligand concentration that results in half of the macromolecule binding sites being occupied by ligand. The *Hill coefficient*, r, is a measure of the cooperativity: if $r > 1$, then when a ligand binds to the macromolecule, the binding affinity for other ligands increases (bound ligands encourage additional binding of ligands, called "positive cooperativity"); if $r < 1$, then when a ligand binds to the macromolecule, the binding affinity for other ligands decreases (bound ligands discourage additional binding of ligand, called "negative cooperativity"); if $r = 1$, the binding affinity of a ligand to the macromolecule is independent of whether or not other ligands are already bound (called "noncooperativity"). In our model, $r = 3$, meaning that once P_n binds to B_m, even more P_n is likely to bind.

4.1 Role of Technology

Clearly the nonlinear, seven-dimensional model in Eqs. (31)–(37) is too complicated to solve analytically. By solving numerically, we find approximately 24-h oscillations in the mRNA and proteins, as expected in healthy circadian oscillations (see Exercise 12).

Exercise 12 Use technology to reproduce Figure 3A from [4]. Required parameter values are available in [4] and in the Appendix of this chapter.

To investigate the effect alcohol has on this system, we make one small change to Eq. (31):

$$\frac{dP_m}{dt} = \frac{av_P(B_a + c)}{k_{MP}(1 + (P_n/k_i)^s) + (B_a + c)} - d_{Pm}P_m. \tag{38}$$

Here, a represents the effect of alcohol in reducing transcription of *Per* and *Cry*. If $a = 1$, we have the standard, maximal transcription rate given in Becker-Weimann et al.; if $a = 0$, the effect of alcohol is so strong that transcription has been completely shut off. Values of a between 0 and 1 represent varying degrees of reduction in transcription. Using technology, we can study the effect parameter a has on the system.

To understand the effect of alcohol on circadian rhythms, we numerically solve the model using a values less than 1 and compare the results to the Becker-Weimann et al. model, which does not contain alcohol. Sample code is provided in the Appendix, but the idea is very similar to that presented in Sect. 3: input the model equations, parameters, time steps, and initial conditions, then solve the equations. As in Sect. 3, we can solve the model for a range of values of a particular parameter (in this case a instead of ω), and save information about how the model variables behave for each parameter value. In this project, however, we find oscillations rather than equilibrium values; our mRNA and protein concentrations are not leveling out at certain values, but rather oscillating as time goes on.

4.2 Model Dynamics

The dynamics of the model (Eqs. (32)–(38)) are shown in Fig. 9. In the left panel of Fig. 9, the concentrations of PER2/CRY protein and BMAL1 protein in the absence of alcohol are shown as gray solid and black dashed curves, respectively.

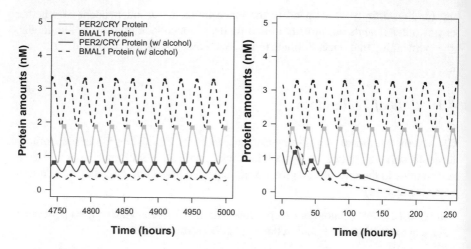

Fig. 9 Dynamics of model with parameter values and initial conditions specified in the sample code in the Appendix. Parameter values that differ between plots are (Left) $a = 0.2$; (Right) $a = 0.1$. Only late time points are displayed in the figure on the left to show that oscillations are sustained, even in the presence of alcohol

The same proteins in the presence of a strong alcohol effect ($a = 0.2$) are shown in red. We see that alcohol reduces both the peak concentration and the period of oscillation. A shorter period of oscillation means a less-than-24-h circadian rhythm, which would negatively affect the health of the individual. In the right panel of Fig. 9, the concentrations of PER2/CRY and BMAL1 in the absence of alcohol are still shown in gray and black, and in red are the concentrations in the presence of a stronger alcohol dependency ($a = 0.1$). In this case, oscillations disappear completely, meaning there is no longer a circadian rhythm. Studies have linked the disruption of the circadian oscillator to the development of cancer [18], heart disease [41], and depression [36], so our theoretical patient whose oscillations disappear could be in trouble.

4.3 Bifurcation

When simply changing a parameter value results in a drastic qualitative switch in behavior (e.g., moving from a regime in which oscillations persist into one with no oscillations), we say that a *bifurcation* has occurred. (See Ref. [44] for a nice introduction to bifurcations.) We can get a clearer picture of the effect parameter a has on the system by conducting a numerical experiment. To obtain Fig. 10 we numerically solve the model equations 30 different times for 30 different a values, equally spaced between $a = 0.0$ and $a = 1.0$. For each value of a, we numerically calculate the peak protein concentration (using findpeaks() in the pracma

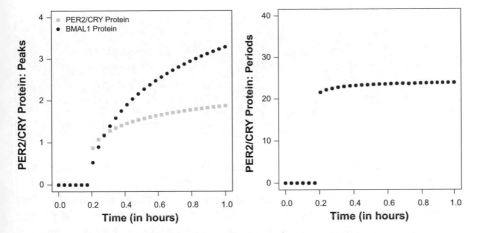

Fig. 10 Effect of varying a, the alcohol sensitivity parameter, on (Left) peak values and (Right) period of oscillation, with other parameter values specified in the sample code. Note that a values near 0 indicate a large alcohol effect; $a = 1$ indicates no alcohol effect

package [6]) and use the peak times to calculate the period of oscillations, then summarize the results in Fig. 10. From this computational work, we see that there must be a bifurcation that occurs at an a close to 0.2; for $a \geq 0.2$, oscillations exist, but for $a < 0.2$, there is a stable steady state without oscillations. We also see that alcohol seems to have a bigger effect on peak BMAL1 protein than on peak PER2/CRY protein. Peak PER2/CRY protein levels are not that different for a-values between 0.3 and 1.0, but peak BMAL1 protein levels are reduced by about half in that same a range. Alcohol seems to affect the period of oscillation of PER2/CRY and BMAL1 proteins in the same way (gray squares and black circles are almost indistinguishable in Fig. 10 (Right panel)).

It is possible to make *bifurcation diagrams* with specialized software like XPP AUTO [16]. A bifurcation diagram plots the stable and unstable steady states, as well as the periodic orbits (regions of oscillations), of the system as a function of a parameter value. Making bifurcation diagrams look "nice" can take quite a lot of finessing, but we can learn a lot of the same information using the numerical technique we employed to make Fig. 10. We will not explicitly discuss the construction of bifurcation diagrams here, but we turn the interested reader to the XPP AUTO tutorial for detailed information and sample codes [14, 15].

Research Project 5 (Bifurcation Diagram) Use XPP AUTO to make a bifurcation diagram using a as the bifurcation parameter. This could take quite a bit of fine-tuning, as well as decision-making. For example, which variable should you focus on as a varies? Perhaps start with PER2/CRY or BMAL1

(continued)

Research Project 5 (continued)
protein. Is it also worth looking at how mRNA levels are affected by a? After creating the bifurcation diagram, analyze it to understand how alcohol affects the concentration of your chosen variable. What insight does this give you into possible targets for treating patients with alcohol-dependent circadian rhythm problems?

We made a fairly simple modification (Eq. (38)) to an already-published model [4]. This is a great way to get started on a research project. Read a published model, looking carefully at the model assumptions. Is there an assumption you could relax, or another factor that you could add? Asking questions will get you thinking about the problem in new ways, and could open up avenues of exploration. For example, in Eqs. (31)–(37), proteins PER2 and CRY were treated as a single protein; no distinction was made between the two. One extension of the model would be to include both proteins (and their corresponding mRNA), in order to understand the relative importance of each individual protein (see Project 7).

Care must be taken not to make the new model overly complicated (assumptions should still be made in order to keep the model "as simple as possible but no simpler"), but modifying an existing model is a good way to get your feet wet with research. Another benefit of modifying an existing model is that you can compare the results of your new model to the results of the published model. You can see if the changes you made had a large effect on the system or not. Even if your results look similar to the original, something important has been learned: whatever changes you made likely *aren't* critical to the underlying problem (or you did not model these critical components appropriately).

Research Project 6 (Time-Dependent Alcohol Parameter) In Eqs. (32)–(38), the effect of alcohol was modeled as a constant, a, or as a piecewise constant. Modify the model so that the alcohol parameter is a time-dependent function, $a(t)$. The help file for the lsoda differential equations solver contains examples of how to accomplish this in general, but for simple choices of forcing function, the term can just be added in-line in the differential equation. Think of functions that might reasonably model alcohol consumption over time, and implement them one at a time into the code. Compare and contrast the results from each case. What insight does this give you into possible targets for treating patients with alcohol-dependent circadian rhythm problems?

Research Project 7 (Modified Model) Modify the model to account separately for PER2 and CRY (rather than including them in a single variable). Do background reading to decide how best to do this. Compare results of this model to the model presented in Eqs. (32)–(38). Was it important to account for the two proteins separately, or was it reasonable to make the assumption that we could treat them as the same protein?

5 Influenza Dynamics, Healthcare Options, and Associated Costs

Our final research example returns to larger-scale, population-level *infectious disease dynamics*. In this project, we study the dynamics of a common pathogen and the healthcare costs associated with its treatment.

At some point in your life, you or someone you know has likely contracted influenza, commonly called "the flu." Symptoms vary, but are often bad enough to keep the sufferer home from work or school. In the USA alone it has been estimated that the annual impact of seasonal influenza infections is approximately 31.4 million outpatient visits, 3.1 million days of hospitalization, and over 600,000 life-years lost [39]. Mathematical models of influenza seek to explain person-to-person transmission dynamics at the population level [10, 38], the immune response to infection at the individual level [3], and even the evolutionary dynamics of the virus within and across epidemics [33]. Broadly speaking, infections by seasonal influenza viruses pose serious risk to the elderly, the young, and to those with compromised immune systems.

The project described in this section was the result of a year-long undergraduate research effort aimed at analyzing traditional epidemiological models from an actuarial science perspective; that is, understanding the healthcare costs associated with a variety of public health approaches to managing seasonal influenza. We were especially interested in how assumptions or decisions about treatment influenced the total healthcare costs accumulated across one or more seasons. This model presents an attempt at specifying a few of the complex medical, economic, and ethical factors that meet at the intersection of infectious disease dynamics and public health and policy. You are encouraged to explore the consequences of our assumptions, as well as some of your own, in a few of the projects at the end of this section.

The diagram in Fig. 11 can be transcribed into Eqs. (39)–(45). Implicit in the model is the assumption that all births create susceptible individuals and that there is no vertical transmission of infection (infection from mother to newborn) or transfer of acquired immunity to offspring. Births occur at a constant rate of p individuals per year and all individuals are subject to natural mortality at a rate m per year. The transmission parameter β characterizes the rate at which new cases occur in

Fig. 11 Compartments and transitions for the influenza model. We consider populations that are healthy or susceptible S, infectious I, vaccinated V, dead D, recovered naturally N, or recovered following treatment R. Though the transition is not shown to simplify the diagram, vaccinated, naturally recovered, and treated individuals are allowed to lose immunity and return to the susceptible group

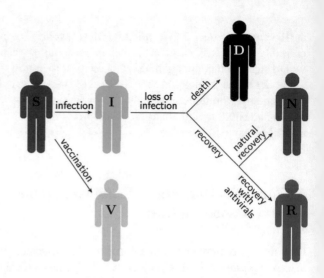

response to interactions between susceptible and infectious individuals. We have modeled the loss of infection (and infectiousness) as a process whereby individuals are lost to death or recovery at a rate γ. A fraction δ die and the remaining fraction $(1 - \delta)$ are assumed to have recovered in response to treatment with probability ρ, or naturally with probability $(1 - \rho)$. Individuals are vaccinated at a rate of ζ per year and all individuals with immunity, including those who have recovered from infection, lose immunity at a rate ω.

$$\text{Susceptible} \qquad \frac{dS}{dt} = p - mS - \beta SI - \zeta S + \omega(V + N + R) \quad (39)$$

$$\text{Vaccinated} \qquad \frac{dV}{dt} = \zeta S - mV - \omega V \quad (40)$$

$$\text{Infectious} \qquad \frac{dI}{dt} = \beta SI - \gamma I - mI \quad (41)$$

$$\text{Natural recovery} \qquad \frac{dN}{dt} = (1 - \delta)(1 - \rho)\gamma I - mN - \omega N \quad (42)$$

$$\text{Recovery after treatment} \qquad \frac{dR}{dt} = (1 - \delta)\rho\gamma I - mR - \omega R \quad (43)$$

In an uncommon twist, we append Eq. (44) to help explicitly track the number of deaths associated with infection

$$\text{Dead} \qquad \frac{dD}{dt} = \delta\gamma I. \quad (44)$$

The distinction between natural recovery, recovery following treatment, and death as outcomes of infection in the model is somewhat unique. In particular, by certain choices of parameters (e.g., $\delta = 0$ and/or $\rho = 1$), we can conveniently experiment with different outcomes of infection using the same model (see Project 12).

Depending on the goal of a particular modeling project, deaths may be ignored or lumped in with members of a "removed" category—those individuals for some reason unable to participate in dynamics. Since significant financial costs are associated with death, we made a decision to track influenza-related deaths. Ignoring Eqs. (44)–(45), the model presented would be a fairly generic and flexible model of infectious disease transmission allowing for vaccination, recovery with and without medical intervention, and infection-associated deaths [2, 8, 26]. We assign healthcare costs associated with vaccination by z, with infection by b, with death by d, with natural recovery by n, and with hospitalization-aided recovery by r. Specifically, Eq. (45) allows us to track the healthcare costs associated with many of the transitions recorded in the other model equations (e.g., cost of vaccination, cost of curative antiviral treatment).

$$\text{Cost} \quad \frac{dC}{dt} = z\zeta S + b\beta S I + d\delta\gamma I + n(1-\delta)(1-\rho)\gamma I + r(1-\delta)\rho\gamma I \quad (45)$$

Many of the healthcare cost parameters listed in the sample code were calculated by scaling national estimates to a population of size and density reflective of Oklahoma County, Oklahoma (roughly one thousand per square mile) [39].

5.1 Role of Technology

The model described in Eqs. (39)–(45) is a seven-dimensional nonlinear system of differential equations, for which numerical approaches are appropriate and quite helpful. As in the other projects presented in this chapter, we begin with computer code for numerically solving the two-dimensional predator–prey model (Sect. 2.2) and update the code to reflect the new variables and equations. See Appendix for new code.

Again, we are interested in both the dynamics of the model and how the model results change as a function of a specific parameter. In this case, we are interested in the vaccination rate, ζ. We conduct a similar experiment to that described in Sect. 3.3, where we solve the model for several values of ζ, and then plot the equilibrium values of the state variables as a function of ζ. The vaccination rate is a parameter that we have some control over in real-life situations, which makes this project particularly interesting. We can use our numerical experimentation to suggest ideal (but realistic) vaccination rates that can minimize the cost of an influenza outbreak.

Finally, technology plays an essential role in an extension to this project: the development of an analogous stochastic model of these dynamics. We will discuss

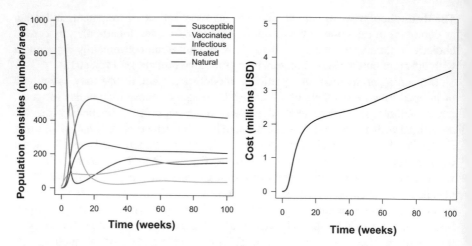

Fig. 12 Sample results for influenza transmission dynamics (Left) and cost (Right) as described in Fig. 11 from initial conditions $S(0) = 980$, $V(0) = 10$, $I(0) = 10$ with other variables initialized to zero

this model and its analysis more carefully in Sect. 6.4, so we only mention here that studying a model of this complexity is really only possible by using numerical approaches.

5.2   Model Dynamics

We present sample dynamics in Fig. 12 (Left panel) for initial conditions and parameters stated in the code samples in the Appendix. We start from a mostly susceptible population with small groups of infected and vaccinated individuals. For the parameters specified in the included computer code (see Appendix), we find a large spike in the infectious group followed by increases in the sizes of the two recovered groups: those that recover naturally outnumber those that undergo treatment. Fortunately, for this parameterization, influenza-related deaths are rather rare. Since much of the population has some form of immunity (following recovery or vaccination), the second outbreak, which is centered near $t \approx 60$ weeks, is relatively small. Additionally, the accumulation of healthcare and associated economic costs is plotted in Fig. 12 (Right panel). The initial rise in the cost function is largely due to costs related to infection and treatment, while the slower, sustained rise beyond $t = 20$ weeks is attributed largely to the cost of vaccination.

Exercise 13 Investigate the cost of vaccination by plotting the right graph in Fig. 12 with and without vaccination included in the cost. Then plot with and without treatment included in the cost. Is it more costly to vaccinate or treat influenza?

5.3 Stability Analysis

After studying the behavior of this model for a variety of parameterizations, we gained some insight into which parameters were of interest biologically or dynamically. Specifically, we chose to numerically solve the model for a range of vaccination parameter values, ζ, then to plot the long-term behaviors as a function of ζ. The graph that this approach yields is similar to a bifurcation diagram generated using by-hand analysis or more specialized software such as XPP AUTO (Fig. 13). We see that for vaccination rates above $\zeta \approx 0.09 \, \frac{1}{\text{week}}$, the infection is eliminated and the population is comprised solely of individuals from the susceptible and vaccinated groups.

These results can be verified by a direct calculation of equilibrium values and an analysis of stability, but we started with the numerical results in order to gain intuition about the model. Now that numerical experimentation seems to suggest that ζ is an important parameter, we can calculate by hand the inequality for which infection is eliminated

$$\zeta > \frac{(\omega + m)(\beta p - m(\gamma + m))}{m(\gamma + m)}. \tag{46}$$

Whether this inequality is satisfied for a given vaccination rate, ζ, depends on the two host demographic parameters (natural mortality rate, m, and population birth rate, p) and three epidemiological parameters (the rate at which infection ends, γ, the transmission parameter, β, and the rate at which immunity wanes, ω).

Research Project 8 (Oscillations) Plan and carry out well-organized numerical experiments with the transmission parameter β, the vaccination rate ζ, the rate at which immunity wanes ω, and perhaps the recovery rate γ, to determine if the short-lived, damped oscillations illustrated in Fig. 12 can ever be sustained.

Research Project 9 (Host Demography: The Birth Function) We chose a constant birth rate p for only susceptible individuals, which has certain important implications for the dynamics. Explore the effects of other choices of birth terms, for example $p(S + V + I + R + N)$. To start, ask yourself how the units of p differ between each form. You might also ask what happens under the popular assumption of a logistically growing population.

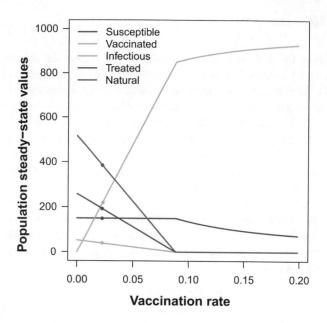

Fig. 13 Long-term numerical solution values plotted against the vaccination rate. The labels "Treated" and "Natural" both refer to recovered populations, with and without antivirals, respectively. Points on the graph illustrate the equilibrium values of the solution plotted in Fig. 12, where $\zeta = 0.023076923 \frac{1}{\text{week}}$. While knowing ζ to so many decimal places is not biologically feasible, we report the decimal places here for mathematical reproducibility

Research Project 10 (Host Behavior: The Transmission Function) The paper "How should pathogen transmission be modelled?" [37] discusses a variety of possible models of pathogen transmission. Modify the influenza healthcare cost model by choosing an alternative form of pathogen transmission as described in the paper. Identify any new assumptions about pathogen transmission or host behavior that come with this new form of transmission function. Alter the model code and explore the dynamics of your modified model.

Research Project 11 (Quarantine) Imagine an alternative where we consider the role of quarantine—that is, the removal of infectious individuals to prevent further transmission. How do the diagram and equations change? Are these individuals given antivirals that may lead to recovery, or simply supportive care that may allow them to recover naturally?

Research Project 12 (To Treat or Not to Treat) According to cost estimates made from the literature, natural recovery is relatively inexpensive
when compared to recovery following treatment, and especially when compared with the case where infection ends in death. With this in mind,
experiment with the parameter δ (the probability that infection ends with
death, where larger values indicate a particularly lethal infection) and/or ρ
(the probability that an infection ends as the result of treatment). Imagine your
goal is to keep people healthy and costs low. What can you do to reduce costs
over time by preventing infection through vaccination and providing treatment
to those infected? From your experiments, what model parameters (i.e., public
health strategies) might you strive for and what real-world constraints might
you face?

6 Stochastic Models

The models presented thus far can be interpreted as giving the average outcome of
the systems. If we are interested in individual, specific outcomes, in which chance
or randomness can play a role, we need a new modeling technique.

Numerical experimentation is particularly useful when studying problems that
involve randomness. All the models discussed up to this point were *deterministic*;
that is, given initial conditions, the models predict the exact, unique outcome
with complete certainty. Now we will transition to *stochastic* (or random) models.
Stochastic models give the probability of an outcome; given the same initial
conditions, a stochastic model will produce a different final result every time it is
solved. Each of these solutions represents a sample path from the unique underlying
probability distribution. In this section we will discuss stochastic models and
stochastic extensions of deterministic models discussed earlier in the chapter. For a
thorough overview of stochastic processes and their relation to biology, see [1, 19].

6.1 Simple Birth–Death Process

A *stochastic process* is a system that describes the evolution of a time-dependent
random variable, for example, $N(t)$, the number of individuals in a population. The
probability that the random variable N takes on the value of n at time t is given by

$$P(N = n, t) = P(n, t) = P_n(t) \tag{47}$$

where in general it is more convenient to subscript by the value of n than to include n as an argument to the function [1, 19]. A *master equation* of a stochastic process is an equation for the time evolution of probability distributions. For example,

$$\frac{dP_n(t)}{dt} = \mu(n+1)P_{n+1}(t) + r(n-1)P_{n-1}(t) - (\mu+r)nP_n(t) \tag{48}$$

could describe the time evolution of the probability of a population having n individuals at time t. Here we let r ($\frac{1}{\text{time}}$) denote the birth rate and μ ($\frac{1}{\text{time}}$) denote the death rate. The left-hand side of Eq. (48) represents the change in probability (given by the derivative), and the right-hand side lists all the ways the population could transition to or from the n state. The first two terms on the right-hand side say that we can increase the probability of the population having n individuals in two ways: (1) the population could have contained $n + 1$ individuals, but one of them died; (2) the population could have contained $n - 1$ individuals, and a new one was born. These terms are multiplied by $(n + 1)$ and $(n - 1)$, respectively, because any one of the $n + 1$ or $n - 1$ individuals could have died or given birth. The probability the population has size n is decreased if the population was already size n, and one individual died or was born. This type of model is called a birth-death process.

Solving master Eq. (48) analytically is possible, but requires the use of moment generating functions to derive a partial differential equation that can be solved using the method of characteristics (see [19] for details of the solution techniques). These are nontrivial mathematical concepts. The probability the population has size n at time t, $P_n(t)$, is obtained by Taylor expanding the probability generating function $G(t, z)$ below about $z = 0$,

$$G(t, z) = \left(\frac{(z-1)\mu e^{(r-\mu)t} - rz + \mu}{(z-1)r e^{(r-\mu)t} - rz + \mu} \right)^N, \tag{49}$$

where N is the size of the population at time $t = 0$. Though there is certainly nothing wrong with it, we imagine you agree that this is not a particularly illustrative solution. How many individuals are we likely to end up with at any given time? It is not at all clear from the analytical solution. We could use technology to Taylor expand then plot the resulting function in order to gain more intuition into what is happening. However, in practice, in our undergraduate research programs, we would approach this problem numerically from the get-go. We explain our approach in the following section.

6.2   Gillespie Algorithm

"We" (read: someone else) solved the relatively simple master equation for the birth–death process in Eq. (48), to obtain the probability distribution of population size as a function of time. However, for even slightly more complicated systems, the master equation can become very difficult (or even impossible) to solve analytically. In this case, numerical computation is critical. The *Gillespie algorithm* generates statistically exact realizations of the master equation. In other words, a result obtained by solving the master equation using the Gillespie algorithm is not an approximation; it is a true solution (out of many possible solutions, because of the randomness) of the master equation. This means we can use a computer to generate exact statistics, instead of having to analytically solve a big, potentially ugly, master equation.

The Gillespie algorithm explicitly simulates every possible reaction between the individuals in the system; the algorithm randomly chooses which reaction happens next, and the time at which the reaction happens, using exponential distributions and parameter values from the model [20, 21]. The basic steps of the Gillespie algorithm are as follows:

0. Initialize the system so $t = 0$ and $x =$ number of individuals.
1. Compute the propensity functions (based on the reaction rates), $a_i(x)$ for $i = 1, \ldots$ total number of reactions, and the sum, $a_0(x) = \sum_i a_i(x)$.
2. Pick 2 uniformly distributed random numbers, r_1 and r_2. To determine which reaction happens next, we calculate r_1 and then look for the first j for which

$$r_1 < \frac{1}{a_0(x)} \sum_{i=1}^{j} a_i(x).$$

Compute the time to the next reaction, τ, via

$$\tau = -\ln(r_2)/a_0(x).$$

3. Update the system by increasing the time by τ, and by updating the number of individuals in the system according to the chosen reaction.
4. Return to step 1 and repeat, until either there are no individuals left, or we have reached a pre-determined final time or final number of events.

The propensity functions, $a_i(x)$, can be thought of as the stochastic reaction rates. Given that the system contains x individuals at time t, the probability that reaction i occurs somewhere in the system in the infinitesimal time interval $[t, t + dt)$ is given by $a_i(x)dt$ [22]. For our purposes, we can think of these propensity functions as the number of individuals at time t, x, times the rate at which that number changes. For example, the propensity function describing a birth would be $a_1(x) = rx$, because each of the x individuals could birth a new individual at rate r.

6.2.1 A Sample Calculation

Let us walk through one iteration of a specific example to see how this works in practice. Assume we have a birth-death process and we start with a population with eight individuals. Assume the birth rate is $r = 0.4 \frac{1}{time}$, the death rate is $\mu = 0.1 \frac{1}{time}$, and our initial population contains eight individuals.

0. Initialize by setting $t_1 = 0$, $x_1 = 8$.
1. Set $a_1 = rx_1 = 0.4(8)$ and $a_2 = \mu x_1 = 0.1(8)$, so $a_0 = 0.5(8) = 4$.
2. Pick $r_1 = 0.3077661$, $r_2 = 0.2576725$ from a uniform distribution.

   (a) Now start summing up the propensity functions, checking to find the first for which the sum is greater than $r_1 a_0 = (0.3077661)(4) = 1.2310644$:
   $a_1 = 3.2 > 1.2310644$. Since we have a number bigger than 1.2310644, we stop here (there is no need to add the second propensity function). This means the first reaction is the one that occurs next.
   (b) Calculate the time to the next reaction:
   $\tau_1 = -\ln(0.2576725)/4 = 0.3390165$.

3. Update $t_2 = t_1 + \tau_1 = 0 + 0.3390165 = 0.3390165$, and $x_2 = x_1 + 1 = 8 + 1 = 9$ (because the reaction that was chosen in step 2 corresponds to birth).
4. Repeat, with the new "initial" conditions $t_2 = 0.3390165$, $x_2 = 9$.

6.2.2 A Computer Implementation

To ensure that the code exactly reproduces the first few results presented in the worked example above, we will set the seed of the random number generator using set.seed(100), where there is nothing special about our choice of the number 100. Any time you run such a model it is good to set the seed in the file so that you can reliably reproduce the sequence of values, possibly for debugging purposes— was your result the result of random chance or a programming error?

```
set.seed(100)  ## for reproducibility
## parameters
   (r <- 0.4)
   (mu <- 0.1)
## transition probabilities
   (a <- c(r, mu))
## transitions
   (trans <- c(1, -1))
## initialization
   (Xs <- X <- 8)
   taus <-  0
```

Below we choose to run a maximum of 199 steps of the method (if it is even allowed to go that far). The loop starts at i=2 since the step corresponding to `i=1` was taken care of during the initialization above. By default, the `for`-loop prints no output to the screen. Below the assignments are wrapped with parentheses for testing purposes. This way we could manually set a few values for `i` and step through the calculation one line at a time and see printed output. Equally useful is the included, but commented out, line that prints the reaction selected in the method. Uncommenting this allows the user to detect if the same reaction is repeatedly, and perhaps erroneously, selected due to the misspecification of the probability values or other coding errors. One important line, `if(Xs[i]==0)break;` ensures that the calculation stops when the population reaches zero. Notice that if we continued the calculation with a population size of zero that we would divide by zero in the following iteration.

```
for(i in 2:200){
    if(X==0)break;
    (rn <- runif(2))
    (rate <- a*Xs[i-1])
    (pr <- cumsum(rate)/sum(rate))
    (rxn <- min(which(pr > rn[1])))
    ## print(rxn) ##(useful to debug)
    (X <- X + trans[rxn])
    (Xs <- c(Xs, X))
    (taus[i] <- - log(rn[2])/sum(rate))
}
```

For reasons we will discuss in the exercises, we have chosen to store the time increments τ rather than the progression of time t itself. Before plotting we should calculate the cumulative sum of the `taus` to calculate the actual clock time. The remaining two lines allow us to specify "nice" upper axes limits based on the results of the simulation.

```
## calculate actual time
    (times <- cumsum(taus))
    (tmax <- ceiling(max(times)))
    (xmax <- ceiling(max(Xs)))
```

The results can be plotted using the following code which displays the integer-valued solution to the stochastic model as points, and the calculated exact solution to the corresponding deterministic model as a smooth curve (Fig. 14). In general, plotting a function in R is not quite as easy as it is in other "symbolic"-oriented languages—here it requires the specification that we are plotting a `function(t)` so that the values of the independent variable can be provided in the specification of the x-axis limits.

```
plot(function(t)Xs[1]*exp((r-mu)*t), col=2, lwd=2,
    xlim=c(0, tmax), ylim=c(0, xmax),
    xlab='Time', ylab='Population size', las=1)
points(times, Xs, pch=19, lwd=2)
```

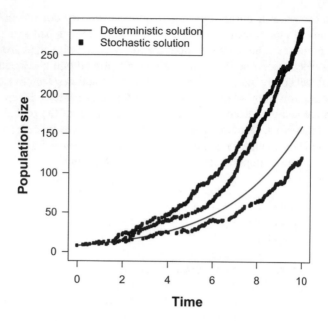

Fig. 14 Three sample stochastic solutions plotted with the corresponding deterministic solution (from an appropriately parameterized Eq. (1)). In all cases, the initial population size is 8

To really get a sense of what stochastic models are all about, at this point we should rerun the entire code (except for the `set.seed` line, since that would lead to the identical randomization and result, and the `plot` line, since that would open a new plotting window). Running just the final line of the plotting commands (`lines`) will superimpose the results of the new simulation on top of the existing plot that contains the deterministic solution and first stochastic sample solution (Fig. 14).

Exercise 14 Use the Gillespie algorithm code provided to run ten simulations of a birth-death process with $r = 0.1 \; \frac{1}{\text{time}}$ and $\mu = 0.4 \; \frac{1}{\text{time}}$, for initial population sizes of 8 and 80. Set the final time to be $t = 100$. What happens in the ten cases? Does the population always go extinct by time 100? Does the population sometimes go extinct and sometimes survive? How do your results change if $r = 0.4 \; \frac{1}{\text{time}}$ and $\mu = 0.1 \; \frac{1}{\text{time}}$? Experiment by changing other values and see what results you get. Can you justify your results biologically?

Even for simple stochastic models, the analysis of master equations becomes quite involved. Perhaps more important than the relative ease of studying stochastic models by simulation is the realization that the stochastic analog of a deterministic model can allow the researcher to approach vastly different questions. In even the simplest population model (such as that from Eq. (1)), and depending on parameters, the solution may indicate that a population exponentially decays *towards* a size of zero, but never reaches zero. In the stochastic analog to this model, extinction is indeed possible. The stochastic model allows for calculations about the time to extinction or how the time to extinction depends on initial population size.

6.3   Stochastic Extension of Predator–Prey Model

Here we formulate the stochastic analog of the predator–prey model. The deterministic predator–prey model can be made into a stochastic model if we replace the rates by their underlying event probabilities. Practically, we think of replacing the interactions that occur in the deterministic model with the corresponding interactions and transitions that define the stochastic model. A very rewarding exercise is to write the master equation for this model, describing the probabilities of having $H = h$ hares and $L = \ell$ lynxes at time t. Instead of writing the master equation, here we will specify the interactions and transitions in tabular form in Table 2.

Exercise 15 Write the master equation for the stochastic predator–prey model.

With these transitions, we can formulate a transition matrix (shown with states as labeled rows and transitions as labeled columns in Eq. (50)).

$$
\begin{array}{c|cccc}
 & \alpha H & \beta HL & \gamma HL & \delta H \\
\hline
\text{Hare} & 1 & -1 & 0 & 0 \\
\text{Lynx} & 0 & 0 & 1 & -1
\end{array} \tag{50}
$$

To implement the Gillespie algorithm for this problem we store the reaction rates, initialize the state vector X and solution Xs, initialize the vector of time increments, and define the transition matrix.

```
parms <- c(alpha=1.1, beta=0.05, gamma=0.01, delta=0.5)
Xs <- X <- t(c(H=50, L=8))
taus <- 0
trans <- rbind(c(1, -1, 0, 0), c(0, 0, 1, -1))
```

Table 2 Interactions and terms from the model in Eqs. (20) and (21) with the associated stochastic transitions

Interaction	Term	Transitions
Hare production	αH	$h \leftarrow h + 1, \ell \leftarrow \ell$
Lynx-hare interaction (predation)	βHL	$h \leftarrow h - 1, \ell \leftarrow \ell$
Lynx production	γHL	$h \leftarrow h, \ell \leftarrow \ell + 1$
Lynx mortality	δH	$h \leftarrow h, \ell \leftarrow \ell - 1$

Arrows are used to indicate the assignment of a new value. For example, the first row corresponds to prey birth so the new h value is $h + 1$ and ℓ value is unchanged (set to itself)

Now we step through the process for an arbitrary number of steps (here up to 10,000), all the while ensuring that there is at least one individual in the population. Should both populations go extinct, the calculation halts. Our rate vector includes 4 possible events (see Table 2 and Eq. (50)), and once a "reaction" is selected that column of the transition matrix is selected in order to update the vector of states. After the reaction is selected, the corresponding time increment is stored as `taus[i]`.

```
for(i in 2:10000){
## stops at extinction
   if(sum(X) == 0)break;
## same as before
   rn <- runif(2, 0, 1)
## rate contains more possibilities
   rate <- with(as.list(parms),
      c(alpha*X[1], beta*X[1]*X[2],
      gamma*X[1]*X[2], delta*X[2]))
## same as before
   prop <- cumsum(rate)/sum(rate)
   rxn <- min(which(rn[1] < prop))

## apply transition and store state
   X <- X + t(as.matrix(trans[, rxn], 2, 1))
   Xs <- rbind(Xs, X) ## other ways to do this
## store time increment
   taus[i] <- -log(rn[2])/sum(rate)
}
```

To this point we have only stored the time increments; from here calculating the vector of event times is as simple as taking the cumulative sum of the increments. We bind the event times (`times`) to the model states (which were stored in the two-column matrix `Xs`), then assign names to the columns to facilitate use.

```
## calculate event times
   times <- cumsum(taus)
## organize output
   dat <- as.data.frame(cbind(times, Xs))
   names(dat) <- c("time", "H", "L")
```

A sample solution to the stochastic model, generated with initial conditions and parameters matching those from the deterministic model and using the code above, is plotted with the code below. This solution is plotted with the corresponding deterministic solution in Fig. 15 (Left panel). The sample stochastic solution is the thinner, jagged curve and the deterministic solution is the thicker, smoother curve. Additional 9 solutions to the stochastic model are shown in Fig. 15 (Right panel). This can be accomplished by rerunning the stochastic code and adding solutions to the plot of the first solution using `matlines()`.

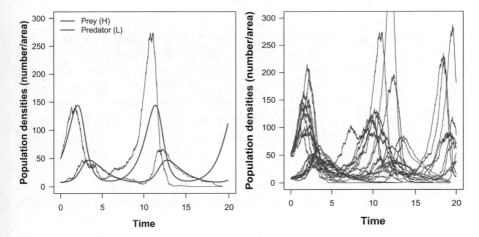

Fig. 15 Comparison of stochastic and deterministic results to the predator–prey model. (Left) Smooth curves show deterministic solution and jagged curves show stochastic solution. (Right) Ten sample stochastic solutions with identical parameters and initial conditions. In all cases, the initial conditions are 50 hares and 8 lynxes

```
## a fairly generic plot, one solution
matplot(dat$time, dat[, c("H", "L")], type='l', lwd=1,
    col=c('blue', 'red'), lty-1, xlab-"Time",
    ylab-"Population densities (number/area)", las=1)
```

Figure 15 (Right panel) shows that the oscillatory nature of the solutions often persists, but variability in the second peak (and its timing) for each species is quite striking compared to the corresponding solution of the deterministic model.

By using the Gillespie algorithm to numerically solve the stochastic predator–prey model, we were able to quickly generate many sample solutions, which would have been unattainable through analytical techniques alone. Additionally, we could run the Gillespie algorithm thousands of times more and generate distributions of time to the extinction of either species, of the time lag between peaks, and of peak population sizes. In the deterministic model, extinction is impossible and peak heights and periods do not vary. So we are able to obtain more (and different) information from a stochastic model, but only through the use of numerical computation. From a practical perspective, numerically solving this more complex stochastic model requires only slight modification to the approach used in the simpler one-dimensional case. Scaling up pen-and-paper analysis in a similar manner would be more work, and at some point, practically impossible.

Research Project 13 (Three-Dimensional Predator–Prey Model) In Exercises 10 and 11 in Sect. 2.2, you were asked to write down three-dimensional deterministic models involving either two predator and one prey species, or one predator and two prey species. Write the stochastic extensions of these models, and code up the Gillespie algorithm to solve the models. Think of a real-world system that could be described by either of these models. Read about those species, and try to estimate some of the reaction rates. Gather some data from the model output (e.g., extinction time distributions for several values of a parameter of interest). How do the results change with changing parameter values?

6.4 Stochastic Extension of Influenza Dynamics Model

Similar to the stochastic extension of the predator–prey model, it is possible to list all reactions in the influenza dynamics model (see Eqs. (39)–(45)). To keep Table 3 as simple as possible, we only list one example of the possible reactions associated with mortality and with loss of immunity.

The reactions are ordered in a vector similar to, but much bigger than, the one used in Sect. 6.2.2, and we construct a transition matrix to help update the model state vector once a reaction is selected in the Gillespie algorithm. In this model, some reactions have an associated healthcare cost which accumulates with the selection of a reaction. As we saw in the earlier stochastic models, the stochastic influenza model can be used to study a variety of unique questions involving extinctions or rare outcomes.

In Fig. 16 (Left panel), we show the result of a stochastic simulation of the model compared with the associated deterministic model results obtained by numerically solving Eqs. (39)–(42). Black dots indicate points in time for which

Table 3 Interactions and terms from the model in Eqs. (39)–(44) with the associated stochastic transitions

Interaction	Term	Transitions
Susceptible increase	p	$S \leftarrow S + 1$
Mortality	e.g., mS	$S \leftarrow S - 1$
Infection	βSI	$S \leftarrow S - 1, I \leftarrow I + 1$
Vaccination	ζS	$S \leftarrow S - 1, V \leftarrow V + 1$
Immunity loss	e.g., ωV	e.g., $V \leftarrow V - 1$ and $S \leftarrow S + 1$
Infection-associated mortality	$\delta \gamma I$	$I \leftarrow I - 1, D \leftarrow D + 1$
Natural recovery	$(1 - \rho)(1 - \delta)\gamma I$	$I \leftarrow I - 1, N \leftarrow N + 1$
Recovery with antivirals	$\rho(1 - \delta)\gamma I$	$I \leftarrow I - 1, R \leftarrow R + 1$

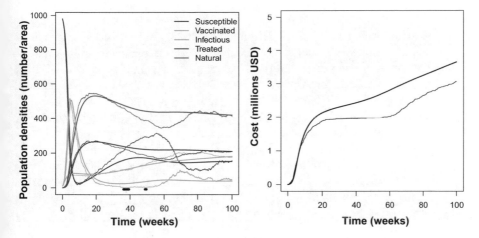

Fig. 16 Comparison of stochastic and deterministic results. (Left) Full sample stochastic and deterministic solution over 100 weeks. (Right) Evolution of cost for stochastic model. Initial conditions are the same as in Fig. 12

only two infectious individuals are present in the stochastic solution. Remarkably, the infection persists through these periods (which range in length from 2 to 4 days); in other words, influenza was on the cusp of eradication, but was able to bounce back. In a different stochastic solution, it is possible that influenza would go extinct. These types of results are impossible to obtain from the deterministic model, which does not reach zero in finite time (solutions can asymptotically approach zero, but never actually reach it).

We could use a collection of simulation results, say 1000 or more, to calculate the distribution of the times at which the infection vanishes, of the peak infection sizes, or of a number of other characteristics. A hallmark of deterministic models is that using identical initial conditions and identical parameters, we would expect identical results from two analyses (analytical or numerical) for the same model. In contrast, as few as two solutions to an identically parameterized and initialized stochastic simulation illustrate the surprising variability inherent in these models.

Research Project 14 (Stochastic Influenza Model) Write code to solve the stochastic influenza model given in Table 3. Fix the parameter values and initial conditions and run at least 1000 simulations, saving the data from each simulation. Calculate how many times the infection vanished, and investigate how this percentage changes as certain parameter values are varied. Think carefully about which parameters can be controlled. If you find a parameter set for which the infection often vanishes, calculate the distribution of times at which the infection vanishes. Repeat this type of experiment, but now looking at cost of infection. Continue to ask (and answer) questions like this: what would you recommend to someone who is trying to reduce the effect of influenza in a population?

7 Conclusion

With computers we can numerically solve systems of equations that are challenging, if not impossible, to solve analytically. Since we can then generate numerical solutions quite quickly and easily, we can carry out numerical experiments by varying different parameter values to learn how model dynamics are affected. As research progresses, comparing results to real-world data can help identify likely values of unknown parameters. We can use numerical results to help decide which avenues to follow for additional experimentation, or to set up a more targeted pen-and-paper analysis. Through this chapter, we hope we have convinced you of the value of programming and computation as research tools and demonstrated that numerical experimentation is an incredibly important approach when analyzing mathematical models. To encourage your pursuits, notice that we have provided the information and the code necessary to start experimenting on your own.

To start a mathematical modeling research project, you can find a published model and modify it somehow (like Sect. 4), or learn the relevant biology needed to comfortably write your own model (like Sects. 3 and 5). Writing a model is a great way to organize your thoughts and assess your understanding of how various parts of the biological system are connected. Once the model has been written, you can quickly solve it numerically to see if results make sense. If the dynamics are really different than expected, you know you need to modify the model before proceeding. Perhaps you left out an important variable, or are using unrealistic parameter values, or simply have a typo in your code. Once you obtain reasonable results, you can start to experiment in earnest, or perhaps reach out to experts in the field. What happens if you change one parameter value? Are there bifurcations in your model, where qualitative behavior changes as you change a parameter value? What do you learn about the biological system you are studying? Can you propose experiments to biologists to help confirm your model predictions?

As you pursue research, it is important to keep asking questions. After all, your project began with a question, one asked by you or suggested by someone else. As you generate results in the search for answers, keep an open mind to the new questions that emerge.

Acknowledgements The authors wish to acknowledge undergraduate students Emily Kelting, Grace Kelting, and Chaissone Moore whose research helped inspire many of the projects in this chapter. Though their work is not explicitly featured in the collection of projects described in the chapter, Sean's work with two former students, Katie Link and Bryan Dawkins, had an important impact on the development of many of the programming-infused approaches to the pursuit of undergraduate mathematical biology research and the role of numerical experimentation. Many of these projects have benefitted from research and travel funding from the University of Central Oklahoma's Office of High-Impact Practices and the RCSA program, and the College of Mathematics and Science's CURE-STEM program. We appreciate the supportive research environment provided for undergraduate students at University of Central Oklahoma.

Appendix 1: Notes on Programming

Where possible we have attempted to simplify code in such a way to allow you to more simply reproduce the figures in this Chapter. We expect that the results you are able to produce from running code snippets will match those that we have included in the chapter, but please do not be alarmed by slight aesthetic differences. In particular, we prefer to use the mtext(...) command in R to generate axis labels which can be adjusted slightly more easily than can the defaults. For example, to produce many of our figures in the text, by plotting the model output . . ., we recommend the code

```
plot(..., xlab='Time', ylab='Population size')
```

For our own use, we prefer the slightly more complex code below, which results in slightly better looking output, though this is a matter of preference.

```
par(mfrow=c(1, 1), mar=c(4.1, 4.1, 1.1, 1.1))
plot(..., xlab='', ylab='') # empty axis labels
mtext('Time', 1, line=2.5, font=2, cex=1.25)
mtext('Population size', 2, line=2.5, font=2, cex=1.25)
```

The first line of this code sets the layout of the plot, here one row by one column, and the second sets the plot margins in the order bottom-left-top-right. We add our plot (with blank labels) to a window of that shape and configuration and add text to the margins. The ambitious reader may be interested in exploring the graphics capabilities of the ggplot2 package.

Figures can be saved from menu selections or using the following approach

```
pdf('filename.pdf', height=5, width=5)
## insert arbitrary figure code
dev.off()
```

Table 4 List of useful commands and operations included in the base installation of R

Code	Name	Example	Notes
`<-`	Assignment operator	`x <- 5`	Assigns the value 5 to the object x
`c()`	Concatenation	`x <- c(1, 2, 3)`	Assigns the vector to the object x
`+ - * / ^`	Arithmetic operations	`((1 + 2)/(3*4))^2`	* Is needed for multiplication
`plot()`	Plot	`plot(x, y)`	Plots vector y against x (for equally sized vectors)
`mtext()`	Margin text	`mtext('x-axis', 1)`	Adds 'margin text' to side 1 (bottom), with other options available
`legend()`	Plot legend	See text	Adds a legend with custom text and line styles and colors
`lines()`	Adds lines (or curves) to an existing plot	`lines(x2, y2)`	This function is not 'strong enough' to open its own window and can only be used to add to existing plots

Table 5 List of differential equations solvers and commands included in `deSolve`

Code	Name	Example	Notes
`euler()`	Euler's method	`euler(xstart, times, <<model>>, parms)`	Poor choice, requires small time step
`rk()`	Runge–Kutta methods	`rk(..., method='rk4')`	Same arguments as above can specify `method=` to be one of `rk2`, `rk4` or others in documentation
`lsoda()`	'Livermore' solver	`lsoda(...)`	Robust solver, default for `ode(...)` function, can be called directly

Entering `?pdf` into the R console will show how the `pdf(...)` command works and related commands like `png(...}` or `jpeg(...)`.

A few examples of syntax and useful commands for basic calculations or plotting are given in Table 4. A brief overview of available differential equations solvers from the `deSolve` package is given in Table 5.

Appendix 2: Circadian Rhythm Code

Start by defining the model. We have included some `if` statements that allow us to turn the alcohol parameter on or off at a given time (`t.switch1` and `t.switch2`), to test how the system responds. Earlier we mentioned having to find and replace in order to replace the printed single quote used to specify many arguments in R if you copy code from an electronic source. Here we use double asterisks (**) to enter exponents rather than the more familiar caret (^), since the latter does not copy well from electronic text.

```
## the model
CLOCKmod <- function(t, x, parms)   {
  with(as.list(c(parms, x)), {
    ## alcohol value switch
    if(t > t.switch1){
    a <- 1
    }
    if(t > t.switch2){
    a <- 1
    }
```

```
#conc. of Per2/Cry mRNA
    dY1 <- (a*v1b*(Y7+c))/(k1b*(1+(Y3/k1i)**p+Y7+c))
       -k1d*Y1
# conc. of PER2/CRY complex in the cytoplasm
    dY2 <- k2b*(Y1)**q - k2d*Y2 - k2t*Y2 + k3t*Y3
# conc. of PER2/CRY complex in the nucleus
    dY3 <- k2t*Y2 - k3t*Y3 - k3d*Y3
# conc. of Bmal1 mRNA
    dY4 <- (v4b*(Y3)**r)/((k4b)**r + (Y3)**r) - k4d*Y4
# conc. of BMAL1 protein in the cytoplasm
    dY5 <  k5b*Y4   k5d*Y5    k5t*Y5 i k6t*Y6
# conc. of BMAL1 protein in the nucleus
    dY6 <- k5t*Y5 - k6t*Y6 - k6d*Y6 + k7a*Y7 - k6a*Y6
# conc. of transcriptionally active form BMAL1
    dY7 <- k6a*Y6 - k7a*Y7 - k7d*Y7
# the output
    list(c(dY1, dY2, dY3, dY4, dY5, dY6, dY7))
})}
```

Again to verify our assignments, we print the parameters and initial conditions to the screen, but since times refers to a *very* long vector, we assign it but do not print it to the screen.

```
(parms <- c(v1b = 9, k1b = 1, k1i = 0.56, c = 0.01,
    p = 8, k1d = 0.12, k2b = 0.3, q = 2, k2d = 0.05,
    k2t = 0.24, k3t = 0.02, k3d = 0.12, v4b = 3.6,
    k4b = 2.16, r = 3, k4d = 0.75, k5b = 0.24,
    k5d = 0.06, k5t = 0.45, k6t = 0.06, k6d = 0.12,
    k6a = 0.09, k7a = 0.003, k7d = 0.09,
    a = 1, t.switch1 = 5000, t.switch2 = 5000))
times <- seq(0, 5000, by=0.1)
(xstart <- c(Y1 = 0.25, Y2 = 0.3, Y3 = 1.15,
```

```
    Y4 = 0.85, Y5 = 0.72, Y6 = 1.35, Y7 = 1.075))
```

To assess the role of the alcohol sensitivity parameter on dynamics, we first numerically solve the model for the original parameter value $a = 1$, store, and plot the solution over the interval of interest.

```
## calculate solution
    out   <- lsoda(xstart, times, CLOCKmod, parms)
    out   <- as.data.frame(out)
## plots
    plot(out$time, out$Y3, type='l', col='gray',
        lty=1, lwd=2, ylim=c(0, 5), xlim=c(0, 250),
        xlab='Time (hours)',
        ylab='Protein amounts (nM)', las=1)
    lines(out$time, out$Y5 + out$Y6 + out$Y7,
        type='l', col='black', lty=2, lwd=2)
```

Then, we change the value of a in the list of parameters by manually resetting it. Since the initial conditions, solution times, and the structure of the model itself are unchanged, we can simply calculate a new solution with the updated parameters and add the desired results to the existing plot. This realization allows us to greatly simplify our computer code by leaving out large, redundant chunks that would otherwise be tempting to repeat.

```
## fix parameter value, repeat as necessary
    parms['a'] <- 0.1
    out2   <- lsoda(xstart, times, CLOCKmod, parms)
    out2   <- as.data.frame(out2)
```

After computing the solution for the parameter of interest, we can add these solutions to our existing plot to better understand the role of the parameter.

```
## add corresponding solutions
    lines(out2$time, out2$Y3, type='l', col='red',
        lty=1, lwd=2)
    lines(out2$time, out2$Y5 + out2$Y6 + out2$Y7,
        type='l', col='red4', lty=2, lwd=2)
```

Finally, since the plot is now quite complex, we add a nice legend.

```
## add legend
    legend('topleft', c('PER2/CRY Protein',
        'BMAL1 Protein', 'PER2/CRY Protein (w/ alcohol)',
        'BMAL1 Protein (w/ alcohol)'), col=c('gray',
        'black', 'red', 'red4'), lty=c(1, 2, 1, 2), lwd=2,
        bg='white')
```

For space purposes, we have not included the code in which peak concentrations were calculated and stored. This was done using the pracma package [6], and can be found in the project's GitHub page.

Appendix 3: Influenza Code

First we define the model, since the equation for the change in cost is long, the + at the end of the first two parts of the equation ensures that each subsequent line is counted as continuation of the previous line.

```
## the model
FLUmod <- function(t, x, parms)  {
  with(as.list(c(parms, x)), {
#for all, m is parameter for natural mortality
#Susceptible
    dS <- p - m*S - beta*S*I - zeta*S + omega*(V + N + R)
#Vaccination
    dV <- zeta*S - m*V - omega*V
#Infectious
    dI <- beta*S*I - gamma*I - m*I
#Recovered Naturally
    dN <- (1 - delta)*(1 - rho)*gamma*I - m*N - omega*N
#Treated, coded as R to avoid confusion with T for TRUE
    dR <- (1 - delta)*(rho)*gamma*I - m*R - omega*R
#Mortality
    dD <- delta*gamma*I
#Added cost equation
    dC <- z*zeta*S + b*beta*S*I + d*delta*gamma*I +
      + n*(1 - delta)*(1 - rho)*gamma*I   +
      + r*(1 - delta)*(rho)*gamma*I
    list(c(dS, dV, dI, dN, dR, dD, dC)) # the output
})}
```

Then we define parameters, the list of times for which to calculate the solution, and initial conditions. The factor of $1/52$ included in many of the parameter values was used to scale all values to be measured in weeks. To provide example dynamics in Fig. 12, we solve the model over a period of 2 years.

```
## parameters
   parms <- c(p = (13)*(1/52), m = (0.013)*(1/52),
       beta = (0.08)*(1/52), zeta = (1.2)*(1/52),
       gamma = (12)*(1/52), omega = (0.8)*(1/52),
       delta = 0.0005, rho = (1/3), z = 52.92,
       b = 3.00, d = 542430.70, n = 110.28, r = 5613.65)
## time points
   times <- seq(0, 52*2, by=0.1)
## initial values
   xstart <- parms['p']/parms['m']*c(S = 0.98,V = 0.01,
       I = 0.01, N = 0, R = 0, D = 0, C = 0)
```

Finally, we use the numerical solver `lsoda` to find the model solution.

```
out <- lsoda(xstart, times, FLUmod, parms)
out <- as.data.frame(out)
```

The `matplot(...)` command lets us plot multiple columns simultaneously (even allowing for different line styles or widths in addition to colors). We can access multiple columns of output, each of which corresponds to a model variable, by referring to their numbers (e.g., `out[, c(2, 4, 6)]`), a range of numbers (e.g., `out[, 2:6]`, which is all of the entries 2 through 6), or by name (as shown in the sample code below). Since the vector of custom colors is rather long, we define the `cols` at the start, which can be referenced anytime the variables are plotted.

```
## define colors
cols <- c("red", "gray", "green", "blue", "magenta")
## panel 1
matplot(out$time, out[, c('S', 'V', 'I', 'R','N')],
    type='l', lty=1, lwd=2, col=cols,
    xlab='Time (weeks)', ylab='Population
    densities (number/area)', las=1)
legend('topright', c('Susceptible', 'Vaccinated',
    'Infectious', 'Treated', 'Natural'),
    col=cols, lty=1, lwd=2, bg='white')

## panel 2 (cost scaled to millions)
plot(out$time, out$C/1e6, type='l', lty=1,
    xlab='Time (weeks)', ylab='Cost (millions USD)',
        las=1)
```

To solve the model for a range of ζ parameter values, then plot the long-term behaviors as a function of ζ as it varies, we use the code below.

```
## time vector
    times <- seq(0, 52*200, by=0.1)
## range of zeta values
    zetas <- c(seq(0.0, 0.2, by=0.01))
## object to store model output
    vals <- NULL

for(i in 1:length(zetas)){
    parms['zeta'] <- zetas[i]
    out  <-  lsoda(xstart, times, FLUmod, parms)
    out  <- as.data.frame(out)
    if(i/10==floor(i/10))print(i) ## progress bar
    vals <- rbind(vals, out[nrow(out), ])
}
```

The above code assigns the parameter value of interest, calculates the corresponding solution, and stores the final output of the solution in an object called vals. Here we take advantage of the power of the computer to generate solutions over a *very* long span of time, just to ensure that solutions have equilibrated. Inspecting plots of solutions against time (as in Fig. 12) offers some explanation, as the dynamics of the two recovered groups are especially slow.

```
matplot(zetas, vals[, c('S', 'V', 'I', 'R', 'N')],
    type='l', lty=1, lwd=2, col=cols,
    xlim=c(0, 0.2), ylim=c(0, 1000), xlab='Vaccination
    rate', ylab='Population steady-state values', las=1)
legend('topleft', c('Susceptible', 'Vaccinated',
    'Infectious', 'Treated', 'Natural'),
    col=cols, lty=1, lwd=2, bg='white')
```

References

1. L. J. S. Allen. *An Introduction to Stochastic Processes with Applications to Biology.* CRC Press, second edition, 2010.
2. R. M. Anderson and R. M. May. *Infectious Diseases of Humans: Dynamics and Control.* Oxford University Press, 1992.
3. C. A. Beauchemin and A. Handel. A review of mathematical models of influenza a infections within a host or cell culture: lessons learned and challenges ahead. *BMC Public Health,* 11(1): S7, 2 2011.
4. S. Becker-Weimann, J. Wolf, H. Herzel, and A. Kramer. Modeling feedback loops of the mammalian circadian oscillator. *Biophysical Journal,* 87(5):3023 – 3034, 2004.
5. P. Blanchard, R. L. Devaney, and G. R. Hall. *Differential Equations.* Cengage Learning, 2012.
6. H. W. Borchers. *pracma: Practical Numerical Math Functions,* 2019. URL https://CRAN.R-project.org/package=pracma. R package version 2.2.5.
7. W. E. Boyce, R. C. DiPrima, and D. B. Meade. *Elementary Differential Equations and Boundary Value Problems.* Wiley, 2017.
8. F. Brauer. Mathematical epidemiology: Past, present, and future. *Infectious Disease Modelling,* 2(2):113–127, 2017.
9. R. L. Burden, J. D. Faires, and A. M. Burden. *Numerical Analysis.* Cengage Learning, 2016.
10. B. J. Coburn, B. G. Wagner, and S. Blower. Modeling influenza epidemics and pandemics: insights into the future of swine flu (H1N1). *BMC Medicine,* 7(1):30, 6 2009.
11. O. Diekmann, J. A. P. Heesterbeek, and J. A. J. Metz. On the definition and the computation of the basic reproduction ratio R_0 in models for infectious diseases in heterogeneous populations. *Journal of Mathematical Biology,* 28(4):365–382, 6 1990.
12. O. Diekmann, J. A. P. Heesterbeek, and M. Roberts. The construction of next-generation matrices for compartmental epidemic models. *Journal of the Royal Society, Interface,* 7(47): 873—885, 6 2010.
13. L. Edelstein-Keshet. *Mathematical Models in Biology.* Classics in Applied Mathematics. Society for Industrial and Applied Mathematics (SIAM, 3600 Market Street, Floor 6, Philadelphia, PA 19104), 1988.
14. B. Ermentrout. Bifurcation calculations with AUTO, 1995. URL http://www.math.pitt.edu/~bard/bardware/tut/xppauto.html.
15. B. Ermentrout. *Simulating, Analyzing, and Animating Dynamical Systems.* Society for Industrial and Applied Mathematics, 2002.

16. B. Ermentrout. XPPAUT 8.0, 2016. URL http://www.math.pitt.edu/~bard/xpp/xpp.html.
17. Z. Feng, J. Velasco-Hernandez, and B. Tapia-Santos. A mathematical model for coupling within-host and between-host dynamics in an environmentally-driven infectious disease. *Mathematical Biosciences*, 241:49–55, 01 2013.
18. L. Fu and N. M. Kettner. Chapter nine - the circadian clock in cancer development and therapy. In M. U. Gillette, editor, *Chronobiology: Biological Timing in Health and Disease*, volume 119 of *Progress in Molecular Biology and Translational Science*, pages 221 – 282. Academic Press, 2013.
19. C. Gardiner. *Stochastic Methods: A Handbook for the Natural and Social Sciences*. Springer Series in Synergetics. Springer Berlin Heidelberg, 2009.
20. D. T. Gillespie. A general method for numerically simulating the stochastic time evolution of coupled chemical reactions. *Journal of Computational Physics*, 22(4):403 – 434, 1976.
21. D. T. Gillespie. Exact stochastic simulation of coupled chemical reactions. *The Journal of Physical Chemistry*, 81(25):2340–2361, 1977.
22. D. T. Gillespie. A diffusional bimolecular propensity function. *The Journal of Chemical Physics*, 131(16):164109, 2009.
23. G. C. González-Parra, A. J. Arenas, D. F. Aranda, R. J. Villanueva, and L. Jódar. Dynamics of a model of toxoplasmosis disease in human and cat populations. *Computers & Mathematics with Applications*, 57(10):1692 – 1700, 2009.
24. N. J. Gotelli. *A Primer of Ecology*. Sinauer, 2008.
25. J. A. P. Heesterbeek. The law of mass-action in epidemiology: A historical perspective. In B. Beisner and K. Cuddington, editors, *Ecological paradigms lost: routes of theory change*, Theoretical Ecology Series, chapter 5, pages 81–106. Elsevier Science, 2005.
26. H. Hethcote. The mathematics of infectious diseases. *SIAM Review*, 42(4):599–653, 2000.
27. D. E. Hill and J. P. Dubey. Toxoplasma gondii. In Y. R. Ortega and C. R. Sterling, editors, *Foodborne Parasites*, pages 119–138. Springer International Publishing, Cham, 2018.
28. M.-C. Huang, C.-W. Ho, C.-H. Chen, S.-C. Liu, C.-C. Chen, and S.-J. Leu. Reduced expression of circadian clock genes in male alcoholic patients. *Alcoholism: Clinical and Experimental Research*, 34:1899–904, 11 2010.
29. J. Keener and J. Sneyd. *Mathematical Physiology: I: Cellular Physiology*. Interdisciplinary Applied Mathematics. Springer New York, 2008.
30. J. Keener and J. Sneyd. *Mathematical Physiology: II: Systems Physiology*. Interdisciplinary Applied Mathematics. Springer New York, 2009.
31. E. K. Kelting. Toxoplasma gondii: A mathematical model of its transfer between cats and the environment. *SIAM Undergraduate Research Online*, 11, 2018.
32. W. O. Kermack and A. G. McKendrick. A Contribution to the Mathematical Theory of Epidemics. *Proc. R. Soc. A Math. Phys. Eng. Sci.*, 115(772):700–721, 1927.
33. K. Koelle, S. Cobey, B. Grenfell, and M. Pascual. Epochal evolution shapes the phylodynamics of interpandemic influenza a (H3N2) in humans. *Science*, 314(5807):1898–1903, 2006.
34. H. Kokko. *Modelling for Field Biologists and Other Interesting People*. Cambridge University Press, 2007.
35. M. Kot. *Elements of Mathematical Ecology*. Elements of Mathematical Ecology. Cambridge University Press, 2001.
36. L. Lyall, C. A Wyse, N. Graham, A. Ferguson, D. Lyall, B. Cullen, C. Celis-Morales, S. Biello, D. Mackay, J. Ward, R. Strawbridge, J. Gill, M. Bailey, J. P Pell, and D. J Smith. Association of disrupted circadian rhythmicity with mood disorders, subjective wellbeing, and cognitive function: a cross-sectional study of 91,105 participants from the UK Biobank. *The Lancet Psychiatry*, 5, 05 2018.
37. H. McCallum, N. Barlow, and J. Hone. How should pathogen transmission be modelled? *Trends in Ecology & Evolution*, 16(6):295 – 300, 2001.
38. J. McVernon, C. McCaw, and J. Mathews. Model answers or trivial pursuits? the role of mathematical models in influenza pandemic preparedness planning. *Influenza and Other Respiratory Viruses*, 1(2):43–54, 2007.

39. N.-A. M. Molinari, I. R. Ortega-Sanchez, M. L. Messonnier, W. W. Thompson, P. M. Wortley, E. Weintraub, and C. B. Bridges. The annual impact of seasonal influenza in the US: Measuring disease burden and costs. *Vaccine*, 25(27):5086 – 5096, 2007.
40. J. Montoya and O. Liesenfeld. Toxoplasmosis. *The Lancet*, 363(9425):1965 – 1976, 2004.
41. F. Portaluppi, R. Tiseo, M. H Smolensky, R. Hermida, D. E Ayala, and F. Fabbian. Circadian rhythms and cardiovascular health. *Sleep Medicine Reviews*, 16:151–66, 06 2011.
42. R Core Team. R: A language and environment for statistical computing, 2018.
43. K. Soetaert, T. Petzoldt, and R. W. Setzer. Solving differential equations in R: Package deSolve. *Journal of Statistical Software*, 33(9):1–25, 2010.
44. S. H. Strogatz. *Nonlinear Dynamics and Chaos: With Applications to Physics, Biology, Chemistry, and Engineering*. Studies in Nonlinearity. Avalon Publishing, 2014.

Simulating Bacterial Growth, Competition, and Resistance with Agent-Based Models and Laboratory Experiments

Anne E. Yust and Davida S. Smyth

Abstract In this chapter, we will introduce the concepts of growth, competition, and adaptation using bacteria. We will use both simulation and laboratory-based exercises and provide practice with critical skills to assess your understanding throughout the chapter. The chapter ends by focusing on the capacious and global problem of antibiotic resistance, considered to be one of the most important public health threats of the twenty-first century. We will review the basic biological concepts underlying the phenomenon, and introduce the mathematical content necessary to begin to create agent-based models in order to simulate and analyze antibiotic resistance and its effect on the planet. we will present a selection of research projects throughout the later portion of the chapter for you to further explore antibiotic resistance and other contexts of bacteria.

Suggested Prerequisites *This chapter is intended to be accessible to a wide range of undergraduates. The only necessary prerequisite is a strong interest and desire to learn about simulation, bacteria, and antibiotic resistance. We do recommended previous exposure to microbes and introductory biology, programming—especially with* NETLOGO, *mathematical modeling, and data analysis.*

1 Introduction

Bacteria are microscopic, unicellular, prokaryotic organisms that can exist in the form of spheres, rods, and spirals. The major category which differentiates bacteria is the structure of their cell walls, which are called *gram-positive* or *gram-negative*. The prototypical gram-negative organism, *Escherichia coli*, has an inner and outer

A. E. Yust (✉) · D. S. Smyth
Department of Natural Sciences and Mathematics, Eugene Lang College of Liberal Arts
at The New School, New York, NY, USA
e-mail: yusta@newschool.edu; smythd@newschool.edu

© Springer Nature Switzerland AG 2020
H. Callender Highlander et al. (eds.), *An Introduction to Undergraduate Research in Computational and Mathematical Biology*, Foundations for Undergraduate Research in Mathematics, https://doi.org/10.1007/978-3-030-33645-5_5

217

membrane, between which is found a periplasmic space with a thin layer of peptidoglycan. Gram-positive bacteria have a much thicker layer of peptidoglycan surrounding their inner membrane and have no outer membrane. The prototype gram-positive organism is *Staphylococcus aureus*.

Bacteria are noted for their ubiquity; they are some of the most adaptable and resilient organisms on the planet. Different species of bacteria can survive and grow at temperatures ranging from -10 to $100\,°C$. They can happily reside in cold salty ponds and in the frigid waters of the polar regions to the boiling water of hot springs. They can even be found growing in the vicinity of thermal volcanic vents at the bottom of the oceans where temperatures exceed $300\,°C$. Bacteria are also resilient to pH; they have been isolated in acid wastes from mines and in the alkaline waters of soda lakes [63]. They have been isolated in black anaerobic silts of estuaries and in the purest waters of biologically unproductive or oligotrophic lakes. This means that, in whatever situation, bacteria can find a way of adapting and surviving, regardless of the environment they find themselves [63].

Bacteria have proven a useful model system in which to investigate many cellular functions and processes. They have simple genomes, many of which are amenable to genetic modification. They can also be readily propagated in the laboratory, and they have fast generation times. Knowledge gained when studying bacterial systems can often be applied to homologous proteins in more complex higher organisms. Bacteria are used in industry and are critical for the production of yoghurt, cheese, sour cream, pickles, sauerkraut, and kombucha to name but a few. Bacteria can also be engineered to make useful products such as human proteins and drugs, and importantly they can be used in bioremediation and to detoxify poisonous substances. It is their adaptability and resilience that not only makes bacteria one of the world's greatest allies but also one of the world's greatest foes. Bacteria are responsible for many different diseases and a major cause of morbidity and mortality across the globe. One of the most important and clinically relevant bacterial adaptations is antibiotic resistance.

1.1 Introduction to Antibiotic Resistance

The World Health Organization has named antibiotic resistance as one of the most important public health threats of the twenty-first century [90]. Importantly, infections caused by antibiotic-resistant organisms are associated with significant mortality [3] and are an important economic burden, estimated to cost over $20 billion per year in the USA alone [35, 39, 113]. At least 23,000 people die from infections with antibiotic-resistant bacteria annually in the USA as estimated by the Centers for Disease Control and Prevention [3]. By 2050, it is estimated that antibiotic resistance will cause around 300 million premature deaths, and result in a loss of up to $100 trillion by the global economy [89]. More worryingly still, the World Health Organization has warned that there is a serious lack of new antibiotics under development to combat the growing threat of antibiotic resistance [8], with only eight of the 51 new antibiotics and biological agents currently in clinical development to treat antibiotic-resistant pathogens adding value to the current drugs on offer [62].

Antibiotic resistance is an ancient phenomenon, a consequence of long-term biowarfare among organisms in their natural environments. Most antimicrobials are natural molecules. In fact, many are secreted by bacteria and other environmental organisms. When organisms growing in communities were exposed to the effects of antibiotics secreted by members of the community, they evolved counter-mechanisms to overcome their action in order to survive. Some organisms that are naturally resistant to antibiotics are called *intrinsically resistant*. A great example of intrinsic resistance occurs in multi-drug resistant gram-negative bacteria such as *E. coli*, which are insensitive to many types of antibiotics used to effectively treat gram-positive bacteria. Their resistance is due to the presence of the outer membrane, which is the differentiating factor between gram-negative and gram-positive bacteria. The outer membrane is impermeable to many molecules. Additionally, these strains possess a variety of efflux pumps that can effectively pump out the antimicrobial from the cells [84]. Many environmental organisms are prolifically and intrinsically resistant to many different classes of antibiotics, and in particular, those that dwell in the soil have many uncharacterized mechanisms of resistance [38]. What is more intriguing is that antibiotic resistance often predates the clinical use of antibiotics and has emerged independently of the selective pressure imposed by using antibiotics [25, 37]. Scientists consider environmental organisms, such as those found in soil or in the urban environment, to be important reservoirs of novel resistance genes that could be transferred to pathogens. This presents a major health concern [59].

While environmental organisms often have intrinsic mechanisms of resistance, antibiotic-resistant bacteria in the hospital or clinical setting often exhibit resistance that has been acquired. Acquired resistance emerges in a bacterial population that was originally susceptible to the antibiotic. In response to exposure, resistance is often acquired by mutations in the chromosome of the bacteria or the acquisition of external resistance-encoding genes, i.e., *horizontal gene transfer (HGT)*.

1.1.1 Genetic Basis of Antibiotic Resistance

Antibiotic resistance emerges genetically in two main ways:

- **Resistance by mutation.** Cells within a susceptible population of bacteria develop mutations in genes that ameliorate the activity of the drug. These cells thus survive the antimicrobial agent, multiply, and proliferate, while the susceptible cells succumb to the agent. Depending on the type of mutation, there can be a fitness cost, and as a consequence, these mutations are only selected for in the presence of the antibiotic. Interestingly, the use of antibiotics has been shown to increase the mutation rate of bacteria [68] and even to select for mutants with higher mutation rates in the microbial flora of patients treated with antibiotics [53].

- **Resistance by horizontal gene transfer.** HGT is a major driver of bacterial evolution. This process involves the acquisition of foreign DNA, which could contain genetic sequences that transfer the antibiotic resistance. HGT is frequently responsible for acquired resistance to antibiotics and antimicrobials. The most common mechanisms used by bacteria to acquire external genetic material are transformation (incorporation of naked DNA), transduction (mediated by phages—viruses that infect bacteria), and conjugation (when bacteria have "sex" mediated by a pilus). The simplest type of HGT, transformation, is demonstrated by a small number of species of high clinical relevance including the pneumococcus or *Streptococcus pneumoniae* [86], and *Neisseria meningitidis* and *Neisseria gonorrhoeae* [116]. Transduction by phages is a very important mode of HGT and has been shown to be an important vehicle for resistance genes in the environment [19]. The most common and most efficient form of HGT is conjugation. This type of transfer needs cell-to-cell contact and is mediated by the presence of conjugative elements in the genome of the donor cell. Tetracycline resistance is readily transferred among *N. gonorrhoeae* and *Enterococcus faecalis* strains by means of a conjugative plasmids, circular forms of DNA [71, 116]. Other types of mobile DNA such as integrons and transposons also play important roles in the dissemination of antibiotic resistance genes, such as carbapenamases [97]. Genes encoding resistance to streptomycin, spectinomycin, and sulfonamides as well as metals such as mercury have been found on complex transposons and plasmids in members of the Enterobacteriaceae [23].

1.1.2 Mechanistic Basis of Antibiotic Resistance

There are several categorizations of antibiotic resistance mechanisms:

- **Modifying the antimicrobial molecule itself.** One mechanism found in both gram-positive and gram-negative bacteria is to produce enzymes that modify the chemical composition of the antimicrobial molecule by phosphorylation, acetylation, and adenylation. Chloramphenicol resistance is mediated by chloramphenicol acetyltransferases known as CATs, widespread among bacteria [103]. Alternatively, some bacteria produce enzymes that can destroy the antibiotic itself. One of the most famous examples is the family of beta-lactamases. Beta-lactamases were identified before the introduction of penicillin to the market [12] and are considered to be ancient. More than 1000 different beta-lactamases have been described to date.
- **Blocking the action of the antibiotic against its target.** The first line of defense used by gram-negative bacteria to prevent antimicrobials from reaching their intracellular or periplasmic targets is their outer membrane as we have

discussed. In addition, they can prevent hydrophilic (water soluble) molecules from traversing the membrane (which is not water soluble) using porins (channels in the membrane) by altering the types of porins present, the expression of the porin genes and by impeding porin function [85]. Efflux is another widespread mechanism to avoid antibiotic action. *E. coli* can actively pump the antibiotic tetracycline out of the cell using an efflux pump. Efflux pump mechanisms are found in both gram-positive and gram-negative bacteria and can be antibiotic specific such as *mef* which encodes macrolide resistance in pneumococci or can be broadly specific facilitating the multi-drug resistance (MDR) phenotype [98].

- **Changing the target site or bypassing it entirely.** This occurs through two main mechanisms, which include protection of the target and modifications of the target site, decreasing affinity for the antibiotic. Tet(M) first described in *Streptococcus* spp., interacts with the ribosome and actively dislodges tetracycline from the target site [33]. Linezolid resistance involves mutation of the binding site in the ribosome and results in decreased affinity of the drug for its ribosomal target [78]. Lastly, bacteria can evolve entirely new target structures that have the same function but bypass the antibiotic entirely, such as methicillin resistance in *S. aureus* due to the acquisition of an exogenous PBP (PBP2a) and vancomycin resistance in enterococci through modifications of the peptidoglycan structure mediated by the *van* gene clusters [17, 31].

- **Changing regulatory networks which control important metabolic pathways.** An important example of this type of resistance is resistance to daptomycin (DAP) and vancomycin (low level in *S. aureus*). In these cases, the bacteria make systematic changes to fundamental systems such as their cell wall structure to withstand the action of the drug. An example in both enterococci and *S. aureus*, YycFG (WalKR), an essential two-component regulatory system, has been implicated in cell wall synthesis and homeostasis, is important for resistance to daptomycin. The exact mechanism is unknown, but it appears to involve alteration in cell wall metabolism resulting in changes in surface charge which repulses the positively charged calcium-DAP complex from the cell envelope [22, 115]. High-level vancomycin resistance in *S. aureus* was the result of acquisition by a methicillin-resistant *S. aureus* (MRSA) strain of the *vanA* gene cluster from a vancomycin-resistant enterococcus (*E. faecalis*) isolate [107]. Thankfully such high-level resistance to the last available drug for treatment, vancomycin, is rare in Staphylococci. However, low level resistance, called vancomycin intermediate *S. aureus* (VISA), is much more prevalent and involves several systematic changes that reduce peptidoglycan cross-linking (in the cell wall) which results in a thicker cell wall. Additional changes in VISA cells include an increase in fructose utilization and fatty acid metabolism, as well as an increase in the expression of cell wall synthesis genes [56].

1.2   Spread and Severity of Antibiotic-Resistant Infections

Infections due to antibiotic-resistant bacteria are already widespread in the USA and across the planet [118]. In 2011, the Infectious Diseases Society of America (IDSA) Emerging Infections Network survey of national infectious disease specialists concluded that more than 60% of participants had seen a pan-resistant, untreatable bacterial infection within the prior year [109]. The gram-positive pathogens, *S. aureus* and Enterococcus species, are responsible for a global pandemic, which poses the biggest threat [3]. MRSA kills more Americans each year than HIV/AIDS, murder, Parkinson's, and emphysema combined [52]. Vancomycin-resistant enterococci are developing resistance to many common antibiotics [50]. Health care settings are seeing serious gram-negative infections due to resistant Enterobacteriaceae (mostly *Klebsiella pneumoniae*), *Pseudomonas aeruginosa*, and *Acinetobacter* [3], with multi-drug resistant gram-negative strains, including extended-spectrum beta-lactamase-producing *E. coli* and *N. gonorrhoeae* emerging in the community and non-health care settings [102]. A review in 2014 indicated that an estimated 700,000 deaths globally were caused by infections caused by antibiotic-resistant organisms, and predicted this number rise to 10 million per year by 2050 [4, 89].

Antibiotic-resistant bacteria and the infections they cause are having an impact on every field of medicine and have a significant impact on morbidity and mortality. It has been estimated that infections caused by antibiotic-resistant bacteria have two-fold higher rates of adverse outcomes compared with similar infections caused by susceptible strains [36]. The impacts of negative outcomes include treatment failure and/or death as well as economic impacts such as increased cost of care and length of stay due to treatment failure of the antibiotic [44]. Serious infections due to MRSA have a significantly higher case fatality rate when compared with methicillin-susceptible *S. aureus* infections [36]. Enterobacteriaceae that produce extended-spectrum beta-lactamases are associated with greater treatment failure and mortality than non-ESBL producing strains [77]. Infections due to *K. pneumoniae* with resistance to carbapenems demonstrate a two- to five-fold higher risk of death than infections caused by carbapenem-susceptible strains [29]. Forty-five percent of bacteremia cases due to carbapenem-resistant *Acinetobacter baumannii* are associated with a 14-day mortality [87].

1.3   The Economic, Social, and Civic Impacts of Antibiotic Resistance

It is also important to consider the impact that antibiotic-resistant bacteria have on local and global economies, individuals, communities, and populations and on policies, regulations, and future planning pertaining to healthcare, food production, and agriculture. The emergence of antibiotic-resistant bacteria has been called a "crisis" or "nightmare scenario" that could have "catastrophic consequences" and in recent years has been recognized as a global threat. As a consequence it is now being recognized by governments and worldwide organizations as a target for policy generation and implementation [83]. The federal Interagency Task Force on Antimicrobial Resistance founded in 1999 succeeded in documenting collaboration and communication among the 11 agencies working on resistance issues, but it failed to set an agenda for federal response [1]. In 2013, the CDC declared that the human race is now in the "post-antibiotic era," and in 2014, the World Health Organization (WHO) warned that the antibiotic resistance crisis is becoming dire [80], stating that the problem "threatens the achievements of modern medicine. A post-antibiotic era—in which common infections and minor injuries can kill—is a very real possibility for the 21st century." Antibiotic resistance poses a substantial threat to US public health and national security according to the IDSA and the Institute of Medicine. In March 2015, the Obama administration released a National Action Plan for Combating Antibiotic-Resistant Bacteria [6] and the 2016 federal budget almost doubled the amount of federal funding for combating and preventing antibiotic resistance to more than $1.2 billion [1, 7].

1.3.1 Economic Burden of Antibiotic Resistance

Antibiotic-resistant infections pose an economic burden as well. Patients with antibiotic-resistant infections spend longer in the hospital, from 6.4 to 12.7 days, collectively adding an extra eight million hospital days [50]. The medical cost per patient infected with an antibiotic-resistant strain is estimated to be in the range of from $18,588 to $29,069 [21, 50]. The US economy faces a total economic burden estimated to be as high as $20 billion in health care costs and $35 billion a year in lost productivity due to antibiotic resistance [50]. Individual families and communities lose wages and have higher health care costs [80]. Staggeringly, the global gross domestic product could be reduced by 2–3.5% by 2050 due to the mortality from antibiotic-resistant infections, about $60 and $100 trillion [4, 89].

1.3.2 Increased Impact on Subpopulations

Many subpopulations are affected by the rise in antibiotic-resistant pathogens considerably more than others. We outline a few specific subpopulations below.

- **Developing populations.** For people in the developing world, a post-antibiotic era has already arrived. In parts of Africa, studies have shown that as many as 97% of S. aureus are caused by MRSA [11] and high levels of resistance to amoxicillin and penicillin in S. pneumoniae and Haemophilus influenzae have been observed, causing concern given that pneumonia is a leading cause of death in children [114]. In India and Pakistan, up to 95% of adults carry bacteria that are resistant to β-lactam antibiotics including carbapenems, where by comparison, only 10% of adults in the Queens area of New York carry such bacteria [101]. Worryingly, not all countries are collecting data on the prevalence of these bacteria and their infections. According to the WHO only 129 of 194 member countries provided any national data on drug resistance in bacteria with only 22 countries tracking the organisms and resistance that pose the greatest threat including *S. aureus* and methicillin, *E. coli* and cephalosporins, and *K. pneumoniae* and carbapenems [101].

- **Underserved and impoverished communities.** According to 2016 US census data, the official poverty rate was 12.7%, down from 13.5% in 2015. Since 2014, the poverty rate has fallen 2.1 percentage points from 14.8 to 12.7%. This means that in 2016, 40.6 million people were living in poverty, 2.5 million fewer than in 2015 and 6.0 million fewer than in 2014 [105]. For most demographic groups, the number of people in poverty decreased from 2015. Adults aged 65 and older were the only population group to experience an increase in the number of people in poverty [105]. Many factors associated with poverty contribute to the development of antibiotic-resistant organisms, some of which impact affecting resistance in the USA [95]. Studies have shown that seniors and low-income patients obtain antimicrobials from other countries and may engage in the sharing of medications while others will save antibiotics from a regimen they did not complete and self-treat [95]. Self-treating can drive antimicrobial resistance because of the inappropriate use of antibiotics for viral illness, the antimicrobials may not work for the specific organism type and the dosage may be incorrect [95]. The high cost of healthcare and lack of access to healthcare for those who are uninsured, prevents many from seeking necessary and lifesaving treatment. The WHO has cited the provision of universal healthcare as a means to

  > improve access to appropriate and affordable treatment of infections, especially for the poor through enactment and enforcement of regulations, dissemination of treatment guidelines based on antibiotic resistance surveillance data, along with awareness raising on the responsible use of antimicrobials and the challenge of antibiotic-resistant bacteria [26].

  It is imperative to remove financial barriers and allow access to antimicrobial treatment of infections.

- **At-risk populations.** While antibiotic-resistant bacteria pose a threat to the population as a whole they are likely to cause illness in populations with greater overall risk of contracting infectious diseases. These at-risk populations include the military [32], the homeless [5], children attending daycare [9], immunocompromised persons [42], and the elderly [18]. Using prisoners as an example, community-associated MRSA outbreaks in the USA have been reported among persons incarcerated in prisons and jails with estimates of MRSA colonization in prisons as high as 80–90%. Crowding and sharing of contaminated personal items may contribute to MRSA spread among incarcerated persons [69]. Of considerable global concern, Russian prisons are said to be driving resistance among strains of TB [61].

1.3.3 Impact on the Food Supply and Agriculture

Antibiotics have been widely used in agriculture and in some countries for growth promotion [64, 89]. This practice was discontinued in the European Union in 2006 [2]. In the Americas and Asia this practice is still in use, where large scale husbandry systems contribute to infection with these bacteria. Treatment is generally delivered via the feed or water to all animals regardless of their infection status [73]. Data from the US Food and Drug Administration shows that in 2015, 74% of farm animal antibiotics were administered via feed and 21% in drinking water, for mass medication. It is estimated that the use of medically important antibiotics in food animals in the USA is approximately three times higher than human use [74]. As a consequence, antibiotic use in animals is thought to be an important selective pressure for antibiotic resistance globally [64]. Sales in the USA in 2015 of the critically important fluoroquinolones antibiotics was 20 tonnes, a 16% increase over 2014 and a 33% increase over 2013 [74]. However, in the USA since 2005, the use of fluoroquinolones has been banned in poultry due to scientific evidence that this use was leading to fluoroquinolone resistance in human Campylobacter infections. In 2016, the FDA showed that sales and distribution of all antimicrobial drugs approved for use in food-producing animals rose by 1% from 2014 to 2015 [75]. An FDA policy named FDA Guidance for Industry #213, asked that drug sponsors voluntarily remove growth promotion from the labels of all medically important antibiotics used in food animals from 2017 onwards [76]. Thankfully, major US food companies such as McDonald's and Tyson Foods have reduced and in some cases eliminated antibiotics in their products [79]. The success or failure of #213 will not be known for a number of years.

1.4 Introduction to Agent-Based Models of Bacteria

Agent-based models (ABMs) vary from differential equation (DE) models by differentiating individual agents acting within the world, instead of treating populations as homogeneous. Though it is common for DE models to capture heterogeneity in a population by partitioning it into subpopulations (compartment models), probabilistically perturbing parameters or rates of flow from one state to another (stochastic differential equations (SDEs)), or using spatial characteristics of the world to influence proportions of the population (partial differential equations (PDEs)), these techniques all maintain anonymity of the agents within the population. There are pros and cons to both types of modeling approaches. Used in conjunction, ABMs and DEs can help us better understand the dynamics of a complex system than if we used one method of modeling in isolation. There has been much discussion in the modeling community about the individual benefits of each [28, 92, 94, 99, 104] and the creation of hybrid models that incorporate both methods [27, 30, 119]. We recommend that you read through the portions of these papers that quite elegantly describe the utility of ABMs, often grounded in biological contexts.

Recently, ABMs have been used in addition to DEs to explore the spread of infectious disease throughout a population. Agents (e.g., bacteria, people, bears, sharks, coral, hurricanes, houses) act as identifiable entities that make a series of decisions, choosing between a prescribed set of options. For instance, the agents may move throughout the world by moving in a uniformly random direction, then potentially transmitting a disease to another agent with a probability that depends on the distance between the agents. Using ABMs allows us to more easily integrate spatial components and randomness into our model than formulating and analyzing PDE or SDE models.[1] Ultimately, we are interested in the behavior of the system that emerges when the agents continue to make stochastic or deterministic decisions based on their current state, which may be affected by other agents in the world or their environment. In the models of bacterial behavior studied throughout this chapter, we will often want to use an ABM to study the emergent behavior of the system. For instance, we may want to predict the proportion of a population that will be affected by an infectious disease or investigate the effects of an intervention (e.g., antibiotics).

Though we include a research project on modeling the spread of infectious disease, most of this chapter will be dedicated to creating, using, and interpreting ABMs that simulate bacterial growth, competition between bacteria within a system, and the ways that bacteria can gain resistance to an antibiotic. The ability to allow probabilistic interactions between various agents in space will help us mimic the biological mechanisms inherent in the processes to better understand the reasons for emergent behavior that we witness, predict trends we expect to see in the future, and reconcile the output of our models with the data collected through laboratory experiments.

To create our ABMs, we will be using NETLOGO throughout this chapter—a common environment for programming ABMs. Created by Uri Wilensky in 1999, the platform is free to download with a plethora of texts to assist with the basics of model creation and is currently one of the standards for ABMs [120]. See [100] and [121] for thorough descriptions of ABMs, NETLOGO, and their use and capabilities. Though we will spend time building a knowledge-base and familiarity with common techniques and commands with tutorial-style exercises while designing models of bacterial growth, we recommend that students who wish to pursue the challenge problems and research projects outlined later in this chapter use supplemental resources to gain additional experience and assistance with NETLOGO. Working through Chapters 2, 4, and 5 of [100] would be particularly useful. If you feel comfortable creating simulations in NETLOGO, you can likely move through the next section relatively quickly, focusing your attention on the biological content. If this is your first experience with programming and/or using NETLOGO, we strongly recommend (at the least) completing the introductory exercises and three tutorials created by Wilensky that are available for free on the NETLOGO website [120].

---

[1]This is especially notable for students that would like to research a complicated biological and/or social phenomenon without a mathematical background that includes advanced topics like PDEs and SDEs.

Throughout the following sections, the exercises, challenge problems, and projects are meant to prompt your own discovery of concepts associated with bacteria and ABMs through guided inquiry. There are few responses that require computations. Most will necessitate trial-and-error-type processes using the simulations you will create, followed by thoughtful reflection on why an action "worked" or "failed." For this reason, solutions are not provided—though practically all versions of the code are freely available on the QUBES (Quantitative Undergraduate Biology Education and Synthesis) website [41], with some complete code included in the Appendix.

2 Bacteria Growth

To gain some familiarity with modeling in NETLOGO, we will begin by creating a model of simple bacterial growth. However, before we begin to dive into the code, we need to distill down the microbiological background given in the previous section to the basic processes integral to bacterial growth. Bacteria reproduce asexually by a process called binary fission. Typically, bacteria divide into two identical daughter cells, containing identical genetic material. Depending on the strain of bacteria and the environmental conditions (e.g., temperature, nutrients), the rate of cell division, called generation time, can vary. In a laboratory, *E. coli*, a type of bacteria, divide every 15–20 min in nutrient-rich media. The same bacterium will divide every 12–24 h in the human intestine, where the environment is less friendly and nutrients are limited [49]. Certain disease-causing bacteria, or pathogens, have especially long generation times even when measured in the laboratory. *Mycobacterium tuberculosis*, the causative agent of TB, has a generation time of 15–20 h. Long generation times are thought to play an advantage in their capacity to cause disease or virulence [66].

Exercise 1 (Theory) The bacterium *E. coli* reproduce approximately once every 20 min in a near-optimal environment [106]. If you begin with one *E. coli* how many bacteria would you expect to see after

1. 20 min?
2. 1 h?
3. 2 h?
4. 1 day?
5. *n* divisions (i.e., generation times)?

If you were to plot the number of *E. coli* cells against time, what type of curve would you expect?

An initial amount of bacteria, N_0, with a generation time of t_g, will theoretically grow to

$$N(t) = N_0 2^{\frac{t}{t_g}},\qquad (1)$$

after t units of time.

Exercise 2 (Theory) Check your responses to Exercise 1 with the formula given in Eq. (1). In terms of the variables and parameters used in Eq. (1), how would you calculate the number of divisions or generation times n?

When recording and graphing data generated in a laboratory experiment, biologists plot log (logarithm) counts of the bacteria, due to the large numbers of bacteria.[2] By the time bacteria have saturated the culture, there can be as many as 8×10^8 cells per ml. You likely encountered this in the previous exercises when calculating the number of bacteria after relatively few generations, even when beginning with only one bacterium. Imagine trying to plot the numerical counts you calculated in the previous exercises using a linear scale. You would either have to use a very inaccurate scale, or you would run out of paper (or at least table space). Using a log scale makes the very large cell counts typically found easier to visualize. Since bacterial growth is exponential, the log transformation will appear linear. Additionally, a log transformation allows us to estimate the generation time of bacteria by simply finding the slope of the curve.

Exercise 3 (Theory) Show that the function $\log(N(t))$ is linear. Then, determine the generation time, t_g, by only using the slope the linear function, $\log(N(t))$.

Exercise 4 (Lab) Many of the following concepts can be demonstrated in the laboratory with the minimum of resources and equipment. They can be accomplished in a microbiology teaching lab under the supervision of your biology or microbiology instructor. We have provided examples of experiments (with commercial kits) that can reinforce your learning about the simulations, which can be found here: [108]. In addition virtual labs and online resources are supplied to aid with student learning.

Exercise 5 (Theory) Suppose we have discovered a new as of yet unidentified, deadly bacterial pathogen, *Morbum malum*. Through extremely careful laboratory experimentation, we closely monitored its growth. Beginning with a bacterial count of approximately 1.0×10^8 *M. malum* cells, we estimated approximately 4.2×10^9 cells 66 h and 6 min later by plating serial dilutions of the culture on agar and

---

[2]There is a debate over using log transformations with data that expresses counts when performing advanced statistical analyses [88]. However, most of the arguments against this type of transformation come from ecologists who frequently encounter zero counts.

counting the numbers of colonies that form (these are called colony-forming units or cfu). Find an estimate for the generation time (or doubling time) of *M. malum*. Compare this generation time with that of *E. coli*. Which bacteria would you expect to pose more of a threat to humanity? Why?

Now, in NETLOGO, let us begin with a single bacterium centered in the world, simulating a bacterium placed in the center of a nutrient-rich, continuous-culture solution in a flask. In continuous culture, nutrients are added and waste is removed continuously. In the Code tab, type the following in order to create a `setup` procedure:

```
to setup
  clear-all
  create-turtles 1 [set shape "circle 2" set color pink]
  reset-ticks
end
```

When this procedure is called, NETLOGO will clear any graphs and erase any variable values it once knew, create one pink circle at the center of the world, then reset the `ticks` to zero. A `tick` typically represents one iteration through the code. Notice the primitive command to generate an agent is `create-turtles`. Though we are creating bacteria, the default name for the agentset is turtles.[3] We set the shape as circles to approximate the form of the bacteria.[4] Below the `setup` procedure, create a `go` procedure. Use the `hatch` command in the code below to model binary fission. Every time you click the `go` button this procedure will "hatch" (create) a clone of each turtle in the agentset `turtles`. We will consider each tick to be the estimated generation time of the bacteria.

```
to go
  ask turtles [
    hatch 1 [right random 360 forward 1]
  ]
  tick
end
```

Do not forget to create a `setup` and `go` button in the interface area.

---

[3] In the NETLOGO language, all movable agents are called turtles. A parent programming language, LOGO, could be used to program robots that moved in the physical world based on the commands given. The original robots had shells that gave the appearance of a turtle. A pen could be attached to them so they would draw their path on paper. Visualizing the agents as these roaming robots may help you better understand the natural language programming conventions and syntax.

[4] If you are reading an electronic version of this text, we strongly recommend typing the sample code into the Code tab of NETLOGO instead of copying and pasting. This will help with your understanding of the flow and syntax. Also, and possibly more importantly, NETLOGO will not understand the formatted quotation marks, and you will receive an error. The naming conventions, spacing, hyphenation, and lack of capitalization used throughout the code are standard style choices for NETLOGO.

Exercise 6 (Code) The primitive procedures `right`, `random`, and `forward` take a single number as an input. Vary the numeric inputs. Explain what each of these primitives do.

Use the model to confirm your responses to Exercises 1 and 2. For now, leave the *Forever* checkbox unclicked. When the *Forever* checkbox is clicked, a circular arrow symbol appears in the bottom right-hand corner of the button. This makes the `go` button run *Forever*—meaning NETLOGO will continue to loop through the code until you unclick the `go` button. By default, the *Forever* option is turned off. This allows you to view the evolution of the simulation after each tick by a manual click. You may notice that it may be difficult to count the number of bacteria even after a few ticks (and for your computer to generate such large numbers of circles). In order to better understand and analyze the results of our simulation, we will want a count of the bacteria. We can create a monitor that displays the exact number present at any given iteration.

In the Interface tab, choose *Monitor* from the drop-down menu to the right of the *Add* button. Click within the interface area to place your monitor. In the *Reporter* text box, we want to report the count of the turtles. Type `count turtles`, then click *Ok*. Now, click the `go` button a few times and note the count of bacteria.

Exercise 7 (Theory) Do the counts that appeared in the simulation match the number of bacteria you calculated in Exercises 1 and 2? You found a closed-form formula for the number of bacteria present after n generations. Can you find an iterative formula for the number of bacteria present after n generation times given the bacterial count at $n - 1$ generations?

It is difficult to understand the growth rate of the bacteria by just examining the counts at each step. A graphical display of the counts will provide more insight. In the Interface tab, choose *Plot* from the drop-down menu to the right of the *Add* button. Click within the interface area to place your graph. The default *Pen update commands* already contain the appropriate command, `plot count turtles`. A *pen* is a line created by plotting the reporter over time, in this case, the total count of bacteria over time. Following best practices, give the plot a more appropriate title (e.g., Bacteria Growth Curve) and label the axes (e.g., Number of Bacteria, Concentration of Bacteria (cfu/ml), Optical Density of the Bacteria (OD600) or Absorbance[5] versus Time[6]). Note, there are other modifications you can make to the plot, like changing the color of the pen or adding more pens. Then click *Ok*.

---

[5]In the laboratory setting, spectrophotometry is often used to estimate the concentration of bacteria (cells per ml or colony-forming units (cfu) per ml). In essence, light is shone through a bacterial suspension and the spectrophotometer records the amount of light that makes it all the way through to the sensor. This is called the optical density or OD. For instance an OD600 (600 refers to the wavelength of the light used) of 1.0 is roughly the same as 8×10^8 cells/ml depending on the strain. In the laboratory experiments provided in this chapter, this method is used to approximate the count of the bacteria cells present at a given time [108].

[6]Note that we are simplifying our model so that each unit on the horizontal is a generation time of the bacteria. This means the scale could have a wide range of standard time units. In the case of

Now when you run the simulation, you should see a curve, showing the total amount of bacteria over time. The slope of this curve is the growth rate. You can find the complete code up to this point at [125]: Model 1.0.

Exercise 8 (Theory) What type of curve do you see? Is this what you expected? Describe how the growth rate changes over time. Explain why this change in growth rate occurs.

Exercise 9 (Code) As previously mentioned, biologists often plot bacterial counts on a log scale. Create an additional plot in the interface area to show the growth curve on a log scale.[7] Does this curve appear as you expected? (See [125]: Model 1.1 for sample code.)[8]

At this point, our bacteria have not moved from their initial positions, with clusters grown around the original, single bacterium. Though there are non-motile bacteria (e.g., *S. aureus*), many types of bacteria are motile such as strains of *E. coli*. Though *E. coli* are known to move in a coordinated fashion (which is an entire area of study in itself),[9] we will simplify this motion to allow free and random movement of the bacteria [24].

We will take this opportunity to reorganize the model structure since we will be asking the bacteria to do two separate sub-procedures: move and divide. Our goal is to create a straightforward go procedure that will just ask the bacteria to move then divide. Often when designing a simulation and determining the order of actions, modelers will create flow diagrams to visually map an agent's path through a single tick. Figure 1 illustrates a simple example that we will use to build the code for Model 1.2.0.

Then, we can translate the visual map we created into the go procedure that will replace our previous go procedure:

```
to go
  ask turtles [
    move
    divide
  ]
  tick
end
```

E. coli, one tick could represent 20 min. However, if simulating *M. tuberculosis* growth, the same change in the horizontal axis could represent 20 h.

[7]The NETLOGO Dictionary found on the NETLOGO website is extremely useful for browsing available primitive commands and their necessary syntax. In this case, you may want to check the NETLOGO Dictionary for help with the log primitive.

[8]It is good practice to save a model with a new name whenever you make a substantial addition or change in the code. Also, make comments with text following a semicolon or two in the code to help others understand what each line does. Often this will even be useful to remind yourself what you were thinking when you wrote the code.

[9]Though not the focus of this chapter, cell motility is a very interesting topic of study. The NETLOGO Model Library contains some models that explore this (e.g., Bacteria Food Hunt, Bacteria Hunt Speeds). You could create a research project on this topic by performing a literature search and designing a model that simulates the current understanding of the way bacteria move on and in different media.

Fig. 1 This flow diagram shows a visual plan for Model 1.2.0. The initialization provided in the setup procedure to create one bacteria cell is the starting point of the diagram. Then the cell will move in a way prescribed by the move sub-procedure we will create. Then, the cell will divide in a way prescribed by the divide sub-procedure we will create

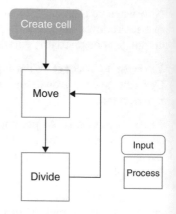

We must write a procedure for each of those actions. We have already written the divide procedure; we just wrote it straight into the go procedure (hatch 1 [right random 360 forward 1]). We can copy and paste that into its own procedure like so:

```
to divide
  hatch 1 [right random 360 forward 1]
end
```

Note, we called the divide procedure within the turtle context in the go procedure (ask turtles [...]). Therefore, we do not need to "ask" the turtles again in the divide procedure. Always be mindful of the context of each procedure. Who is being asked to act? Who is asking?

Now, we will insert a move procedure, giving the cells the same directions that we gave to the "hatched" cells:

```
to move
  right random 360
  forward 1
end
```

At this point, we could simplify the code even more and instruct the daughter cell that was "hatched" to follow the move procedure; however, we may want the flexibility to change the way that the bacteria move in further additions and revisions of this model. If you are interested in building a model that emulates the observed movement of particular motile bacteria, you will certainly need to enhance the move procedure. (See [125]: Model 1.2.0 and the Appendix for sample code.)

Exercise 10 (Theory) If we start with the same amount of bacteria in ten different nutrient-rich flasks, would you expect to count the exact same number of bacteria after 1 h in each of the flasks? Why?

Example 10 is referring to deterministic versus probabilistic or stochastic system dynamics. If we repeat a process over and over again, are we sure to get the same result (deterministic) or will there be some (or possibly a lot of) variability in the result (probabilistic)? For instance, our current model uses both deterministic and probabilistic processes. The number of cells exactly doubles with each click; this is deterministic. Each cell moves forward one step with each click (deterministic), but in a uniformly random direction (probabilistic). Though we will always create the same number of cells after n clicks, the cells will be located in different places because of the probabilistic move procedure.

Exercise 11 (Theory) In reality, would you expect to see bacteria to continue growing in this manner? Why?

One of the many reasons that you will witness variability in the growth (and not just placement) of bacteria in laboratory experiments (and in nature) is that bacterial cells die naturally, like all living organisms. Since bacterial growth rates are found from empirical counts seen in the lab or in patients, the generation times used in the model already account for cell death.[10] In future models, we will consider many other reasons for why the death rate may rise (or the growth rate will decrease), as well as other environmental conditions that lead to variability in bacterial growth.

Challenge Problem 1 (Code) Update the flow diagram and code to add a probabilistic natural death rate into Model 1.2.0. There are multiple ways this could be done. A simplified version of natural death could be coded by asking each bacterium at every tick to roll a (non standard, many sided) die to determine if they will die. In other words, you would be asking each bacterium at every tick to die with some probability p, where p is likely quite small. It is important to consider the order of the sub-procedures. For instance, what is the difference in the effect if you allow bacteria to die before versus after they reproduce? You will use the built-in random and die commands in NETLOGO. (See [125]: Model 1.2.1 for sample code.)

When growing in a flask or on an agar plate in the laboratory, bacteria do not demonstrate unrestricted exponential growth *ad infinitum*. A flask or an agar plate would represent a closed system. In a closed system, the growth of an organism is limited by the available resources and many other possible factors. This is comforting, as otherwise the world would have been overrun by *E. coli* and many other species of bacteria by now.

Exercise 12 (Theory) Consider some environmental conditions that might lead to a reduction of the growth rate of bacteria. Why do you think these conditions would lead to a reduction in growth rate?

---

[10]We have included a project on quantifying bacteria death at the end of the chapter. The idea uses genetic sequencing techniques used more commonly when studying viruses, but could be applied to estimate changes in bacterial death rates to identify genetic mutations that affect fitness [124].

2.1 *Bacterial Growth on Agar Plates*

Let us assume the spatial aspects of the world is a limitation in the bacteria's growth. Suppose the bacteria is growing on a two-dimensional agar plate and that it is non-motile. If space is not available for the bacteria to divide, then the bacteria will not divide. We can model this in NETLOGO by using an if-statement. We check the bacteria's eight closest neighboring patches to see if any are empty. If so, we will allow the bacteria to divide and occupy one of the available neighboring patches. Model 1.1 simulated unrestricted growth of non-motile bacteria; alter the go procedure of that model in following way:

```
to go
  ask turtles [
    if any? neighbors with [count turtles-here = 0] [
      hatch 1 [
        move-to one-of neighbors with [count turtles-here = 0]
      ]
    ]
  ]
  tick
end
```

Now, the limitation on growth is the total number of patches in the simulated plate. No longer will each bacterium divide at each tick. Observing the plot, notice the growth now appears logistic. Instead of unlimited growth, we now see the total bacteria count not only approaching but also achieving a carrying capacity. Figure 2 shows a flow diagram of the potential paths for each bacteria cell for the code as it is written.[11] (See [125]: Model 2.0.0 for sample code.)

Exercise 13 (Theory) Examining the graph of the log of the counts of bacteria present at each generation time, why is the curve no longer linear?

Exercise 14 (Theory) Since the time it takes for each cell to divide now varies based on its environment, would you expect the average generation time to be higher or lower than when there were no spatial restrictions?

Exercise 15 (Theory) If you modify the necessary condition for cell division in Model 2.0.0 to be `count turtles-here <= 1`, how do you think the model output would change? Try these modifications in the code, making sure to change both instances of the logical expression. Were you right? Compare the carrying capacity to that of the previous model. Right-click on a patch. What do you notice in the menu? In terms of the bacterial growth on a plate, what is the difference in how the bacteria will form?

---

[11]If we were concerned about computational power, we may want to improve the efficiency of this code. In its current form, the loop that continually asks bacteria that had already failed the unoccupied-neighbors test is in vain; there is no mechanism in the current code to transform occupied patches into unoccupied patches. We always want to be mindful of efficiency by eliminating redundancy in the code and/or using structures that minimize computation time.

Fig. 2 This flow diagram shows a visual plan for Model 2.0.0. The initialization provided in the `setup` procedure to create one bacteria cell is the starting point of the diagram. Next, the cell must determine if it has any unoccupied neighboring patches. If so, the cell divides, and the newly formed daughter cell is placed in one of the unoccupied neighboring patches. If not, the cell will continue to look for open neighboring patches

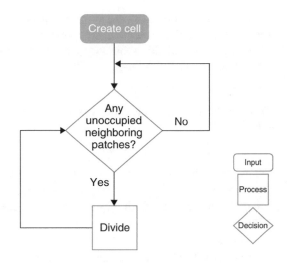

Challenge Problem 2 (Code) Even on plates, some bacteria (like certain strains of *E. coli*) are motile on the thin film of fluid spanning the agar surface [34, 40, 60, 112]. Though they still express coordinated behavior (swimming, swarming, twitching, etc.) in their motion on an agar plate, we will simplify this to add movement in a uniformly random way like we used when modeling movement in a fluid. Create two different models that introduce movement on a plate. One can be created by inserting the move procedure from Model 1.2 into Model 2.0.0 ([125]: Model 2.0.1). The other method could restrict movement in the same way that we restricted the cell division, by only allowing a cell to divide if there is a neighboring patch that contains no bacteria cells ([125]: Model 2.0.2). Describe the difference in the bacterial growth between these two models. Why does the first model appear to grow without bound, though with a significantly reduced growth rate?

2.2 Effects of Energy Source Availability on Bacterial Growth on an Agar Plate

In the previous exercises, you explored the theoretical idea of the vertical, or three-dimensional, growth of bacteria on an agar plate by allowing multiple cells to exist on the same patch. This simulates stacked bacteria growing out from the plate. Experimental results combined with modeling have shown that the growth rate of bacteria is almost identical whether they are grown in broth or on agar with similar nutrients [45]. This leads us to posit that there are other environmental factors that cause bacteria growth to subside before spatial constraints could possibly cause a statistically significant effect.

You may have considered the absence of an energy source as a condition likely to reduce bacteria growth. Indeed, bacteria cannot survive forever on an unmodified surface; without food from which to obtain energy, they would die. Let us modify our model of bacteria growth on a plate, Model 2.0.0, and introduce an initial food source (e.g., sugar, in this case, glucose) that will be consumed by the bacteria over time. First, we will add to the code a patch variable called sugar. At the very top of the code in Model 2.0.0 type,

```
patches-own [sugar]
```

Then, in the setup procedure, we need to specify the initial amount of sugar on each patch.

```
ask patches [set sugar 50]
```

We use 50 units here in order for the bacteria to grow to capacity before the food source becomes limited. At every tick, we will ask each bacteria to consume one unit of sugar from the patch they are on. Let us create a consume procedure in the following way:

```
to consume
  if [sugar] of patch-here = 0 [die]
  ask patch-here [
    if sugar > 0 [
      set sugar sugar - 1
    ]
  ]
end
```

This procedure first checks if the sugar on the patch is depleted. If so, the bacterium dies. In this case, the bacterium no longer exists to ask... anything. Therefore, the consume procedure would be exited if there is no sugar on the patch. However, if there is sugar left on the patch, the bacterium does not die and continues to consume one unit of sugar (simulated by the incremental decrease in the value of the sugar variable). Do not forget to add the new consume procedure into the go procedure, like so:

```
to go
  ask turtles [
    consume
    if any? neighbors with [count turtles-here = 0] [
      hatch 1 [
        move-to one-of neighbors with [count turtles-here = 0]
      ]
    ]
  ]
  tick
end
```

We want to ask the bacteria to consume the sugar prior to dividing, as energy is required for binary fission. Once we instruct the bacteria to consume the sugar, the simulation will produce a logistic growth curve with a relatively steep decay rate beginning after 52 iterations until all of the bacteria die. (See [125]: Model 2.1 for sample code.)

Exercise 16 (Code) When you run the simulation, the graph of the log growth curve stops plotting and the title turns red. Investigate this. What is the error? Edit the `go` procedure to force the simulation to terminate using the `stop` primitive command to correct the error, while also eliminating unnecessary iterations. You may find an alternative method than the solution provided in the sample code for Model 2.2 in the Appendix and in [125].

Exercise 17 (Lab + Code) Bacterial growth on different substrates supplemented with different nutrients can be easily demonstrated in the laboratory. We have provided ideas for experiments where students could use agar or broth, or vary the conditions including the pH, the sugars incorporated into the media, or vary the oxygenation [108]. Students would prepare serial dilutions of bacteria and plate for counts. The counted values could be logarithmically transformed and the log values graphed against units of time.

After collecting the data, vary the parameters in your model to best mimic the trends you witnessed in the bacteria counts. Think about adjusting numeric values like the initial amount of sugar, the maximum amount of sugar a single bacteria consumes in one tick, or the restrictions on how closely packed the bacteria can be. Devise methods for determining what makes your simulation best reflect the data. Additionally, as you learn more about environmental effects on bacteria growth, you can return to this exercise or the future lab protocols provided to continue to improve the replication of the trends you discovered by collecting data in the lab.

2.3 Effects of Energy Source Availability on Bacterial Growth in a Flask

Now, in a three-dimensional flask, we would not expect spatial restrictions to affect bacterial growth as much as the availability of an energy source. Additionally, the food would be relatively uniformly dispersed throughout the solution through mixing. We will return to our simulation that considers motile bacteria growing unrestricted by space in a flask (Model 1.2.0), but now we will add a restriction that they must have enough energy to perform binary fission and to generally survive. To the two sub-procedures we already included (`move` and `divide`), let us add two new sub-procedures to execute within the `go` procedure: `consume` and `expire`.[12] In Model 2.2, we added a `consume` procedure to ask the bacteria

---

[12]It seems that the obvious name for this sub-procedure would be "die." However, as you previously saw, `die` is a primitive command hard-coded into NETLOGO that performs the opposite action as `hatch`—`hatch` creates a turtle, and `die` removes a turtle.

to consume spatially fixed sugar on an agar plate. We will use a similar idea for the consume procedure in a flask, while keeping the idea in mind that the sugar will be well-mixed in the solution. We will begin by declaring sugar as a global variable (globals [sugar]). We will also need the bacteria to have an energy variable (turtles-own [energy]) that will determine whether the bacteria can divide or if it will expire. Both of those declarations will need to be added to the top of the code.

Exercise 18 (Code) In this new model, we will want to keep track of the amount of sugar that remains in the system as it will be depleted over time and not replenished (like agar in a plate or broth in a flask). Create a monitor in the interface area that reports the amount of sugar that has not yet been consumed. Note, sugar is a variable, not an agentset. The syntax is slightly different than the monitor that shows the number of bacteria.

In the consume procedure, we will want each bacterium to consume a sugar unit if it is available. This means the sugar will get depleted and the energy of the bacterium will increase. However, if the sugar has been completely consumed, the bacterium's energy will remain the same. We could use the following code:

```
to consume
  if sugar > 0 [
    set sugar sugar - 1
    set energy energy + 2
  ]
end
```

Now, we will alter the move procedure written in Model 1.2.0 to use energy:

```
to move
  right random 360
  forward 1
  set energy energy - 1
end
```

So for every movement, the bacteria have their energy reduced by one unit. However, while sugar is available, the bacteria will be steadily increasing their energy levels at a rate of one unit per tick.

The divide sub-procedure introduced in Model 1.2.0 will be transformed into a conditional process. For the purposes of our model, we will assume that in order for bacteria to perform binary fission, they require a sufficient amount of energy. Then, when the division occurs and the two daughter cells remain, they will each have half of the original cell's energy. Though this may not be the precise method of energy redistribution, the even split will be a reasonable proxy. Therefore, we could use the following code:

```
to divide
  if energy >= 20 [
    set energy energy / 2
    hatch 1 [right random 360 forward 1]
  ]
end
```

Note that the hatched bacterium is a clone of the original, and so has the same energy level. Also, we have semi-arbitrarily set the energy threshold for binary fission to be 20 energy units. This parameter was chosen in order to approximately reproduce the growth patterns we see in the scientific studies that plot bacteria amounts or concentrations over time. See [13, 45] for graphic and verbal explanations examples of slightly more sophisticated curve-fitting techniques with experimental bacteria growth data. Although the authors of these papers are using DE models, the basic ideas are the same—use an underlying set of relationships and behaviors, then find relative parameter values that best imitate the trends you see in the data.

The bacteria will die when their energy levels are completely depleted, and so we can create a simple expire procedure in the following way:

```
to expire
  if energy = 0 [die]
end
```

Exercise 19 (Code) The order that the sub-procedures are called within the go procedure matters. What order makes sense to you? Change the order of the sub-procedures in the code. Do you notice any changes in the emergent behavior? Repeat this multiple times.

In this model, we will also begin with a larger number of randomly placed initial bacteria to simulate a well-mixed solution. The energy level of each bacterium is determined by a uniform distribution. Therefore we will alter the setup procedure in this way:

```
to setup
  clear-all
  create-turtles 25 [
    set shape "circle 2"
    set color pink
    set energy random 20
    setxy random-xcor random-ycor
  ]
  set sugar 100000
  reset-ticks
end
```

The total amount of sugar is set to 100,000 units in order to visualize the exponential growth, a plateau, then a sharp decay. The simulation will need to run for approximately 100 ticks to witness the growth and decay. Just as we observed in Model 2.1 (bacteria growth on an agar plate), the log plot turns red when all of the bacteria have died. Therefore, we may fix this error in the same way by using the stop command.

Exercise 20 (Code) If you do not wish to click the go button 100 times, create another go button that runs the simulation until you unclick the button using the *Forever* option. Note, in order to witness the growth and decay at a visually processable speed, you may use the slider near the top of the Interface tab to slow down the tick rate. (See [125]: Model 3.0.0 and the Appendix for sample code.)

Exercise 21 (Code) What do you expect will happen in the simulation if the amount of energy required for the bacteria to divide is decreased? Increased? Try this in the model. Were you correct? What changed in the simulation output?

Challenge Problem 3 (Code) Instead of following the count of sugar by observing the monitor, use the color of the world to indicate the amount of sugar that remains in the solution. I suggest using the `pcolor` and `scale-color` primitive commands. (See [125]: Model 3.0.1 for sample code.)

Challenge Problem 4 (Code) Return to Model 3.0 once more to incorporate a spatial component of the nutrients in the way we used in the agar plate example in Model 2.2. Do this by attaching quantities of sugar to the patches and only allowing the bacteria to consume the sugar if it is on a patch that has remaining sugar. (See [125]: Model 3.0.2 for example code if you get stuck.)

You may notice in Model 3.0.0 the plateau signifying the stationary phase is quite abrupt and crudely models logistic growth. If you think about the nutrients available in the flask, there would certainly be a spatial component; once a substantial portion of the sugar has been broken down, not all bacteria would be in the proximity of an energy source. Let us consider a way to simulate a spatially dependent food source by using breeds in NETLOGO. Breeds allow us to designate classes of agents that can have their own variables and actions. We will alter Model 3.0.0 that introduced the four main sub-procedures: consume, move, divide, expire. The bacteria will be a class of agents (breed) that are required to be in the proximity of a sugar (another breed) in order to consume it and gain energy for binary fission. First, we must define our breeds at the top of the code:

```
breed [sugars sugar]
breed [bacteria bacterium]
```

Note, the `breed` primitive requires both a plural and singular form of the agentset. Instead of designating the energy variable to all agents by using `turtles-own`, we can restrict the assignment of energy to only the bacteria using `bacteria-own`. Essentially, we now use the plural name we gave to the agentset anywhere we would have previously used `turtles`. The singular form is still reserved for addressing a particular agent. Moreover, if we would like to address all agents (in this case bacteria and sugar), we would still use the entire agentset of `turtles`. Now, we must make some additions and slight alterations to the `setup` procedure. We must populate the solution with sugar, so we will add:

```
create-sugars 2000 [
    set shape "dot"
    set color white
    setxy random-xcor random-ycor
]
```

This creates 2000 randomly placed small white circles, representing sugar. Notice we use `create-sugars` instead of `create-turtles` since we have two distinct breeds. The initial amount of 2000 sugars was chosen in order to model all phases of growth and decay. Next, we need to modify the bacteria initialization to designate the agents as bacteria:

```
create-bacteria 15 [
    set shape "circle 2"
    set color pink
    set energy random 21
    setxy random-xcor random-ycor
  ]
```

Again, the numbers were chosen to produce an accurate simulation. In the laboratory, bacteria growing in liquid are rotated quickly so as to agitate the contents, ensuring mixing of nutrients, the cells themselves and good aeration. For our simulation, we will assume the flask is shaken regularly to ensure the sugars are mixing evenly in the solution. Thus, our simulated sugars (and bacteria) will move randomly within the world. We add the movement of the sugar to the go procedure in the following way:

```
ask sugars [
    right random 360
    forward 5
  ]
```

A forward movement of 5 ensures the remaining sugars are well-mixed and not remaining far away from the bacteria clusters. In the move procedure for the bacteria, increase the forward movement to 5 as well, as it seems as if the mixing would have the same effect on the positional change of the sugar and bacteria.

The divide and expire sub-procedures for the bacteria will remain the same; however, we must modify the consume procedure. Just as we considered the spatial proximity of the bacteria and sugar in Challenge Problem 4, we will use a similar concept here. The bacteria can only consume the sugar if they are adjacent to it. So, we will use the built-in in-radius command to ask the bacteria to look with a certain range around themselves and consume a sugar if they find one. If there are no sugars within the designated area, the bacteria lose energy.

```
to consume
  if any? sugars in-radius 2 [
    ask one-of sugars in-radius 2 [die]
    set energy energy + 15
  ]
end
```

Note, we have increased the energy gain from a sugar unit to 15, enough for the bacteria to divide after consuming two sugars within a few ticks. This change simulates conditions where bacteria can easily and quickly perform binary fission. (See [125]: Model 4.0 for sample code.)

Exercise 22 (Code) Click the go button a few times and observe the growth curves. Why do they appear flat? And, why is the sugar monitor reporting an error (indicated by the red font color)? Compare to the number of bacteria. Does this make sense? Edit the *Pen update commands* within the plots in order to plot the

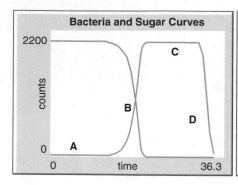 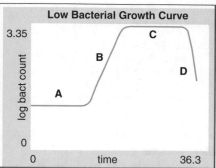

Fig. 3 The labeled curves show all four phases of bacterial growth: lag (A), log (B), stationary (C), and death (D). The increasing then decreasing bacterial counts (left) are displayed in pink and then shown on a log scale (right). The additional decreasing gray curve (left) displays the amount of sugar remaining at each time step. The plots were generated with Model 4.3 found in [125]

actual bacterial growth curves. Then add an additional pen to the plot to display the amount of sugar that remains. Modify the monitors for bacteria and sugar so that they show the proper totals. (See [125]: Model 4.1 for sample code.)

Exercise 23 (Code) Run the simulation for 50 ticks. What is not happening? Inspect a bacterium by right-clicking on it, choosing one of the bacteria on the patch, and clicking *Inspect* from the menu shown. Carefully consider the values of the variables associated with the bacterium. Find the logical error in the code that is not producing the desired results. Then, alter the code in order to fix the simulation. (See [125]: Model 4.2 for sample code.)

Model 4.2 simulates three of the four phases of bacterial growth: the log (or exponential) phase, the stationary phase, and the death phase, shown in Fig. 3 by B, C, and D, respectively. The remaining phase, indicated by A, occurs at the beginning of the growth process, called the lag phase. This is when the bacteria have just been placed in the growth medium and are preparing for the necessary processes that must take place in order to consume the energy source. The plots shown in Fig. 3 are copies of simulated bacteria growth trajectories that were generated in NETLOGO.[13] Though the resolution of the data visualizations is lacking, these images are shown here to validate your own output (with some variation, given the variability in the simulation).[14]

---

[13]Though the horizontal axis is labeled as time, recall, the time unit used in this simulation is arbitrary. The purpose of building this model was to replicate the phases of growth of an unspecified bacteria in a closed system; we did not try to use parameter values that would approximate growth with empirical cell counts or concentrations of a particular bacteria over time. Adjusting the generation time to fit data you collected through a laboratory experiment or using a known generation time to observe the emergent behavior of the system are components of many of the proposed research projects later in the chapter.

[14]The letter labels in Fig. 3 were added in post-production. You should not expect to see those when you create your plots in NETLOGO.

Exercise 24 (Code) Modify your model to include a lag phase. Note, even when the bacteria are adjusting to their new environment in the flask, they will continue to move and use energy. (See [125]: Model 4.3 and the Appendix for sample code.)

Not only does the depletion of energy sources cause bacterial decay, but paradoxically, the consumption of an energy source can contribute to bacteria death as well. For instance, when *E. coli* consume glucose , they are fermenting the glucose in a process called *mixed-acid fermentation*. This process results in the production of acetate as well as other acids and gas (carbon dioxide) which acidifies the medium, rendering the environment sub-optimal for the *E. coli* [16, 96]. Note, if the bacteria were inhabiting a continuous-culture solution, the acidic waste would be removed.

Let us add this unfortunate, yet natural, effect of consumption in a closed system into our model. We will begin by including a new global variable, acidity, defined at the very top of the code with globals [acidity]. Whenever we use a global variable, it is important to initialize its value in the setup procedure. Since we assume the initial medium is near-optimal, we will set the acidity to be zero, set acidity 0. Since acid is added to the solution when sugar is consumed by the bacteria, we need to incorporate an increase in acidity (or a decrease in pH) to the consume procedure. After a sugar is consumed, we can simulate the acidity level rising in this way:

```
to consume
  if any? sugars in-radius 2 [
    ask one-of sugars in-radius 2 [die]
    set energy energy + 15
    set acidity acidity + 1
  ]
end
```

In order to model the adverse effect of the acidity on the *E. coli*, we will add another cause of death on top of starvation. One way to simulate the continual decay in the quality of the solution due to acidity is to transform acidity into a death rate. We will use the following code as an example:

```
to expire
  if energy <= 0 [die]
  if random-float 1 < acidity / 200000 [die]
end
```

Here, random-float is introduced as an alternative way to model probabilistic behavior. This chooses a uniformly random non-negative floating point number that is strictly less than 1. Then, because 200,000 is 100 times greater than the initial amount of sugar specified in the setup procedure—and, therefore, much greater than the maximum acidity value—this creates a monotonically increasing death rate due to consumption with a minimum of 0 and a maximum of 0.01. In other words, the acidity will begin by killing 0% of the bacteria with every tick and slowly increases to killing 1% once all the sugar has been consumed. (See [125]: Model 5.0 and the Appendix for sample code.)

In the versions of Model 1, we used one tick as a proxy for generation time. In the versions of Models 2, 3, and 4, we restricted cell division by other environmental conditions, e.g., occupation of neighboring patches, availability of an energy source. We pose a research project to assess how environmental changes affect the average generation time.

Research Project 1 We noted early in Sect. 2 that biologists have witnessed significantly different generation times for *E. coli* when observed in a laboratory versus in a human body. Use the sources provided above, along with your own literature review, to determine potential causes of the variation of generation times, e.g., availability of efficient energy sources, ability to thrive in extreme temperatures. Create multiple models that incorporate those environmental differences. Within each model, track the number of ticks between every cell division. Record those disaggregated generation times in a list for export. You may want to investigate the use of lists in the Programming Guide found in the NETLOGO User Manual hosted on the NETLOGO website [120]. You will want to analyze the simulated data by considering measures of center and spread of generation times in a given environmental setting, ultimately, reconciling the emergent behavior with data you gathered by completing one of the laboratory protocols we provided or with experimentally produced data found in the literature.

In the following steps to design the next iteration of the model, we will create a slider so the user can input a bacteria's empirically estimated generation time. This will lay a framework that will allow simulated competition between two strains of bacteria. However, we should be mindful that we are hard-coding a minimum generation time into the simulation. Therefore, the average generation time that emerges will be longer than the one specified.

At this point, we will adjust the code in order to allow for user-defined variability in generation time. We will convert ticks to time units and create an internal counter to track the time required before the cell division is completed. In making these changes, we can also make some adjustments to improve the sample code given for Exercise 24, which simulates the lag phase. First, we need to add a counter variable to the bacteria. Within the variable declaration for the bacteria breed we will now include:

```
bacteria-own [
  energy
  generation-time-counter
]
```

The `generation-time-counter` will increase by one for every tick, triggering a division when the counter has reached the specified generation time. Of course, we still need to maintain the energy requirement for cell division as well. To keep track of each bacterium's clock, we will insert a standard counter at the end of the `consume-move-expire-divide` cycle within the `go` procedure:

```
ifelse ticks > 10
  [
    ask bacteria [
      consume
      move
      divide
      expire
      set generation-time-counter generation-time-counter + 1
    ]
  ]
  [
    ask bacteria [ move ]
  ]
```

Note, when the `ifelse` conditional statement checks if the number of ticks is greater than 10, this forces the lag phase to continue for 10 ticks (a possible solution to Exercise 24 to include a lag phase in the growth curve). In the `divide` sub-procedure, we must restrict the cell division even further than before. Not only is an energy threshold required, but enough time must have passed for the bacteria to complete binary fission. As an example, we will use the 20-min average generation time of *E. coli*, where a tick represents 1 min in the following code:

```
to divide
  if energy >= 100 and generation-time-counter >= 20 [
    set energy energy / 2
    set generation-time-counter 0
    hatch 1 [right random 360 forward 1]
  ]
end
```

Here we remember to reset the `generation-time-counter` to 0 to begin the binary fission process again. Recall, the "hatched" bacterium will have the same variable values as its parent bacterium, resulting in the `generation-time-counter` for both daughter cells to be 0.[15] Notice that the energy threshold has increased. This is due to the increase in the number of sugars that the bacteria can consume between cell divisions and for the purpose of simulating all phases of bacterial growth. We will also change the amount of energy gained from the consumption of a sugar to 5 (`set energy energy + 5`) in the `consume` procedure, in order to maintain the relatively arbitrary choice to have the maximum amount of energy gained over the generation time to be equal to the threshold needed for division.

Finally, we need to initialize the bacteria with a `generation-time-counter` value in the `setup` procedure. We will allow this value to be a uniformly random value between 0 and 20 using `set generation-time-counter random 21`

---

[15] It is important to recognize that in this construction, the 20-min generation time used here is a minimum generation time; this will not result in an average generation time of 20 min.

and placing this in the setup procedure in the creation of the bacteria. This way, all bacteria will begin at different stages of preparedness for cell division. After the lag phase is over, some of the bacteria will be able to split immediately, while others will need a bit more time. This simulates the empirically driven concept that individual bacteria will adjust to their environment at different rates. The stochasticity of the cell division times is part of what gives us variability in the bacterial counts after a fixed time has passed. Much work has been done to model with stochastic differential equations and agent-based models backed with experimental evidence on methods to measure the variability caused by multiple stochastic processes during all phases of bacterial growth [15, 47].

Because of the additional ticks that result in additional consumed sugars, we will increase the initial number of sugars. Instead of hard-coding the amount of sugar added into the model, let us create a slider to allow us to change this value in the interface without needing to alter the code. In the Interface tab, find *slider* in the drop-down menu. Then click within the interface area to place the slider. Sliders are named with the global variable that you plan to alter. Conventionally, we use a descriptive noun or phrase with hyphens between words. In this case, we will name the variable initial-sugars. Then, we will set the range and granularity of the slider. In this case, we will allow the slider to go from 0 to 20,000 by 1000 sugar increments. We set the default value to be the center, 10,000, since this results in an outcome that displays all phases of bacterial growth.

Now, we need to incorporate this variable into our code. This user-determined value should replace the previous hard-coded numeric value for the initial number of sugars created in the setup procedure. Substitute create-sugars 2000 with create-sugars initial-sugars in the code. Now, we can use the slider to change this value easily. Note, we did not need to define this global variable using globals within the code. In fact, if you do this, you will receive an error. In allowing the initial amount of sugar to be user-defined, we must alter the death-by-acidity conditional statement. We will make this adjustment:

```
if random-float 1 < acidity / (500 * initial-sugars) [die]
```

By multiplying the user-determined global variable, initial-sugars, by 500, we are restricting the death rate even more than in the previous version, now simulating an even slower monotonic increase from 0 to 0.5%.

Finally, all that remains is to change the axis titles on the graphs to reflect the appropriate units. Time in standard units (minutes, hours, etc.) is now measured by the horizontal axis instead of a unit of time indicating a generation time.[16] You may

---

[16]No units have been specified for time because any time unit could be used in the model as long as we are consistent. For instance, in the case of *E. coli*, we use 20 for our generation-time variable. The units on this variable are now minutes, and so the x-axis would show time in minutes. If we were modeling bacteria with a longer generation time, like, *M. tuberculosis*, we may choose to still use 20 for generation-time. However, the x-axis would be a measure of time in hours. Alternatively, we could stick with minutes and set generation-time to be 1200, which would mean the x-axis units would be minutes again. You may realize that such a large value for the

also wish to remove the pen that displays the amount of sugar that was added in Model 4.1, as we did in the sample code, so you can view the bacterial growth curve in more detail. (See [125]: Model 6.0 for sample code.)

Exercise 25 (Code) Vary the parameter determining the maximum death rate due to fermentation and increased acidity of the environment.

(a) What happens to the growth phases when the parameter is decreased? Why do you think this occurs?
(b) Use the methods described in Research Project 1 to find the average generation time for various values of the parameter. Is there an effect?

Exercise 26 (Code) Vary the slider for the initial value of added sugar.

(a) What do you notice in the time it takes to execute the simulation once? Why does this occur?
(b) What do you notice in the graphical output when you run the simulation to completion? Why does this occur?
(c) Use the methods described in Research Project 1 to find the average generation time for various values of the sugar-slider. Is there an effect?

Exercise 27 (Code) Vary the amount of energy gained from consuming a sugar and the threshold needed for cell division.

(a) What changes occur in the bacterial growth curve? Inspect the bacteria. Why does this occur?
(b) Use the methods described in Research Project 1 to find the average generation time for various values of the parameter. Is there an effect?

Exercise 28 (Code) At this point our simulation allows the bacteria to continue to consume sugar and amass stores of energy. In reality, the bacteria would genetically regulate this, so we will in our model as well. Include an additional condition in the consume procedure to only allow bacteria to consume sugar if their energy is below a fixed threshold, say 100 units. (See [125]: Model 6.1 for sample code.)

Exercise 29 (Theory) Instead of using the simulated data to find an average generation time, how could you estimate the generation (doubling) time of the bacteria from the graphs produced? You may want to consider Eq. (2). Use this method to estimate and confirm the expected generation time of the E. coli example. Now devise a method to find the generation time from the log bacteria growth curve. Estimate and confirm the expected generation time with this method.

Exercise 30 (Code) Change the code to model *Lactobacillus acidophilus*, a probiotic (i.e., "good bacteria") that is part of healthy gut flora. *L. acidophilus* has an average generation time of approximately 2.86 h when in a glucose solution [67].

---

generation time would lead to a lengthier simulation, as each tick or iteration through the go procedure is one time unit, likely leading to a preference for using a specific unit of time.

Use the methods from above to support that the changes you made in the code actually reflect the new generation time.

Exercise 31 (Code) Create a slider in the interface for generation time to more easily model different strains of bacteria. (See [125]: Model 6.2 and the Appendix for the complete sample code.)

Research Project 2 DIY Lab: Let's make kombucha!

We can witness the mixed-acid fermentation first-hand with a delicious experiment you can conduct at home. Kombucha is a carbonated acidic beverage made by fermenting sweet tea with a symbiotic culture of bacteria and yeast (SCOBY). The SCOBY contains multiple strains of bacteria and yeast. You can acquire this stiff, jelly-like culture by purchasing one online, getting one from a friend who makes kombucha, or growing your own with a plain, unflavored kombucha that you purchase at a store. The simplified process of fermentation that creates the finished kombucha product is as follows: the yeast consume the sugar, which produces carbon dioxide and ethanol. The bacteria then convert the ethanol into acid [72]. Note, because of the symbiotic nature of the yeast and bacteria, the ethanol that is produced is mostly metabolized by the bacteria, causing the alcohol content of the beverage to typically be kept to under 0.5% ABV (alcohol by volume).[17] The longer the SCOBY is thriving in the sweet tea solution, the more carbonated and acidic the beverage becomes. Additionally, the acid-producing bacteria are genetically predisposed to thrive in relatively high-acid environments—unlike many bacteria, like *E. coli*, that would die in these conditions. For this project, you will combine DIY scientific experimentation with computer simulation to better understand the fermentation process and predict the outcomes of your kombucha dependent on varied parameters.

1. Create a laboratory protocol to systematically measure the amount of sugar, acid, and SCOBY present in the solution.[18] If you are doing this at home—and not in a lab—you could qualitatively measure the sugar with a sweetness scale, use litmus strips to measure the pH, and either weigh the SCOBY with a kitchen scale or estimate its volume by using a ruler to find its dimensions. Note, temperature can

---

[17]So, for those of you under the legal age for drinking most of the fermented beverages currently on the market, you may legally consume the beverage produced after the project is complete (unless your kombucha was produced in the lab—never consume anything that was produced/modified/brought into the lab. There could be (and probably are) nasty hitchhiking pathogens swimming around in there!).

[18]See [81] for an example of a chemistry-focused laboratory experiment that uses technical equipment to perform measurements of more variables.

affect growth rates and the fermentation process, so be sure to consider this in your data collection. For a more advanced experiment, repeat your experiment multiple times to collect more data.[19]

2. Use the data collected to find relationships between the three variables you measured. Create a simulation that mimics the fermentation process. You may want to create a few more breeds to model the additional organisms and byproducts found in and necessary for the fermentation process. You should also do additional research on the types of yeast and bacteria present in a standard SCOBY to better model their interactions. Keep in mind that the bacteria used are not affected by the rise in acidity—unlike our previous model that was made with *E. coli* in mind.

3. Use the relationships you found to determine parameters and units for your model.

4. Compare the results of your experiment with the results of your simulation to assess your model's validity. Adjust parameter values to maximize your confidence in your model's ability to predict the quantities or concentrations of the model variables over time.

5. Consider several scenarios and use your model to predict the outcomes. Then, test your predictions by making kombucha under conditions that are as close to the hypothetical scenario you chose. For instance:

   a. You are fermenting your sweet tea in an apartment in New York City over the summer, and your air conditioner breaks. Alternatively, it is the winter, and your heat gets shut off.

   b. You order a SCOBY from a company online. You make kombucha as directed, but the end product has an acidity level close to that of apple cider.

   c. Oh no! Visible mold growth has appeared on the surface of your kombucha.[20]

Exercise 32 (Lab) The ability of bacteria to ferment sugars and to change the environment can be demonstrated several ways. The simplest would be to use solid media that changes color as the bacteria grow, ferment and produce acid, the acid changes the pH, thus the agar changes color. Fermentation tubes can also be used to demonstrate the ability of bacteria to ferment (this is indicated by a color change) and to produce gas (a bubble) and pH indicator strips could be inserted (cautiously) and the pH recorded. Triple sugar iron slants can also be used to demonstrate fermentation and gas production but

(continued)

---

[19] Each fermentation process requires significant time (on the order of a month). You could consider running multiple experiments simultaneously to produce more data in less time. You could also vary parameters (type of tea, incubation temperature, light exposure) and see what happens.

[20] You will likely want to read through the next section on competition before tackling this scenario.

Exercise 32 (continued)
also the production of H_2S gas. All these procedures would require access to the laboratory. You can find lab protocol examples here: [108].

3 Competition Between Bacteria Strains

Now that we have a simplified model of the growth of a single strain of bacteria, let us consider a simulation of the competition between two types of bacteria that consume the same energy source. There are many ways the bacteria could vary; they could have different generation times, death rates, energy consumption rates, etc. For instance, *E. coli* and *Salmonella enteritidis* both consume glucose for energy. However, *E. coli* has a much shorter generation time than *S. enteritidis*, ~20 min and ~30 min, respectively. Therefore, we would expect when competing for the same resource in a closed system, *E. coli* would grow at a faster rate than *S. enteritidis* initially, while the environment is still habitable. Since there will eventually be many *E. coli* to ferment the glucose and increase the acidity of the medium, this will cause the environment to be much too acidic for both strains of bacteria in the solution.

Exercise 33 (Lab) Competition between bacteria as well as their ability to adapt and evolve is one of the most fascinating labs to perform but it takes more time than a standard lab. This may be best suited to a long-term project or independent study. Find a mentor for this [108].

Challenge Problem 5 Use the main ideas from Model 6.2 to create a simulation of two strains of bacteria competing over the same energy source. When analyzing the results of the simulation, consider changes in the initial amount of sugar, initial amount of each bacteria strain, and generation time of each strain. How do these changes affect the amount of each bacteria strain over time?

Lactic acid bacteria, like the previously introduced *Lactobacillus acidophilus*, are characterized by their production of lactic acid and their propensity for thriving in high-acid environments that most bacteria cannot tolerate. However, *L. acidophilus*' 2.86 h generation time is significantly longer than that of *E. coli*'s—only 20 min. Both strains can thrive independently in milk and consume lactose for energy [13].

Exercise 34 (Theory) Suppose there are equal numbers of *L. acidophilus* and *E. coli* in milk. Formulate a hypothesis for the trajectory of the quantity of each strain over time.

Challenge Problem 6 Create a simulation to test the hypothesis you formulated in Exercise 34.

Now, suppose we want to simulate competition in a solution containing two energy sources: glucose and lactose. Let us return to competition between *E. coli* and *S. enteritidis*. *E. coli* can metabolize glucose and lactose. However, in media containing both glucose and lactose, *E. coli* favors glucose; it will metabolize all of the glucose first, ignoring the lactose until the glucose has been consumed. The genes needed for the breakdown of lactose are being repressed by the bacteria in a tightly regulated process called catabolite repression [82]. While the bacteria are breaking down the glucose, they produce a glucose breakdown product which inhibits an important enzyme called adenylate cyclase. This enzyme is responsible for the conversion of ATP (adenosine triphosphate—the energy currency of the cell) into cAMP. When the *E. coli* have metabolized all the glucose, the breakdown product is no longer created, and adenylate cyclase is activated, which forms cAMP. When this happens, the cell relieves the repression on the genes for lactose breakdown, the genes are transcribed, the proteins translated and the breakdown of lactose as a source of energy can begin. On the other hand, *S. enteritidis* can only metabolize glucose; it is a non-lactose fermenter.

Exercise 35 (Theory) Suppose we begin a culture with an equal number of cells of *E. coli* and *S. enteritidis*. Assume they consume all energy sources at roughly the same rate. Sketch the curves that represent the bacteria count of each type over time when

1. no food source is added;
2. only glucose is added;
3. only lactose is added; and
4. glucose and lactose are added.

Challenge Problem 7 Modify the competition simulation you previously made to include glucose and lactose, making sure to abide by the metabolic and fermentation constraints outlined above. Use your simulation to confirm your expectations for the curves in the previous exercise.

Lactose usually is fermented rapidly by *Escherichia*, *Klebsiella*, and some Enterobacter species and more slowly by *Citrobacter* and some *Serratia* species. In the clinical lab, this is the mechanism used to distinguish between pathogenic and non-pathogenic Enterobacteriaceae—non-lactose fermenters are usually pathogens, e.g., Salmonella and Shigella [14]. The ability of bacteria to ferment different sugars can be demonstrated using a variety of procedures in the laboratory (See [108].).

Exercise 36 (Lab) The ability of bacteria to ferment lactose is of particular importance in clinical microbiology where non-lactose fermenters are commonly pathogenic. Students can determine the ability of bacteria to ferment this sugar by simply streaking them onto MacConkey agar [14, 108].

4 Genetic Mutations

Every time a cell divides, there is a chance—albeit, typically quite a small chance—for a genetic mutation. For instance, researchers have used whole-genome sequencing to determine that *E. coli* have a mutation rate of approximately 2.2×10^{-10} mutations per nucleotide per generation or 1×10^{-3} mutations per genome per generation [65].[21] However, this method is impractical for determining mutation rates for bacteria with larger genomes due to the computationally expensive process.

Exercise 37 (Theory) Use the equivalent rates of mutation given above to approximate the number of nucleotides in the genome of the strain of *E. coli* studied in [65].

Challenge Problem 8 Use the approximate number of mutations per genome per generation to create a simulation of *E. coli* growth with mutations. Track the total number of mutations over time and the total number of mutated bacteria. Note, when a mutated bacterium divides, it replicates the mutation.

Some genetic mutations affect the fitness of the bacteria—the bacteria's ability to survive and divide. For example, a mutation could improve the fitness of a bacterium: it could enable the bacterium to consume a new energy source; it could increase the range of temperatures in which the bacterium would thrive; it could render a chemical compound ineffective whose purpose is to kill the bacteria (more on this later!). However, genetic mutations can also decrease the fitness of a bacterium: it could render the bacteria unable to complete the binary fission process; it could prevent the bacteria from consuming a vital energy source; it could decrease the range of pH levels that the bacteria could thrive. Another possibility is that the mutation could have no effect on the fitness level of the bacterium at all.

Challenge Problem 9 Create a simulation that allows mutations to affect the fitness level of bacteria in a positive, negative, or neutral direction. Incorporate effects on survival and division.

---

[21]Though, it does appear that not all nucleotides are equally likely to mutate under selective pressures due to evolutionary processes [65].

Research Project 3 Create a model that allows the user to change the environment (e.g., temperature, energy source, pH level). Simulate mutations that directly affect the fitness of a bacterium to the potential changes in the environment. Run your simulation many times with many different parameter settings to identify combinations of parameter values where certain types of mutated bacteria prevail. Read about the BehaviorSpace tool in NETLOGO to assist with running this type of experiment.

Recall from the laboratory experiments and simulation exercises on bacterial growth that biologists estimate the generation time of a bacteria strain in a fixed environment by determining the number of bacteria present over multiple time steps during the log growth phase. You may remember that we ignored natural cell death (for the most part) in our growth models. This is due to the difficulty in experimentally determining the number of bacteria that have died; our laboratory procedure only records living cells as we count them on the agar. We cannot count dead cells. Because there are no bacterial corpses to count, this makes estimating a natural death rate difficult. In the study of viruses (virology), researchers often use next generation DNA sequencing to determine viral death and genetic diversity distributions [124]. Simply put, this technology allows you to determine the proportions of a given DNA sequence as a fraction of all the DNA sequences in a sample at a given time. If you can determine the number of each type of mutated virus strains that exist in every sample over time, then when a strain no longer exists in the sample, you know that (at least) all of those viruses have been lost. The same method could conceivably be used in determining bacterial strain death, particularly for bacteria with higher mutation rates.

Research Project 4 Create a model that includes (significantly) reduced genome sequence of a bacteria and include a natural death rate. Force mutations to the genome at a fixed rate. Isolate the different sequences made by the mutations and track each strain's count over time. From the data you generate, attempt to recover the natural death rate you had hard-coded into the model. Then, incorporate the effect that mutations can have on fitness levels into your model. Could you use the data you generate to identify specific portions of the genome that have particular effects on fitness? If you allow changes in the environment, could you recover more information about the mutations? Even though your findings will only be recovering information you had to specify in the simulation, how could you use this process in a laboratory setting to better understand bacterial mutations?

5   Antibiotic Intervention and Resistance

Though most bacteria inside and outside of our bodies are harmless—or even helpful, there are strains of bacteria that are detrimental to the health of a human. When inside a human body, these disease-causing, or pathogenic, bacteria are often treated with antibiotics in order to hasten the elimination of the bacteria, thereby relieving the host of undesirable symptoms more quickly than their natural immune response. As mentioned in the introduction, antibiotics are not only used to treat humans but are widely used to promote growth in food-producing animals. Though legislative efforts have been successful in decreasing the overall use of "medically important" antibiotics in food-producing animals, FDA data has shown that in 2017, 62% of farm animal antibiotics were administered via feed, and 30% in drinking water, for mass medication [10]. The widespread and non-discriminate use of antibiotics in animal populations has been implicated in the rapid increase and spread of antibiotic-resistant bacteria.

Challenge Problem 10 Modify and add an antibiotics breed to Model 6.2 to create a simulation of the effect of antibiotics on bacterial growth. Determine a typical prescribed dosing schedule of an antibiotic and incorporate this into the simulation to track the bacterial growth over time. Consider alternative dosing schedules to witness the effectiveness of antibiotics when a non-standard dosing schedule is followed.[22]

Suppose you have a urinary tract infection (UTI), and your medical doctor prescribes you Cipro (ciprofloxacin)—a broad-spectrum antibiotic, which indiscriminately kills all strains of bacteria—based on your description of your symptoms. These broad-spectrum antibiotics will kill the pathogenic bacteria, but will also eliminate "good bacteria" such as lactobacilli found in a healthy digestive tract [55]. Alternatively, if your medical doctor first tests the bacteria present in a urine sample, they could prescribe you with a narrow-spectrum antibiotic, like Primsol (trimethoprim) that would only target the pathogenic bacteria causing the unwanted symptoms of a UTI [51].

Challenge Problem 11 Simulate the effect that antibiotics have on bacteria. Consider the varied affect that broad- versus narrow-spectrum antibiotics would have on multiple strains of bacteria contained in a system.

> **Research Project 5** Expand the simulation from above to specifically consider the use of broad-spectrum antibiotics for a UTI, which will kill the "good" and "bad" bacteria. One of the probiotics ("good" bacteria) found

(continued)

---

[22]See the existing model, Bacterial Infection, located in the NETLOGO Model Library for an example of this.

> **Research Project 5** (continued)
> in the urinary tract is *Lactobacillus acidophilus*, which was introduced in
> a previous section. *E. coli* are a common cause of urinary tract infections.
> *L. acidophilus* strains secrete antibacterial substances with activity against
> *E. coli*, along with other bacteria and yeast [48]. *L. acidophilus* can also
> thrive at much lower pH levels than *E. coli*. Use the ideas of bacterial
> growth, competition, and antibiotic intervention to simulate the trajectory of
> the populations of bacteria. In your model, consider the difference between
> antibiotics introduced in a closed system (e.g., an agar plate in a laboratory)
> or in the human (or livestock) body. For instance, how would the immune
> system contribute to the elimination of the pathogens? Additionally, what if
> narrow-spectrum antibiotics were used instead of broad-spectrum?

Under laboratory conditions, some antibiotics kill the targeted bacteria (bactericidal), while others prevent the bacteria from dividing (bacteriostatic) [91]. For first line therapy of patients, bactericidal rather than bacteriostatic agents are recommended because the eradication of microorganisms serves to limit, although not completely avoid, the development of bacterial resistance [111]. Resistance has been shown to occur more rapidly with bacteriostatic agents such as tetracyclines, sulfonamides, and macrolides than it does with bactericidal agents such as beta-lactams and aminoglycosides [111].

The use of antibiotics increases the chance of creating antibiotic-resistant bacteria.[23] Some bacteria gain resistance through mutations. Recall, mutations occur at some rate per nucleotide per genome per generation. When antibiotics are present, bacteria that are resistant will survive, while the others will die. This forces the selection of this mutation in the genome.

> **Research Project 6** Modify the previous simulation that tracked genetic
> mutations and their effect on fitness to focus only on mutations that result
> in antibiotic resistance. Perform a literature search to find an estimate for an
> average rate of acquiring resistance for a particular bacteria to a particular

(continued)

---

[23] Antibiotics can increase the mutation rate in bacteria [68]. Antibiotics not only impose a selective challenge to all bacteria but also accelerate the rate of adaptation by magnifying the rate at which advantageous mutations arise (any type of mutations, not just for resistance). In addition, mutation rates have been shown to increase in the bacterial flora of patients treated with antibiotics, not only for the targeted bacteria [53]. Strains that have acquired antibiotic resistance mutations often have a lower growth rate and are less invasive or transmissible initially than their susceptible counterparts [126]. The fitness costs of resistance mutations can be ameliorated by secondary site mutations. These so-called compensatory mutations may restore fitness in the absence and/or presence of antimicrobials.

Research Project 6 (continued)

antibiotic. Then model the introduction of the antibiotic to the bacteria. Run the simulation many times to analyze the outcomes. How frequently is resistance gained and selected for? Consider the differences between the effect of the antibiotic in a controlled laboratory setting and in the human body. What other environmental factors may influence the rate of selection for resistance?

Tuberculosis (TB) is often treated with a drug cocktail; in other words, a collection of narrow-spectrum antibiotics that target *Mycobacterium tuberculosis*—the causative agent of TB. Drug cocktails are prescribed in order to lessen the chance of creating a resistant strain. The cocktail is often made of four bactericidal drugs: isoniazid (INH), rifampin (RIF), streptomycin (STM) or ethambutol (EMB), and pyrazinamide (PZA). The individual use of any of these drugs to treat TB has a relatively high chance to cause resistance, but the risk is significantly reduced when taken together.

Exercise 38 (Theory) Suppose an individual with sensitive TB was prescribed a cocktail of INH, PZA, RIF, and EMB. The rate at which TB gains resistance to each antibiotic in the cocktail has been previously determined—when introduced in isolation. Resistance is acquired to INH at a rate of 2.56×10^{-8} mutations per bacterium per division, PZA at a rate of 10^{-5}, RIF at a rate of 3.32×10^{-9}, and EMB at a rate of 1.0×10^{-7} [57, 110]. Assuming each mutation is independent of the other, at what rate would we expect TB to mutate to gain resistance to all four antibiotics?

Research Project 7 Using the rates given above (and your own literature review), create a simulation that supports your calculation. Use the estimate for the generation time of *M. tuberculosis* (provided earlier in the chapter) and the simulation to determine the typical length of time resistance would emerge when taking any combination of the drugs in the cocktail. Find the standard dosing schedule used for sensitive strains of TB. Use your model to simulate the dosing schedule. Alter the schedule and/or dosages to change the chances of resistance. What if you were just given one antibiotic at a time and were only prescribed the next if the strain gained resistance to the current drug? What if you used the same cocktail for a strain of TB that is already resistant to at least one of the antibiotics?

6   Spread of Antibiotic-Resistant Bacteria in Humans

Another way bacteria gain resistance is through horizontal gene transfer (HGT). This is the process where bacteria acquire DNA from similarly-designed bacteria. For instance, suppose you have acquired an innocuous strain of *E. coli* that has gained resistance to carbapenem—antibiotics that are often considered to be the last stand against bacteria. However, you have a healthy immune system that prevents this strain of *E. coli* from reaching high enough levels to make you sick. Perhaps you visit your elderly, diabetic grandfather in a nursing home, and you pick up an antibiotic-sensitive strain of *K. pneumoniae*. The *K. pneumoniae* begins to make you quite ill. You go to a walk-in clinic, and they prescribe you a broad-spectrum antibiotic. The antibiotic begins to wipe out all of the bacteria in your body, but the stress causes *K. pneumoniae* to conjugate with *E. coli* and in so doing to acquire the carbapenamase plasmid that confers resistance [54]. After a few generations, there are enough carbapenem-resistant *K. pneumoniae* to take hold. A healthy immune system may be able to respond to this attack, but you did just wipe out your entire gut flora with the broad-spectrum antibiotic. Unfortunately, the resistant *K. pneumoniae* will not respond to another antibiotic now.

Exercise 39 (Lab) There are several commercially available kits to demonstrate the ways that bacteria can exchange genetic information. Though they require access to a microbiology lab, the resources needed are minimal and the experiments are simple to perform and interpret [108].

Research Project 8 Create a simulation of HGT. Formulate a model of the intracellular mechanisms and dynamics involved in the process, then consider a human population-scale model of the implications of HGT on planetary health.

Research Project 9 In the introduction, we discussed the societal causes and implication of antibiotic resistance. There are many agent-based models created in NETLOGO that address the spread of infectious disease. In the NETLOGO Model Library you can find the epiDEM models, which give basic models that simulate an epidemic [122, 123]. There are also NETLOGO ABMs that consider the effect of antibiotic resistance [20], the spread of vector-borne diseases [46], and the effect of vaccinations [58]. Other ABMs use

(continued)

Research Project 9 (continued)

many of the above strategies and network structures to illustrate physical and social connections between individuals [43, 70]. Other researchers are using more advanced ABM systems that can utilize immense amounts of data. For instance, many are incorporating Geographic Information Systems (GIS) data to incorporate real city-structures and census data to better understand the spread of infectious disease [93, 117]. We challenge you to create an agent-based model that considers and analyzes the social implications of antibiotic resistance that were addressed in the introduction. For instance, incarceration of people with limited healthcare and physical space, regulation of antibiotic use on livestock, use of broad- versus narrow-spectrum antibiotics, changes in the prescription of antibiotics prompted by the regulation of healthcare, and increased mobility of people and populations. Standard infectious disease models could be adapted to incorporate these social effects. A thorough analysis of the simulation could help inform policy or support further research in the area you studied.

Appendix

Model 1.2.0

```
to setup
  clear-all
  create-turtles 1 [set shape "circle 2" set color pink]
  reset-ticks
end

to go
  ask turtles [
    move
    divide
  ]
  tick
end

to move
  right random 360
  forward 1
end


to divide
  hatch 1 [right random 360 forward 1]
end
```

Model 2.2

```
patches-own [sugar]

to setup
  clear-all
  create-turtles 1 [set shape "circle 2" set color pink]
  ask patches [set sugar 50]
  reset-ticks
end

to go
  ask turtles [
    consume
    if any? neighbors with [count turtles-here = 0] [
      hatch 1 [move-to one-of neighbors with [count turtles-here
          = 0]]
    ]
  ]
  if count turtles = 0 [stop]
  tick
end

to consume
  if [sugar] of patch-here = 0 [die]
  ask patch-here [
    if sugar > 0 [
      set sugar sugar - 1
    ]
  ]
end
```

Model 3.0.0

```
globals [sugar]

turtles-own [energy]

to setup
  clear-all
  create-turtles 25 [
    set shape "circle 2"
    set color pink
    set energy random 20
    setxy random-xcor random-ycor
  ]
  set sugar 100000
  reset-ticks
end

to go
  ask turtles [
```

```
    consume
    move
    divide
    expire
  ]
  if count turtles = 0 [stop]
  tick
end

to consume
  if sugar > 0 [
    set sugar sugar - 1
    set energy energy + 2
  ]
end

to move
  right random 360
  forward 1
  set energy energy - 1
end

to divide
  if energy >= 20 [
    set energy energy / 2
    hatch 1 [right random 360 forward 1]
  ]
end

to expire
  if energy = 0 [die]
end
```

Model 4.3

```
breed [sugars sugar]
breed [bacteria bacterium]

bacteria-own [energy]

to setup
  clear-all
  create-sugars 2000 [
    set shape "dot"
    set color white
    setxy random-xcor random-ycor
  ]
  create-bacteria 15 [
    set shape "circle 2"
    set color pink
    set energy random 21
    setxy random-xcor random-ycor
  ]
  reset-ticks
```

```
end

to go
  ifelse ticks > 10
  [
    ask bacteria [
      consume
      move
      divide
      expire
    ]
  ]
  [
    ask bacteria [ move ]
  ]
  if count bacteria = 0 [stop]
  ask sugars [
    right random 360
    forward 5
  ]
  tick
end

to consume
  if any? sugars in-radius 2 [
    ask one-of sugars in-radius 2 [die]
    set energy energy + 15
  ]
end

to move
  right random 360
  forward 5
  set energy energy - 1
end

to divide
  if energy >= 20 [
    set energy energy / 2
    hatch 1 [right random 360 forward 1]
  ]
end

to expire
  if energy <= 0 [die]
end
```

Model 5.0

```
globals [acidity]

breed [sugars sugar]
breed [bacteria bacterium]
```

```
bacteria-own [energy]

to setup
  clear-all
  create-sugars 2000 [
    set shape "dot"
    set color white
    setxy random-xcor random-ycor
  ]
  create-bacteria 15 [
    set shape "circle 2"
    set color pink
    set energy random 21
    setxy random-xcor random-ycor
  ]
  set acidity 0
  reset-ticks
end

to go
  ifelse ticks > 10
  [
    ask bacteria [
      consume
      move
      divide
      expire
    ]
  ]
  [
    ask bacteria [move]
  ]
  if count bacteria = 0 [stop]
  ask sugars [
    right random 360
    forward 5
  ]
  tick
end

to consume
  if any? sugars in-radius 2 [
    ask one-of sugars in-radius 2 [die]
    set energy energy + 15
    set acidity acidity + 1
  ]
end

to move
  right random 360
  forward 5
  set energy energy - 1
end
```

```
to divide
  if energy >= 20 [
    set energy energy / 2
    hatch 1 [right random 360 forward 1]
  ]
end

to expire
  if energy <= 0 [die]
  if random-float 1 < acidity / 200000 [die]
end
```

Model 6.2

```
globals [acidity]

breed [sugars sugar]
breed [bacteria bacterium]

bacteria-own [
  energy
  generation-time-counter
]

to setup
  clear-all
  create-sugars initial-sugars [
    set shape "dot"
    set color white
    setxy random-xcor random-ycor
  ]
  create-bacteria 15 [
    set shape "circle 2"
    set color pink
    set energy random 100
    setxy random-xcor random-ycor
    set generation-time-counter random generation-time + 1
  ]
  set acidity 0
  reset-ticks
end

to go
  ifelse ticks > 10
  [
    ask bacteria [
      consume
      move
      divide
      expire
      set generation-time-counter generation-time-counter + 1
    ]
```

```
  ]
  [
    ask bacteria [move]
  ]
  if count bacteria = 0 [stop]
  ask sugars [
    right random 360
    forward 5
  ]
  tick
end

to consume
  if any? sugars in-radius 2 and energy < 100 [
    ask one-of sugars in-radius 2 [die]
    set energy energy + 5
    set acidity acidity + 1
  ]
end

to move
  right random 360
  forward 5
  set energy energy - 1
end

to divide
  if energy >= 100 and generation-time-counter >=
      generation-time [
    set energy energy / 2
    set generation-time-counter 0
    hatch 1 [right random 360 forward 1]
  ]
end

to expire
  if energy <= 0 [die]
  if random-float 1 < acidity / (500 * initial-sugars) [die]
end
```

References

1. U.S. Action to Combat Antibiotic Resistance. https://www.cdc.gov/drugresistance/federal-engagement-in-ar/index.html.
2. Ban on antibiotics as growth promoters in animal feed enters into effect. Press Release, European Commission, December 2005.
3. Antibiotic Resistance Threats in the United States, 2013. Technical report, Centers for Disease Control and Prevention, April 2013.
4. The global economic impact of anti-microbial resistance. Technical report, KPMG LLP, December 2014.
5. Multidrug-resistant Shigellosis Spreading in the United States. Press Release, Centers for Disease Control and Prevention, April 2015.

6. National action plan for combating antibiotic-resistant bacteria. Technical report, The White House, March 2015.
7. President's 2016 Budget Proposes Historic Investment to Combat Antibiotic-Resistant Bacteria to Protect Public Health. Fact Sheet, The White House, January 2015.
8. Antibacterial Agents in Clinical Development: An analysis of the antibacterial clinical development pipeline, including tuberculosis. Technical report, World Health Organization, Geneva, Switzerland, 2017.
9. Antibiotic Resistance in Nursing Homes and Day Care Centers. https://www.cdc. gov/healthcommunication/toolstemplates/entertainmented/tips/AntibioticResistance.html, September 2017.
10. 2017 Summary Report On Antimicrobials Sold or Distributed for Use in Food-Producing Animals. Technical report, Food and Drug Administration, Center for Veterinary Medicine, December 2018.
11. Nasiru Abdullahi and Kenneth Chukwuemeka Iregbu. Methicillin-Resistant *Staphylococcus aureus* in a Central Nigeria Tertiary Hospital. *Ann. Trop. Pathol.*, 9(1):6, January 2018.
12. E. P. Abraham and E. Chain. An enzyme from bacteria able to destroy penicillin. 1940. *Rev. Infect. Dis.*, 10(4):677–678, 1988 Jul-Aug.
13. P Ačai, L' Valík, A Medved'ová, and F Rosskopf. Modelling and predicting the simultaneous growth of *Escherichia coli* and lactic acid bacteria in milk. *Food Sci. Technol. Int.*, 22(6):475–484, September 2016.
14. Mary E. Allen. MacConkey Agar Plates Protocols. http://www.asmscience.org/content/ education/protocol/protocol.2855, September 2005.
15. Antonio A. Alonso, Ignacio Molina, and Constantinos Theodoropoulos. Modeling Bacterial Population Growth from Stochastic Single-Cell Dynamics. *Appl Environ Microbiol*, 80(17):5241–5253, September 2014.
16. Natalie Angier. A Population That Pollutes Itself Into Extinction (and It's Not Us). *The New York Times*, June 2018.
17. Cesar A. Arias and Barbara E. Murray. The rise of the Enterococcus: Beyond vancomycin resistance. *Nat. Rev. Microbiol.*, 10(4):266–278, March 2012.
18. S. Augustine and R. A. Bonomo. Taking stock of infections and antibiotic resistance in the elderly and long-term care facilities: A survey of existing and upcoming challenges. *Eur J Microbiol Immunol (Bp)*, 1(3):190–197, September 2011.
19. Jose Luis Balcazar. Bacteriophages as Vehicles for Antibiotic Resistance Genes in the Environment. *PLoS Pathog*, 10(7), July 2014.
20. Sean L. Barnes, Daniel J. Morgan, Anthony D. Harris, Phillip C. Carling, and Kerri A. Thom. Preventing the transmission of multidrug-resistant organisms (MDROs): Modeling the relative importance of hand hygiene and environmental cleaning interventions. *Infect Control Hosp Epidemiol*, 35(9):1156–1162, September 2014.
21. John G. Bartlett, David N. Gilbert, and Brad Spellberg. Seven ways to preserve the miracle of antibiotics. *Clin. Infect. Dis.*, 56(10):1445–1450, May 2013.
22. Arnold S. Bayer, Tanja Schneider, and Hans-Georg Sahl. Mechanisms of daptomycin resistance in *Staphylococcus aureus*: Role of the cell membrane and cell wall. *Ann. N. Y. Acad. Sci.*, 1277:139–158, January 2013.
23. P M Bennett. Plasmid encoded antibiotic resistance: Acquisition and transfer of antibiotic resistance genes in bacteria. *Br J Pharmacol*, 153(Suppl 1):S347–S357, March 2008.
24. Howard C. Berg. *E. coli in Motion*. Springer Science & Business Media, January 2008.
25. Kirandeep Bhullar, Nicholas Waglechner, Andrew Pawlowski, Kalinka Koteva, Eric D. Banks, Michael D. Johnston, Hazel A. Barton, and Gerard D. Wright. Antibiotic resistance is prevalent in an isolated cave microbiome. *PLoS ONE*, 7(4):e34953, 2012.
26. Gerald Bloom, Gemma Buckland Merrett, Annie Wilkinson, Vivian Lin, and Sarah Paulin. Antimicrobial resistance and universal health coverage. *BMJ Glob Health*, 2(4), October 2017.

27. Georgiy V. Bobashev, D. Michael Goedecke, Feng Yu, and Joshua M. Epstein. A Hybrid Epidemic Model: Combining The Advantages Of Agent-Based And Equation-Based Approaches. In *2007 Winter Simulation Conference*, pages 1532–1537, Washington, DC, USA, 2007. IEEE.

28. E. Bonabeau. Agent-based modeling: Methods and techniques for simulating human systems. *Proc. Natl. Acad. Sci.*, 99(Supplement 3):7280–7287, May 2002.

29. Abraham Borer, Lisa Saidel-Odes, Klaris Riesenberg, Seada Eskira, Nejama Peled, Ronit Nativ, Francisc Schlaeffer, and Michael Sherf. Attributable mortality rate for carbapenem-resistant *Klebsiella pneumoniae* bacteremia. *Infect Control Hosp Epidemiol*, 30(10):972–976, October 2009.

30. Andrei Borshchev and Alexei Filippov. From system dynamics and discrete event to practical agent based modeling: Reasons, techniques, tools. In *Proceedings of the 22nd International Conference of the System Dynamics Society*, volume 22. Citeseer, 2004.

31. Henry F. Chambers and Frank R. Deleo. Waves of resistance: *Staphylococcus aureus* in the antibiotic era. *Nat. Rev. Microbiol.*, 7(9):629–641, September 2009.

32. Ruvani M. Chandrasekera, Emil P. Lesho, Uzo Chukwuma, James F. Cummings, and Paige E. Waterman. The State of Antimicrobial Resistance Surveillance in the Military Health System: A Review of Improvements Made in the Last 10 Years and Remaining Surveillance Gaps. *Mil. Med.*, 180(2):145–150, February 2015.

33. Sean R. Connell, Dobryan M. Tracz, Knud H. Nierhaus, and Diane E. Taylor. Ribosomal protection proteins and their mechanism of tetracycline resistance. *Antimicrob. Agents Chemother.*, 47(12):3675–3681, December 2003.

34. Matthew F. Copeland, Shane T. Flickinger, Hannah H. Tuson, and Douglas B. Weibel. Studying the Dynamics of Flagella in Multicellular Communities of *Escherichia coli* by Using Biarsenical Dyes. *Appl. Environ. Microbiol.*, 76(4):1241–1250, February 2010.

35. Sara E. Cosgrove. The relationship between antimicrobial resistance and patient outcomes: Mortality, length of hospital stay, and health care costs. *Clin. Infect. Dis.*, 42 Suppl 2:S82–89, January 2006.

36. Sara E. Cosgrove, George Sakoulas, Eli N. Perencevich, Mitchell J. Schwaber, Adolf W. Karchmer, and Yehuda Carmeli. Comparison of mortality associated with methicillin-resistant and methicillin-susceptible *Staphylococcus aureus* bacteremia: A meta-analysis. *Clin. Infect. Dis.*, 36(1):53–59, January 2003.

37. Vanessa M. D'Costa, Christine E. King, Lindsay Kalan, Mariya Morar, Wilson W. L. Sung, Carsten Schwarz, Duane Froese, Grant Zazula, Fabrice Calmels, Regis Debruyne, G. Brian Golding, Hendrik N. Poinar, and Gerard D. Wright. Antibiotic resistance is ancient. *Nature*, 477(7365):457–461, August 2011.

38. Vanessa M. D'Costa, Katherine M. McGrann, Donald W. Hughes, and Gerard D. Wright. Sampling the antibiotic resistome. *Science*, 311(5759):374–377, January 2006.

39. Carlos A. DiazGranados, Shanta M. Zimmer, Mitchel Klein, and John A. Jernigan. Comparison of mortality associated with vancomycin-resistant and vancomycin-susceptible enterococcal bloodstream infections: A meta-analysis. *Clin. Infect. Dis.*, 41(3):327–333, August 2005.

40. Willow R. DiLuzio, Linda Turner, Michael Mayer, Piotr Garstecki, Douglas B. Weibel, Howard C. Berg, and George M. Whitesides. *Escherichia coli* swim on the right-hand side. *Nature*, 435(7046):1271–1274, June 2005.

41. Sam Donovan, Carrie Diaz Eaton, Stith T. Gower, Kristin P. Jenkins, M. Drew LaMar, DorothyBelle Poli, Robert Sheehy, and Jeremy M. Wojdak. QUBES: A community focused on supporting teaching and learning in quantitative biology. *Lett. Biomath.*, 2(1):46–55, January 2015.

42. Donald M. Dumford and Marion Skalweit. Antibiotic-Resistant Infections and Treatment Challenges in the Immunocompromised Host. *Infect. Dis. Clin. North Am.*, 30(2):465–489, June 2016.

43. Joshua M Epstein and Robert Axtell. *Growing Artificial Societies: Social Science from the Bottom Up*. Brookings Institution Press, 1996.

44. N. D. Friedman, E. Temkin, and Y. Carmeli. The negative impact of antibiotic resistance. *Clinical Microbiology and Infection*, 22(5):416–422, May 2016.
45. Hiroshi Fujikawa and Satoshi Morozumi. Modeling surface growth of *Escherichia coli* on agar plates. *Appl. Environ. Microbiol.*, 71(12):7920–7926, December 2005.
46. Holly Gaff. Preliminary analysis of an agent-based model for a tick-borne disease. *Math. Biosci. Eng.*, 8(2):463–473, April 2011.
47. Míriam R. García, José A. Vázquez, Isabel G. Teixeira, and Antonio A. Alonso. Stochastic Individual-Based Modeling of Bacterial Growth and Division Using Flow Cytometry. *Front Microbiol*, 8, January 2018.
48. Ralitsa Georgieva, Lyubomira Yocheva, Lilia Tserovska, Galina Zhelezova, Nina Stefanova, Akseniya Atanasova, Antonia Danguleva, Gergana Ivanova, Nikolay Karapetkov, Nevenka Rumyan, and Elena Karaivanova. Antimicrobial activity and antibiotic susceptibility of Lactobacillus and Bifidobacterium spp. intended for use as starter and probiotic cultures. *Biotechnol Biotechnol Equip*, 29(1):84–91, January 2015.
49. Beth Gibson, Daniel J. Wilson, Edward Feil, and Adam Eyre-Walker. The distribution of bacterial doubling times in the wild. *Proc Biol Sci*, 285(1880), June 2018.
50. Zhabiz Golkar, Omar Bagasra, and Donald Gene Pace. Bacteriophage therapy: A potential solution for the antibiotic resistance crisis. *J Infect Dev Ctries*, 8(2):129–136, February 2014.
51. G. Gopal Rao and Mehool Patel. Urinary tract infection in hospitalized elderly patients in the United Kingdom: The importance of making an accurate diagnosis in the post broad-spectrum antibiotic era. *J Antimicrob Chemother*, 63(1):5–6, January 2009.
52. Michael Gross. Antibiotics in crisis. *Curr. Biol.*, 23(24):R1063–1065, December 2013.
53. I. Gustafsson. Bacteria with increased mutation frequency and antibiotic resistance are enriched in the commensal flora of patients with high antibiotic usage. *J. Antimicrob. Chemother.*, 52(4):645–650, September 2003.
54. C. A. Hardiman, R. A. Weingarten, S. Conlan, P. Khil, J. P. Dekker, A. J. Mathers, A. E. Sheppard, J. A. Segre, and K. M. Frank. Horizontal Transfer of Carbapenemase-Encoding Plasmids and Comparison with Hospital Epidemiology Data. *Antimicrob. Agents Chemother.*, 60(8):4910–4919, August 2016.
55. Peera Hemarajata and James Versalovic. Effects of probiotics on gut microbiota: Mechanisms of intestinal immunomodulation and neuromodulation. *Therap Adv Gastroenterol*, 6(1):39–51, January 2013.
56. Benjamin P. Howden, John K. Davies, Paul D. R. Johnson, Timothy P. Stinear, and M. Lindsay Grayson. Reduced vancomycin susceptibility in *Staphylococcus aureus*, including vancomycin-intermediate and heterogeneous vancomycin-intermediate strains: Resistance mechanisms, laboratory detection, and clinical implications. *Clin. Microbiol. Rev.*, 23(1):99–139, January 2010.
57. David L. Hugo. Probability Distribution of Drug-Resistant Mutants in Unselected Populations of Mycobacterium tuberculosis. *APPL MICROBIOL*, 20:5, 1970.
58. Winfried Just and Hannah Callender Highlander. Vaccination Strategies for Small Worlds. In Aaron Wootton, Valerie Peterson, and Christopher Lee, editors, *A Primer for Undergraduate Research*, pages 223–264. Springer International Publishing, Cham, 2017.
59. Kang Kang, Yueqiong Ni, Jun Li, Lejla Imamovic, Chinmoy Sarkar, Marie Danielle Kobler, Yoshitaro Heshiki, Tingting Zheng, Sarika Kumari, Jane Ching Yan Wong, Anand Archna, Cheong Wai Martin Wong, Caroline Dingle, Seth Denizen, David Michael Baker, Morten Otto Alexander Sommer, Christopher John Webster, and Gianni Panagiotou. The Environmental Exposures and Inner- and Intercity Traffic Flows of the Metro System May Contribute to the Skin Microbiome and Resistome. *Cell Rep*, 24(5):1190–1202.e5, July 2018.
60. Daniel B. Kearns. A field guide to bacterial swarming motility. *Nat Rev Microbiol*, 8(9):634–644, September 2010.
61. Georgina Kenyon. Russia's prisons fuel drug-resistant tuberculosis. *The Lancet Infectious Diseases*, 9(10):594, October 2009.
62. Zosia Kmietowicz. Few novel antibiotics in the pipeline, WHO warns. *BMJ*, 358:j4339, September 2017.

63. A. A. A. Kwaasi. MICROBIOLOGY | Classification of Microorganisms. In Benjamin Caballero, editor, *Encyclopedia of Food Sciences and Nutrition (Second Edition)*, pages 3877–3885. Academic Press, Oxford, January 2003.

64. Timothy F. Landers, Bevin Cohen, Thomas E. Wittum, and Elaine L. Larson. A review of antibiotic use in food animals: Perspective, policy, and potential. *Public Health Rep*, 127(1):4–22, 2012 Jan-Feb.

65. H. Lee, E. Popodi, H. Tang, and P. L. Foster. Rate and molecular spectrum of spontaneous mutations in the bacterium *Escherichia coli* as determined by whole-genome sequencing. *Proc. Natl. Acad. Sci.*, 109(41):E2774–E2783, October 2012.

66. Helen C. Leggett, Charlie K. Cornwallis, Angus Buckling, and Stuart A. West. Growth rate, transmission mode and virulence in human pathogens. *Philos. Trans. R. Soc. B Biol. Sci.*, 372(1719):20160094, May 2017.

67. Jennifer A. P. Liu and Nancy J. Moon. Commensalistic Interaction Between *Lactobacillus acidophilus* and *Propionibacterium shermanii*. *Appl. Environ. Microbiol.*, 44(3):715–722, September 1982.

68. Hongan Long, Samuel F. Miller, Chloe Strauss, Chaoxian Zhao, Lei Cheng, Zhiqiang Ye, Katherine Griffin, Ronald Te, Heewook Lee, Chi-Chun Chen, and Michael Lynch. Antibiotic treatment enhances the genome-wide mutation rate of target cells. *Proc. Natl. Acad. Sci.*, 113(18):E2498–E2505, May 2016.

69. Bianca Malcolm. The Rise of Methicillin-Resistant Staphylococcus aureus in U.S. Correctional Populations. *J Correct Health Care*, 17(3):254–265, July 2011.

70. Carrie A. Manore, Kyle S. Hickmann, James M. Hyman, Ivo M. Foppa, Justin K. Davis, Dawn M. Wesson, and Christopher N. Mores. A network-patch methodology for adapting agent-based models for directly transmitted disease to mosquito-borne disease. *J. Biol. Dyn.*, 9(1):52–72, January 2015.

71. Janet M. Manson, Lynn E. Hancock, and Michael S. Gilmore. Mechanism of chromosomal transfer of *Enterococcus faecalis* pathogenicity island, capsule, antimicrobial resistance, and other traits. *Proc. Natl. Acad. Sci. U.S.A.*, 107(27):12269–12274, July 2010.

72. Alan J. Marsh, Orla O'Sullivan, Colin Hill, R. Paul Ross, and Paul D. Cotter. Sequence-based analysis of the bacterial and fungal compositions of multiple kombucha (tea fungus) samples. *Food Microbiol.*, 38:171–178, April 2014.

73. Scott A. McEwen and Paula J. Fedorka-Cray. Antimicrobial Use and Resistance in Animals. *Clin Infect Dis*, 34(Supplement_3):S93–S106, June 2002.

74. Center for Veterinary Medicine. CVM Updates - FDA Annual Summary Report on Antimicrobials Sold or Distributed in 2015 for Use in Food-Producing Animals. https://www.fda.gov/AnimalVeterinary/NewsEvents/CVMUpdates/ucm534244.htm.

75. Center for Veterinary Medicine. CVM Updates - FDA Releases Annual Summary Report on Antimicrobials Sold or Distributed in 2016 for Use in Food-Producing Animals. https://www.fda.gov/AnimalVeterinary/NewsEvents/CVMUpdates/ucm588086.htm.

76. Center for Veterinary Medicine. FDA Announces Implementation of GFI #213, Outlines Continuing Efforts to Address Antimicrobial Resistance. WebContent.

77. Mark Melzer and Irene Petersen. Mortality following bacteraemic infection caused by extended spectrum beta-lactamase (ESBL) producing *E. coli* compared to non-ESBL producing *E. coli*. *J. Infect.*, 55(3):254–259, September 2007.

78. Rodrigo E. Mendes, Lalitagauri M. Deshpande, and Ronald N. Jones. Linezolid update: Stable in vitro activity following more than a decade of clinical use and summary of associated resistance mechanisms. *Drug Resist. Updat.*, 17(1-2):1–12, April 2014.

79. Zlati Meyer. Tyson Foods will eliminate antibiotics in chicken. *USA TODAY*, May 2017.

80. Carolyn Anne Michael, Dale Dominey-Howes, and Maurizio Labbate. The antimicrobial resistance crisis: Causes, consequences, and management. *Front Public Health*, 2:145, 2014.

81. Breanna Miranda, Nicole M. Lawton, Sean R. Tachibana, Natasja A. Swartz, and W. Paige Hall. Titration and HPLC Characterization of Kombucha Fermentation: A Laboratory Experiment in Food Analysis. *J. Chem. Educ.*, 93(10):1770–1775, October 2016.

82. Annik Nanchen, Alexander Schicker, Olga Revelles, and Uwe Sauer. Cyclic AMP-Dependent Catabolite Repression Is the Dominant Control Mechanism of Metabolic Fluxes under Glucose Limitation in *Escherichia coli*. *J. Bacteriol.*, 190(7):2323–2330, April 2008.
83. Carl Nathan and Otto Cars. Antibiotic Resistance — Problems, Progress, and Prospects. https://www.nejm.org/doi/10.1056/NEJMp1408040?url_ver=Z39.88-2003&rfr_id=ori%3Arid%3Acrossref.org&rfr_dat=cr_pub%3Dwww.ncbi.nlm.nih.gov, November 2014.
84. H. Nikaido. Prevention of drug access to bacterial targets: Permeability barriers and active efflux. *Science*, 264(5157):382–388, April 1994.
85. Hiroshi Nikaido. Molecular basis of bacterial outer membrane permeability revisited. *Microbiol. Mol. Biol. Rev.*, 67(4):593–656, December 2003.
86. Eric L. Nuermberger and William R. Bishai. Antibiotic Resistance in *Streptococcus pneumoniae*: What Does the Future Hold? *Clin Infect Dis*, 38(Supplement_4):S363–S371, May 2004.
87. A. Nutman, R. Glick, E. Temkin, M. Hoshen, R. Edgar, T. Braun, and Y. Carmeli. A case-control study to identify predictors of 14-day mortality following carbapenem-resistant *Acinetobacter baumannii* bacteraemia. *Clin. Microbiol. Infect.*, 20(12):O1028–1034, December 2014.
88. Robert B. O'Hara and D. Johan Kotze. Do not log-transform count data. *Methods Ecol. Evol.*, 1(2):118–122, June 2010.
89. Jim O'Neill. Antimicrobial resistance: Tackling a crisis for the health and wealth of nations. *Rev. Antimicrob. Resist*, 20:1–16, 2014.
90. World Health Organization, editor. *Antimicrobial Resistance: Global Report on Surveillance*. World Health Organization, Geneva, Switzerland, 2014. OCLC: ocn880847527.
91. G. A. Pankey and L. D. Sabath. Clinical Relevance of Bacteriostatic versus Bactericidal Mechanisms of Action in the Treatment of Gram-Positive Bacterial Infections. *Clin Infect Dis*, 38(6):864–870, March 2004.
92. H Van Dyke Parunak, Robert Savit, and Rick L Riolo. Agent-based modeling vs. equation-based modeling: A case study and users' guide. In *International Workshop on Multi-Agent Systems and Agent-Based Simulation*, pages 10–25. Springer, 1998.
93. Liliana Perez and Suzana Dragicevic. An agent-based approach for modeling dynamics of contagious disease spread. *International Journal of Health Geographics*, 8(1):50, August 2009.
94. Gael Pérez-Rodríguez, Martín Pérez-Pérez, Daniel Glez-Peña, Florentino Fdez-Riverola, Nuno F. Azevedo, and Anália Lourenço. Agent-Based Spatiotemporal Simulation of Biomolecular Systems within the Open Source MASON Framework. *BioMed Res. Int.*, 2015:1–12, 2015.
95. Margaret B. Planta. The Role of Poverty in Antimicrobial Resistance. *J Am Board Fam Med*, 20(6):533–539, January 2007.
96. P. Pletnev, I. Osterman, P. Sergiev, A. Bogdanov, and O. Dontsova. Survival guide: Escherichia coli in the stationary phase. *Acta Naturae*, 7(4):22–33, 2015.
97. Laurent Poirel, Johann D. Pitout, and Patrice Nordmann. Carbapenemases: Molecular diversity and clinical consequences. *Future Microbiol*, 2(5):501–512, October 2007.
98. Keith Poole. Efflux-mediated antimicrobial resistance. *J. Antimicrob. Chemother.*, 56(1):20–51, July 2005.
99. Hazhir Rahmandad and John Sterman. Heterogeneity and Network Structure in the Dynamics of Diffusion: Comparing Agent-Based and Differential Equation Models. *Manag. Sci.*, 54(5):998–1014, May 2008.
100. Steven F. Railsback and Volker Grimm. *Agent-Based and Individual-Based Modeling: A Practical Introduction*. Princeton University Press, Princeton, 2012.
101. Sara Reardon. Antibiotic resistance sweeping developing world. *Nat. News*, 509(7499):141, May 2014.
102. Gian Maria Rossolini, Fabio Arena, Patrizia Pecile, and Simona Pollini. Update on the antibiotic resistance crisis. *Curr Opin Pharmacol*, 18:56–60, October 2014.

103. Stefan Schwarz, Corinna Kehrenberg, Benoît Doublet, and Axel Cloeckaert. Molecular basis of bacterial resistance to chloramphenicol and florfenicol. *FEMS Microbiol. Rev.*, 28(5):519–542, November 2004.
104. Shelby M. Scott, Casey E. Middleton, and Erin N. Bodine. An Agent-Based Model of the Spatial Distribution and Density of the Santa Cruz Island Fox. In *Handbook of Statistics*, volume 40, pages 3–32. Elsevier, 2019.
105. Jessica L Semega, Kayla R Fontenot, and Melissa A Kollar. Income and Poverty in the United States: 2016. Technical report, US Census Bureau, September 2017.
106. G. Sezonov, D. Joseleau-Petit, and R. D'Ari. Escherichia coli Physiology in Luria-Bertani Broth. *J. Bacteriol.*, 189(23):8746–8749, December 2007.
107. DM Sievert, ML Boulton, G Stoltman, D Johnson, MG Stobierski, FP Downes, PA Somsel, JT Rudrik, W Brown, W Hafeez, T Lundstrom, E Flanagan, R Johnson, J Mitchell, and S Chang. *Staphylococcus aureus* Resistant to Vancomycin. *Morb. Mortal. Wkly. Rep.*, 51(26):565–567, July 2002.
108. Davida S. Smyth. Laboratory exercises on microbial growth, horizontal gene transfer, competition and evolution. *QUBES Educ. Resour.*, 2018.
109. Brad Spellberg and David N. Gilbert. The future of antibiotics and resistance: A tribute to a career of leadership by John Bartlett. *Clin. Infect. Dis.*, 59 Suppl 2:S71–75, September 2014.
110. Karolien Stoffels, Vanessa Mathys, Maryse Fauville-Dufaux, René Wintjens, and Pablo Bifani. Systematic Analysis of Pyrazinamide-Resistant Spontaneous Mutants and Clinical Isolates of *Mycobacterium tuberculosis*. *Antimicrob Agents Chemother*, 56(10):5186–5193, October 2012.
111. Charles W. Stratton. Dead Bugs Don't Mutate: Susceptibility Issues in the Emergence of Bacterial Resistance. *Emerg Infect Dis*, 9(1):10–16, January 2003.
112. Jean-Marie Swiecicki, Olesksii Sliusarenko, and Douglas B. Weibel. From swimming to swarming: *Escherichia coli* cell motility in two-dimensions. *Integr Biol (Camb)*, 5(12):1490–1494, December 2013.
113. Emily R. M. Sydnor and Trish M. Perl. Hospital epidemiology and infection control in acute-care settings. *Clin. Microbiol. Rev.*, 24(1):141–173, January 2011.
114. Birkneh Tilahun Tadesse, Elizabeth A. Ashley, Stefano Ongarello, Joshua Havumaki, Miranga Wijegoonewardena, Iveth J. González, and Sabine Dittrich. Antimicrobial resistance in Africa: A systematic review. *BMC Infect Dis*, 17, September 2017.
115. Truc T. Tran, Diana Panesso, Hongyu Gao, Jung H. Roh, Jose M. Munita, Jinnethe Reyes, Lorena Diaz, Elizabeth A. Lobos, Yousif Shamoo, Nagendra N. Mishra, Arnold S. Bayer, Barbara E. Murray, George M. Weinstock, and Cesar A. Arias. Whole-genome analysis of a daptomycin-susceptible *enterococcus faecium* strain and its daptomycin-resistant variant arising during therapy. *Antimicrob. Agents Chemother.*, 57(1):261–268, January 2013.
116. Magnus Unemo and William M. Shafer. Antimicrobial resistance in *Neisseria gonorrhoeae* in the 21st century: Past, evolution, and future. *Clin. Microbiol. Rev.*, 27(3):587–613, July 2014.
117. Srinivasan Venkatramanan, Bryan Lewis, Jiangzhuo Chen, Dave Higdon, Anil Vullikanti, and Madhav Marathe. Using data-driven agent-based models for forecasting emerging infectious diseases. *Epidemics*, 22:43–49, March 2018.
118. C. Lee Ventola. The antibiotic resistance crisis: Part 1: Causes and threats. *P T*, 40(4):277–283, April 2015.
119. Gudrun Wallentin and Christian Neuwirth. Dynamic hybrid modelling: Switching between AB and SD designs of a predator-prey model. *Ecol. Model.*, 345:165–175, February 2017.
120. Uri Wilensky. NetLogo. Center for Connected Learning and Computer-Based Modeling, Northwestern University, 1999.
121. Uri Wilensky and William Rand. *An Introduction to Agent-Based Modeling: Modeling Natural, Social, and Engineered Complex Systems with NetLogo*. The MIT Press, Cambridge, Massachusetts, 2015.
122. C Yang and Uri Wilensky. NetLogo epiDEM Basic model, 2011.
123. C Yang and Uri Wilensky. NetLogo epiDEM Travel and Control model, 2011.

124. Fengjiao Yu, Yujie Wen, Jibao Wang, Yurong Gong, Kaidi Feng, Runhua Ye, Yan Jiang, Qi Zhao, Pinliang Pan, Hao Wu, Song Duan, Bin Su, and Maofeng Qiu. The Transmission and Evolution of HIV-1 Quasispecies within One Couple: A Follow-up Study based on Next-Generation Sequencing. *Sci. Rep.*, 8(1):1404, January 2018.
125. Anne E. Yust. Sample NetLogo Code of Bacterial Growth. *QUBES Educ. Resour.*, October 2018.
126. Pia Schulz zur Wiesch, Jan Engelstädter, and Sebastian Bonhoeffer. Compensation of Fitness Costs and Reversibility of Antibiotic Resistance Mutations. *Antimicrob. Agents Chemother.*, 54(5):2085–2095, May 2010.

Agent-Based Modeling in Mathematical Biology: A Few Examples

Alexandra L. Ballow, Lindsey R. Chludzinski, and Alicia Prieto-Langarica

Abstract Agent Based Modeling is a computer based modeling approach that allows undergraduate students to dive into research with very little mathematical background. This chapter aims to give students and faculty a better idea of what ABM is and a few examples of where and how it can be used to model real-life problems in mathematical biology. An example project is explained in detail for the reader to use as a guide. Also, in the afterword, the student co-authors will discuss their experience and share their advice to other students interested in doing research in mathematical biology.

Suggested Prerequisites *We are writing this chapter with the assumption that student readers are either comfortable with programming or willing to learn and have basic knowledge of probability. This chapter is set up as inquiry-based research. The premise is that exercises start basic enough that any college student willing to put the time can work through the chapter exercises (this could form the basis of an independent study, for example) and be ready to undertake an agent-based modeling project. A student who has all suggested prerequisites should be able to more rapidly complete the exercises and then dive quickly into the research projects. In Sect. 3, the author expands on how to approach this chapter for students with different backgrounds.*

1 Preface

The faculty author, Prieto-Langarica, would like to begin by discussing her personal philosophy in leading undergraduate research, as it is with this philosophy in mind that the following projects and exercises are written. In her relatively short career, the author has led over fifty students in research projects, ranging from mathematical biology to data science applied to psychology and with students ranging from high

A. L. Ballow · L. R. Chludzinski · A. Prieto-Langarica (✉)
Youngstown State University, Youngstown, OH, USA
e-mail: alballow@student.ysu.edu; lrchludzinski@student.ysu.edu; aprietolangarica@ysu.edu

© Springer Nature Switzerland AG 2020   273
H. Callender Highlander et al. (eds.), *An Introduction to Undergraduate Research in Computational and Mathematical Biology*, Foundations for Undergraduate Research in Mathematics, https://doi.org/10.1007/978-3-030-33645-5_6

school and all the way to master's degree. However, the way in which the author leads research might not be the most common.

When meeting a potential research student, the highest importance is given to the student's professional goal and particular needs. This way, research is viewed as a mechanism to acquire the tools and skill to succeed at such goal, with research topic and research outcome becoming less important. For this reason, research is treated as an opportunity to experiment, to led students take risks, develop their own ideas and their own models, make their own mistakes and learn how to recognize shortcomings and improve on their models. Most of the students the author works with start as freshmen, or before. They take their time to discover concepts on their own, then eventually articulate their own research ideas, many times far surpassing the author's knowledge on the topic.

This kind of research approach might sound odd and even scary as the professor will no longer be the expert, but just another collaborator for the student. However, giving students ownership of their work comes with high reward. Students report being better prepared for future research endeavors, they develop confidence and excitement about mathematics, they become comfortable with the idea of thinking mathematics in every aspect of their lives. In addition, students who decide to enter the workforce after an undergraduate degree, have reported success in making operations and procedures at their current job better, faster, cheaper, more optimal giving credit to the skills learned while undertaking research as undergraduates.

One last very important recommendation is for the faculty and the students to seek collaborations. From students working in groups to working with faculty in other departments, the power of collaborating within and across disciplines provides the student with invaluable skills required in academia and the industry. When working with a team of biology and mathematics majors, and partnering with biology faculty, we expose the students to different ways of thinking and to knowledge in different disciplines. In addition, faculty in these departments may offer great research problems and/or data.

2   Introduction

Agent-Based Modeling, from here on referred to as ABM, is a relatively new mathematical/computational way of modeling real-life phenomena that allows stochasticity. Historically, mathematical models have often been limited to analytic models and our ability to "solve" analytically such models [33]. However, with computer simulations, we are not limited by mathematical tractability and are free to create models as complex and as close to reality as data collection and understanding of the system permit.

Since ABM started, it has been used to model many types of situations in biology and the social sciences. In general, ABM is an excellent tool to model and experiment with real-life situations where outcomes are highly dependent on randomness. For mathematical biology research, it has evolved into an effective method to mathematically investigate biological phenomenon [5, 20, 24]. In addition, ABM is a supremely useful tool to introduce deep theoretical, mathematical concepts, especially when used in tandem with other more classical modeling techniques. It is especially valuable as a way for mathematicians and experts in other fields to communicate. In a time where multi-disciplinary work is becoming more and more the standard, ABM is quickly becoming an indispensable tool [14, 17, 34].

So what exactly is Agent Based Modeling? ABMs are models where individuals or agents are described as unique and autonomous entities that usually interact with each other and their environment locally. These agents can be cells, animals, humans, businesses, institutions, and any other entity that has a goal and makes decisions mostly based on that goal [33]. In practice, ABMs are computational programs that follow the evolution of a collection of individuals making decisions based on their neighbors and the state of the system.

A particular kind of Agent-Based Model is a Cellular Automata [CA] Model [2, 11, 43]. A CA model is based on a grid. Each cell in the grid is in one of a set of possible states at each point in time. Each time step, the state of each cell is updated based on the previous state, as well as the previous state of the *"neighboring"* cells. The way neighborhoods are defined depends on the problem and will be explored in future exercises. CA models are easy to start but can get extremely complex. The study of the theory behind CA models has rapidly grown in the last two decades, and it is a rich field of study for mathematicians as well as computer scientists.

It is worth noting that ABM is considered inferior by some mathematicians, particularly in the pure mathematics field. Leaving aside the question whether this argument has any validity, the author of this manuscript defends the great value ABM offers in training undergraduates in applied mathematics. ABM has huge potential simply to teach students about collaborative work in mathematical biology and/or the social sciences. In order to satisfy the more pure at heart, the author recommends using this type of modeling in tandem with more classical types of modeling and/or analyzing the agent based model fully with the introduction of heavy probabilistic and analytic tools.

This chapter is organized as follows. Section 3 suggests previous knowledge (or concurrent learning) that students should acquire before or while working on the exercises and/or suggested problems in this chapter. Recall, Prieto-Langarica's styles of advising tends to leave students free to experiment with ideas and construct their own hypothesis. Many of the exercises suggested throughout this work are very open ended. This is intentional as a way to get students comfortable with open ended questions, very similar to those found in the real world, but often lacking in the classroom. The author encourages students and faculty to work through these exercises in teams, to talk about your results, to formulate hypotheses, and to work on expanding the algorithms suggested here.

3 Background for Students and Faculty

One of the main reasons Prieto-Langarica has used and continues to use this type of modeling with undergraduate students is how little initial background students need, and how much they can learn while constructing their models. A student with no programming experience, who might not have the time to learn how to program, can use already well-established ABM languages such as NetLogo [42], which are easy to learn and allow students to start running simulations within a few weeks. Here are a few NetLogo tutorials that might be of use to students and faculty [7, 15, 33]. However, the author highly recommends taking the time to learn more complex programming languages such as C++, Python, or even Matlab, which will most likely be useful to the student beyond their undergraduate training. The examples in this chapter aim to stay as general as possible when it comes to programming language.

As for mathematical background, a general high school level probability knowledge will be more than enough to start working with ABMs. Once the model is running, higher level mathematical tools can be used to analyze the ABM and to extract population level information. Advanced students with a background in mathematical analysis and maybe even partial differential equations may want to write discrete probabilistic equations for the system and attempt to upscale the system into a continuous population level. This will be beyond the scope of this chapter but interested students can look at books [4], or articles [32].

This chapter starts by detailing introductory exercises, followed by more in-depth exercises. It then describes a current undergraduate project as an example for students to follow, detailing each step of the research process and what will be done/can be done in order to expand on it. The last section includes some interesting applications, projects, or modeling choices that can be made by other faculty/undergraduate teams, as well as sources where faculty can find other applications.

The faculty author wishes to re-emphasize the importance of collaboration, as a skill faculty need to teach students, in order to advance science in general. Most of the projects the author has been involved with began by attending Biology Department Seminars and asking Biology faculty for possible collaborations. This not only helps the faculty member, but is an invaluable lesson for the students who gain experience with real-life problems and with real-life data. The author encourages the faculty reading this chapter to do the same, and seek out problems and collaborations within their own institutions.

4 Introduction to ABM

In this section, several exercises will be given for students who wish to start gaining experience with this type of modeling. These exercises may be done in any programming language the student and professor prefer. Ideally, the faculty can work with multiple students at the same time who can try to program on their own but who will then work together comparing their approach and collaborating in improving on solutions.

4.1 Spatial Transmission

The first set of exercises will help construct a CA Model, and therefore take place on a grid. These exercises aim to familiarize the researcher with situations in which something (for example, a disease, or gossip, or a political position) spreads in time and space. Every cell in the grid has a value, but its value can change at every time step, depending on the values of its geographical neighbors.

Note In most programming languages, you can represent a grid by a matrix. Be mindful of different ways in which elements of the matrix are indexed depending on the language you are using.

Problem 1 Start with a matrix and randomly fill it out with zeros and ones. Define *Neighborhood Type 1* as the element on top, below, on the right, and on the left. In Table 1, the element labelled *Indi* has four neighbors marked with the letter N.

At each time step, every element in the matrix can change or remain the same, based on the values of the immediate neighbors, that is, the more zeros around the neighborhood, the higher the probability that the element either stays or becomes a zero. Let $p^0_{i,j}$ = the probability of the number in cell (i, j) in the grid changing (or remaining) a 0 and $p^1_{i,j}$ = the probability of the number in cell (i, j) in the grid changing (or remaining) a 1. Create an experiment with different rules for how these numbers will change. Here is an example of one possible such rule for the grid shown in Table 2.

Table 1 Neighborhood Type 1

	N	
N	Indi	N
	N	

The center cell in the grid labelled *Indi* contains the individual and the cells with N represent that individual's neighbors. In the literature, this type of neighborhood is known as *Von Neumann* Neighborhood [36]

Table 2 Sample grid the center cell in the grid labelled *Indi* has three neighbors with value 0 and one neighbor whose value is 1

	0	
0	Indi	0
	1	

Table 3 Neighborhood Type 1

N	N	N
N	Indi	N
N	N	N

The center cell in the grid labelled *Indi* contains the individual and the cells with N represent that individual's neighbors. In the literature, this type of neighborhood is known as *Moore* Neighborhood [36]

$$p_{i,j}^0 = \frac{1}{4}(3) = \frac{3}{4}$$

$$p_{i,j}^1 = \frac{1}{4}(1) = \frac{1}{4}.$$

Since the cell has 3 "zero" neighbors and 1 "one" neighbor.

Let T be the number of steps you run the program for. Run the simulation for $T = 10,\ 100,\ 1000$ times. Use the same initial conditions for all your simulations.

Problem 2 Define Neighborhood Type 2 as in Table 3. Repeat the exercise using the same initial conditions but with the different neighborhood. What changes?

Challenge Problem 1 Define Neighborhood Type 3 as in Table 4. Repeat the exercise using the same initial conditions but with the different neighborhood. What changes? Consider two types of neighbors D and N. What if the influence of D is less than N? How would you implement this? After running simulations of both problems describe how your results change.

Challenge Problem 2 What happens if you add a number? For example, what would happen if you fill the matrix with zeros, ones, and negatives ones? Repeat the Challenging Problem 1 using ones, zeros, and negative ones.

Table 4 Neighborhood
Type 3

D	D	D	D	D
D	N	N	N	D
D	N	Cell	N	D
D	N	N	N	D
D	D	D	D	D

The center cell contains the individual and the cells
with and N represent the center cell's neighbors

4.2 Random Walks

In most biological situations, individuals will move either randomly [39] or directed
by environmental cues [23]. The following set of problems addresses situations in
which the goal is to track the movement of one or more individuals. Since even
directed movement almost always has a random component, you will be asked to
run your program many times and report the average behavior of individuals.

Problem 3 (One Dimensional) Start with n particles placed in the same point in a
one dimensional grid. At each time step, each particle will randomly decide to move
one step to the left or one step to the right to the neighboring point in the grid. Run
the program for T time steps. After T time steps, record the number of particles at
each point in the grid. Run the simulation for $N = 100$, $N = 1000$, $N = 10,000$,
and finally for $N = 100,000$ times. Record the average number of particles in each
point in the grid. How do things change when you vary T? How do things change
from $N = 10,000$ and $N = 100,000$? How do things change if you allow particles
to choose between moving left, moving right or staying?

Problem 4 (Two Dimensional) Repeat the exercise above with particles in the
middle of the plane and let particles choose between moving up, down, right, and
left. How do things change if you let the particle move diagonally? (i.e., there are
eight directions of movement). Compare the results of both experiments.

Challenge Problem 3 (Three Dimensional) Repeat exercise above with particles
in the middle of a three dimensional space. Let particles move in 1 out of 6
directions. How does your simulations change when you move from one to
two dimensions and then to three dimensions? What challenges do you find
in working with three dimensions? What do you think will change if you add
diagonal directions? (i.e., how many directions will there be? How will adding
these directions change your results?)

Problem 5 (Directed Walks) Start with two different kinds of particles. Think of the particles as predator and a prey. Predators sense and chase prey. In practice, that means that a certain percent of the time, $0 \leq q \leq 1$, which is fixed and given at the beginning of the simulation, predators will move towards prey and the rest of the time $(1 - q)$, they will just move randomly. Start in one dimension with one predator and one prey and for now let the prey move completely randomly. Vary initial condition and report the time it takes for the predator to reach the prey. This exercise is far more interesting in two and three dimensions; do both, and pick how many directions of movement you want to use.

Problem 6 (Many Predators) Add more predator and prey particles but make all the predators start at one point in the plane (or space) and all the prey in a different point in the plane (or space). Vary initial conditions and report the average number of particles in each point in space after running your simulations $N = 100, 1000, 10{,}000, 100{,}000$ times for a fixed set of initial conditions. What are "interesting" initial conditions? How does the final distribution of particles change if you vary the "percent" of the time the movement is directed as opposed to random? Do this same exercise on two and three dimensions (how will you report averages in three dimensions?).

Challenge Problem 4 In reality, most of the time, q depends on how far the predator is from the prey. How would you implement this?
Note to Faculty: this is an excellent time to introduce the concept of metrics and different metrics to students. A great exercise here would be to run this program using different metrics and making observations and hypothesis of how the results of the program would change under different metrics.

4.3 Reproduction

Another very important biological process that can be modeled using ABM is reproduction. Biologically, reproduction can be either sexual or asexual. The following series of exercises prepares students to model those kind of situations.

Problem 7 (Asexual Reproduction) Start with a large rectangular domain and a few circular particles randomly placed on it. Assign a randomly selected positive integer to each particle t_i. Each particle will reproduce, by division, at $T = t_i$. Every time a particle divides, the daughter particles should be placed in the same spot as the mother particle with probability $p = 1/2$ and in adjacent grid points with probability $1 - p$. After division, assign new integers t_i to each particle. Run the simulation for long enough to fill out the grid with particles and record how long it took to fill out the grid.

Problem 8 (Asexual Reproduction 2) Start this exercise using the same conditions as Exercise 7; however, the reproduction clocks on the particles should be randomly chosen with a Poisson distribution, with $\lambda = 20$ the average waiting time until reproduction. Just as in the previous exercise, run the simulation for long enough to fill out the grid with particle and record how long it took to fill out the grid.

Note to Faculty: This is an excellent opportunity to discuss probability and queue theory with students. A great book to facilitate learning is: [16].

Challenge Problem 5 (Asexual Reproduction 3) Repeat the same exercise as above, with all the characteristics discussed included and one simple change: when reproduction happens, the particles pile up in towers. As soon as a tower reaches over 10 particles length, the towers will crumble and spill out randomly into neighboring grid points. Towers can start building up from there. Run this simulation until all points in the grid have at least one particle. This is also known as the sandpile problem [6].

Problem 9 (Sexual Reproduction 4) Start with a large grid and two different randomly placed particles, one male, one female. Allow the particles to move randomly. When the particles are a certain predefined distance d from each other, they will move towards each other. Once the particles find each other, a new particle will be produced and a random gender will be assigned. The particles then will continue moving randomly for a fix number of steps T, when they become eligible to reproduce again, at which time they can move randomly until they run into another particle of the opposite gender. Run this simulation until there are half the particles than grid points.

4.4 Movement in Graphs

Sometimes it makes sense to look at a biological process as a process happening in graphs. Some applications of this can be found in epidemiology [18]. The following series of exercises aim to familiarize students with working with graphs. For all these exercises, students will need to be familiar with the adjacency matrix associated with a graph. The authors suggest students who wish to learn about graphs use any of the following books [10, 21, 41].

Problem 10 (Randomly Creating Graphs) For $n = 10, 100, 1000$ create a random undirected graph with n vertex and n, $2n$, n^2 randomly placed edges. For each graph calculate:

- The vertex degree
- The average vertex degree
- The graph distance matrix
- The vertex connectivity
- The closeness centrality
- The betweenness centrality

Problem 11 (Movement in a Graph) Randomly create a 100 vertex graph and 1000 edges. Randomly place a particle in one of the vertices. Each time step, the particle can move to any other connected vertex. Follow the particle for at least 1000 time steps and record the number of times each vertex was visited.

Problem 12 (Movement in a Graph 2) Randomly create a 100 vertex graph and 1000 edges. Randomly place a particle in one of the vertices. Each time step, the particle can move to any other connected vertex. This time the particle can only visit each vertex once. For the same graph and different initial placement, record how many vertices were visited.

Problem 13 (Particle Chasing Particle) Randomly create a 100 vertex graph and 1000 edges. Make sure your graph is completely connected. Randomly place two different particles (particle a and particle b) in two different vertices. Each time step, particle a will choose to move closer to particle b by moving to any other connected vertex. Particle b will move randomly. Play this game until particle a catches up with particle b. Let $T_i =$ the number of time steps it took a to catch b on simulation i. Run this game 100,000 times and plot the distribution of T. Is there anything you can say hypothesis about T and how it depends on any of the measurements you calculated on the first problem in this section?

Note that this system is similar to the predator–prey system we studied in Problem 5, this time with particles moving on a graph.

Challenge Problem 6 (Particle Chasing Particle 2) Randomly create a 100 vertex graph with 1000 edges. Make sure your graph is completely connected. Randomly place two different particles (particle a and particle b) in two different vertices. Each time step, particle a will choose to either move randomly q_a fraction of the time, or move closer to particle b by moving to any other connected vertex $1 - q_b$ fraction of the time with $0 \leq q_a \leq 1$. Particle b will move away from particle a $1 - q_b$ fraction of the time and randomly q_b fraction of the time with $0 \leq q_b \leq 1$. Play this game until particle a catches up with particle b. Let $T_i =$ the number of time steps it took a to catch b on simulation i Run this game 100,000 times and plot the distribution of T. What happens when $q_a > q_b$? How about when $q_b > q_a$? Are there any hypothesis you can come up with about what you think T will be equal to? Do you think T depends on any of the measurements you calculated on the first problem in this section? If so, is there a way in which you could relater such measurements to T?

Challenge Problem 7 (Particle Chasing Particle 3) Same problem than last one, however, in reality, q_a and q_b should depend on the distance between a and b. How would you implement this?

5 Research Projects

In Sect. 5.1, the authors will describe in detail a particular problem they are currently working on using agent-based modeling as an example ABM research project for readers to follow. This project at the stage described was conducted by two sophomores (co-authors Ballow and Chludzinski) in collaboration with Dr. Diana Fagan from the Biology Department at Youngstown State University. By this point, the students had met with Dr. Fagan on multiple occasions, and she has given them material relevant to the context to read. The project team had also met throughout the semester to report on the progress of our model and to calibrate and improve in the biological understanding. The final project goal is to have a computational simulation that can be used as a tool to test different medical interventions. Collaborating with experts in other disciplines has provided us not only with real-life data for model realism and parameterization, but has greatly helped the students develop the skills to communicate outside mathematics.

The students started working on ABMs for modeling wounds as Freshmen, a little under a year ago. This was informed by some prior knowledge of Prieto-Langarica and a literature review. However, after a few months of experimenting with building models the group established a collaboration with the Biology Department and were able to start working with Dr. Fagan in modeling subcutaneous infections.

This is a particular approach the faculty author recommends: as much as possible, one should take advantage and start collaborations with the Biology/Ecology department (or other departments in the university) Not only does this potentially advance the faculty members' own research agenda, but it provides a unique opportunity for students to be involved in multi-disciplinary research. This is an invaluable experience and particularly important skill to have in the current state of academia and industry.

5.1 Modeling Subcutaneous Infections

Subcutaneous infections may be only a small subset of an impressive realm of immunological topics, yet they hold a very significant impact on the human population. By one count, over 1400 species of pathogens capable of causing disease in mankind have been identified in literature [38]. Under the right conditions, a single bacterium is capable of producing upwards of 20 million new bacteria in only 24 h [3]. With such staggering numbers, it is scarcely surprising that humans are well equipped with a number of protective mechanisms.

The human immune system has three main layers of defense: physical barriers, such as the skin, innate immunity, and the adaptive immune response. Our skin is responsible for preventing a practically innumerable number of pathogens from entering the body; innumerable simply because it is nearly impossible to count all of the diseases an individual did not catch but may have been exposed too.

When a pathogen does enter the body, it is almost immediately faced with a battalion of cells fighting against it. Among the first responders are the macrophages, an important kind of phagocyte that resides in tissue. One of the many jobs of the macrophage is to begin attacking pathogens while sounding an alarm by releasing several different kinds of effector molecules. Many of these molecules lead to the inflammation often associated with infections and are thus collectively referred to as inflammatory mediators. These include a number of proteins called cytokines and chemokines, such as TNF-α and CXCL8 (also known as IL-8). CXCL8 binds with very specific receptors on the neutrophil, a short-lived phagocyte whose sole purpose is to kill pathogens. The binding of CXCL8 changes the adhesive properties of the neutrophil, enabling it to slow down and begin to interact with the blood vessel walls. TNF-α is responsible for allowing neutrophils and similar cells to cross the blood vessel in a multi-step process called extravasation. This involves the neutrophils slowing and rolling along the wall of the blood vessel, squeezing through, and migrating towards the infection [29].

Once a neutrophil has reached the infection it will engulf and essentially consume pathogens. This process is called phagocytosis, a term accurately derived from the Greek root "phagein" meaning "to eat" [1]. Like before, the process is facilitated by receptors. This time the receptors are specific for proteins generated by the pathogens. As soon as a pathogen has been ingested it is attacked by a host of enzymes and quickly degraded. As neutrophils are particularly designed for phagocytosis, they commit most of their resources to creating and containing these enzymes [29]. Once their supply of digestive material is depleted, they go through apoptosis, or pre-programmed death. The resulting pus is later cleaned up by the more versatile macrophages [26]. This attack requires an increased need for oxygen, referred to as the respiratory burst. This activity is also accompanied by the production of additional enzymes designed to protect surrounding tissue from the destructive effects of the toxins in use [13].

Although this process programmed into people usually gets the job done, there are times that the innate immune response fails to completely control an infection. When phagocytosis alone cannot overpower pathogens, specific antibodies are produced to combat the problem. This incredible process, the adaptive immune response, takes time when the body is first introduced to a certain variety of pathogen. However once a given pathogen has been introduced, a small amount of its corresponding antibody is kept on hand at all times and future infections can be vanquished much more rapidly [3]. It is this response that vaccinations seek to strengthen, by introducing extremely small amounts or dead samples of common problematic pathogens.

The goal of the subcutaneous infection modeling begun by Prieto-Langarica and three undergraduate researchers has been to create an accurate and adaptable representation of the processes of innate immunity first initiated when a pathogen enters the body. Such a model is important to further the understanding of why some infections can be effectively countered by innate immunity and why others require medical assistance such as antibiotics or vaccinations. Patients with deficient numbers of neutrophils or otherwise compromised immune systems suffer from frequent infections and their devastating effects. It could be that a visual simulation of what occurs during the inflammatory response, coupled with more traditional methods of modeling, will allow researchers to more efficiently develop treatments of subcutaneous infections.

To model the response to infection, the researchers considered only a small section of the human body including a segment of a blood vessel and tissue. Initially, the neutrophils move through the blood vessel in one direction. An infection starts to bloom in the area designated as tissue, starting off as a singular cell. This is all represented by a matrix. Each number in the matrix is represented by a different color in a grid, allowing the user to fully visualize the system. The empty cells are zeroes, the infection is made up of threes, the blood vessel wall fives, and the neutrophils are twos.

Each of these cells moves in different ways in the human body and some, such as the neutrophils, move differently depending on where they are in the body. Each of these stipulations has to be considered in the model. In order to do so, *for loops* are used to locate different cells and the part of the body they are in. For example, a 2 (neutrophil) outside of the blood vessel can move in all 8 directions. Each of these directions is given a number representation, shown in Table 5. Then by means of a user defined function one of these directions is randomly chosen, taking into consideration only possible moves (i.e., a neutrophil cannot be placed outside the bounds of the matrix without an error so any move which causes the neutrophil to move out of bounds is not considered). Then the program makes sure the neutrophil is only moving to an empty spot or somewhere with infection. Finally the system subtracts 2 from that cell of the matrix and adds 2 to the new random location of the neutrophil. This process is repeated for each neutrophil outside of the blood vessel and similar processes are repeated for other particles; for instance, a 2 (neutrophil) within the bounds of the blood vessel can only move three directions so only 1 through 3 are direction options. After each particle is moved within its parameters

Table 5 This table represents the different ways particles can move

5	4	3
6	Cell	2
7	8	1

The numbers are given to indicate the different directions. During the simulations, particles can move in any of those 8 directions with different probabilities the colored grid is shown. When this process is repeated several times the particles on the grid appear to move and give a semi-accurate visual representation of the movement of the particles (Fig. 1).

Now that the system is set up, it is time to introduce the characteristics of the body when an infection hits. The one initial cell of infection grows randomly outward one cell for every two movements of the cells, using the same system while skipping the step where the previous location of the cell is deleted, giving the appearance of growth instead of movement. Inflammatory mediators should now start moving through the system from infected cells. One inflammatory mediator is added to a random spot bordering the boundary of the infection using the same methods as above. These inflammatory mediators move randomly upward until they move into the blood vessel. When a neutrophil comes into contact with an inflammatory mediator, the whole system of neutrophils slows down and begins to move down toward the infection. The particles sometimes move randomly and other times move toward the closest infection point. When the neutrophils make it to the infection, they combine to create a new number in the matrix for one cycle and then that number is removed to represent the killing of the infection (phagocytosis), the death of the neutrophil (apoptosis), and the removal of the dead cells.

The current program is a working model of infection in the human body. Factors like the size of the system, the size of the initial infection, and the growth rate of the infection can easily be manipulated. Great care has been taken to closely parallel the biological processes represented, yet there is still a lot that can be done to this model to make it more accurate. Ideally, the model should be made three dimensional and the initial count of neutrophils should be based on biological data. The communication between particles and the decision making involved in their movements will eventually need to incorporate more mathematically precise methods. Currently all cells appear to be the same size, when in reality there is great variation in both size and shape of the cell types. These variations are biologically very significant, and likely will prove to have a great impact on the appearance of the model and the probability of the infection being successfully removed. The movement of particles across the blood stream, known as extravasation, is presently shown in a very simplified manner. Currently neutrophils simply slow down and cross the line representing the vessel wall, while in reality the process involves several distinct steps, each of which involves unique interactions between the different cells. Macrophages, which play two very important roles of initially releasing the chemical signals that initiate the inflammatory response and cleaning up dead material are not included in the present model. The processes of phagocytosis, apoptosis, and the removal of pus are simplified to one stage but may eventually be expanded into the three distinct steps that would technically be correct.

Although some of these potential adjustments may prove to be insignificant to the mechanisms being reproduced, the researchers hope to begin to implement each of them in time and investigate their potential impacts. Both undergraduates plan to continue working with the model throughout their undergraduate career, hopefully creating an effective tool for other researchers and potentially moving towards an industry leading model.

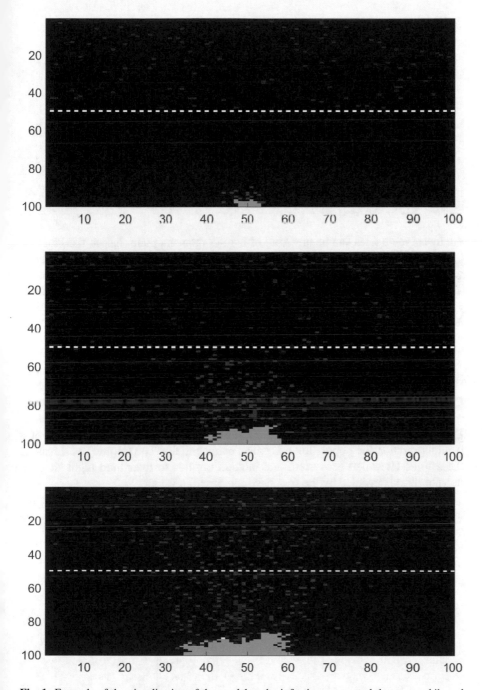

Fig. 1 Example of the visualization of the model as the infection grows and the neutrophils and inflammatory mediators move throughout the system

5.2 Research Projects for ABM in Biology

> **Research Project 1** In general, **ABM** that simulate the invasion of a non-native species will deal with the geographical spread of the species, populations growth. It can also include predator–prey dynamics and consumption of natural resources. The exact details will depend on the particular problem of interest. Once a model is constructed based on ecological information, the model can be used to answer a wide variety of important questions. One particular place to look for research projects is in management of invasive species.

An **invasive species** can be any kind of living organism, plant, insect, fish, fungus, bacteria, or even an organism's seeds or eggs that are not native to an ecosystem and causes harm. They can harm the environment, the economy, or even human health. This problem has been well studied in the field of ecology and in mathematical ecology [30, 31, 35, 37]. However, there are many questions open for investigation. Most existing models aim to be as general as possible. Even models dealing with a specific species invading a specific region, try to be general and offer general answers to their own proposed questions.

ABMs are effective tools to deal with very specific invasive species problems. For example, the faculty author has worked with an undergraduate student, Meghan Chambers, in using ABM to test different control strategies to deal with the invasion of the Coqui Frog in the Hawaii islands. The Coqui Frog is a native species of Puerto Rico and was brought to the Hawaiian islands, where in the absence of a natural predator, the population has grown exponentially. The Coqui Frog threatens the natural Hawaiian ecosystem and in addition, due to their loud night singing, neighborhoods invaded by the frog have decreased in value.

Due to her participation in the PURE Math program (a research experience for undergraduate students in Hawaii), Megan was able to make connections with the ecology department at the University of Hawaii and was able to get data that we used to create the ABM. This is another great argument for faculty seeking to engage in this type of research with undergraduates to pursue collaborations with biologists to get access to knowledge, data and to very specific questions they need answered.

> **Research Project 2** A **CA Model** can be created to demonstrate the effects an over population of geese can have over the environment and the health of other animal species. These kind of visualization can be an invaluable tool for educating the public. More important than that, however, is the use the CA model can have in testing other population control strategies, such as sterilizations, and test scheduling routines for such interventions.

Another set of possible projects can be found in **population control** problems. As an example, the city of Youngstown Ohio is lucky to have one of the largest, most beautiful metropolitan parks in the nation. During the summer of 2014, the populations of Canadian Geese in the park exploded. Overpopulation of a particular species can bring all sorts of ecological in-balances and habitat depletion. In order to mitigate the damage of the growing geese population, the United State Department of Agriculture had to euthanize large groups of geese several times through the year. Here is an article on the local paper [25]. As one can imagine, this produced a very negative reaction from the Youngstown populations who did not want to see the geese being killed. The city has a strong relationship with the park and there are numerous educational programs for students of all ages and even for adults. The park management wishes to find other none lethal ways to control the population and the city is very willing to learn about the effects to the environment of overpopulation of a species.

In this particular CA model, each cell can represent a small park area containing information about the number of geese, the quality of that area as a breeding place, availability of resources, whether this area is accessible to humans (to model the effect that humans feeding the wild life), any information on natural predators, among other markers of interest. Most of this information can be given by the Youngstown Metropolitan Parks authorities. Since it is of interest to model one season, the time step can be set as small as a day or maybe a week. The CA can show the movement of geese through the park, births and natural deaths. Once a CA is constructed that mimics the current park situation, the model can be used to test many different population control strategies, teach park attendees the consequences of feeding the wild life, and help educate the public to why it is necessary to control the populations.

Research Project 3 A fascinating project for students interested in ecological problems is to model migration of a specific bird of interest. Students can create a **CA Model** in which each element in the matrix saves information on all measurable factors that determine migration patterns. Once a model has been created, the model can be used to predict migration changes due to changes in the terrain, such as night lights from cities, changes in climate and deforestation, among others.

Migratory bird species rely on environmental cues to travel in response to the seasons. Migration is controlled by internal mechanisms that ensure arrival and departure times coincide with the right temperatures and peak food availability [8]. Despite this, rapid **climate change** has altered environmental cues over the recent decades; as a result, these birds are forced to shift not only their time of flight but also their flight patterns to make up for variation in food supply and temperature [8]. According to [22], migratory birds are declining at much higher rates compared to non-migratory birds. Changes in weather and habitat are among the primary drivers of bird population decline [22]. Additionally, collisions with man-made structures kill thousands of birds each year and are expected to kill millions by 2030 [12].

As an example, in collaboration with the Geography department at Youngstown State University, a group of students have been working on mathematically modeling the recent shifts in migratory patterns from land use change and population density for one bird species, the Yellow-rumped Warbler, and to predict how these migratory patterns will look in the future. Since shifting environmental cues due to climate change have been driving shifts in bird migration patterns over recent decades, an alteration in land type and population density coupled with prolonged start of migration will contribute to a change in the most efficient route for the Yellow-rumped warbler.

The geography department has helped the team get detailed information on the state of the terrain under the migration route for the Yellow-rumped Warbler and have guided predictions on changes in terrain due to climate change. Students are working on incorporating this information into the model and will use it to predict future migration patterns and health of the overall population.

There are two ways in which this project can be developed. If the reader would like to focus on the environment itself, this could be a great project to develop a **CA Model**. However, if the reader wants to consider the birds as agents, an ABM might be a better way to model this. The article [19] contains a lot of information of different migratory species and how climate change is perturbing their migration patterns. The author also recommends using NetLogo [42] and its GIS data capabilities.

Research Project 4 The field of **microbiology** offers many opportunities for **Agent Based Modeling** and for close collaborations with biologist who are currently experimenting at the cellular level. ABM done in tandem with experiments offer the opportunity to create data driven models which can then guide further experimentation. The authors will describe two past projects she worked on with undergraduate students and with biologists in an effort to inspire similar projects.

The importance of collaborations with biologists cannot be overstated. In addition, ABMs are an incredible asset as a visualization tool to communicate with other scientists, so collaborations that start with ABM models can evolve into the development of more complex experiment-driven models. Below are two examples of two different projects in microbiology.

Bone Formation and Metabolism
At the macroscopic scale bone is a static, hard tissue, however, at the microscopic level, bone consists of different types of cells, bone matrix cells (Osteocytes), cells that create new bone (Osteoblast) and cells that destroy old or defective bone (Osteoclast).

This project was done in very close collaboration with Dr. Marnie Saunders at the Biomedical Engineering Department at Akron University. Dr. Saunder's lab was experimenting with bone cells and trying to understand how cells communicate with each other. They developed small area experiments in which they isolated a few bone cells and observed their behavior under different stimulus.

The faculty author has done some preliminary work in this area with students, but there are still open questions available, started with a **Cellular Automata Model** which aimed to replicate their exact lab experiments. In the lab, cells that create bone, Osteoblasts, are placed in a matrix and induced to begin bone formation. Measurement of bone formation is taken after one, two, four, eight, sixteen, and 26 days. The data were used to create a CA model and we validated the model using a permutation test. This CA model can then be modified to improve bone formation and this will lead to new possible experiments Dr. Saunder's lab can perform. Most of the CA model was developed by a mathematics student during her Junior year. Preliminary results for this project can be found in [40].

In this case, as the cells are not moving but mineralizing and creating bone, a CA was better suited to model the locations where minearlization was happening. Other examples of possible applications of cellular biology in which movement is negligible are:

- Angiogenesis. As tumors grow, they require more oxygen and other nutrients to grow. By sending signals to nearby blood vessels, tumors are able to extend those blood vessels towards themselves, providing all the oxygen and nutrients necessary to grow. Processes such as this can be modeled using a Cellular Automata Model, where blood vessels extend to neighboring squares in the grid, dependent on concentrations of chemoatractants [9].
- Wound healing is a process well studied and yet many questions remain. New uses of STEM cells to enhance healing are being tested in different ways in many Biology or Biomedical Engineering Departments at universities all across the country. Using data provided by biologist or immunologist, a CA model can be constructed to simulate the wound area, the immune response, bacteria, and normal tissue. This models can start simple, with just a few types of cells, and increase in complexity with time [44].

Pre-Natal Muscle Formation
The Mathematics Department at the University of Texas at Arlington (UTA) was interested in starting collaborations with the Nursing Department at UTA, and in particular, with a Nursing faculty whose lab studies muscle formation in the fetus. As in many cases, the biologists do not necessarily see the point of a collaborations with mathematicians, so by creating a ABM that simulates their *in vitro* experiments and using it as a visualization tool for the Nursing professor, the lab was able to see the potential for a collaboration. This collaboration has led to grant proposals and will produce peer reviewed publications as well.

The ABM that helped this collaboration start was done by a team of four freshmen students, two mathematics majors and two pre-med. The ABM tracked a few initial cells before differentiation. Each cell would then differentiate into a myocite (muscle cell), reproduce, move towards neighboring cells, align with them, and merge to form muscle fibers. Students learned to program (in Matlab), learned quite a bit about fetus formation, learned cellular biology, and in mathematics, they learned about probability and a bit about measures.

In the project described above, the movement of cells and their orientation is a vital component of what needs to be better understood. Therefore, an ABM was developed which tracks cells position, orientation, length, and polarity (since cell fuse at the poles).

Research Project 5 With the right collaboration, students create ABM which model the competition for resources that naturally arise between cancerous and non-cancerous cells. The problem of competing populations is a well-studied problem and a wide variety of standard models exist in the literature [27, 28]. One real-world application of a competing population system that has become increasingly important is the modeling of cancerous and non-cancerous cells as competing populations. Since ABM is good for modeling very complex systems, the model can start simple and complexity can be added as students learn more about the biology or get access to lab data. They can start by modeling competing cells and add the different growth rates, the immune response, angiogenesis (the formation of arteries around a tumor), tumor formation, the necrotic nucleus of tumors, etc.

6 Afterword

In this section, the two student authors whose work was feature in Sect. 5.1 of this chapter will talk about their personal experience being involved in the subcutaneous example research project. By sharing their experience, the authors hope that other students and faculty thinking about doing research in this area can learn what we did right, what needs improvement and ultimately, why enabling students to do research early in their undergraduate career is a worthwhile endeavor.

6.1 Lindsey Chludzinski

As a first semester Freshman, I knew very little about college or research projects, and essentially nothing about mathematical biology research, agent-based modeling, or programming. When we embarked on this project we had no experience and almost no background to draw from. We honestly did not know what the end goal of the project would be, why we were doing it, or how long it was going to take. But it was an opportunity to try something new and it ended up being a good one.

Although it is obviously not ideal to have no idea what one is doing, the nature of our research project ended up being well suited for our lack of knowledge. We started by learning the basics of Matlab and a very simplified general overview of our biological topic. Our research professor gradually added more details to the exercises she assigned us, and the Biology professor that we are working with gradually provided us with more information and reading materials. The problems that seemed insurmountable at the beginning of our project, things as basic as making a for loop or creating a matrix, are now more or less routine for us. I would imagine that our current impossible challenges will quickly lead to new knowledge and new obstacles.

As a student that has always preferred to work through confusing topics in solitude one of the biggest initial challenges for me was learning how to problem solve with a partner. I had never really struggled with group work, but I had also only ever collaborated on topics that I was already confident in. Learning to admit that I had no clue what was going on or where to go next took some time. Learning to make suggestions that might be absolutely wrong was also an adjustment. Thankfully, Alex and Dr. Prieto-Langarica have been incredibly patient partners.

We have grown a lot in our ability to code—a skill I never really expected to have but that is obviously applicable in a lot of areas. We have made a lot of mistakes, some of which we are still facing the consequences for. When we first embarked on this project every half line of code we wrote was painful and required at least two Google searches. Once we got the hang of it we were able to crank out about 1600 lines of working code in just a few months, which to an experienced coder is likely insignificant, but to our novice perspectives seemed incredible. Once we finally had a program that looked cool and appeared to get the job done we decided to review what we had written and quickly discovered how disorganized, unreadable, and inefficient our technique was. Debugging was nearly impossible and finding any specific line took so long that we would forget what we were looking for before we found it. We have spent the last semester of our research completely rewriting the code for the program and in order to bring it down to about 300 hundred lines, doing essentially the same thing that it did before. In the process we have encountered a lot of new (or previously hidden) problems, but we are slowly addressing and eliminating them. To someone of any degree of genuine skill our current work would probably be embarrassing, but to us it feels like an accomplishment. It has been a constant cycle of problem finding, problem solving, problem fixing, and problem creating, but we are making progress.

If I had to start the project over, I do wish that we had taken more time to learn the basic mechanisms of how MatLab reads through a program, how the syntax works, how to choose variable names that make sense and can be remembered, or what separates efficient code from a disastrous mess. Although we learned as we went, I would suggest to other students to spend as much time as possible reading about their topics before jumping in. It was also difficult to actually understand and envision the end goal of our project, largely because it was and still is evolving as our abilities and interests and progress evolves. If you choose to start a project with as little background as we had, be patient. Occasionally take the time to compare what you have accomplished to what your original understanding was. Even slow progress has significance.

We had the opportunity to present a talk about our research at the MathFest in Denver this past summer. Our advisor helped us apply for travel grants and made the trip to our first academic conference possible. It was thrilling to hear the amazing things other undergraduates are working on, and it was very rewarding to share our own efforts. As has been the recurring theme in this experience, I learned a lot about research and a lot of practical life skills.

Mathematics and Biology are beautiful on their own, but any time the two can be combined is nothing short of awesome. Our particular project has not yet required very advanced levels of math, but I think it has exercised our critical thinking and problem solving abilities. I look forward to delving into new ways to apply deeper math to our topic and to potentially creating a useful research tool for medical professionals.

As a whole, my research experience has often been frustrating and confusing and slow moving, but it has been extremely constructive and altogether worthwhile. I have grown as a student, as a presenter, and as a partner. I have learned how to collaborate with others, how to problem solve, and how to use Matlab. What I find most exciting is that we still have a lot of room to continue with this project. As an undergrad, I highly advise other students to get involved in research as early as possible.

6.2 Alexandra Ballow

My story is a little different than most. I came into college wanting a career in academia and anxious to start research. I sought out professors who would work with me and was referred to Dr. Prieto-Langarica. She was trying to put together a team to research wound healing computationally. I had no interest in biology and no experience in coding, but I was happy to accept the opportunity doing math research, even if the focus was different from the projects I dreamed of doing.

For me, the people I am working with are invaluable. My ability to work well can be impeded by bad partners and severely improved by good partners. Because I knew this about myself, I wanted to make sure I had a good partner to experience research for the first time with. To that end, I recommended Lindsey, who was in my English and Calculus classes and a budding friend. All I knew was she was a biology major, so I hoped she could fill in the gaps of my lacking biology knowledge. Dr. Prieto encouraged me and promised to start small, but it still was comforting to have someone with me almost as clueless about how research worked. Dr. Prieto happily brought Lindsey into the project and the two of us were able to start. These details probably feel insignificant, a few days compared to the weeks we have spent rescarching, but looking back, I feel like these little decisions were the reasons we have been able to succeed. We were a small team who wanted to work together, the undergraduates involved had equal knowledge and were eager to learn, and the professor was careful to preserve that eagerness with challenges just hard enough to cause a struggle, but not too hard to crush spirits. Lindsey and I learned everything together, we encouraged each other, pushed each other, and struggled through the hard times together.

Our complete lack of understanding at the beginning of this project did not matter at the end, because Dr. Prieto-Langarica helped teach us coding and the infection healing process as we built the model. We struggled through one task at a time starting as simple as building a matrix and eventually, after many hours brainstorming in the library and hundreds of Google searches later, we had a working model. I do not mean to belittle the struggles and accomplishments and triumphs that happened during this time. We both learned so much, grew so much. There are no words to describe what this experience has given me or how strongly I recommend all undergraduates start research as early as possible. The project does not matter. I had no interest in biology, but knowing how to do biomath research and understanding how to work within other fields has been invaluable. It does not matter how young you are, how inexperienced you are, or how little you know about what you are researching. Find a good group of people and a patient professor and everyone, freshmen too, can do research.

6.3 Lindesy and Alex

For both of us, the modeling process has been well suited for our limited background. When we first began we had extremely limited knowledge of the biology behind infections and essentially no background in MatLab or coding. Throughout this project we have learned a tremendous amount every step of the way. In the beginning Dr. Prieto gave us very simple exercises to complete in Matlab, seemingly unrelated to the overall goal. These exercises are very similar to those included here. The first task was to create a matrix of zeroes with ones randomly placed around the edges. This may sound like a simple undertaking, but to someone who had never coded before it was extremely difficult. We tried different tactics and fiddled with the program for over a month before we finally created the matrix as described.

Giving us a month to do something so simple probably sounds a bit crazy when Dr. Prieto-Langarica could have given them the simple two lines of code. However, what we learned in that month was well worth the time invested. We learned more than coding. Dr. Prieto-Langarica was teaching us how to complete a research project, not how to complete this specific project. We learned the process, the frustration, the ups and downs of actually discovering something. We began to develop a new approach to problem solving and established how to collaborate on challenges neither of us initially understood. With our new knowledge we started to tackle the next problem; making the particles move. This problem took even longer, the rest of the semester. This was another good use of time. By the time we had completed the assignment we had a deeper understanding of coding and a more efficient approach to debugging. We were learning the possibilities and complexities of a new language.

All of these discoveries were made while on the side, we were focused on wounds. Once we had enough experience to create a rough representation of something a partnership with an immunology professor in the biology department was established. With her guidance, it was decided to transfer the research focus to infections, due to a slightly simpler pattern of movement and a significantly smaller number of cell types involved in the basic processes. Because we had learned the basics of coding, the second time around was much easier. We were able to apply what we learned to set up the system quickly. Dr. Prieto directed their progress by breaking the big picture into simplified pieces, yet she only offered specific advice after the undergraduates began to approach each step independently. This allowed us to take ownership of OUR project. Despite the slower pace and the increased amount of mistakes made resulting from the trial and error methods often used, the experience gained through the progress achieved so far has proven invaluable.

Acknowledgements The authors would like to thank the reviewers and the editors of this book, Carrie Diaz Eaton and Hannah Highlander, for their guidance and help in making this chapter better.

References

1. *Merriam Websters Collegiate Dictionary*, volume Eleventh Edition. Merriam-Webster, Incorporated, 2009.
2. A. I. Adamatzky. *Identification of cellular automata*. CRC Press, 2014.
3. B. Alberts, A. Johnson, J. Lewis, M. Raff, K. Roberts, and P. Walter. *Molecular biology of the cell: Reference edition*. Garland Science, 4 edition, 2002.
4. L. J. Allen. *An introduction to stochastic processes with applications to biology*. Chapman and Hall/CRC, 2010.
5. R. L. Axtell, C. J. Andrews, M. J. Small, et al. Agent-based modeling and industrial ecology. *Journal of Industrial Ecology*, 5(4):10–14, 2001.
6. P. Bak, C. Tang, and K. Wiesenfeld. Self-organized criticality: An explanation of the 1/f noise. *Physical review letters*, 59(4):381, 1987.
7. M. J. Berryman and S. D. Angus. *Tutorials on agent-based modelling with NetLogo and network analysis with Pajek*. World Scientific, 2010.
8. C. Carey. The impacts of climate change on the annual cycles of birds. *Philosophical Transactions of the Royal Society B: Biological Sciences*, 364(1534):3321–3330, 2009.
9. P. Carmeliet and R. K. Jain. Angiogenesis in cancer and other diseases. *nature*, 407(6801):249, 2000.
10. G. Chartrand and P. Zhang. *Chromatic graph theory*. Chapman and Hall/CRC, 2008.
11. B. Chopard. *Cellular automata modeling of physical systems*. Springer, 2012.
12. A. B. Conservancy. American bird conservancy website, 2018.
13. C. Dahlgren and A. Karlsson. Respiratory burst in human neutrophils. *Journal of immunological methods*, 232(1-2):3–14, 1999.

14. J. S. Dean, G. J. Gumerman, J. M. Epstein, R. L. Axtell, A. C. Swedlund, M. T. Parker, and S. McCarroll. Understanding Anasazi culture change through agent-based modeling. *Dynamics in human and primate societies: Agent-based modeling of social and spatial processes*, pages 179–205, 2000.
15. M. Dickerson. Multi-agent simulation and NetLogo in the introductory computer science curriculum. *Journal of Computing Sciences in Colleges*, 27(1):102–104, 2011.
16. R. Durrett. *Probability: theory and examples*, volume 49. Cambridge university press, 2019.
17. J. M. Epstein. Modeling civil violence: An agent-based computational approach. *Proceedings of the National Academy of Sciences*, 99(suppl 3):7243–7250, 2002.
18. F. Fadel, M. Khalil El Karoui, and B. Knebelmann. Spread of a novel influenza a (H1N1) virus via global airline transportation. *Nat Rev Immunol*, 8:153–60, 2008.
19. A. S. Gallinat, R. B. Primack, and D. L. Wagner. Autumn, the neglected season in climate change research. *Trends in Ecology & Evolution*, 30(3):169–176, 2015.
20. V. Grimm, E. Revilla, U. Berger, F. Jeltsch, W. M. Mooij, S. F. Railsback, H.-H. Thulke, J. Weiner, T. Wiegand, and D. L. DeAngelis. Pattern-oriented modeling of agent-based complex systems: lessons from ecology. *science*, 310(5750):987–991, 2005.
21. N. Hartsfield and G. Ringel. *Pearls in graph theory: a comprehensive introduction*. Courier Corporation, 2013.
22. A. Howard, K. Challis, J. Holden, M. Kincey, and D. Passmore. The impact of climate change on archaeological resources in Britain: a catchment scale assessment. *Climatic change*, 91(3-4):405–422, 2008.
23. C.-M. Lo, H.-B. Wang, M. Dembo, and Y.-l. Wang. Cell movement is guided by the rigidity of the substrate. *Biophysical journal*, 79(1):144–152, 2000.
24. A. J. McLane, C. Semeniuk, G. J. McDermid, and D. J. Marceau. The role of agent-based models in wildlife ecology and management. *Ecological Modelling*, 222(8):1544–1556, 2011.
25. P. Milliken. 238 geese euthanized in mill creek park, Jun 2014.
26. I. Mukundan, J. I. Odegaard, C. R. Morel, J. E. Heredia, J. W. Mwangi, R. R. Ricardo-Gonzalez, Y. S. Goh, A. R. Eagle, S. E. Dunn, J. U. Awakuni, et al. PPAR δ senses and orchestrates clearance of apoptotic cells to promote tolerance. *Nature medicine*, 15(11):1266, 2009.
27. J. D. Murray. *Mathematical biology. II Spatial models and biomedical applications Interdisciplinary Applied Mathematics*, volume 18. Springer-Verlag New York Incorporated New York, 2001.
28. J. D. Murray. Mathematical biology: I. an introduction (interdisciplinary applied mathematics), 2007.
29. P. Parham. *The immune system*. Garland Science, 2 edition, 2005.
30. A. T. Peterson. Predicting the geography of species' invasions via ecological niche modeling. *The quarterly review of biology*, 78(4):419–433, 2003.
31. E. C. Pielou et al. An introduction to mathematical ecology. *An introduction to mathematical ecology.*, 1969.
32. A. Prieto-Langarica, H. V. Kojouharov, and B. M. Chen-Charpentier. Upscaling from discrete to continuous mathematical models of two interacting populations. *Computers & Mathematics with Applications*, 66(9):1606–1612, 2013.
33. S. F. Railsback and V. Grimm. *Agent-based and individual-based modeling: a practical introduction*. Princeton university press, 2019.
34. W. Rand and R. T. Rust. Agent-based modeling in marketing: Guidelines for rigor. *International Journal of Research in Marketing*, 28(3):181–193, 2011.
35. A. K. Sakai, F. W. Allendorf, J. S. Holt, D. M. Lodge, J. Molofsky, K. A. With, S. Baughman, R. J. Cabin, J. E. Cohen, N. C. Ellstrand, et al. The population biology of invasive species. *Annual review of ecology and systematics*, 32(1):305–332, 2001.
36. J. L. Schiff. *Cellular automata: a discrete view of the world*, volume 45. John Wiley & Sons, 2011.

37. C. M. Taylor and A. Hastings. Finding optimal control strategies for invasive species: a density-structured model for Spartina alterniflora. *Journal of Applied Ecology*, 41(6):1049–1057, 2004.

38. L. H. Taylor, S. M. Latham, and M. E. Woolhouse. Risk factors for human disease emergence. *Philosophical Transactions of the Royal Society of London. Series B: Biological Sciences*, 356(1411):983–989, 2001.

39. K. Uriu, Y. Morishita, and Y. Iwasa. Random cell movement promotes synchronization of the segmentation clock. *Proceedings of the National Academy of Sciences*, 107(11):4979–4984, 2010.

40. G. K. Van Scoy, E. L. George, F. O. Asantewaa, L. Kerns, M. M. Saunders, and A. Prieto-Langarica. A cellular automata model of bone formation. *Mathematical biosciences*, 286:58–64, 2017.

41. D. B. West et al. *Introduction to graph theory*, volume 2. Prentice hall Upper Saddle River, NJ, 1996.

42. U. Wilensky. NetLogo, 1999.

43. S. Wolfram. *Theory and applications of cellular automata: including selected papers 1983-1986*. World scientific, 1986.

44. Y. Wu, L. Chen, P. G. Scott, and E. E. Tredget. Mesenchymal stem cells enhance wound healing through differentiation and angiogenesis. *Stem cells*, 25(10):2648–2659, 2007.

Network Structure and Dynamics of Biological Systems

Deena R. Schmidt

Abstract Many biological systems in nature can be represented as a dynamic model on a network. Examples include gene regulatory systems, neuronal networks, food webs, epidemics spreading within populations, social networks, and many others. A fundamental question when studying biological processes represented as dynamic models on networks is to what extent the network structure is contributing to the observed dynamics. In other words, how does network connectivity affect a dynamic model on a network? In this chapter, we will explore a variety of network topologies and study biologically inspired dynamic models on these networks.

Suggested Prerequisites *A standard first course in calculus-based probability is recommended. Some familiarity with graph theory and/or network theory is recommended, but not required. Some familiarity with biology, mathematical modeling, and simulation in* R *will be helpful.*

1 Introduction

This chapter focuses on network structure and dynamics in biological systems. Here we use the term *network* to mean a graph consisting of a set of *nodes* (also called *vertices*) connected by *edges*. The network reflects the structure of the biological system, and the overall dynamics are determined by a dynamic model on the network such that state variables associated with the nodes (and possibly also the network topology) change over time. We use the term *dynamic model* in the broad sense to mean a mathematical model describing how those state variables change over time. We will focus on stochastic process models that evolve on a network, but we can also consider deterministic processes (e.g., differential equation models) on a network within this framework. In this chapter we will often refer to such dynamic

D. R. Schmidt (✉)
University of Nevada Reno, Reno, NV, USA
e-mail: drschmidt@unr.edu

© Springer Nature Switzerland AG 2020
H. Callender Highlander et al. (eds.), *An Introduction to Undergraduate Research in Computational and Mathematical Biology*, Foundations for Undergraduate Research in Mathematics, https://doi.org/10.1007/978-3-030-33645-5_7

models, either stochastic or deterministic, simply as processes on a network. Formal definitions of fundamental network concepts are given in Sect. 3 and examples of processes on a network are given in Sect. 5.

The aim of this chapter is to introduce the topic of network structure and dynamics for modeling a variety of biological systems, and to guide students toward interesting research projects on the topic. This chapter is not intended to be an exhaustive resource that includes all necessary background to do research in this area. Indeed, there are many books and a vast literature on each of the examples and projects given in this chapter. The goal, rather, is to introduce the basic concepts and motivation as well as to provide resources (e.g., R code, further references) that students can use in collaboration with their advisor to work on the suggested (or related) research projects in this area.

Suggested research projects are outlined in Sect. 6. These projects involve modeling and analyzing different biological processes using the tools of network science and applied probability. In particular, students will learn about a variety of network models (i.e., models of network structure) which allow for different types of network connectivity. Which network structure is most appropriate for modeling a particular biological system depends on that particular biological setting. Students will then learn about processes (dynamic models) on networks, and explore what types of biological systems can be modeled as such. The level of projects envisioned ranges from a course term project or a small summer REU project up through projects that could span multiple semesters as part of, for example, the undergraduate McNair Program.

A fundamental question when studying biological systems represented as dynamic models on networks is to what extent the network structure is contributing to the observed dynamics? In other words, how does network connectivity affect a given process on the network? This will be the focus of the first part of this chapter, and one of the most important problems in network science today [33]. A more difficult problem is to connect these modeling ideas to real biological processes. For example, we could study how (physical) social contacts affect disease progression in a population. We could look at how neuronal connectivity affects the collective behavior of a neuronal network within the brain. These examples have a fixed (possibly randomly generated) network structure and look at a process evolving on the network. We could also study how changing the network structure over time might influence the process on the network.

In order to build up to the research projects at the end of this chapter, we present the material along with selected exercises intended to deepen the understanding of each topic. It is strongly recommended that students attempt these exercises before moving on to the next topic. Knowledge of basic probability, as in a standard first course in calculus-based probability, is recommended. Some familiarity with graph theory and/or network theory is recommended, but not required to work through this chapter. Some familiarity with biology, mathematical modeling, and simulation in R will be helpful, but not required. We will define all necessary terms and concepts needed to work on the research projects.

The remainder of the chapter is organized as follows: Section 2 gives an overview of biological systems represented as dynamic models on networks along with some examples. Section 3 presents the fundamentals of network theory which include mathematical descriptions of networks, network metrics and properties used to characterize network structure. In Sect. 4, we introduce network models which are used to explore different connection topologies, and in Sect. 5 we discuss various dynamic models or processes on networks. Lastly, Sect. 6 describes the suggested research projects on this topic.

1.1 Further Reading

This chapter will only scratch the surface of the vast topic of biological networks in terms of structure and dynamics. For comprehensive references on the basics of probability and stochastic processes, see Sheldon Ross' books [39, 40]. A nice reference for stochastic processes in biology is given by Allen [7]. For a more complete introduction to network theory, see Mark Newman's book [33]. For more details on dynamical processes on networks, see [11, 36, 37]. A comprehensive reference on adaptive networks is given by Gross and Sayama [26]. For an advanced mathematical treatment of random graphs and random graph dynamics, see [14, 20]. Other references will be given throughout the rest of the chapter.

A significant part of the exercises and projects will involve computation and simulation via the statistical software R [38]. In terms of learning basic R, there are many resources available, particularly online. For example, see the tutorial "Getting Started with R" by the support team at RStudio at https://support.rstudio. com/hc/en-us/articles/201141096 Getting Started with R [1] and the interactive course "Introduction to R" by DataCamp at https://www.datacamp.com/courses/ free-introduction-to-r [2]. At the end of this chapter is an Appendix containing the R code used to create all figures as well as additional definitions and examples for simulating basic stochastic processes.

2 Biological Systems Represented as Dynamic Models on Networks

Networks play a central role in the biological sciences as they form the basis of a wide variety of biological processes across scales, from molecular and cellular processes to species interactions. Particularly at the molecular level, most genes and proteins carry out their functions within a complex network of interactions with other genes and proteins rather than on their own.

After several decades of collecting large-scale data on biological networks, researchers have discovered various network properties that appear to be typical of the structure or *topology* of many biological networks. This has led to considerable research aimed at understanding the origin of such network topologies as well as the corresponding biological relevance or function. Essentially, researchers are interested in linking structure with biological function by studying the interplay of network topology and the dynamic model defined on the network.

Many biological systems in nature can be modeled as a dynamic model on a network. Examples include gene regulatory networks, protein–protein interaction networks, metabolic pathways, neuronal networks, food webs, epidemics spreading within populations, and social networks, just to name a few; see also Table 1.

As a simple example, consider a food web consisting of three species: grass, mice, and snakes. Figure 1 illustrates the food web by showing what each species eats. Grass is the producer, or base of the food web, which grows as a result of sunlight, carbon dioxide, and water. Mice eat the grass and snakes eat mice. In this network, the nodes are the species and the edges have a direction to them (denoted by an arrow) corresponding to the food each species eats. Specifically, an edge from species A pointing to species B means that A is the food consumed by B. In other words, B eats A.

The next example illustrates that sometimes more complicated notions of edges are needed when modeling specific biological interactions. Figure 2 shows a cartoon example of a gene regulatory network for a set of three genes: X, Y, and Z. On the

Table 1 Examples of networks in biology

Network	Nodes	Edges
Brain (neuronal networks)	Neurons	Synapses
Gene regulatory networks	Transcription factors (gene products)	Regulatory interactions
Protein–protein interaction networks	Proteins	Protein interactions
Food webs	Species	Predation
Metabolic pathways	Metabolites	Bio-chemical reactions
Social contact networks	People	Social interactions

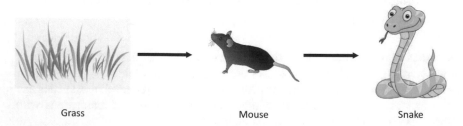

Grass   Mouse   Snake

Fig. 1 A simple example of a food web describing what each species eats. Grass is the producer, mice eat grass, and snakes eat mice. Reproduced with permission from Surya Zaidan ©123RF.com under a standard license

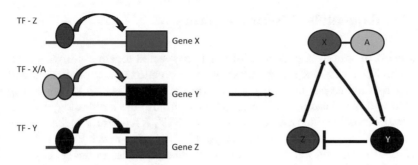

Fig. 2 An example of a gene regulatory network describing the interactions between three genes X, Y, and Z via their gene products, or transcription factors (TF). Transcription factors regulate gene expression by turning genes on or off. The left diagram is the biological representation showing TFs (ovals) binding to the DNA (horizontal lines) and regulating the expression of genes (rectangles). On the right is the associated network diagram where the nodes are TFs and the edges are regulatory interactions. Edges with arrowheads indicate positive (activating) interactions that turn on gene expression, whereas the edge with a hammerhead indicates a negative (inhibiting) interaction that turns off the expression of gene Z

left-hand side is the biological representation. Each gene is transcribed and then translated into a protein product called a *transcription factor* (TF), color coded to match its gene. A TF then binds to the DNA upstream of a (possibly) different gene to control its expression. It is certainly possible for a gene to control its own expression but that is not represented here. These TFs are modeled as ovals and the color coordinated arrows show whether the interaction is positive (activating gene expression) or negative (inhibiting gene expression). The black arrow with the hammerhead is the only negative interaction, and the interpretation is that the TF made from gene Y inhibits the expression of gene Z.

On the right-hand side of Fig. 2 is the network representation of the gene regulatory network example. The nodes are the transcription factors corresponding to genes X, Y, and Z. The two types of edges, denoted by arrowheads and hammerheads, represent the activating and inhibiting interactions described above, respectively. Note that the regulation of gene Y requires a TF complex made up of two TFs: TF-X and TF-A, denoted by TF-X/A. However, the regulation of gene A is not part of the network. This brings up an important point that biological networks are part of a larger complex system of interactions, and typically we only study a small part of that complex system at a time.

Before describing the other biological examples listed in Table 1 and related examples in more detail, we need to introduce some basic concepts from network theory so that we can properly describe network structure (or connectivity), properties, and metrics.

3 Fundamentals of Network Theory

A network is simply a collection of points connected together pairwise by links. In the field of network theory, these points are called *nodes* or *vertices* and the links are called *edges*. In mathematical terms, a network is also known as a *graph* and we will use these terms interchangeably. We use the terminology from graph theory to formally define a network as a graph $G(V, E)$, where V is the set of nodes (or vertices) and E is the set of edges. Typically, the set of nodes $V = \{1, 2, \ldots, N\}$ consists of N natural numbers so that the number of nodes in the set is $|V| = N$. We will define the edge set E after a brief discussion of the different types of edges that can occur.

Mathematically, we typically think of two different types of edges[1]: *directed* and *undirected*, and which type is used depends on the interactions between the nodes. A *directed network*[2] consists of directed edges, or edges with arrowheads that indicate a directed interaction between pairs of nodes, as shown in the motivating example in Fig. 1. In that case, edge (i, j) is an ordered pair representing an interaction from node i to node j, but not vice versa (unless edge (j, i) is also present in the graph). So the edge set E for a directed network $G(V, E)$ is given by the set of ordered pairs identifying which nodes are connected by a directed edge:

$$E = \{(i, j) : \exists \text{ a directed edge from node } i \text{ to } j\}. \tag{1}$$

Alternatively, edges in an *undirected network* $G(V, E)$ assume a bidirectional interaction between pairs of nodes and are illustrated simply by line segments rather than arrows. The edge set E in this case is given by the set of *un*ordered pairs identifying which pairs of nodes are connected by an undirected edge:

$$E = \{\{i, j\} : \exists \text{ an undirected edge between nodes } i \text{ and } j\}. \tag{2}$$

Figure 1 is an example of a directed network. Each directed edge (i, j) has biological meaning: species i is the food consumed by species j. Figure 3 shows two examples of undirected networks. Note that one is connected and one is disconnected; we define these terms below.

A network can be *connected* so that all nodes are connected together in some way by edges. Another way to describe a connected network is one in which no single node or subset of nodes are disconnected from the network. Alternatively, a *disconnected* network is a network that consists of two or more disjoint parts with no edges connecting those parts together. Figure 3 illustrates the difference between a connected and a disconnected undirected network. The left panel shows a connected

---

[1]The term *edge* is frequently reserved for undirected edge, and directed edges are often called *arcs*. *Link* is often used as the general term encompassing all types of such objects.

[2]Directed graphs are also called *digraphs*, and graph could mean either directed or undirected. Often, graph is used to mean an undirected graph.

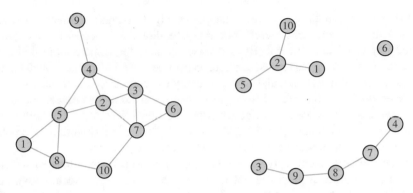

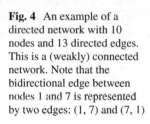

Fig. 3 **Left:** an example of a connected network with 10 nodes and 15 edges. **Right:** an example of a disconnected network consisting of 3 separate parts. These are both undirected networks

Fig. 4 An example of a directed network with 10 nodes and 13 directed edges. This is a (weakly) connected network. Note that the bidirectional edge between nodes 1 and 7 is represented by two edges: (1, 7) and (7, 1)

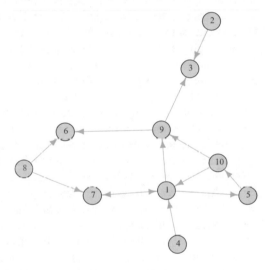

network with 10 nodes and 15 edges, and the right panel shows a disconnected network with three separate parts.

Figure 4 shows a directed network. This network appears to be connected, but is actually only *weakly connected* since not every edge is bidirectional; if it had this property, it would be *strongly connected*. We defer further discussion of weak and strong connectivity for directed graphs to Sect. 3.2, and we give formal definitions of connected and disconnected undirected graphs in Sect. 3.1.

Note that none of the example networks shown here have a node with an edge that loops back to itself; such an edge is called a *self-loop*. In this chapter, we only consider networks with no self-loops.

It is important to understand when to use a directed network and when to use an undirected network. We need to think about edges both in a mathematical context and in the context of the particular biological application in terms of the dynamics on the network. In the food web example in Fig. 1, we use a directed network because those interactions clearly occur in one direction; species B eats species A and not vice versa. In infectious disease models, for instance, the nodes are individuals and the edges represent direct contact between individuals. Such contact is a bidirectional interaction and so we typically use an undirected contact network. However, the infection spreads from person to person along the contact network in a directed manner: Alice may infect Bob, or Bob may infect Alice, but both events do not occur simultaneously in the same outbreak. So here the direction of infection is a property of the disease dynamics, not of the contact network.

When thinking about a dynamic model on a network, sometimes we need to think about additional properties of the edges (in addition to directedness), such as weighted edges or how different directed edges influence node dynamics. For example, the gene regulatory network in Fig. 2 is a directed network with two different types of edges that represent activation and inhibition of gene expression. We can assign positive or negative edge weights, respectively, to account for these interactions in the model. As another example, for the infectious disease example discussed above, we might assign edge weights to represent how frequently different individuals come into contact in a given amount of time. We could do the same thing for the food web example, to represent how much food each species consumes per unit time. These edge weights, then, are treated as properties of the dynamic model on the network; the edge weights agree with the rules for the dynamics. In Sect. 5, we will discuss how to account for different types of directed edges in the context of a specific dynamic model on a network.

In this chapter, we will focus mainly on the simplest type of networks: undirected, unweighted networks with no self-loops, and we will develop most of our theory in this framework. However, the last suggested research project will illustrate the use of other types of networks: directed and weighted (with no self-loops), and these concepts will be specifically developed in Sects. 3.2 and 5.3. In Sect. 3.1, we define the fundamental mathematical concepts used in network theory for undirected networks, and in Sects. 3.3 and 3.4 we introduce a variety of network metrics and centrality measures that can be used to quantify network structure. The R code to generate all figures in this chapter is given in the Appendix [38].

3.1 Mathematics of Undirected Networks

A network can be characterized in various ways. The most fundamental object used to characterize a network is the adjacency matrix, defined formally in Definition 1. This is a matrix of size $N \times N$, where N is the number of nodes in the network, that describes the connections between each pair of nodes in a network.

Definition 1 The *adjacency matrix* $A = (a_{ij})$ of an undirected network $G(V, E)$, where $|V| = N$ is an $N \times N$ matrix such that each entry a_{ij} is either 1 or 0, corresponding to the presence or absence of an edge between nodes i and j, respectively. In other words,

Fig. 5 Left: an undirected
network with 4 nodes and 4
edges. **Right:** corresponding
adjacency matrix

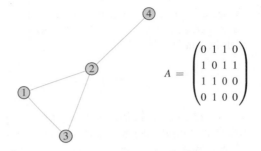

$$A = \begin{pmatrix} 0 & 1 & 1 & 0 \\ 1 & 0 & 1 & 1 \\ 1 & 1 & 0 & 0 \\ 0 & 1 & 0 & 0 \end{pmatrix}$$

$$a_{ij} = \begin{cases} 1, & \text{if } \{i, j\} \in E \\ 0, & \text{if } \{i, j\} \notin E. \end{cases} \tag{3}$$

Recall that the edge set E for an undirected network is a set of unordered pairs; see Eq. (2).

For undirected graphs, A is a symmetric matrix, meaning that $a_{ij} = a_{ji}$ for all $i, j \in V$ and hence $A = A^T$. Figure 5 shows an undirected network (left panel) and its corresponding adjacency matrix A (right panel). We can check that the figure agrees with the matrix. For instance, we see that there is an edge between nodes 1 and 2 and also that entries $a_{1,2}$ and $a_{2,1}$ are both 1. Since this is an undirected network, A is symmetric. Also notice that the diagonal entries of A are zero. This is true for networks with no self-loops.

Another fundamental property of a network is its degree distribution, defined formally in Definition 3. First we define the *degree* of a node in a network.

Definition 2 The *degree* of a node in an undirected network $G(V, E)$ is the number of connections (edges) it has to other nodes in the network. We denote the degree of node $i \in V$ by k_i.

For an undirected network of N nodes, we can compute the degree of each node by summing over the columns of the unweighted adjacency matrix $A = (a_{ij})$ for each row i as

$$k_i = \sum_{j=1}^{N} a_{ij}. \tag{4}$$

Going back to the example network in Fig. 5, since each row of the adjacency matrix A corresponds to a node in the network, if we sum up the 1s in the first row, we get 2 which means that the degree of node 1 is 2. This agrees with the figure since node 1 has two edges. Likewise, the degree of node 2 is 3, and so on.

Definition 3 The *degree distribution* of an undirected network $G(V, E)$ is a function $P(k)$, where $k = 0, 1, 2, \ldots$ that defines the fraction of nodes in the network with degree k.

To see this, note that if there are N nodes in the network and a subset N_k of them have degree k, then $P(k) = N_k/N$. For the example in Fig. 5, $N = 4$ and the degree distribution is

$$P(1) = \frac{1}{4}, \quad P(2) = \frac{2}{4}, \quad P(3) = \frac{1}{4}. \tag{5}$$

The value $P(k)$ can also be thought of as a probability, specifically the probability that a randomly chosen node in the network has degree k. This viewpoint will become useful when we study network models in Sect. 4.

Once we know the degree distribution of a network (i.e., k_i for each $i \in V$), we can compute the mean degree of the nodes in the network. For an undirected network $G(V, E)$ with $|V| = N$ nodes and $|E| = M$ edges, the mean degree is

$$c = \frac{1}{N} \sum_{i=1}^{N} k_i = \frac{2M}{N} \tag{6}$$

since each edge contributes to the degrees of the two nodes it connects. Hence, each edge is counted twice and we get twice the total number of edges in the numerator above.

One last important network property is the notion of a path between any two given nodes, defined below.

Definition 4 A *path*[3] $\mathscr{P}_{i_0,i_n}$ in an undirected network $G(V, E)$ is a sequence of nodes $i_0, i_1, \ldots, i_n \in V$ starting at i_0 and ending at i_n such that every consecutive pair of nodes in the sequence is connected by an edge in the network. Edges can be traversed in either direction.

In general, nodes and edges can occur more than once in a path. The *path length* of a given path is calculated by counting the total number of edges in the path. If an edge is traversed more than once, it is counted each time it is traversed.

We have a *closed path* if the path ends at the same node at which it started, or $i_0 = i_n$ in $\mathscr{P}_{i_0,i_n}$. A closed path can include repeated nodes and edges.

Definition 5 A *cycle* is a closed path in which all nodes and edges in the path are distinct.

Now that we have defined a path in an undirected network, we are able to provide formal definitions of connected and disconnected networks in the undirected case.

Definition 6 An undirected network is *connected* if there exists a path connecting any two nodes in the network. Alternatively, an undirected network is *disconnected* if it is not connected.

---

[3]There are different names for *path* in the literature. The definition given above is also called a *walk*. A *simple path* refers to a path in which all nodes are distinct. Terminology depends on the source.

In Sects. 3.3 and 3.4, we will dive deeper into metrics used to quantify network structure, noting that several of these metrics are based on path length. Further characterization of network structure involves quantifying connectivity properties such as clustering. Intuitively, clustering relies heavily on the notion of a cycle. The more cycles a network has, the more complex it is in terms of structure.

If a network contains no cycles, then it is *acyclic*. A special type of acyclic network is called a tree.

Definition 7 A *tree* is a connected undirected network that is acyclic.

The connected network example in Fig. 3 has several cycles, but the disconnected network example has no cycles. As a result, each piece of this disconnected network is a tree. There are numerous examples of trees in nature, such as river networks and human arteries or veins (but not both or cycles would be present).

The exercises given below and at the end of each section in this chapter are intended to be done first by hand and then by using R (if applicable) as a way to introduce the software by solving simple tasks (and by checking your work). The exercises in this section focus on the example networks given in Figs. 3 and 4; these networks are intentionally simple ($N = 10$ nodes in each case) so that computations can be done by hand. Once students become comfortable using R, these exercises can be repeated in R, perhaps also for larger example networks generated in R or from other sources.

Exercise 1 Find the adjacency matrix for each network shown in Fig. 3.

Exercise 2 Compute the node degrees for each network given in Figs. 3 and 5, and compute their degree distributions. In addition, compute the mean degree for each network.

Exercise 3 Give an example of a path from each of the networks shown in Figs. 3 and 5.

Exercise 4 For the connected network in Fig. 3, list all cycles of length 3 and length 5. List all cycles along with their lengths for the network in Fig. 5.

3.2 Mathematics of Directed Networks

Here we show how the theory developed in Sect. 3.1 changes for directed networks. We start with the adjacency matrix.

Definition 8 The *adjacency matrix* $A = (a_{ij})$ of a directed network $G(V, E)$, where $|V| = N$ is an $N \times N$ matrix such that each entry a_{ij} is either 1 or 0, corresponding to the presence or absence of a directed edge from node i to node j, respectively. In other words,

Fig. 6 **Left:** a directed
network with 4 nodes and 4
edges. **Right:** corresponding
adjacency matrix

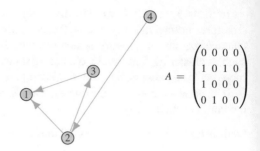

$$A = \begin{pmatrix} 0 & 0 & 0 & 0 \\ 1 & 0 & 1 & 0 \\ 1 & 0 & 0 & 0 \\ 0 & 1 & 0 & 0 \end{pmatrix}$$

$$a_{ij} = \begin{cases} 1, & \text{if } (i, j) \in E \\ 0, & \text{if } (i, j) \notin E. \end{cases} \tag{7}$$

Here, the edge set E is a set of ordered pairs (i, j) such that there is a directed edge
from i to j; see Eq. (1).

Recall that for undirected graphs, A is a symmetric matrix. However, this is not
necessarily true for directed graphs since it is possible to have an edge from node i
to node j but not vice versa. In that case, we would have $a_{ij} = 1$ but $a_{ji} = 0$, and
hence $A \neq A^T$. See Fig. 6 for a directed network along with its adjacency matrix A.
Note that A is not symmetric.

The notation used in Definition 8, where the edge direction runs from the first
index to the second index, is frequently used in the literature; see [11], for instance.
Other references, such as [16, 33], define the adjacency matrix entry a_{ij} to be a
directed edge from node j to node i which is the reverse order of the definition
above. These differences in notation are something to keep in mind when reviewing
the literature or when using software to calculate various network metrics.

Next, node degree has a more complex meaning in directed networks. In a
directed network, each node has two different degrees: the *in-degree*, or the number
of incoming edges to that node, and the *out-degree*, or the number of outgoing edges
from that node. We use the adjacency matrix to compute the in-degrees and out-
degrees for each node i in the network as

$$k_i^{\text{in}} = \sum_{j=1}^{N} a_{ji}, \qquad k_i^{\text{out}} = \sum_{j=1}^{N} a_{ij}. \tag{8}$$

The *total degree* of a node in a directed graph is the sum of its in-degree and out-
degree:

$$k_i^{\text{tot}} = k_i^{\text{in}} + k_i^{\text{out}}. \tag{9}$$

For a directed network with N nodes and M edges, the mean in-degree and mean
out-degree are equal,

$$c^{in} = \frac{1}{N}\sum_{i=1}^{N} k_i^{in} = \frac{1}{N}\sum_{i=1}^{N} k_i^{out} = c^{out} = \frac{c}{2} = \frac{M}{N}. \tag{10}$$

This equality ($c^{in} = c^{out}$) comes from the fact that a directed edge from any node must point to another node in the network, so on average the in- and out-degrees must be equal.

The definition of a path between any two nodes is also modified for directed networks. A *directed path* is a path (see Definition 4) with the additional property that in a directed network, each edge in the path must be traversed in the direction defined by that edge. A cycle in a directed network is given by Definition 5 with directed path instead of path.

Lastly, connectivity is more complex in the directed network setting. Here we introduce two different notions of connectivity for a directed network: weakly connected and strongly connected.

Definition 9 A directed network is *weakly connected* if replacing all directed edges with undirected edges yields a connected undirected network. A directed network is *strongly connected* if there exists a directed path from node i to j for every pair of nodes i, j in the network. Alternatively, a directed network is *disconnected* if it is not weakly connected.

Every strongly connected network is also weakly connected, but not vice versa. Recall the directed network in Fig. 4 from the beginning of Sect. 3. If all edges were undirected, this network would be connected and so it is a weakly connected directed network. It is not strongly connected, however, since there is no directed path from node 1 to node 4, for example.

Exercise 5 Find the adjacency matrix for the network shown in Fig. 4.

Exercise 6 Compute the node in-degrees and out-degrees, mean in-degree, and mean out-degree for the networks given in Figs. 4 and 6.

Exercise 7 Give an example of a path from each of the networks shown in Figs. 4 and 6.

Exercise 8 For the network in Fig. 4, list all cycles along with their lengths.

3.3 Network Metrics

There are numerous metrics used to quantify network structure. The first set of metrics we will discuss are based on path length. In the previous two sections, we defined the notion of a path between any two nodes in a network (undirected or directed), and we talked about computing the path length. We can use path length to compute the distance, in terms of number of edges traversed, between any two nodes in a network. However, we need to be more specific. Which path are we talking about? There may be several different paths between a pair of nodes in a network.

Of broad interest is the *shortest path* between two nodes in a network, also known as the *geodesic path*.

Definition 10 Given a network $G(V, E)$, a *geodesic path* between nodes i and j in the network is the shortest path, in terms of number of edges, between i and j.

A widely used metric is the length of the geodesic path between a pair of nodes in the network, also known as the *geodesic distance* or *shortest distance* in the network.

Definition 11 Given a network $G(V, E)$, the *geodesic distance* between nodes i and j in the network, denoted by $d(i, j)$, is the length (in number of edges) of the geodesic path between i and j.

The geodesic distance between a node i and itself, $d(i, i)$, is defined to be 0. If two nodes are in different parts of a disconnected network such that there is no path connecting them, then the geodesic distance between them is defined to be infinite. Note that geodesic paths are not necessarily unique since it is possible to have two or more paths of the same length. In a directed graph, the geodesic path from node i to node j might not agree with the geodesic path from node j to node i. Hence, it is possible and actually quite likely that $d(i, j) \neq d(j, i)$ in that case. In undirected graphs, however, the geodesic paths between two nodes and their associated lengths must agree.

There are several related metrics that are also used to quantify network structure. First is the *average geodesic distance* between any two nodes in the network, also called the *average path length* and implicitly means the average geodesic path length. This metric is computed by first computing the shortest distance between each pair of nodes in the network and then by taking the mean over these distances. Mathematically, this is given by

$$\bar{d}_i = \frac{1}{N} \sum_{j=1}^{N} d(i, j). \tag{11}$$

This definition includes the possibility that $i = j$, but we know that $d(i, i) = 0$ so it is not contributing to the sum.[4] If the network is disconnected, then there are at least two nodes with no path between them, and hence their path length is infinite. It follows that the average path length in this case is infinite.

The average path length metric tells us, on average, the number of steps (edges) it takes to get from one node in the network to another. One can interpret this metric

---

[4]Some researchers prefer to exclude the case $i = j$ and instead use $\bar{d}_i = \frac{1}{N-1} \sum_{j=1, j \neq i}^{N} d(i, j)$. The difference in the leading factors becomes negligible in large graphs. See [33, chapter 6] for further discussion of these two definitions.

as a measure of efficiency of mass transfer in a metabolic network, for instance. It turns out that many real-world networks have a short average path length.

The second related metric is known as the *network diameter*.

Definition 12 Given a network $G(V, E)$, the *network diameter* is the longest geodesic path between any pair of nodes in the network for which a path exists.

In other words, this is the shortest distance between the two most distant nodes in the network. Of interest is studying how the diameter of a network depends on its size, or number of nodes N.

Exercise 9 Compute both the shortest and longest geodesic paths for the networks given in Fig. 3, and give the geodesic distance for each of these paths. What is the network diameter in each case?

Exercise 10 Repeat Exercise 9 for the directed network in Fig. 4.

3.4 Centrality Measures

The next set of measures defined in this section capture different features of network structure than the metrics given in the previous section, which were based on path length. To date, a lot of research on networks has focused on understanding what nodes are the most important or most central in a network. This idea of centrality has led to the development of an array of *centrality measures* which we describe below. For simplicity, we only consider undirected networks here.

Perhaps the most straightforward measure of centrality in a network is node degree, defined as the number of edges connected to a node in a network (see Sect. 3.1, Definition 2). Node degree is sometimes also called *degree centrality*, and this is a widely used measure in network theory. Degree centrality is a ranking of nodes such that nodes with more connections are ranked higher; the node with the largest degree has the highest degree centrality.

Definition 13 Given a network $G(V, E)$, the *normalized degree centrality* is defined for each node $i \in V$ by

$$C_d(i) = \frac{k_i}{\max_j k_j},$$

where k_i is the degree of node i and the denominator is the maximum node degree in the network.

Other related centrality measures we will consider are betweenness centrality, closeness centrality, and transitivity. There are many other centrality measures that are in the literature, but we will not explore them all here. See [33, chapter 7] for a more complete and detailed discussion of such measures.

Betweenness centrality measures the extent to which a node lies on paths between other nodes. In other words, how often is a node between two other nodes in the sense that it lies on the geodesic path between those two nodes? How often does this happen between any two nodes in the network? Such nodes may act as bridges connecting different regions of the network.

Definition 14 Given a network $G(V, E)$, *betweenness centrality* of node $i \in V$ is defined by

$$C_b(i) = \sum_j \sum_k \frac{g_{jk}(i)}{g_{jk}},$$

where $g_{jk}(i)$ is the number of geodesic paths from node j to node k that pass through node i, and g_{jk} is the total number of geodesic paths from j to k. This measure is defined for all nodes $i \in V$.

We adopt the convention that $g_{jk}(i)/g_{jk} = 0$ if both $g_{jk}(i)$ and g_{jk} are zero.

Nodes with a high degree or nodes that are between other nodes are noteworthy characteristics in terms of network structure. What about nodes that are close to many other nodes in the network? Those nodes may or may not also be between many other nodes. This is the idea of *closeness centrality*. This measure uses the geodesic distance from a node to all other nodes in the network. Recall the average geodesic distance of a node, defined in Eq. (11). Taking the inverse of this measure yields closeness centrality.

Definition 15 Given a network $G(V, E)$ with $|V| = N$, *closeness centrality* of node $i \in V$ is defined by

$$C_c(i) = \left[\frac{1}{N} \sum_{j=1}^N d(i, j) \right]^{-1} = \frac{N}{\sum_{j=1}^N d(i, j)},$$

where $d(i, j)$ is the geodesic distance (in number of edges) from node i to other nodes $j \in V$.

In other words, closeness centrality is the inverse of the average shortest path length between a node and all other nodes in the network. Recall that $d(i, i) = 0$ for each node $i \in V$ by definition. Also recall that $d(i, j) = \infty$ if i and j belong to separate (disconnected) parts of the network. In disconnected networks, we use the convention that $1/\infty \equiv 0$ to compute this metric.

Recalling Eq. (11), there is an alternate version of closeness centrality that excludes the term $i = j$ in the sum. This changes the factor $1/N$ to $1/(N-1)$ which hardly has any influence for large graphs. Also, we are typically only interested in the relative centralities of the different nodes rather than absolute values, so in this chapter, we will use the version given in Definition 15. This is the version commonly used in many references [33].

The last measure we will define is known as *transitivity* which is a measure of clustering within a network, and is widely used in network theory. Transitivity means that if node i is connected by an edge to node j, and node j is connected by an edge to node k, then node k is connected by an edge to node i. This is a *cycle of length three*, the smallest cycle possible in a network with no self-loops. Recall that a cycle is a closed path in which all nodes and edges are distinct; see Definition 5. A *path of length two* is given by the sequence of nodes i, j, k with edges (i, j) and (j, k). A path of length two is also called a *triple* since three nodes are joined together by a path consisting of two edges. Then a cycle consisting of three nodes is called a *closed triple*. These concepts are used to define an important structural metric called the *transitivity coefficient*, more commonly known as the *clustering coefficient* [33].

Definition 16 The *clustering coefficient*[5] C is the ratio of the number of closed triples to the total number of triples:

$$C = \frac{\text{number of closed triples}}{\text{number of all triples (open and closed)}}.$$

This is the frequency of closed triples in the network, or the fraction of triples that are closed.

Moreover, we can calculate the number of closed triples as $3 \times$ the number of *triangles* in the network. The factor of 3 is due to the fact that each triangle is associated with three nodes, and each triangle gets counted three times when we count up the number of closed triples. For instance, the triangle ijk contains the triples ijk, jki, and kij. See Fig. 7 for an illustration of triangles and triples. Therefore, we have an equivalent definition of the clustering coefficient:

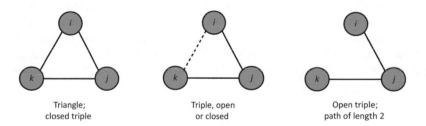

| Triangle; closed triple | Triple, open or closed | Open triple; path of length 2 |

Fig. 7 Illustrations of triples and triangles. **Middle:** a triple is a path of length two connecting three nodes i, j, and k. The presence or absence of edge (i, k) yields a closed or open triple, respectively. **Left:** a triangle ijk consists of three closed triples: ijk, jki, and kij. **Right:** an open triple or a path of length two such that edge (i, k) is absent

---

[5]This is also known as the *global clustering coefficient*. Some authors, most notably Watts and Strogatz [46], use a different definition known as the *network average clustering coefficient* which is the average of node (local) clustering coefficients. This alternate definition does not always give the same value as the global version.

$$C = \frac{3 \times \text{number of triangles}}{\text{number of all triples}}. \tag{12}$$

Since the above two definitions of C are equivalent, it is up to the reader to decide which version to use.

The clustering coefficient ranges between 1 for perfect transitivity (all nodes are connected in triangles) and 0 for a network with no triangles. The latter is the case for networks such as trees, which have no cycles (see Definition 7).

For further reading and a more detailed coverage of network metrics and centrality measures, see [33, chapters 6–7] and [11, chapter 1].

Exercise 11 Compute each measure of centrality (degree, betweenness, and closeness) for the connected network given in Fig. 3 (or Fig. 5 if working by hand).

Exercise 12 Compute the clustering coefficient for the connected network given in Fig. 3.

Exercise 13 Compute the three measures of centrality and the clustering coefficient for the disconnected network given in Fig. 3.

4 Network Models

In this section we will describe a variety of different network models. In Sect. 4.1 we describe some special types of graphs that are fundamental objects in graph theory, and in Sect. 4.2 we describe three classes of random graphs. By *random graph*, we mean that there is a probability distribution describing the family of graphs produced under the random graph model considered, and that any given graph constructed from that model is a *realization* from the family of such graphs. Random graphs are often used in biological applications because they are thought to be more realistic representations of biological systems. Lastly, we describe a class of networks whose structure changes over time, known as adaptive networks. In this case, since the dynamic topology is coupled with the dynamics of the process on the network, we defer this topic to the end of Sect. 5.

4.1 Special Types of Graphs

We focus on undirected graphs here and in the rest of Sect. 4 for simplicity. The first type of graph we will discuss is one in which all nodes have the same degree. This is called a *regular graph* in graph theory. In particular, a *k-regular graph* is a regular graph in which all nodes have degree k. Special cases of a regular graph include a *ring graph* which means the nodes are arranged in a ring and connected to their two nearest neighbors, and a *complete graph* which means that every node is connected

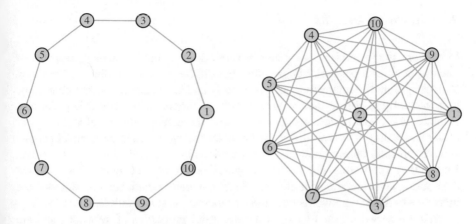

Fig. 8 Left: a ring network with 10 nodes; each node is connected to its 2 nearest neighbors. **Right:** a complete graph with 10 nodes; each node is connected to the other 9 nodes in the graph

to the other $N - 1$ nodes in the graph (where the graph has N nodes total). Figure 8 shows examples of a ring network and a complete network, each with $N = 10$ nodes. Note that in these examples, the ring network is also a 2-regular graph and the complete network is also a 9-regular graph.

Another special case of a regular graph is known as the *r-regular ring graph*, meaning that each node is initially connected only to its r nearest neighbors ($r/2$ nearest neighbors on each side). The r-regular ring network, in particular, will come into play when we discuss small-world networks in the next section. See the left panel of Fig. 10 for an illustration of a 6-regular ring network; each node is connected to its 3 nearest neighbors on each side. In general, the structure of regular networks is well-known and can be used to compare the structure of different random networks.

To generate the networks shown in Fig. 8, use the function `make_ring(N)` to make a ring graph on N nodes and the function `make_full_graph(N)` to make a complete graph on N nodes. Both of these functions are in the `igraph` package in R; see Appendix for the code used in Fig. 8.

Exercise 14 Compute the shortest and longest geodesic paths for the ring and complete networks given in Fig. 8. What is the network diameter in each case?

Exercise 15 Compute each measure of centrality (degree, betweenness, and closeness) for the ring and complete networks given in Fig. 8.

Exercise 16 Compute the clustering coefficient for the ring and complete networks given in Fig. 8.

4.2   Random Graphs

In *random* graph models, we generate networks that have specific properties of interest, such as a particular degree distribution, but which are otherwise random. Random graphs are interesting objects to study and have become an important topic in advanced probability theory as well as in a variety of biological applications. In biological applications, we may not know the actual network of interest. For example, we may not know all details the contact network of a large group of people, and we typically do not know all the connections between all neurons in the brain. But we may know some characteristic properties of these networks, such as degree distribution and clustering coefficients. So we can gain some insight by studying the dynamics on typical networks with these properties, that is, random networks.

Random networks shed light on the structural properties of networks, and are used as a template for modeling processes on networks. In Sect. 5, we study different biologically inspired processes on networks such as epidemics and neuronal activity. In this section, we will explore three important random graph models: the Erdős–Rényi model, the small-world model, and the scale-free model. For simplicity, we only consider undirected graphs in our descriptions below.

Each random graph model describes a family or *ensemble* of graphs and rules for how to generate a representative graph (or a *realization*) from the model. Each model starts with a set of N nodes and then edges are added to connect the nodes according to a random process. For more details on the random graphs described in this section and related network models, see [14, 33, 42].

4.2.1   Erdős–Rényi Random Graph

We begin with one of the most fundamental and widely studied random graph models which was introduced by Erdős and Rényi [21, 22]. The *Erdős–Rényi (ER) random graph model* assumes that, for a given set of nodes, any possible edge occurs independently of all other possible edges with a given probability. More precisely, the model starts with N nodes and the network is generated by connecting each pair of nodes with probability p, independently of all other pairs of nodes in the network. The family of ER random graphs with N nodes and edge probability p is denoted by $G(N, p)$. See Fig. 9 for an illustration of an ER random graph with $N = 30$ nodes and edge probability $p = 0.2$.

In the ER model, a node is equally likely to be connected to each of the $N - 1$ other nodes in the network. As a result, the probability $P(k)$ that a node has degree k is given by the binomial distribution

$$P(k) = \binom{N-1}{k} p^k (1-p)^{N-1-k} \quad \text{for } k = 0, 1, \dots, N-1. \tag{13}$$

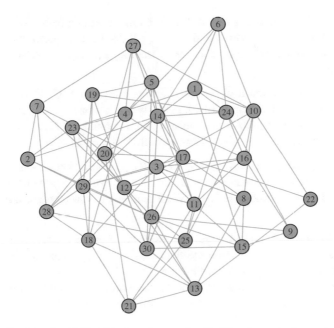

Fig. 9 A realization of an Erdős–Rényi random graph with 30 nodes and edge probability 0.2

In other words, $G(N, p)$ has a binomial degree distribution. Now we can easily compute the expected value of the degree of a node, also known as the mean degree, which is $c = (N - 1)p$. This result is intuitive; the mean number of edges connected to a node is the product of the number of possible edges $(N - 1)$ and the probability that an edge actually exists (p). The mean degree of the ER network shown in Fig. 9 is approximately 5.8 which agrees with the theory.

An interesting mathematical result (see [33, chapter 12] for more details) shows that in the limit as the number of nodes goes to infinity,

$$P(k) = \binom{N - 1}{k} p^k (1 - p)^{N-1-k} \rightarrow e^{-c} \frac{c^k}{k!} \quad \text{as } N \rightarrow \infty, \tag{14}$$

where $c = (N - 1)p$ is the mean degree derived above and c is assumed to be constant. The limiting distribution is the Poisson distribution with parameter c. Thus, if we have a very large ER network, then its degree distribution is approximately Poisson. For more details on the Poisson distribution and other probability distributions, see [39].

To generate an ER random graph in R with the `igraph` package, use either function `sample_gnp(N,p)` or `erdos.renyi.game(N,p)`, where N is the number of nodes in the graph and p is the edge probability. Then use `plot` to visualize the graph. See Appendix for the code used to generate Fig. 9. For further reading on Erdős–Rényi random graphs, see [33, chapter 12] and [11, chapter 3].

Exercise 17 Generate a realization of an ER random graph from $G(N = 30, p = 0.2)$ in R. Use the functions in the `igraph` package to compute the degree distribution of your graph and visualize by making a histogram. Also compute the mean degree.

Exercise 18 Generate ten different realizations drawn from $G(N = 30, p = 0.2)$ and plot their associated degree histograms. What do you notice? Are some nodes more connected than others? Is the entire network connected? How often do you get a realization that is disconnected? How variable is the mean degree?

Exercise 19 Explore realizations of $G(N = 30, p)$ for different choices of the edge probability p. How does changing p affect the shape of the degree distribution, the mean node degree, and connectivity? What happens if you increase the number of nodes N?

Challenge Problem 1 Generate several realizations of an ER random graph from $G(N = 30, p = 0.2)$ in R. Use the functions in the `igraph` package (or write your own code) to compute the following for each realization: the lengths of the shortest and longest geodesic paths in the network, the average path length, the three measures of centrality, and the clustering coefficient for the network. What patterns do you observe?

4.2.2 Small-World Networks

Another class of random networks known as small-world networks were introduced by Watts and Strogatz [46] after they noticed that many naturally occurring networks displayed greater clustering than an Erdős–Rényi random network. Small-world networks are characterized by a high degree of clustering as well as a short average path length between any two nodes in the network. In particular, an edge in a small-world network is more likely to occur if it would complete a triangle, as described in Sect. 3.4.

The *small-world model* by Watts and Strogatz [46] can be described as follows: Start with N nodes initially connected as an r-regular ring network, where each node is initially connected only to its $r/2$ nearest neighbors on each side. The next step introduces a *rewiring probability* q which means that some edges are removed and then rewired to random new positions. Specifically, we look at each edge in the ring network, and with probability q we remove that edge and replace it with one that joins two nodes chosen uniformly at random. This leads to a network with *shortcuts*, also known as *long-range connections*, which is what gives small-world networks their characteristic short average path lengths.

Figure 10 shows a representative small-world network for $r = 6$ in the center panel with rewiring probability $q = 0.05$; this network has about 5% of its edges rewired compared to the initial 6-regular ring network. The left and right panels, respectively, show networks representing the two extreme cases of no rewiring ($q = 0$, ring network with $r = 6$) and complete rewiring ($q = 1$, equivalent to

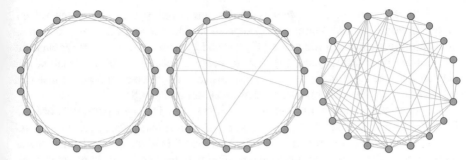

Fig. 10 Small-world networks generated from the Watts–Strogatz model [46] with 20 nodes, each node is initially connected to its six nearest neighbors ($r = 6$) and then the graph is rewired with a specified probability. From left to right, the rewiring probability is $q = 0$, 0.05, and 1 which correspond to a 6-regular ring graph, a characteristic small-world graph, and a graph with complete rewiring (same as ER random graph), respectively

a realization from the ER model with the same edge probability as the small-world network).

A variant of the Watts–Strogatz (WS) small-world model is to not remove any edges during the rewiring step, but only add the shortcut edges to the initial r-regular ring network (as in the original model, with probability q per edge). The two versions are similar in the regime that is of interest to most researchers, the case where q is small. The reason this variant exists is because it is more straightforward to calculate network properties such as degree distribution in this alternate version [33]. We will use the alternate version below when discussing the degree distribution.

In order to describe the degree distribution for the small-world model in which no edges are removed, we need to compute the number of shortcuts that are added to a node. Once shortcuts are added to the ring network, the degree k of a node is r (the number of initial nearest neighbor connections) plus the number of shortcut edges s attached to it. There will be on average rq shortcuts that end at any given node. The number of shortcuts attached to any one node follows a Poisson distribution with mean rq, so that the probability a node has s shortcuts is approximately (for large N and small q)

$$P(s) = e^{-rq} \frac{(rq)^s}{s!} \quad \text{for } s = 0, 1, \ldots \tag{15}$$

Recall that the total degree of a vertex is $k = r + s$. Solving for s yields $s = k - r$ and plugging this into Eq. (15) gives the degree distribution for the small-world model:

$$P(k) = e^{-rq} \frac{(rq)^{k-r}}{(k-r)!} \tag{16}$$

for $k \geq r$ and $P(k) = 0$ if $k < r$. For more details on the derivation of the degree distribution, see [33, chapter 15].

Small-world networks are often used in models of biological systems due to their more realistic structural features as compared to other network models. Examples include brain (neuronal) networks [12, 45] and protein networks that regulate fat metabolism in budding yeast [4], just to name a few. In addition, small-world networks can be used in models of social networks upon which infectious diseases spread [27, 28]. We will consider epidemics on networks in Sect. 5.

To generate a small-world network in R (via the igraph package), use the function sample_smallworld(1,N,h,q) where the parameters correspond to the notation used above and $h = r/2$. Note that this is the original small-world model where each edge is removed and rewired with probability q. In particular, the first parameter is set to 1 so that initially we start with a 1-dimensional r-regular ring graph, and the second parameter is N since the graph consists of N nodes total. The third parameter h is the number of nearest neighbors a node is initially connected to on each side so that it has $2h = r$ initial nearest neighbors. The last parameter q is the rewiring probability discussed above. See Appendix for the code used to generate Fig. 10. For further reading on small-world networks, see [33, chapter 15], [11, chapter 3], and [46].

Exercise 20 Generate a realization of a small-world network with $N = 20$, $r = 6$, and $q = 0.05$ in R. Use the functions in the igraph package to compute the degree distribution of your graph and visualize by making a histogram. Also compute the mean degree.

Exercise 21 Generate ten different realizations from the model in Exercise 20 and plot their associated degree histograms. What do you notice? Are some nodes more connected than others? How variable is the mean degree?

Exercise 22 Explore realizations from the small-world model for different choices of rewiring probability q. How does changing q or r affect the shape of the degree distribution, the mean node degree, and connectivity?

Exercise 23 Compute the clustering coefficient for the network in Exercise 20. Explore how the clustering coefficient changes as q increases or decreases.

Challenge Problem 2 Generate several realizations of a small-world network from the Watts–Strogatz model with parameters $N = 20$, $r = 6$, and $q = 0.05$ in R. Use the functions in the igraph package to compute the lengths of the shortest and longest geodesic paths in the network and the three measures of centrality for each realization. Repeat for small-world networks with rewiring probabilities $q = 0$ and $q = 1$ for comparison.

4.2.3   Scale-Free Networks

The last class of networks that we investigate are known as *scale-free networks*, and they are an important subclass of the small-world networks discussed in Sect. 4.2.2 [8, 10, 33]. Scale-free networks have a degree distribution that follows a *power law* (at least asymptotically). By asymptotically, we mean that as the number of nodes in the network tends to infinity, the degree distribution approaches a power law relationship. That is, the probability that a node has degree k is given by

$$P(k) \sim k^{-\alpha}, \tag{17}$$

where the exponent $\alpha > 0$. In other words, for large values of k the fraction of nodes of degree k is proportional to $k^{-\alpha}$.

Empirical studies have found such distributions with α in the range $2 \leq \alpha \leq 3$ for many important networks that occur in nature [6, 17, 33], including neuronal networks [5, 32] and contact networks of people [29]. As a result, the scale-free nature of various biological networks makes for an interesting topic of research.

The most striking feature of scale-free networks is the presence of nodes that have very high degree, much higher than the average degree of a node in the network. The highest degree nodes are known as *hubs*. The power law degree distribution is a *heavy-tailed* distribution which means that the tail of the distribution is heavier (tends to 0 slower) than an exponential distribution. A heavy tail is what allows hubs to exist. In addition, scale-free networks have short average path length (similar to small-world networks), but the amount of clustering is closer to that of an ER random graph rather than a small-world network. In contrast, small-world networks generally have a higher clustering coefficient than scale-free or ER random networks.

The most popular model for generating scale-free networks is the *preferential attachment model*, developed by Barabási and Albert [10]. This model was proposed to explain the appearance of power-law distributions in real networks. We will also refer to this model as the *Barabási–Albert (BA) scale-free network model*.

The BA scale-free network model proceeds as follows: We start with a small connected network of $m_0 \geq 2$ nodes and add new nodes to the network one at a time until we have a total of N nodes. The probability that the new node will be connected to an existing node i is given by

$$p_i = \frac{k_i}{\sum_j k_j}. \tag{18}$$

This probability is proportional to the degree k_i of node i. High degree hubs tend to quickly accumulate more links, while nodes of small degree are much less likely to gain new connections. Hence, new nodes have a so-called preference to attach themselves to hub nodes, and this explains the name "preferential attachment" in the model.

Figure 11 shows a realization of a scale-free network with $N = 50$ nodes. We can easily see that node 1 has the highest degree in the network, with 16 neighbors. Hence, node 1 is a hub in this network. In larger scale-free networks with $N > 100$, we would likely find several hubs.

To generate a BA scale-free network in R (igraph package), use the function sample_pa(N, directed = FALSE, out.pref = TRUE) making sure

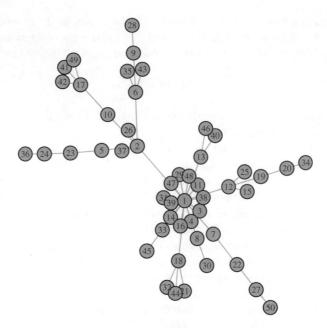

Fig. 11 A realization of a Barabási–Albert scale-free random graph with 50 nodes. Notice that node 1 has a much higher degree than the other nodes, illustrating the hub structure characteristic of scale-free networks

to change the options as indicated. Then use `plot` to visualize the graph. Note that the `sample_pa()` function has the argument `directed = TRUE`, meaning that by default it generates a directed network. Changing this setting to `directed = FALSE` generates an undirected network, and this setting can be changed in ER network generation as well. Alternatively, the function `as.directed(G)` converts any undirected graph G to its strongly connected version (for each component if disconnected). See Appendix for the code used to generate Fig. 11. For further reading on scale-free networks and the preferential attachment model, see [33, chapter 14].

Exercise 24 Generate by hand a realization of a scale-free network with $N = 10$ nodes using the preferential attachment model that starts with $m_0 = 2$ nodes connected together by an edge. Add new nodes one at a time until the network has 10 nodes total, and at each time step, use Eq. (18) (and a suitable device for generating random numbers) to choose which existing node the new node will attach to.

Exercise 25 Generate a realization of a scale-free network with $N = 50$ in R. Use the functions in the `igraph` package to compute the degree distribution of your graph and visualize by making a histogram. Also compute the mean degree.

Exercise 26 Generate ten different realizations from the model in Exercise 25 and plot their associated degree histograms. What do you notice? Are some nodes more connected than others? How variable is the mean degree?

Exercise 27 Use the functions in the `igraph` package to compute the clustering coefficient and the length of the shortest geodesic path for the network you generated in Exercise 25.

5 Processes on Networks

Up to this point, we have focused on the structure of a given system as represented by the network topology. Now we consider the dynamics of the system which are determined by a dynamic model on a network such that the state variables associated with the nodes change over time. A dynamic model is a mathematical model describing how those state variables change over time. We will focus on stochastic process models that evolve on a network, but we can also consider deterministic processes (e.g., differential equation models) on a network within this framework. Here we will often refer to such dynamic models, either stochastic or deterministic, simply as processes on a network.

When studying a process on a network, a key question arises: To what extent is the network structure reflected in the dynamics of the process on the network? This question will be the driving force for the rest of this chapter, and will be of particular interest for some of the research projects detailed in Sect. 6.

In the following sections, we describe a variety of processes that take place on a network, and we do not focus on the specific structure of the underlying network. We are thinking of the network as a static template on which to explore changes in the state of the nodes as the process evolves on the network. The biologically inspired processes we study are percolation, epidemic spreading, and neuronal activity. The last part of this Sect. 5.4 explores adaptive networks in which both the process and the network structure are dynamic and change over time.

The aim here is to introduce students to several different processes on networks by providing the basic background needed to understand more about each topic. Indeed, many books have been written on each of these topics, and this chapter will only scratch the surface. The hope is that one or more of these examples will pique your interest, so at the end of each section, we provide additional resources for students interested in pursuing projects in these areas.

A brief overview of the basics of stochastic processes is given in the Appendix. For a good reference on introductory probability theory, see [39]. For introductory stochastic processes, see [19, 40], and for stochastic processes arising in biology, see [7]. For further reading about processes on networks in biology and other applications, see [11, 30, 33, 37, 42].

5.1 Percolation

Percolation processes are based on a spreading of activity throughout the network and one of the simplest processes to consider on a network [11, 37, 42]. There are various types of percolation processes, but we will only describe one specific type in this chapter. We focus on *bootstrap percolation* which is an activation process such that nodes become active once sufficiently many of their neighboring nodes are active [3, 13]. Bootstrap percolation can also be thought of as an infection process, and is often used in models of epidemics on networks.

More precisely, the process proceeds as follows: Given a network $G(V, E)$ such that $|V| = N$, each of the N nodes can be in one of two states: active or inactive. Initially, each node is active with probability ρ. Typically a small fraction of initially active nodes is used, for instance, 5% of the total number of nodes in the network N which corresponds to $\rho = 0.05$. In that case, for a network of size $N = 100$, we have on average five nodes that are initially active at time $t = 0$. We then define an updating rule for how activity spreads throughout the network as time goes forward in discrete time steps $t = 1, 2, 3, \ldots$. We keep track of the number of active nodes in the network at time t, denoted by $X(t)$. Bootstrap percolation uses a threshold rule which says that once a node has at least k active nearest neighbors, then it becomes active in the next time step. Lastly, once a node becomes active, it stays active and cannot revert back to the inactive state.

Note that this is a deterministic process that is discrete in time and space; the time index $t \in \{0, 1, 2, \ldots\}$ and $X(t)$ takes on non-negative integer values in the set $S = \{0, 1, \ldots, N\}$. In other words, $\{X(t) : t = 0, 1, 2, \ldots\}$ is a discrete-time process taking values in the discrete (i.e., finite) state space S. The only randomness comes from the probabilistic rules governing network generation, the associated degree distribution of the network, and the initial condition of each node starting out in the active state with probability ρ. Once the network is chosen, this is a deterministic dynamic model with trajectories completely specified by the initial conditions.

How can we compute the value of $X(t)$ at each time step t? First, we need to keep track of the state $x_i(t)$ of each individual node $i \in V$ at time t. We do this as follows: Let

$$x_i(t) = \begin{cases} 1, & \text{if node } i \text{ is active at time } t \\ 0, & \text{if node } i \text{ is inactive at time } t. \end{cases} \tag{19}$$

Suppose node i is inactive at time t (for $t \geq 0$); that is, $x_i(t) = 0$. If at least k of node i's neighbors are active at time t, then node i becomes active at time $t + 1$ and $x_i(t + 1) = 1$. Recall that once a node becomes active, it stays in the active state from then on. We can sum over the $x_i(t)$'s to compute the number of active states at time t, likewise for time $t + 1$, and so on. Specifically,

$$X(t) = \sum_{i=1}^{N} x_i(t), \quad \text{for } t = 0, 1, 2, \ldots. \tag{20}$$

This process depends on the threshold value k as well as the number of neighbors each node has. For $k = 1$, bootstrap percolation on a connected network with at least one initially active node will eventually reach the state of all nodes becoming active, meaning that $X(t) = N$ for some time t. Of interest is studying how the process changes for different values of k, and/or for different network topologies. This is the topic of suggested research project 2 in Sect. 6.

Bootstrap percolation on a network as described above can be simulated in R. For the case $k = 1$, the adjacency matrix of the network can easily be used to update the system at each time step. See Exercise 30 below for some details on how to simulate this process.

A related *stochastic process* that can evolve on a network is a *random walk*; see the Appendix for formal definitions of these two terms. A stochastic process is used to model the evolution of a system over time. A random walk describes the path of an individual walking on the nodes of a network over time. The path starts at a (possibly randomly chosen) initial node $i_0 \in V$ at time $t = 0$ for a given network $G(V, E)$ such that $|V| = N$. Now suppose that a node i is occupied at time t, and call this the current state. Then, given the current state, the move to the next state j at time $t + 1$ is chosen randomly, with equal probability, from the nearest neighbors of node i. The stochasticity in this random walk occurs when $k_i > 1$, where k_i is the degree of the currently occupied node i. This means that there is more than one possible state to move to in the next time step, and each state transition happens with probability $p_{ij} = 1/k_i$.

The random walk is a discrete-time process and is defined on the same state space $S = \{0, 1, \ldots, N\}$ as the bootstrap percolation process above. In the Appendix, we define a random walk on the state space of the integers; this process is denoted by $\{Y(t) : t = 0, 1, 2, \ldots\}$. We also provide R code to simulate this random walk, see Exercise 28 and Fig. 15 in the Appendix. See Exercise 29 for details on how to adapt this code to evolve on a finite network. Lastly, we define a related stochastic process $\{\tilde{Y}(t) : t = 0, 1, 2, \ldots\}$ to count the number of distinct nodes that have been visited in the random walk by time t. In other words, we can think of the $\tilde{Y}(t)$ as the number of active nodes at time t and compare this process to bootstrap percolation; see Challenge Problem 3 below.

For further reading on bootstrap percolation, see [3, 13]. For further reading on percolation processes in general or applied to biology, see [33, chapter 16] and [37, chapter 3]. More details on random walks can be found in chapter 4 of [40] and chapter 1 of [19].

Exercise 28 Use the R code in the Appendix (section on Selected Stochastic Processes) to simulate a random walk (RW) on the integers. Set the transition probability $p = \frac{1}{2}$ to simulate a symmetric RW and explore what happens for other values of $p \neq \frac{1}{2}$.

Exercise 29 Modify the random walk R code given in the Appendix (section on Selected Stochastic Processes) to evolve on a given connected network of N nodes, say $N = 10$. Recall that a random walk describes the path of an individual walking on the nodes of the network. Given the current state (say node i is occupied), the move to the next state is chosen randomly, with equal probability, from the nearest neighbors of node i. Run the simulation for 50 time steps and plot the sample path with time on the x-axis and the states on the y-axis.

Exercise 30 Write a script in R to simulate bootstrap percolation on a given network. For instance, generate and fix an Erdős–Rényi random graph $G(N = 10, p)$ (choose p small enough so that you get a disconnected network, say $p = 0.2$), and then simulate bootstrap percolation on the graph using $\rho = 0.1$ and a threshold value of $k = 1$. Run the simulation for 10 time steps or until the process reaches a stable state such that no further state changes can occur. Repeat with $k = 2$ or larger.

Challenge Problem 3 Generate a random network, say an Erdős–Rényi random graph $G(N = 100, p = 0.02)$. Fix the network and explore how bootstrap percolation differs from a random walk on the given network. Specifically, look at how quickly or slowly $\tilde{Y}(t)$ reaches the number of nodes that are eventually turned on by the percolation process (for $k = 1$) and how this number may remain different as $t \to \infty$ when $k > 1$.

5.2   Epidemics on Networks

A very interesting biological process that we can model on a network is the spread of an infectious disease over a so-called contact network [23, 28, 33, 37]. A *contact network* is a network model for how individuals interact with one another in a population. In particular, nodes represent individuals and edges represent interactions (direct contact) between individuals through which an infection can spread from individual to individual.

Before diving into epidemics on networks, we first briefly describe a few classical epidemic models with no associated network structure. These are known as *compartmental models* in which a set of compartments describes the states of the individuals, for instance, "susceptible," "infected," and "recovered," and parameters are used to describe the transition rates for individuals changing states [15]. The traditional modeling approach does not involve contact networks, but rather assumes that individuals in the population are *well-mixed*. This term means that each individual has an equal chance of coming into contact with every other individual in the population, a somewhat unrealistic representation of the real world. Compartmental models are typically described using ordinary differential equations (ODEs) which are deterministic models with continuous-state transitions.

The simplest compartmental model is known as the *Susceptible-Infected (SI) model* which partitions the population into two classes (or compartments): suscepti-

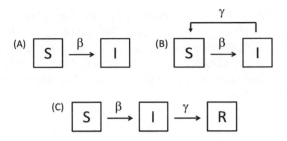

Fig. 12 Flowcharts for three compartmental models described in the text: (**a**) SI model, (**b**) SIS models, and (**c**) SIR model. "S" stands for susceptible, "I" stands for infected, and "R" stands for recovered; parameters β and γ refer to the transmission rate and recovery rate, respectively, for the infectious disease considered

ble (S) and infected (I). The model assumes that susceptible individuals can become infected when they come in contact with an infected individual (given a particular transmission rate β of the infection). Also, once an individual is infected with the disease, they remain infected for all time.

A slightly more complicated version of the SI model allows for individuals to recover (R) from the infection (given a particular recovery rate γ); this model is known as the *Susceptible-Infected-Recovered (SIR) model*. A third model, known as the *Susceptible-Infected-Susceptible (SIS) model* allows for individuals to become susceptible again once they have recovered from the infection. Hence, the SIS model is a good model for infectious diseases that do not yield long-term immunity such as the common cold or norovirus. See Fig. 12 for an illustration of the flowcharts which describe the set of epidemic states and the rates of transition between states (defined formally below) for each of these models.

In reality, most people have a regular set of contacts (e.g., family, neighbors, co-workers) with whom they interact. Therefore, using a contact network to model person-to-person interactions in an infectious disease model is more realistic than the well-mixed assumption given in the classical compartmental models discussed above [11, 33, 37].

As an extension of the classical models, here we consider the influence of network structure on the propagation of an epidemic [11, 23, 29, 30]. To do this, we place a compartmental model on a network so that each node belongs to one of the possible epidemic states, for instance, susceptible or infected corresponding to the SI model, and the nodes have updating rules for how their states change. We start with a random undirected network $G(V, E)$ with $|V| = N$. The nodes in the network represent individuals and each node's set of nearest neighbors are their possible contacts; this is the contact network on which an epidemic can spread.

The dynamics of an epidemic model on a network can be defined in different ways. Typically, one uses a stochastic process to model disease spread on a network due to the inherent randomness in how infectious diseases spread; see the Appendix for more details on stochastic processes. Here will consider a stochastic process in which the infection is transmitted from an infected node to a susceptible neighboring node at *rate* β. Then $\beta \Delta t$ is the *probability* of transmitting the infection within in a small interval of time Δt on a given edge that connects an infected node and a susceptible node. The *infection rate* for a given susceptible node depends on the number of infected neighbors it has, and this term is defined formally below in the SI network model. Note that this rate is used in each of the network epidemic models considered in this section.

5.2.1 SI Network Model

First consider the Susceptible-Infected (SI) model on a contact network $G(V, E)$ with $|V| = N$. The set of nodes V is partitioned into two compartments: susceptible (S) nodes $i \in S$ and infected (I) nodes $j \in I$. Moreover, the sets of nodes in each compartment are disjoint, i.e., $S \cap I = \emptyset$ and $S \cup I = V$. For simplicity, suppose initially that only one node is infected.

Now for a given susceptible node $i \in S$ in the network, we can count the number of infected neighbors it has at any given time. Suppose i has m infected neighbors at time t, where $m \leq k_i$ and k_i is the degree of node i. The probability that node i remains susceptible during a time interval of length Δt is

$$P(i \in S \to i \in S) = \prod_{j=1}^{m}(1 - \beta \Delta t) = (1 - \beta \Delta t)^m, \qquad (21)$$

where the product is taken over node i's m infected neighbors, and each neighbor fails to transmit the infection during that time with probability $1 - \beta \Delta t$.

Alternatively, since the only options in the SI model are for node i to remain susceptible or to become infected, the probability of node i becoming infected during a time interval of length Δt is

$$P(i \in S \to i \in I) = 1 - (1 - \beta \Delta t)^m \to \beta m \quad \text{as} \quad \Delta t \to 0. \qquad (22)$$

Therefore, we define the *infection rate* (or *transmission rate*) for a susceptible node with m infected neighbors to be βm [33, 37]. This is the continuous-time version of the stochastic process described above since we are letting the length of the time step Δt go to 0. One could also formulate this process as a discrete-time stochastic process with a small time step, and this is often done in practice to simulate the process.

Note that β is also the infection rate parameter used in the well-mixed epidemic models. One must take caution when reading about epidemic models on networks versus their well-mixed counterparts; these two β parameters are not equivalent. In the network case, β is the rate of contact that leads to infection between an infected node and a single susceptible node, whereas in the well-mixed case, β is the rate of contacts that lead to infection between an infected individual and all other individuals in the population.

If the contact network is a connected network, the ultimate fate of each node is to become infected after some time so that the entire network is in the infected state. The amount of time it will take for this to happen depends on the network structure and the infection rate describing the inherent randomness in the infection process. If the network is not connected, only the part of the network connected by a path to the initially infected node will become infected. In that case, the behavior of the process will depend on the location of the first infected node. The SI model is a good model for diseases that cannot be cured.

Lastly, we note an important connection between the SI model and bootstrap percolation with $k = 1$. These are different models because the latter is a deterministic process, while the former is a stochastic process. However, the common ground between these models is that the set of eventually infectious nodes will be exactly the same as the set of nodes that gets turned on by bootstrap percolation with $k = 1$.

5.2.2 SIR Network Model

A slightly more complicated model is the Susceptible-Infected-Recovered (SIR) model on a contact network $G(V, E)$ with $|V| = N$. Here V is partitioned into three compartments: susceptible nodes $i \in S$, infected nodes $j \in I$, and recovered (R) nodes $k \in R$. Note that S, I, and R are disjoint sets and that $S \cup I \cup R = V$. For simplicity, suppose initially that only one node is infected and that all other nodes are susceptible. The SIR model uses the same infection rate β (or βm for a susceptible node with m infected neighbors) as the SI model.

The SIR model allows individuals to recover from the infection, so now we need to define the *recovery rate* γ of an infected node. The transition of an infected node to the recovered state is commonly modeled as a stochastic process with rate γ that does not depend on the states of the neighboring nodes. As a result, each infected node moves to the recovered state at a constant rate γ. Hence, in a time interval of length Δt the probability that an infected node becomes recovered is $\gamma \Delta t$, and the probability that it remains infected is $1 - \gamma \Delta t$. Note that once individuals recover from the infection, they are no longer able to become infected. The SIR model is a good model for diseases that confer immunity.

An interesting quantity to measure in SIR network models is the *final size* of the infection which is the total number of individuals who have experienced infection over the course of the epidemic. Mathematically, this is the number of recovered individuals $R(t)$ as $t \to \infty$ [30].

5.2.3 SIS Network Model

Another related model is the Susceptible-Infected-Susceptible (SIS) model on a contact network $G(V, E)$ with $|V| = N$. This model assumes that infected individuals can recover from the disease and then become susceptible again. The set of nodes V is partitioned into two compartments as in the SI model, but we have the additional recovery rate γ given in the SIR model describing the transition of infected nodes back to the susceptible state. As above, we suppose initially that only one node is infected and that all other nodes are susceptible.

As in the SIR model, we use the same two transition rates: β and γ defined above because these parameters describe the rate that individuals leave states S and I, respectively. However, in the SIS model, the recovery rate γ now describes the transition back to the susceptible state. Therefore, in a time interval of length Δt, the probability that an infected node recovers from the infection and is once again susceptible to infection is $\gamma \Delta t$, and the probability that it remains infected is $1 - \gamma \Delta t$. Lastly, note that the SIS model is a good model for diseases that do not confer long-term immunity.

For further reading on epidemics on networks, see chapter 3 in [37], chapter 17 in [33], [27], [30], [11, chapter 9], and [23, section 3.3] as well as references therein.

Exercise 31 Discuss qualitative differences between the SIR, SIS, and SI network models. Assume a random undirected connected contact network. In particular, think about the possible final outcomes of an epidemic in the three models.

Exercise 32 Think about which epidemic model (SI, SIR, SIS) best fits several common infectious diseases, for instance, influenza, norovirus, herpes, and chicken pox. Describe an epidemic model for one of these infectious diseases.

Challenge Problem 4 Simulate the SI model on the ring graph with $N = 10$ nodes in R. Use $\beta = 0.2$ for the infection rate as discussed in the text above, i.e., if a susceptible node has m infected neighbors, then the infection rate is $\beta m = 0.2m$. Recall that in a ring graph each node has exactly two neighbors, thus at any given time for any given node, $0 \leq m \leq 2$. Start with 1 infected individual. How long does it take for all nodes to become infected?

Challenge Problem 5 Modify the simulation described in Challenge Problem 4 to evolve on the complete graph with $N = 10$ nodes.

5.3   Neuronal Activity

Neuronal activity refers to neurons in the brain firing *action potentials*, also called *nerve impulses* or *spikes*; action potentials play a key role in cell-to-cell communication. The activity of a single neuron excites or inhibits the activity of neurons connected to it, and the collective network activity directs the body to perform myriad tasks. It is important to note that neuronal connections are directed, and hence biologically realistic models of neuronal network activity should use directed graphs.

Here we describe a model of neuronal spiking activity on a generic directed network $G(V, E)$ that was developed in [42]. Note that this is more challenging than the previous models in this chapter. We think of the nodes as neurons and the edges as neuronal connections (or synapses), and we only consider excitatory connections. We assume that each node fires action potentials at random times according to a *Poisson process* with a given rate λ. The Poisson process is a well-known continuous-time stochastic processes in probability theory; see the Appendix (section on Selected Stochastic Processes) for a formal definition and R code to simulate the process. Figure 16 (see Appendix) shows an illustration of a sample path for the case $\lambda = 1$. Also see [40] for more details on this process.

Suppose that each node $i \in V$ fires action potentials over time according to a Poisson process with rate r_i. We interpret r_i as the firing rate of node i which can change over time in response to neighboring node activity. We assume that each node has a constant background firing rate r_0 (same for all nodes), and when node i is activated above its baseline, it decays back down to its baseline level according to the firing rate equation

$$\frac{dr_i}{dt} = \frac{1}{\tau_r}(r_0 - r_i), \tag{23}$$

where τ_r is a time constant. To model interactions with neighboring neurons, when node i fires an action potential, the firing rate of each of its neighboring nodes j (i.e., nodes j such that $(i, j) \in E$) gets a boost by a constant parameter $g > 0$. Hence, we formulate the updating rule as

$$r_j \to r_j + g a_{ij}, \tag{24}$$

where a_{ij} is the i, jth entry in the adjacency matrix $A = (a_{ij})$ of the network G. Recall that $a_{ij} = 1$ if there is an edge from node i to node j. Lastly, we impose a refractory period after node i fires an action potential which is a short period of time during which node i cannot fire again.

As described in more detail in Sect. 6, we can simulate this neuronal activity process on analogues of any particular network drawn from one of the random network models described in Sect. 4 for directed graphs. Then we can investigate the influence of network structure on the dynamics of the process on the network. Suggested research project 6 describes a modified version of this model by considering a small number of inhibitory connections between nodes. In that case, we account for how different individual directed edges (excitatory and inhibitory connections) influence node dynamics in the model by assigning a positive or negative sign to those terms, respectively.

Researchers have looked at modeling neuronal activity in different contexts. Recent work in [34, 35] considered a model of two interacting neuronal populations that mimics the dynamics of sleep-wake cycling in developing rats. The spiking model considered in [42] can be viewed as expanding the wake-active population node from the model in [35] into a network of individual spiking neurons where the activity of each node is described by a stochastic differential equation. Other studies have modeled the random gating of ion channels as a stochastic process on specific networks found in neuroscience [43, 44]. There are several other examples in the literature, see [11, chapter 12] for further reading and additional references.

There has also been some work linking neuronal activity to networks with specific topologies. In particular, recent studies have suggested that certain neuronal networks found in the brain have small-world network structure [12, 45]. The authors justify this as an attractive model for brain networks in terms of anatomical structure and function since small-world networks can minimize the costs of wiring (by introducing shortcut edges) while supporting high dynamical complexity.

Exercise 33 See the Appendix (Selected Stochastic Processes) for an R script to simulate a Poisson process $\{N(t) : t \geq 0\}$ with rate $\lambda = 1$, and for more details on the Poisson process. Recall that $N(0) = 0$ by definition. Run the simulation five times and compute the number of events that have occurred by time $t = 10$ for each run.

Exercise 34 Use the R script in the Appendix (Selected Stochastic Processes) to simulate a Poisson process $\{N(t) : t \geq 0\}$ with rate $\lambda = 3$. Repeat with increasing values for λ, such as 5, 10, and 20. Compare to the simulation in Exercise 33. In particular, how does increasing the rate λ influence the average number of events that have occurred by time $t = 10$? What do you notice about the distribution of time between events (also known as the *interarrival times*, see same section in Appendix for definition) in each case?

Challenge Problem 6 Write a script in R to simulate the neuronal spiking process defined by Eqs. (23) and (24) on a given directed network. For instance, generate and fix a Barabási–Albert scale-free network on $N = 50$ nodes and then simulate neuronal activity on the network using the following parameter values: $r_0 = 0.002$ ms^{-1}, $r_i = r_0$ ms^{-1} (initially for all i), $\tau_r = 10$, and $g = 0.01 r_0$ ms^{-1}. Set the refractory period to 20 ms. Run the simulation for 10,000 time steps and plot the average firing rate over time.

5.4 Adaptive Networks

The intent of this short section is to briefly introduce the idea of adaptive networks. If interested, students should be able to modify the project on epidemic spreading to include a particular change in network structure over time. Additionally, we point students to some of the vast literature available on this topic.

Biological systems are dynamic, meaning that they change over time, and the networks that represent them are also dynamic. A recent study reveals the dynamic nature of biological networks [9]. This study focused on how genetic networks rewire themselves in response to DNA damage and showed that a static network of interactions is not realistic for this system. This is not unique to genetic networks. For example, social networks can be highly dynamic, especially in the context of infectious disease transmission. People tend to change their daily interactions (or contacts) once they show symptoms of a disease, for instance, by staying home from work or school. The change in contact network structure might evolve along with the time course of the disease. When modeling biological systems, therefore, such questions naturally arise: How can we model the growth of networks? How can we model networks that change over time, both in terms of the structure and dynamics on the network? As one example, note that the Barabási–Albert model for generating a scale-free network is essentially based on modeling the growth of a contact network.

In many biological applications, the dynamics of the process on the network and the dynamics of the network itself occur on similar time scales. Such networks are known as *adaptive networks* [25, 26, 41]. These networks display a dynamic interplay between the state of the nodes and the structure of the network.

An interesting and realistic example of a dynamic model on a time-dependent network is an adaptive SIS network model [25], in which a susceptible node can break any connection (edge) to an infected node and rewire it to a randomly chosen susceptible node. See also [30] for a comprehensive reference on the mathematics of epidemics on networks and [31] for epidemics specifically on adaptive networks. We will discuss this topic in more detail in Sect. 6 when describing research projects.

For further reading on this topic, see [26, 37]. For the specific application of epidemics on adaptive networks, see [25, 31].

Challenge Problem 7 Extend a standard model, such as a random walk on the complete graph with $N = 10$ nodes, to a context where network connectivity is changing over time. For instance, assume that at each time step, one of the occupied node's neighbors breaks its connection (edge) with probability ρ. Simulate both the standard model and the adapted version with dynamic connectivity in R. In the latter case, explore how the random walk changes for different values of ρ.

Challenge Problem 8 Come up with your own adaptive network model related to a particular biological network. Simulate the model in R and use the `igraph` package to visualize the changing network structure over time.

6 Suggested Research Projects

In this section, we describe six research-level projects centered on analyzing the network structure and dynamics of different biological systems. The first five focus on undirected networks and the last project explores directed networks in the neuronal activity setting. These projects can be studied both analytically and by simulation. We recommend using the `igraph` package in R for network generation and simulation of processes evolving on the network [18, 38]. We also recommend using the Appendix for basic sample code that can be modified to suit the projects.

The first project is a comparative study of network structure for the different network models presented in this chapter.

Research Project 1 Compute and compare the different centrality measures, average path length, and clustering coefficient (defined in Sect. 3) for the three different random network models presented in Sect. 4.

In particular, for research project 1 you want to:

(a) Fix the number of nodes to be the same for each network model, for example, $N = 100$, and then generate realizations from each of the three models: Erdős–Rényi (ER) random graph model, Watts–Strogatz (WS) small-world model, and Barabási–Albert (BA) preferential attachment model for generating a scale-free network; see Sect. 4.2. Focus on undirected networks.
(b) First focus on a representative realization from each of the network models and compute each metric. For average path length, restrict this computation to the largest connected component in the graph, also known as the *giant component*, in the case of a disconnected network. Then compute averages for each metric over 10 (or more) realizations of each network model.
(c) Look for trends, similarities, and differences between the network models. Change parameters (e.g., ER edge probability, small-world rewiring probability) and re-evaluate.
(d) Repeat for larger networks (e.g., $N = 1000$) and/or directed strongly connected networks.

The next project investigates one of the simplest processes on a network: bootstrap percolation.

Research Project 2 Study bootstrap percolation on different networks structures. For example, use the same process on an Erdős–Rényi random network, a small-world network, and a scale-free network. Compute and compare metrics that quantify how quickly activity spreads through the network.

In particular, for research project 2 you want to:

(a) Fix the number of nodes to be the same for each network model, for example, $N = 100$, and then generate representative realizations from: ER model, WS small-world model, BA scale-free model, and complete graph; see Sect. 4.2. Focus on undirected networks.
(b) Plot the observed degree distribution for each of the networks generated in (a).
(c) Set $k = 1$ and $\rho = 0.01$ in bootstrap percolation. Compare the average, minimum, and maximum *time to percolation* (first time step at which all nodes

in the initially active component are active) for many realizations of the process on each of the networks.

(d) Compute the clustering coefficient for each network structure generated above. Explore how the duration of the process changes for the different network structures and relate this back to the differences in clustering. Compare these results to the duration of the process on the complete graph.

(e) Look for trends, similarities, and differences between the process on the different network models. Change parameters (e.g., ER edge probability, small-world rewiring probability, activity threshold k in the process, number of initially active nodes ρ) and re-evaluate.

In the next two projects, you will consider the following questions: How do diseases propagate on networks? What properties of networks encourage or prevent the spread of an epidemic?

Research Project 3 Construct and analyze an epidemic network model for an infectious disease spreading through a population of N individuals.

In particular, for research project 3 you want to:

(a) First identify an infectious disease that you can justifiably model with one of the epidemic network models discussed in Sect. 5.2: SI, SIR, or SIS network model.

(b) Define the contact network using, for instance, the WS small-world network model with moderate rewiring (say $q = 0.05$) or the BA scale-free network model, each with $N = 100$ nodes; see Sect. 4.2. Use an undirected network.

(c) For the infection process, start with a single randomly chosen infected individual and define how the infection spreads from one individual node to the next (e.g., specify the parameters β and γ).

(d) Study the time course of the epidemic via simulation in R. Plot the number of infected, susceptible, and (if applicable) recovered nodes over time. Run many realizations of the process on the same fixed contact network. Compute the maximum number of infected nodes averaged over all realizations. If using an SI model, compute the average time it takes for all nodes in the initially infected component to become infected.

(e) Explore how the centrality score of the initially infected node influences the time course of the epidemic (and also the final size of the epidemic if using an SIR model; recall definition in Sect. 5.2).

(f) Change the network structure (e.g., ring, complete graph, or different random network models) and repeat steps above. Explore how changing the network structure affects the infection process.

Research Project 4 Modify and reanalyze the epidemic model from Project 3 to include an adaptive network framework. That is, the network evolves in time according to some rule which is related to the infectious disease. For instance, an individual (node) changes their contact network by increasing the chance that they stay home after developing symptoms. How does this affect the spread of the epidemic?

For research project 4, careful thought is required to realistically define how the contact network changes over time in response to individuals getting sick. In particular, you want to:

(a) Define the rules for how the adaptive network changes over time. For instance, suppose individuals change their contact network by increasing the chance that they stay home after developing symptoms. This can be modeled by assuming that each infected node will break its connection (edge) with each of its susceptible neighbors with a given probability q, similar to Challenge Problem 7.

(b) Study how quickly the epidemic spreads via simulation in R. Plot the number of infected, susceptible, and (if applicable) recovered nodes over time. Run many realizations of the process and compute the maximum number of infected nodes averaged over all realizations. Compare the adaptive network version of the disease process to the static network version.

(c) Repeat parts (a) and (b) for different values of q or for different adaptive rules.

Alternatively, advisors/students could find a paper that studies dynamics on networks in another biological context, impose or modify the rules for how the network changes over time, and then re-run the analysis presented in that paper.

Research Project 5 With the help of your advisor, find a paper that studies a biological system in a network context (adaptive or not). Identify an assumption the authors make that could be altered to yield a new but similar model. Repeat the analysis presented in the paper on the modified model.

In particular, for research project 5 you want to:

(a) Identify an assumption that could be altered, or a new assumption that could be imposed, to yield a new but similar model. For example, modify the rules (or impose new rules) for how the network gets rewired over time, or change the dynamics of the process on the network.

(b) Pick a particular analysis or question answered in the paper. First re-create that result so that you are confident in your ability to do the analysis; you can check your answer. Then re-run the same analysis on the modified model you develop.

(c) Discuss your results and how they compare to the results in the motivating paper.

For a concrete example, use the paper by Watts and Strogatz [46] on the dynamics of small-world networks. In that paper, they describe a simplified model for the spread of an infectious disease on a family of small-world graphs where the rewiring parameter p varies from 0 to 1. Focus on the simulation results in their Figure 3. Re-run this simple disease model with an adaptive network framework.

The last project explores neuronal activity on different network structures and incorporates a second type of connection between nodes: inhibitory connections. In addition, this project uses directed networks.

> **Research Project 6** Generate a variant of the dynamical process used to model neuronal activity on a random directed network $G(V, E)$ described in Sect. 5.3 to include a small number of inhibitory connections. Simulate the process on different network topologies and explore how network structure affects the dynamics of the process.

In particular, for research project 6 you want to:

(a) Fix the number of nodes to be the same for each network model, for example, $N = 100$, and then generate representative realizations from: ER model, WS small-world model with moderate rewiring, and BA scale free model; see Sect. 4.2. Use as.directed() to generate directed network analogues and focus on strongly connected networks.

(b) Modify the neuronal activity process so that some small proportion p of nodes make inhibitory connections with their neighbors and the remaining nodes make excitatory connections as before. Suppose $p = 0.1$. Then for a network with $N = 100$ nodes, randomly choose $pM = 10$ nodes to be inhibitory. Recall that each node i fires action potentials according to a Poisson process with rate r_i. Now we can modify Eqs. (23) and (24) in different ways to model inhibition. For instance, given an inhibitory node i^* and its neighboring nodes j, the updating rule (Eq. (24)) could be modified by subtracting a constant amount $g > 0$ from the firing rate of i^*'s neighboring nodes. In this case, Eq. (23) would stay the same. This scenario puts excitatory and inhibitory effects on similar same timescales and is given by the modified equations:

$$r_j \rightarrow \max(r_j - g a_{i^* j}, 0) \tag{25}$$

$$\frac{dr_{i^*}}{dt} = \frac{1}{\tau_r}(r_0 - r_{i^*}). \tag{26}$$

Alternatively, we could impose a longer timescale for inhibition by temporarily dropping the baseline firing rate r_0 for i^*'s neighboring nodes j. We introduce a new parameter $r_{0,j}$ to denote this change to node j's baseline firing rate, and then the magnitude of the drop below the original baseline r_0 exponentially decays back to r_0. The modified equations in this case are given by

$$r_{0,j} \to \theta \cdot r_{0,j}, \quad \text{where} \quad \theta \in [0, 1] \tag{27}$$

$$\frac{dr_{0,j}}{dt} = \frac{1}{\tau_{r_{0,j}}}(r_0 - r_{0,j}). \tag{28}$$

For large values of $\tau_{r_{0,j}}$ (i.e., for small $\tau_{r_{0,j}}^{-1}$) and small values of θ, this introduces a slow inhibitory timescale relative to the excitatory node dynamics. See Sect. 5.3 and Challenge Problem 6 for possible parameter values to use.

(c) Choose one of the modifications above and write an R script to run simulations of the process.

(d) Compare the dynamic activity on the three networks. For instance, compute and plot the average firing rate over time. Compute and plot the firing rate for the highest degree node in the network and compare to the average firing rate over time. Repeat for the lowest degree node. Additionally, compute the maximal length of a time interval (in milliseconds) during which there are no gaps in activity of more than $\delta = 1$ ms between subsequent firings. Such periods of activity are called *active bouts*. Explore how the different network topologies affect the dynamics of the process in terms of average firing rate and active bout duration. Change δ and re-evaluate.

(e) Increase the number of inhibitory connections (choose $p > 0.1$) and repeat the steps above. How does this change the dynamics of the process? How many inhibitory connections are needed to yield different dynamics than the fully excitatory case?

(f) Investigate how the refractory period influences the dynamic activity on the three networks. Specifically, increase or decrease the refractory period (set at 20 ms) in the model and repeat steps (d) and (e) above.

Acknowledgements The author thanks two reviewers whose comments greatly helped to clarify and focus the content of this chapter. The author also thanks Paul Hurtado for helpful discussions that influenced the content and scope of the chapter.

Appendix

Here we provide the R code used to generate all network figures in this chapter with the exception of Figs. 1, 2, 7, 12, 13, and 14 which were created from scratch in MS PowerPoint. Networks were generated using the igraph package (version 1.0.1) in R (version 3.5.3) [18, 38]. Note that the code below will generate a realization or sample network from the given model. These are random graph models, and their generation is based on probabilistic rules (e.g., edge probabilities in the Erdős–Rényi model). Therefore, it is unlikely that you will generate the exact same sample network given in this chapter, but rather a different sample network with the same or similar network properties.

We also give a brief overview of the basics of stochastic processes, including formal definitions of stochastic process, random walk, and Poisson process. We provide sample R code to simulate a simple random walk which is intended for use in Exercise 28 and can be adapted for Challenge Problem 3. Lastly, we provide R code to simulate a Poisson process for use in Exercises 33 and 34 and Research Project 3.

R Code for Figures

```
###################################################################
# Figures in Section 3
###################################################################

# First, install the igraph package and load the igraph library
library(igraph)

# FIGURE 3
# Sample connected undirected network, N is number of nodes
N-10;
g <- erdos.renyi.game(N,0.25)
iglayout1 = layout.fruchterman.reingold(g)
plot(g, layout=iglayout1, vertex.size=20, vertex.label.dist=0,
vertex.color="lightblue")

# Sample disconnected undirected network
N=10;
g <- erdos.renyi.game(N,0.1)
iglayout1 = layout.fruchterman.reingold(g)
plot(g, layout=iglayout1, vertex.size=20, vertex.label.dist-0,
vertex.color="lightblue")


# FIGURE 4
# Sample directed network, weakly connected
N=10;
g <- erdos.renyi.game(N,0.1,directed = TRUE)
iglayout1 = layout.fruchterman.reingold(g)
plot(g, layout=iglayout1, vertex.size=20, vertex.label.dist=0,
vertex.color="lightblue")


# FIGURE 5
# Sample undirected network and adjacency matrix
N=4;
g <- erdos.renyi.game(N,0.5)
iglayout1 = layout.fruchterman.reingold(g)
plot(g, layout=iglayout1, vertex.size=20, vertex.label.dist=0,
```

```
vertex.color="lightblue")

# Retrieves the adjacency matrix used in the above graph
adj <- get.adjacency(g)
adj

# FIGURE 6
# Sample directed network and adjacency matrix
N=4;
g <- sample_pa(N, directed = TRUE)
plot(g, layout=iglayout1, vertex.size=20, vertex.label.dist=0,
vertex.color="lightblue")

# Retrieves the adjacency matrix used in the above graph
adj <- get.adjacency(g)
adj


####################################################################
# Figures in Section 4
####################################################################

# FIGURE 8
# Ring network
g <- make_ring(10)
plot(g, layout=layout_with_kk, vertex.color="lightblue")

# Complete graph
g <- make_full_graph(10)
plot(g, layout=layout_with_kk, vertex.color="lightblue")


# FIGURE 9
# Erdos-Renyi network
N=30; p=0.2
g <- sample_gnp(N,p)
plot(g, vertex.size=10, vertex.color="wheat")
# hist(degree_distribution(g), 20, col = "wheat")


# FIGURE 10
# Small-world network: Watts--Strogatz model
N <- 20

# WS network with no rewiring (q=0)
g1 = sample_smallworld(1, N, 3, 0) -- sample realization
mean_distance(g1)
transitivity(g1, type="average")

# > mean_distance(g1)
# [1] 2.105263
# > transitivity(g1, type="average")
# [1] 0.6

# WS network with small rewiring (q=0.05)
g2 = sample_smallworld(1, N, 3, 0.05) -- sample realization
mean_distance(g2)
transitivity(g2, type="average")

# > mean_distance(g2)
```

```
# [1] 1.936842
# > transitivity(g2, type="average")
# [1] 0.4841667

g3 = sample_smallworld(1, N, 3, 1) -- sample realization
mean_distance(g3)
transitivity(g3, type="average")

# > mean_distance(g3)
# [1] 1.831579
# > transitivity(g3, type="average")
# [1] 0.2939432

# Plot all three networks together in one row
# Use layout option to keep nodes in a circle
par(mfrow = c(1,3))
plot(g1, vertex.size-10, layout-layout.circle)
plot(g2, vertex.size=10, layout=layout.circle)
plot(g3, vertex.size=10, layout=layout.circle)


# FIGURE 11
# Scale-free network: Barabasi-Albert model
# This model is directed by default
# Use "directed = FALSE" to make undirected
g <- sample_pa(50, directed = FALSE, out.pref = TRUE)
degree_distribution(g)
plot(g, vertex.size=10, vertex.color="wheat")
```

Basics of Stochastic Processes

Knowledge of introductory probability is assumed. In particular, we assume knowledge of the definitions of random variable and conditional probability as well as familiarity with common probability distributions (e.g., exponential and Poisson distributions). For a review of these basics, see [39].

Definition 17 A *stochastic process* is a sequence of random variables $\{X(t) : t \in T\}$, where T is an index set typically thought of as time. In particular, for each $t \in T$, $X(t)$ is a random variable and we refer to $X(t)$ as the state of the process at time t.

The set of possible values that the random variable $X(t)$ can take on is called the *state space*, S. The time index set T can be discrete or continuous, and the same goes for the state space S. However, here we focus on discrete-state processes since we are talking about processes evolving on the nodes of a finite size network. Often, the set of nodes is the state space S, as in the examples below, but this does not have to be the case. We could, for instance, define a stochastic process to occur on each node of the network; see the neuronal spiking example in Sect. 5.3 and other examples in Sect. 5. We will consider both discrete-time and continuous-time processes in this chapter.

As an example of a discrete-time stochastic process on a finite state space, let $T = \{0, 1, 2, \ldots\}$ and $S = \{1, 2, 3\}$. Then the process $\{X(t) : t = 0, 1, \ldots\}$ starts out in state 1, 2, or 3 and moves around on these three states according to the following rules:

- If the process is in state 1 at time t (for $t \in T$), mathematically this is denoted by $X(t) = 1$, then the process moves to state 2 (with probability 1) in the next time step, or $X(t + 1) = 2$. We can denote this transition by the following conditional probability:

$$P(X(t + 1) = 2 \mid X(t) = 1) = 1. \tag{29}$$

- If $X(t) = 2$, then the process has two choices for where to move in the next time step: move to state 3 with probability $\frac{1}{2}$ or move back to state 1 with probability $\frac{1}{2}$. Using conditional probabilities, we have

$$P(X(t + 1) = 3 \mid X(t) = 2) = \frac{1}{2} \tag{30}$$

$$P(X(t + 1) = 1 \mid X(t) = 2) = \frac{1}{2}. \tag{31}$$

- Finally, if $X(t) = 3$, then the process must move back to state 2 in the next time step. Again, we represent this state transition using a conditional probability.

$$P(X(t + 1) = 2 \mid X(t) = 3) = 1. \tag{32}$$

In general, these conditional probabilities are called *transition probabilities* which describe the probability of a transition from the current state i to next state j in one discrete-time step. This general notation is given by

$$P(X(t + 1) = j \mid X(t) = i) = p_{ij}, \tag{33}$$

where each p_{ij} is the (i, j)th entry in the *transition probability matrix*, P. For this example,

$$P = \begin{pmatrix} 0 & 1 & 0 \\ \frac{1}{2} & 0 & \frac{1}{2} \\ 0 & 1 & 0 \end{pmatrix}. \tag{34}$$

Note that each row of P sums to 1. This is true in general since the sum over all possible one-step transition probabilities from a given state i to any other state $j \in S$ must be 1.

The diagram in Fig. 13 is another way of illustrating this example process: the states are the nodes, the directed edges denote the possibility of a transition from one node to another, and the numbers above the edges denote the probability of such transitions. A sample path or realization of this process up through time $t = 7$ is

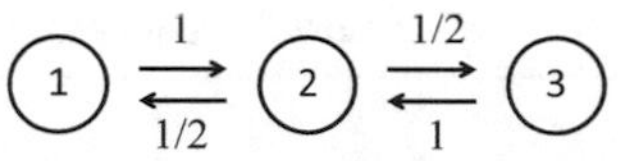

$$1 \underset{1/2}{\overset{1}{\rightleftarrows}} 2 \underset{1}{\overset{1/2}{\rightleftarrows}} 3$$

Fig. 13 An illustration of the transition state diagram corresponding to the discrete-time stochastic process example described above. The state space $S = \{1, 2, 3\}$ is described by the nodes in this 3-state network, and the possible transitions between states are given by directed edges with associated (one time step) transition probabilities

$$\cdots \underset{1-p}{\overset{p}{\rightleftarrows}} -1 \underset{1-p}{\overset{p}{\rightleftarrows}} 0 \underset{1-p}{\overset{p}{\rightleftarrows}} 1 \underset{1-p}{\overset{p}{\rightleftarrows}} 2 \underset{1-p}{\overset{p}{\rightleftarrows}} \cdots$$

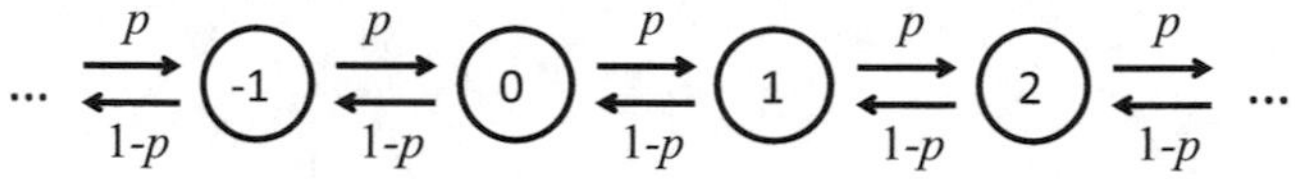

Fig. 14 An illustration of the transition state diagram corresponding to the simple random walk on **Z**. The nodes are the integers and the only possible state transitions (in one time step) are to move by ± 1, also given by the directed edges with associated transition probabilities p or $1 - p$ for some $p \in [0, 1]$

$X(0)=1, X(1)=2, X(2)=1, X(3)=2, X(4)=3, X(5)=2, X(6)=1, X(7)=2, \ldots$.

If we just list the sequence of states starting in state 1 and moving forward in time, this sample path looks like

$$1 \rightarrow 2 \rightarrow 1 \rightarrow 2 \rightarrow 3 \rightarrow 2 \rightarrow 1 \rightarrow 2 \ldots.$$

Another interesting stochastic process called a *random walk* is defined below. We define the discrete-time random walk on a discrete state space only (such as the integers **Z**); see [40] for further reading on other varieties of random walks. Additionally, we focus on the *simple* random walk which can only move in increments of ± 1.

Definition 18 A discrete-time *simple random walk* on state space $S = \mathbf{Z}$ is a stochastic process $\{Y(t) : t = 0, 1, 2, \ldots\}$ such that the state of the process increases or decreases by 1 at each time step with probabilities p and $1 - p$, respectively. These transition probabilities are given by

$$P(Y(t + 1) = i + 1 \mid Y(t) = i) = p \tag{35}$$

$$P(Y(t + 1) = i - 1 \mid Y(t) = i) = 1 - p \tag{36}$$

for some $p \in [0, 1]$.

See Fig. 14 for a graphical representation of the simple random walk process on **Z**. Suppose the simple random walk $\{Y(t) : t = 0, 1, 2, \ldots\}$ starts in state 0 at time 0. Then a sample path of the process is

$Y(0)=0, Y(1)= -1, Y(2)=0, Y(3)=1, Y(4)=2, Y(5)=3, Y(6)=2, Y(7)=1, \ldots$.

We provide R code at the end of the Appendix to generate a simple random walk. Using this code, we generate and plot five sample paths of the symmetric simple

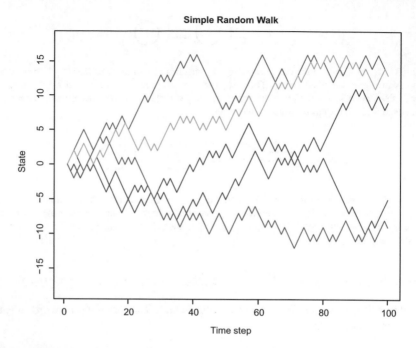

Fig. 15 Sample realizations from the simple random walk simulation on the integers, symmetric case $p = \frac{1}{2}$. Initial state is 0 in all cases

random walk in Fig. 15 below. The symmetric case means that the parameter $p = 1/2 = 1 - p$.

Note that due to the random nature of a stochastic process such as this random walk, each simulation run will be different. In general, in order to draw meaningful conclusions about a given stochastic process, one needs to run multiple simulations with the same parameter settings and look at the distribution of outcomes.

Lastly, we define a *Poisson process* which is a continuous-time stochastic process that takes values in the discrete state space $S = \mathbf{Z}^*$, the non-negative integers. This process is used in the neuronal spiking process described in Sect. 5.3. Poisson processes are commonly used to model a variety of biological processes.

Definition 19 A *Poisson process* on state space $S = \mathbf{Z}^*$ (non-negative integers) is a continuous-time stochastic process $\{N(t) : t \geq 0\}$ that counts events that have occurred up to and including time t. Specifically, $\{N(t) : t \geq 0\}$ is a Poisson process with rate $\lambda > 0$ if

- $N(0) = 0$.
- $\{N(t) : t \geq 0\}$ has independent increments, meaning that for any $0 \leq t_0 < t_1 < \cdots < t_n < \infty$, $N(t_0)$, $N(t_1) - N(t_0)$, ..., $N(t_n) - N(t_{n-1})$ are all independent random variables.

- $N(s + t) - N(s) \sim \text{Poisson}(\lambda t)$ for any $s, t \geq 0$, meaning that the number of events in any interval of length t follows a Poisson distribution with rate λt.

For instance, let $\lambda = 1$. Then the expected number of events that will occur by time t is $\lambda t = t$ since $\lambda = 1$ in this case. In other words, on average, one event will occur per unit time. This follows because $N(t) = N(t) - N(0) \sim \text{Poisson}(\lambda t)$ and $E[N(t)] = \lambda t$, a property of the Poisson distribution [40].

Now let us describe a sample path for a Poisson process $\{N(t) : t \geq 0\}$ with rate $\lambda = 1$. We assume that at time $t = 0$ no events have occurred (since $N(0) = 0$), so the first event will occur after an exponentially distributed amount of time with rate (parameter) λ. Starting at the time of the first event, the amount of time we have to wait until the second even occurs follows an exponential distribution with rate λ. It follows that the length of time between any two successive events in the Poisson process also follows an exponential distribution with rate λ. The times between events are called *interarrival times*, and a key property of the Poisson process is that the interarrival times follow an exponential distribution. Hence, the average length of the interarrival times is given by $1/\lambda$, or the mean of an exponential distribution with rate λ.

A sample path of a Poisson process $\{N(t) : t \geq 0\}$ with rate $\lambda = 1$ is given below.

$$N(0) = 0, N(0.062) = 1, N(0.88) = 2, N(1.02){=}3, N(2.38){=}4, N(3.09){=}5,$$

$$N(5.94) = 6, \ N(8.04) = 7, \ N(8.92) = 8, \ N(9.65) = 9, \ N(10.10) = 10, \ldots.$$

Note that the state of the process increases by 1 at each time step, and the time increments are given by random draws from the exponential distribution with rate $\lambda = 1$. The expected number of events by time $t = 10$ is 10, and in this sample path we have 9 events that occurred by time $t = 10$ since $N(9.65) = 9$ and $N(10.10) = 10$. The tenth event occurred slightly after time $t = 10$, specifically at $t = 10.10$ in this case. See Fig. 16 for a plot of this sample path, and see R code used to generate this plot below. In general, a widely used method for simulating continuous-time discrete-state stochastic processes is known as the Gillespie Stochastic Simulation Algorithm (SSA) [24]; see also the GillespieSSA package in R.

For more reading on introductory stochastic processes, random walks, and Poisson processes, see [19, 40].

R Code for Selected Stochastic Processes

```
####################################################################
# Simulate a random walk
####################################################################

# Simple random walk on the integer line with probability p of
# moving +1 step (right), probability 1-p of moving -1 step (left)
```

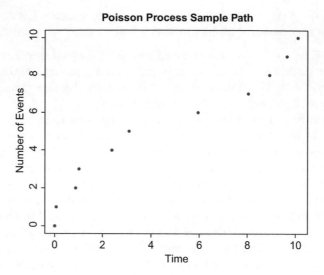

Fig. 16 A plot of the sample path detailed above from a Poisson process with rate $\lambda = 1$

```
# Parameters: n=number of time steps, p=prob increase state by 1,
# x1=initial state

RW.sim <- function(n,p,x1) {
sim <- as.numeric(n)
if (missing(x1)) {
sim[1] <- 0  # initial condition: start at state 0
}
else {
sim[1] <- x1
}
for (i in 2:n) {
newstate <- sample(c(-1,1),1,prob=c(1-p,p)) + x1
sim[i] <- newstate
x1 = newstate
}
sim
}

# Generate 5 sample paths and plot all on the same figure
p = 0.5  # symmetric case

run <- RW.sim(100,p,0)
plot(run, type="l", col="blue", ylim=c(-18,18), xlab="Time step",
ylab="State", main = "Simple Random Walk")

run <- RW.sim(100,p,0)
points(run, type="l", col="red")

run <- RW.sim(100,p,0)
points(run, type="l", col="darkgreen")
```

```
run <- RW.sim(100,p,0)
points(run, type="l", col="orange")

run <- RW.sim(100,p,0)
points(run, type="l", col="magenta")


###################################################################
# Simulate a Poisson process
###################################################################

# Install poisson package and load library
library(poisson)

# Simulate a homogeneous Poisson process with a given rate

# Parameters: rate=rate of Poisson process, num.events=number
# of events to simulate, num.sims=number of times to run the
# simulation (default is 1)

PP.sim <- function (rate, num.events, num.sims = 1, t0 = 0)
{
if (num.sims == 1) {
x = t0 + cumsum(rexp(n = num.events, rate = rate))
return(c(t0,x))
}
else {
xtemp = matrix(rexp(n=num.events*num.sims, rate=rate), num.events)
x = t0 + apply(xtemp, 2, cumsum)
return(rbind(rep(t0, num.sims), x))
}
}

# Run the function PP.sim
rate = 1
PP.sim(rate, num.events = 10)
PP.sim(rate, num.events = 10, num.sims = 5)

# Run the function and plot the time series for 1 simulation
rate = 1
num.events = 10
run1 <- PP.sim(rate, num.events)
plot(run1, 0:num.events, xlab="Time", ylab="Number of Events",
main = "Poisson Process Sample Path", pch = 19, col="blue")
run1
```

References

1. Getting Started with R. URL https://support.rstudio.com/hc/en-us/articles/201141096-Getting-Started-with-R
2. Introduction to R. URL https://www.datacamp.com/courses/free-introduction-to-r
3. Adler, J.: Bootstrap percolation. Physica A: Statistical Mechanics and its Applications **171**(3), 453–470 (1991)
4. Al-Anzi, B., Arpp, P., Gerges, S., Ormerod, C., Olsman, N., Zinn, K.: Experimental and computational analysis of a large protein network that controls fat storage reveals the

design principles of a signaling network. PLoS Computational Biology **11**(5), e1004,264 (2015)

5. Albert, R.: Scale-free networks in cell biology. Journal of Cell Science **118**(21), 4947–4957 (2005)
6. Albert, R., Barabási, A.L.: Statistical mechanics of complex networks. Reviews of Modern Physics **74**(1), 47 (2002)
7. Allen, L.J.: An Introduction to Stochastic Processes with Applications to Biology, 2nd edn. Chapman and Hall/CRC (2010)
8. Amaral, L.A.N., Scala, A., Barthelemy, M., Stanley, H.E.: Classes of small-world networks. Proceedings of the National Academy of Sciences **97**(21), 11,149–11,152 (2000)
9. Bandyopadhyay, S., Mehta, M., Kuo, D., Sung, M.K., Chuang, R., Jaehnig, E.J., Bodenmiller, B., Licon, K., Copeland, W., Shales, M., et al.: Rewiring of genetic networks in response to DNA damage. Science **330**(6009), 1385–1389 (2010)
10. Barabási, A.L., Albert, R.: Emergence of scaling in random networks. Science **286**(5439), 509–512 (1999)
11. Barrat, A., Barthelemy, M., Vespignani, A.: Dynamical Processes on Complex Networks. Cambridge University Press (2008)
12. Bassett, D.S., Bullmore, E.: Small-world brain networks. The Neuroscientist **12**(6), 512–523 (2006)
13. Baxter, G.J., Dorogovtsev, S.N., Goltsev, A.V., Mendes, J.F.: Bootstrap percolation on complex networks. Physical Review E **82**(1), 011,103 (2010)
14. Bollobás, B.: Random Graphs, 2 edn. Cambridge Studies in Advanced Mathematics. Cambridge University Press (2001). https://doi.org/10.1017/CBO9780511814068
15. Brauer, F., Castillo-Chavez, C.: Mathematical Models in Population Biology and Epidemiology, vol. 40. Springer (2001)
16. Caswell, H.: Matrix Population Models: Construction, Analysis, and Interpretation, 2nd edn. Oxford University Press (2006)
17. Clauset, A., Shalizi, C.R., Newman, M.E.: Power-law distributions in empirical data. SIAM Review **51**(4), 661–703 (2009)
18. Csardi, G., Nepusz, T.: The igraph software package for complex network research. InterJournal, Complex Systems p. 1695 (2006). URL http://igraph.org
19. Durrett, R.: Essentials of Stochastic Processes, vol. 1. Springer (1999)
20. Durrett, R.: Random Graph Dynamics, vol. 200. Cambridge University Press, Cambridge (2007)
21. Erdos, P., Rényi, A.: On random graphs. Publicationes Mathematicae **6**, 290–297 (1959)
22. Erdos, P., Rényi, A.: On the evolution of random graphs. Publ. Math. Inst. Hung. Acad. Sci **5**(1), 17–60 (1960)
23. Estrada, E.: Introduction to complex networks: Structure and dynamics. In: Evolutionary Equations with Applications in Natural Sciences, pp. 93–131. Springer (2015)
24. Gillespie, D.T.: Exact stochastic simulation of coupled chemical reactions. The Journal of Physical Chemistry **81**(25), 2340–2361 (1977)
25. Gross, T., D'Lima, C.J.D., Blasius, B.: Epidemic dynamics on an adaptive network. Physical Review Letters **96**(20), 208,701 (2006)
26. Gross, T., Sayama, H.: Adaptive networks: Theory, models and applications. In: Understanding Complex Systems. Springer (2009)
27. Just, W., Callender, H., LaMar, M.D.: Algebraic and Discrete Mathematical Methods for Modern Biology, chap. Disease transmission dynamics on networks: Network structure vs. disease dynamics., pp. 217–235. Academic Press (2015)
28. Just, W., Highlander, H.C.: Vaccination strategies for small worlds. In: A. Wootton, V. Peterson, C. Lee (eds.) A Primer for Undergraduate Research: From Groups and Tiles to Frames and Vaccines, pp. 223–264. Springer, New York (2018)
29. Keeling, M.J., Eames, K.T.: Networks and epidemic models. Journal of the Royal Society Interface **2**(4), 295–307 (2005)

30. Kiss, I.Z., Miller, J.C., Simon, P.L.: Mathematics of Epidemics on Networks: From Exact to Approximate Models, vol. 46. Springer (2017)
31. Masuda, N., Holme, P.: Temporal Network Epidemiology. Springer (2017)
32. Milo, R., Shen-Orr, S., Itzkovitz, S., Kashtan, N., Chklovskii, D., Alon, U.: Network motifs: simple building blocks of complex networks. Science 298(5594), 824–827 (2002)
33. Newman, M.: Networks: An Introduction. Oxford University Press (2010)
34. Patel, M.: A simplified model of mutually inhibitory sleep-active and wake-active neuronal populations employing a noise-based switching mechanism. Journal of Theoretical Biology 394, 127–136 (2016)
35. Patel, M., Joshi, B.: Switching mechanisms and bout times in a pair of reciprocally inhibitory neurons. Journal of Computational Neuroscience 36(2), 177–191 (2014)
36. Porter, M.A., Gleeson, J.P.: Dynamical systems on networks: a tutorial. arXiv preprint arXiv:1403.7663 (2014)
37. Porter, M.A., Gleeson, J.P.: Dynamical systems on networks: A tutorial. In: Frontiers in Applied Dynamical Systems: Reviews and Tutorials, vol. 4. Springer-Verlag, Heidelberg, Germany (2016)
38. R Core Team: R: A Language and Environment for Statistical Computing. R Foundation for Statistical Computing, Vienna, Austria (2018). URL https://www.R-project.org/
39. Ross, S.: A First Course in Probability, 9th edn. Pearson (2012)
40. Ross, S.M.: Introduction to Probability Models, 11th edn. Academic Press (2014)
41. Sayama, H., Pestov, I., Schmidt, J., Bush, B.J., Wong, C., Yamanoi, J., Gross, T.: Modeling complex systems with adaptive networks. Computers & Mathematics with Applications 65(10), 1645–1664 (2013)
42. Schmidt, D., Blumberg, M.S., Best, J.: Random graph and stochastic process contributions to network dynamics. Discrete and Continuous Dynamical Systems, Supp 2011 2, 1279–1288 (2011)
43. Schmidt, D.R., Galán, R.F., Thomas, P.J.: Stochastic shielding and edge importance for Markov chains with timescale separation. PLoS Computational Biology 14(6), e1006,206 (2018)
44. Schmidt, D.R., Thomas, P.J.: Measuring edge importance: a quantitative analysis of the stochastic shielding approximation for random processes on graphs. The Journal of Mathematical Neuroscience 4(1), 6 (2014)
45. Watts, D.J.: Small worlds: the dynamics of networks between order and randomness, vol. 9. Princeton University Press (2004)
46. Watts, D.J., Strogatz, S.H.: Collective dynamics of 'small-world' networks. Nature 393, 440–442 (1998)

What Are the Chances?—Hidden Markov Models

Angela B. Shiflet, George W. Shiflet, Mario Cannataro, Pietro Hiram Guzzi, Chiara Zucco, and Dmitriy A. Kaplun

Abstract Hidden Markov Models (HMMs) can be used to solve a variety of problems from facial recognition and language translation to animal movement characterization and gene discovery. With such problems, we have a sequence of observations that we are not certain is correct—we are not sure our observations accurately reveal the corresponding sequence of actual states, which are hidden— but we do know some important probabilities that will help us. In this chapter, we will develop the probability theory and algorithms for two types of problems that HMMs can solve—calculate the probability that a particular sequence of observations occurs and determine the most likely corresponding sequence of hidden states. The chapter will end with a collection of research projects appropriate for undergraduates.

Suggested Prerequisites *Previous programming experience; a basic understanding of probability; knowledge of Markov models, such as in Module 13.4, "Probable Cause: Modeling with Markov Chains," from [19]; understanding of parallel programming for optional Sect. 9, "Parallel Forward Algorithm," and optional Sect. 10.3, "Parallel Viterbi Algorithm" and the corresponding projects.*

1 Introduction

We can employ **Hidden Markov Models (HMMs)** to solve a variety of problems from facial recognition and language translation to animal movement characteriza-

A. B. Shiflet (✉) · G. W. Shiflet · D. A. Kaplun
Wofford College, Spartanburg, SC, USA
e-mail: shifletab@wofford.edu; shifletgw@wofford.edu; kaplunda@email.wofford.edu

M. Cannataro · P. H. Guzzi · C. Zucco
University "Magna Græcia" of Catanzaro, Catanzaro, Italy
e-mail: cannataro@unicz.it; hguzzi@unicz.it

© Springer Nature Switzerland AG 2020
H. Callender Highlander et al. (eds.), *An Introduction to Undergraduate Research in Computational and Mathematical Biology*, Foundations for Undergraduate Research in Mathematics, https://doi.org/10.1007/978-3-030-33645-5_8

tion and gene discovery (see "Further Reading"). In such a problem, we know a sequence of observations, but we are not certain that our observations are accurate; the actual sequence of states is hidden, or unknown. However, we often know for each state the probability that it is the initial one, the probabilities of transitioning from any one state to every other state, and the probabilities of a state resulting in each type of observation. Hidden Markov Models can attack three types of problems:

- Likelihood problem: Calculate the probability of a particular sequence of observations
- Decoding problem: For a particular sequence of observations, determine the most likely underlying sequence of states
- Training problem: For a sequence of observations and a sequence of hidden states, discover the most likely HMM parameters

In this article, we cover the basics of Hidden Markov Models, algorithms, and applications involving the first two types of problems. We begin with an example of how HMMs are used to advance our understanding of genetics and the workings of the human body.

1.1 Case in Point

Nina was 15 years old and 50 inches tall. Her mother was concerned that her daughter was so short. The Centers for Disease Control and Prevention growth charts indicate a range for normal height for 15-year-old females to be between 60 and 68 inches [2]. Testing done from her physical examination indicated that Nina has growth hormone deficiency. Children diagnosed with this deficiency have inadequate secretion of **growth hormone (GH)**, a hormone produced and stored by **somatotropic cells** of the pituitary gland at the base of the brain. Growth hormone molecules are stored in the pituitary until **growth hormone-releasing hormone (GHRH)** is secreted by the brain. The secreted GHRH binds specifically to **growth hormone-releasing hormone receptors**, found within the cell membranes that form the surface of the somatotropic cells. The receptor is associated with a protein, called a **G-protein**, which transduces the external signal into internal chemical signals (second messengers). The second messengers induce the cell to release GH.

Because Nina's brother was given the same diagnosis, both patients were referred for genetic evaluation. The results of these tests indicated that both children had a genetic mutation in the gene that codes for the GHRH receptor. This mutation may alter the receptor enough that it no longer binds to the signal as well, and therefore there is less internal response to the signal. Hence, less GH is released into the bloodstream. Children with growth hormone deficiency are generally treated with periodic injections of GH.

The mechanism for the release of growth hormone from the pituitary is very complex. Much is still unknown about G-proteins and their receptors, so scientists are working to unravel these mysteries. So, how do G-proteins and receptors relate to hidden Markov models? One example is found in experiments conducted at the University of Birmingham, UK [22]. Using sophisticated imaging studies of individual G-proteins and receptors, researchers could track and map the diffusion of particles (proteins) along the membrane. Using HMM, they assumed that particles shift among discrete diffusive states that follow a randomly determined course. They found that G-proteins and receptors move through four discernable, diffusive states, varying from immobile to fast diffusing, and have association with four different diffusion coefficients, which are the possible observations. Their results were consistent with results of the imaging studies. From these experiments and analyses, the scientists concluded that G-proteins and their receptors are sequestered into small membrane compartments. Such restriction of diffusion allows the two particles to bind more easily, although the binding is very short-lived.

From these results we understand more about how these important components in cell signaling work. As we figure out the intricacies of cell signaling, we may discover ways to modify or correct defects in the components.

2 Example Model

Although the studies of human growth factor are quite interesting, there are many other more approachable problems. Moreover, we will consider other genetic applications of HMMs later in this chapter. Thus, we begin our study of the mathematics of HMMs with a simpler hypothetical example involving animal behavior.

With a primary diet of leaves, which are not very nutritional and are hard to digest, red howler monkeys spend most of their time eating and resting. Suppose scientists in a simplified appraisal considered the monkey to be in two primary states, eating (E) or resting/sleeping (R), so that the **set of possible states** is $S = \{E, R\}$. Moreover, suppose the biologists observe that, on average, the monkeys spend 30% of their time eating and 70% sleeping or resting. In this case, where u_0 is the state at time 0 and P indicates probability, the **initial state probability**, π, for E is $\pi(E) = P(u_0 = E) = 0.30$, and the initial state probability for R is $\pi(R) = P(u_0 = R) = 0.70$. (As we will see later, the choice of variable, $\boldsymbol{u}$, represents the "underlying," or hidden, state.)

Suppose also the biologists determined that if a monkey is eating at hour k, so that its state is $u_k = E$, then there is a 60% chance that the animal will be eating the next hour $(k + 1)$, or $u_{k+1} = E$. Thus, in terms of conditional probability, $P(u_{k+1} = E \mid u_k = E) = 0.6$ and $P(u_{k+1} = R \mid u_k = E) = 1 - 0.6 = 0.4$; that is, given $u_k = E$, the probability of $u_{k+1} = E$ is 0.6, and the probability of $u_{k+1} = R$ is 0.4. Suppose the scientists also discovered that $P(u_{k+1} = E \mid u_k = R) = 0.2$, so that $P(u_{k+1} = R \mid u_k = R) = 1 - 0.2 = 0.8$. If resting at hour k, the animal has a 20% chance of eating and an 80% chance of resting the next hour. Figure 1 presents a state diagram of the findings, with probabilities of transitioning from one

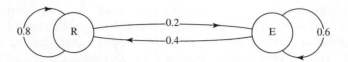

Fig. 1 State diagram for hypothetical study on red howler monkeys

state to another on the arrows. Thus, the following **transition matrix** summarizes their findings:

$$u_k/u_{k+1} \quad \text{E} \quad \text{R}$$

$$T = \begin{matrix} \text{E} \\ \text{R} \end{matrix} \begin{bmatrix} 0.6 & 0.4 \\ 0.2 & 0.8 \end{bmatrix}.$$

We can also denote these **transition probabilities**, t, as $t(previous\ state,\ next\ state$ so $t(\text{E}, \text{E}) = 0.6$, $t(\text{E}, \text{R}) = 0.4$, $t(\text{R}, \text{E}) = 0.2$, and $t(\text{R}, \text{R}) = 0.8$. The transition probabilities (or, comparably, matrix T) form a **Markov model** with each state, u_{k+1}, only depending on its previous state, u_k, and no other states.

Suppose, however, scientists want to study the behavior of a red howler monkey in a more remote area. Knowing they will have limited opportunities of making visual observations, they attach a small microphone to one of the monkeys, whom they name Holly. The biologists discern that when hearing munching (M), the monkey is probably eating; but when hearing breathing (B) noises, the animal is likely to be at rest. Thus, the **set of possible observations** is $O = \{M, B\}$. These hypothetical researchers have developed a computer program to analyze the sounds and record B or M once an hour. However, the microphone/computational results are not completely accurate. Besides background noises, such as from rain or other monkeys, a sleeping Holly might be moving her mouth, perhaps dreaming of luscious leaves. For a while, the scientists are able to observe Holly personally and with their computer-enhanced microphone. In doing so, they discover that there is a 90% chance that if Holly is resting ($u_k = \text{R}$), then their monitoring system indicates breathing noises ($v_k = \text{B}$). (We select the symbol v to represent the "visible," or observed, symbol obtained from the monitoring system). Thus, the **emission probability**, e, of B at state R is $e(\text{B} \mid \text{R}) = P(v_k = \text{B} \mid u_k = \text{R}) = 0.9$, so that $e(\text{M} \mid \text{R}) = P(v_k = \text{M} \mid u_k = \text{R}) = 1 - 0.9 = 0.1$. However, the scientists discover that the system is only 80% accurate in detecting eating; given that Holly is eating ($u_k = \text{E}$), their computer program interprets the audio signal as munching noises ($v_k = \text{M}$) 80% of the time. Thus, the emission probability of M given state E is $e(\text{M} \mid \text{E}) = P(v_k = \text{M} \mid u_k = \text{E}) = 0.8$ and $e(\text{B} \mid \text{E}) = P(v_k = \text{B} \mid u_k = \text{E}) = 1 - 0.8 = 0.2$. The HMM property of **output independence** states that regardless of the situation, the probability of an observation, v_k, *only* depends on the corresponding underlying state, u_k, that leads to the observation and no other states or observations. For example, the probability of the system displaying an output of munching (M) depends exclusively on underlying state of Holly eating (E).

After Holly scampers into the jungle, where the scientists cannot make visual observations, their monitoring system records an observation of B or M each hour. Using this **sequence of observations**, or **observed symbol sequence** ($v = v_1, v_2, v_3, \ldots, v_n$, abbreviated $v_1 v_2 v_3 \ldots v_n$), and the measures for initial, transition, and emission probabilities, (π, t, e), the scientists can answer a number of questions. Such problems generally fall in one of three categories: likelihood, decoding, and learning problems. For example, the scientists might want to determine the probability, or likelihood, of obtaining an observed symbol sequence, $P(v)$, to discern if v is an unusual sequence of observations or not. As another **likelihood-type problem**, the scientists might want to determine the probability that a particular **sequence of states**, or **underlying state sequence** ($u = u_1 u_2 u_3 \ldots u_n$) would generate a particular v; so that they need to evaluate $P(v \mid u)$. Additionally, given a particular observation, v_k, the scientists might want to know the probability of an underlying state, u_k, written $P(u_k \mid v_k)$. Determining $P(u_k \neq u_{k+1} \mid v)$ represents a **change-detection problem**. In this case, given a sequence of monitoring system readings, we are determining the probability of Holly eating 1 h but resting the next, or vice versa. Perhaps in earlier studies we observed that usually the monkeys eat for exactly two time periods before sleeping deeply for at least 3 h, and we might want to use our system to estimate Holly's sleeping habits. In a **decoding-type problem**, they might be interested in finding the most likely u to generate a particular observed sequence, v. In a **learning-type problem** or **training-type problem**, for an observation sequence, v, and a set of states, we would be interested in determining the parameters for the system. For example, suppose we do not know the numbers for the initial state probabilities and in the transition and output matrices of the HMM associated with Holly. Given a long sequence of observations, such as $v = \text{MMMBBBBBBMM} \ldots \text{BB}$, a learning-type problem is to derive those numbers for the HMM that maximize the likelihood of observing the sequence, v. Determination of these parameters is called **training the HMM**.

In all cases, we are using observations and probabilities, which include a Markov model, to estimate something that is hidden. Hence, the name of this system is **Hidden Markov Model (HMM)**. The HMM for Holly consists in the following parameters, which Fig. 2, an expansion of Fig. 1, diagrams:

Holly's HMM

State space, or set of possible states, $S = \{E, R\}$, with elements representing eating and resting/sleeping, respectively.

Observation space, or set of possible observations, $O = \{M, B\}$, with elements representing munching and breathing noises, respectively.

Initial state probabilities, $\pi(E) = 0.30$ and $\pi(R) = 0.70$

Transition probabilities, $t(E, E) = 0.6$, $t(E, R) = 0.4$, $t(R, E) = 0.2$, and $t(R, R) = 0.8$, summarized by the following transition matrix:

(continued)

$$u_k/u_{k+1} \quad E \quad R$$

$$T = \begin{matrix} E \\ R \end{matrix} \begin{bmatrix} 0.6 & 0.4 \\ 0.2 & 0.8 \end{bmatrix}$$

Emission probabilities, $e(M \mid E) = 0.8$, $e(B \mid E) = 0.2$, $e(M \mid R) = 0.1$, and $e(B \mid R) = 0.9$, summarized by the following output, or emission, matrix:

$$hidden/observable \quad M \quad B$$

$$\begin{matrix} E \\ R \end{matrix} \begin{bmatrix} 0.8 & 0.2 \\ 0.1 & 0.9 \end{bmatrix}$$

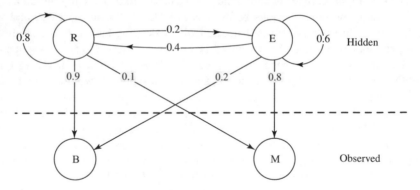

Fig. 2 Diagram of HMM for hypothetical study on red howler monkeys

Answers to Quick Review Questions appear at the end of the module, after "Projects."

Quick Review Question 1

Consider the HMM diagram in Fig. 3 where the initial state probability for A is 0.2 and for B is 0.1. Determine each of the following:

a. The set of possible states, S
b. The set of possible observations, O
c. $t(B, C)$
d. $t(B, A)$
e. The transition matrix with column headings being u_{k+1}
f. $\pi(A)$
g. $\pi(C)$
h. $e(G \mid B)$

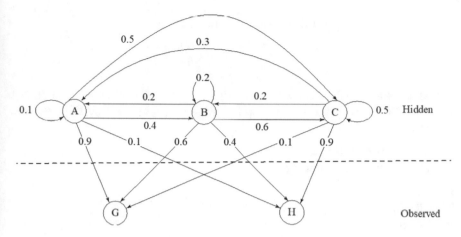

Fig. 3 HMM diagram for Quick Review Question 1

i. $e(H \mid C)$
j. The output (emission) matrix with column headings being observable values

3 Probability Equalities

Before we can start solving some of the problems with the HMM, we need to consider several probability equalities.

3.1 Joint Probability

Joint probability evaluates the probability of the simultaneous occurrence of two events, A and B, which are not necessarily independent as follows:

$$P(A \text{ and } B) = P(A, B) = P(A \mid B) \cdot P(B) \tag{1}$$

and

$$P(A \text{ and } B) = P(A, B) = P(B \mid A) \cdot P(A), \tag{2}$$

where the comma means "and," or "intersection."

For example, consider Holly's HMM. Specifically, $P(E) = \pi(E) = 0.30$, $P(R) = \pi(R) = 0.70$, and the output, or emission, matrix is

$$\begin{array}{cc} hidden/observable & \text{M} \quad \text{B} \\ \begin{array}{c} \text{E} \\ \text{R} \end{array} & \begin{bmatrix} 0.8 & 0.2 \\ 0.1 & 0.9 \end{bmatrix} \end{array}.$$

Knowing their probabilities, we can condition on the hidden states, E or R. For example, $P(hidden =\text{E}$ and $observable =\text{M}) = P(\text{E}, \text{M}) = P(\text{M}\,|\,\text{E}) \cdot P(\text{E}) = 0.8 \cdot 0.3 = 0.24$.

Quick Review Question 2

Calculate each of the following:

a. $P(\text{E}, \text{B})$
b. $P(\text{R}, \text{M})$
c. $P(\text{R}, \text{B})$
d. $P(\text{E}, \text{M}) + P(\text{E}, \text{B}) + P(\text{R}, \text{M}) + P(\text{R}, \text{B})$
e. $\displaystyle\sum_{x\in\{\text{M}, \text{B}\}} P(\text{E},\ x) = P(\text{E},\ \text{M}) + P(\text{E},\ \text{B})$
f. $\displaystyle\sum_{x\in\{\text{M}, \text{B}\}} P(\text{R},\ x)$
g. The sum of the answers to Parts e and f
h. $\displaystyle\sum_{x\in\{\text{E}, \text{R}\}} P(x,\ \text{M})$
i. $\displaystyle\sum_{x\in\{\text{E}, \text{R}\}} P(x,\ \text{B})$
j. The sum of the answers to Parts h and i

To obtain an intuition for why the joint probability relationship Eq. (1) is true, consider the Venn diagram in Fig. 4, where the area for A consists in the areas a and c, while B contains areas B and c. Let us evaluate each term and verify that the left-hand side equals the right-hand side in Eq. (1). $P(A,\ B)$, the probability that an item is simultaneously in A and B, is the area of the intersection, c, divided by the diagram's entire area, $(a + b + c + d)$, or

$$P(A,\ B) = \frac{c}{a + b + c + d}.$$

Fig. 4 Venn diagram

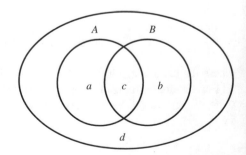

Asking for $P(A \mid B)$ means, "Given that an item is in B, what is the probability that the item is also in A?" B's area is $b + c$, while the part of A that is simultaneously in B is c, so

$$P(A \mid B) = \frac{c}{b+c}.$$

The probability that an item is in B considers the area of B, $(b + c)$, in relationship to the whole diagram, whose area is $(a + b + c + d)$, so that

$$P(B) = \frac{b+c}{a+b+c+d}.$$

Substituting in Eq. (1) and simplifying, as follows, we verify that the left-hand side does indeed equal the right-hand side:

$$P(A, \ B) \overset{?}{=} P(A|B) \cdot P(B)$$

$$\frac{c}{a+b+c+d} \overset{?}{=} \frac{c}{\cancel{(b+c)}} \cdot \frac{\cancel{b+c}}{a+b+c+d}$$

$$\frac{c}{a+b+c+d} = \frac{c}{a+b+c+d}.$$

This argument is not a proof of Eq. (1) but does provide intuition into the equality.

Quick Review Question 3
Using a similar argument as for Eq. (1), provide an intuitive argument for Eq. (2).

If A and B are independent, so that A does not depend on B and vice versa, $P(A \mid B) = P(A)$ and $P(B \mid A) = P(B)$. Thus, for independent events, the equality reduces to the following:

$$P(A \text{ and } B) = P(A, \ B) = P(A) \cdot P(B) \text{ for independent } A \text{ and } B. \tag{3}$$

3.2 Marginal Probability

We can summarize the calculations from the last section, particularly Quick Review Question 2, for the joint probabilities in a matrix as follows:

$$
\begin{array}{cc}
 & \begin{array}{cc} M & B \end{array} \\
\begin{array}{c} E \\ R \end{array} & \begin{bmatrix} 0.24 & 0.06 \\ 0.07 & 0.63 \end{bmatrix} \begin{array}{c} 0.30 \\ 0.70 \end{array} \\
 & \begin{array}{cc} 0.31 & 0.69 \end{array} \quad 1.00
\end{array}
\tag{4}
$$

Each matrix element gives a joint probability. For example, the probability that the monkey is eating (E) while the monitoring system simultaneously indicates munching (M) is $P(E, M) = 0.24$.

As developed in the quick review question, the sum of the probabilities in each row is to the right, and the sums of the column probabilities are below. When the monkey is eating (E), the equipment will either record munching (M) or breathing (B) with probabilities $P(E, M) = 0.24$ and $P(E, B) = 0.06$, respectively. Thus, the probability that the monkey is eating is $0.24 + 0.06 = 0.30$, a value indicated in Holly's HMM by $\pi(E) = 0.30$ and calculated in Quick Review Question 2 e for $\sum_{x \in \{M, B\}} P(E, x) = P(E, M) + P(E, B)$. That is, the probability of E is the sum of the probabilities of hidden value E with each of the possible observable values, M and B; or $P(E) = \sum_{x \in \{M, B\}} P(E, x)$.

Additionally, the values below Matrix (4) are the sums of the probabilities in the columns. Here, we have new information. The probability of the equipment recording munching is 0.31; Holly is either eating (with probability 0.24) or resting/sleeping (with probability 0.07). Quick Review Question 2 h determined this value for $\sum_{x \in \{E, R\}} P(x, M) = P(E, M) + P(R, M)$, which is $P(M)$. We see that the probability of observable value M is equal to the sum of the joint probabilities of M and every possible hidden state, E and R; or $P(M) = \sum_{x \in \{E, R\}} P(x, M)$.

These row and column sums are called **marginal probabilities** because they are written in the "margins" of the matrix. For example, the marginal probability of M is $P(M) = \sum_{x \in \{E, R\}} P(x, M) = P(E, M) + P(R, M) = 0.31$, and we are **marginalizing** over $\{E, R\}$. Notice also that the sum of the marginal probabilities on the right (or below) is 1.00, which is the sum of all the probabilities in the matrix. In general, the marginal probability of y is as follows:

$$P(y) = \sum_{x \in X} P(x, y) = \sum_{x \in X} P(y, x). \tag{5}$$

Quick Review Question 4

Consider the following matrix of joint probabilities:

$$\begin{array}{c} \\ J \\ K \end{array} \begin{array}{ccc} F & G & H \\ \left[\begin{array}{ccc} 0.05 & 0.18 & 0.20 \\ 0.36 & 0.11 & 0.10 \end{array} \right] \end{array}$$

a. Determine $P(J)$.
b. Determine the marginal probability of K.

c. For Part b, give the set over which we are marginalizing.

d. Using the matrix of joint probabilities, the answer for $P(J)$ in Part a, and joint probability formula (1) or (2), determine the conditional probability $P(G \mid J)$.

e. Using the answer for Part b, determine $P(G \mid K)$.

f. Write $P(K)$ using sigma notation.

g. Determine $P(F)$.

h. Using the answer for Part g, determine $P(J \mid F)$.

i. Using the answer for Part g, determine $P(K \mid F)$.

j. Determine the marginal probability of G.

k. For Part j, give the set over which we are marginalizing.

l. Write $P(G)$ using sigma notation.

m. Determine $P(H)$.

3.3 Bayes' Theorem

Bayes' Theorem, or **Bayes' Law**, which follows, is useful in manipulating conditional probabilities:

$$P(A \mid B) = \frac{P(B \mid A) \cdot P(A)}{P(B)}. \tag{6}$$

Sometimes, we do not know $P(A \mid B)$ directly, but we do have values for each of the probability terms on the right. In this case, using Bayes' Theorem we can easily evaluate $P(A \mid B)$. For example, suppose we need to evaluate the probability of Holly eating (E) given that the equipment registers munching (M), $P(E \mid M)$. From Holly's HMM, we know $P(M \mid E) = e(M \mid E) = 0.8$ and $P(E) = \pi(E) = 0.30$. Moreover, in the "Marginal Probability" section, we calculated $P(M) = 0.31$. Thus, we can calculate $P(E \mid M)$ using Bayes' Law as follows:

$$P(E \mid M) = \frac{P(M \mid E) \cdot P(E)}{P(M)} = \frac{0.8 \cdot 0.3}{0.31} \approx 0.77.$$

Exercise 1 provides an intuitive justification for Bayes' Law.

Another version of Bayes' Theorem in which the comma means "and," or "intersection," is as follows:

$$P(A \mid B, C) = \frac{P(B \mid A, C) \cdot P(A \mid C)}{P(B \mid C)}. \tag{7}$$

Notice that this form is the same as the original version, except C is always in the condition. Using a Venn diagram, Exercise 2 provides intuition into Eq. (7).

Quick Review Question 5

Evaluate each of the following using $P(J) = 0.43$, $P(K) = 0.57$, $P(F) = 0.41$, $P(G) = 0.29$, $P(H) = 0.30$, $P(G \mid J) = 0.42$, $P(G \mid K) = 0.35$, $P(J \mid F) = 0.12$, $P(K \mid F) = 0.88$, and the matrix of joint probabilities from Quick Review Question 4:

$$
\begin{array}{c} \\ J \\ K \end{array}
\begin{array}{ccc} F & G & H \end{array}
\begin{bmatrix} 0.05 & 0.18 & 0.20 \\ 0.36 & 0.11 & 0.10 \end{bmatrix}
$$

a. $P(J \mid G)$
b. $P(K \mid G)$
c. $P(F \mid J)$
d. $P(F \mid K)$

4 Probability of a State Given an Observation

Using Bayes' Theorem, other probability equalities, and HMM, we can determine the probabilities of a variety of situations. For example, suppose after Holly scampers off into the jungle, at hour k the monitoring equipment registers $v_k = M$, munching. The scientists might wonder if she really is eating, or, symbolically, if it is true that $u_k = E$. They are asking, "For this single munching reading and without reference to earlier readings, what is the likelihood, or probability, that Holly is eating?" This question falls in the category of an HMM likelihood problem. In notation, they want to discover $P(u_k = E \mid v_k = M)$, abbreviated $P(E \mid M)$; that is, given a reading of M, what is the probability that Holly is in state E.

Examining the HMM model, we do not find the answer directly. However, using Bayes' Theorem, we can rewrite the question as follows:

$$
P(u_k = E \mid v_k = M) = \frac{P(v_k = M \mid u_k = E) \cdot P(u_k = E)}{P(v_k = M)}
$$

or, alternatively,

$$
P(E \mid M) = \frac{P(M \mid E) \cdot P(E)}{P(M)}.
$$

The advantage of this equality is that we can evaluate each of the terms on the right. Consulting Holly's HMM, $P(M \mid E) = e(M \mid E) = 0.8$, and $P(E) = \pi(E) = 0.30$. The denominator, $P(M)$ does take a bit more thought. If the monitoring system records $v_k = M$, Holly could either be eating ($u_k = E$, $v_k = M$) or resting ($u_k = $

R, $v_k = M$). Thus, marginalizing over E, R, we have

$$P(v_k = M) = P(u_k = E,\ v_k = M) + P(u_k = R,\ v_k = M)$$

or

$$P(M) = P(E,\ M) + P(R,\ M).$$

Using a joint probability to determine the first summand, we have

$$P(E,\ M) = P(M,\ E) = P(M\,|\,E) \cdot P(E).$$

That is, the probability of Holly eating leaves and the system recording munching is the same as the probability that the equipment correctly records munching when Holly is eating *and* Holly really is enjoying her dinner. Similarly,

$$P(R, M) = P(M\,|\,R) \cdot P(R).$$

The probability of Holly resting and the system indicating munching is identical to the probability of an incorrect recording of munching when she is resting and Holly is actually inactive. Thus, putting the pieces together determined from joint and marginal probabilities, we have

$$P(M) = P(E,\ M) + P(R,\ M)$$
$$= P(M\,|\,E) \cdot P(E) + P(M\,|\,R) \cdot P(R).$$

Fortunately, from Holly's HMM, we know each of the terms on the right: $P(M\,|\,E) = e(M\,|\,E) = 0.8$, $P(E) = \pi(E) = 0.30$, $P(M\,|\,R) = e(M\,|\,R) = 0.1$, and $P(R) = \pi(R) = 0.70$. Incorporating all calculations, we have the following:

$$P(E\,|\,M) = \frac{P(M\,|\,E) \cdot P(E)}{P(M)} = \frac{P(M\,|\,E) \cdot P(E)}{P(M\,|\,E) \cdot P(E) + P(M\,|\,R) \cdot P(R)}$$

$$P(E\,|\,M) = \frac{0.8 \cdot 0.3}{0.8 \cdot 0.3 + 0.1 \cdot 0.7} \approx 0.77.$$

There is a 77% chance of Holly eating when the system records munching.

In general, for Holly's situation, the following equation determines the probability of an underlying state, u_k, given an observation, v_k:

$$P(u_k\,|\,v_k) = \frac{P(v_k\,|\,u_k) \cdot P(u_k)}{P(v_k)} = \frac{P(v_k\,|\,u_k) \cdot P(u_k)}{P(v_k\,|\,E) \cdot P(E) + P(v_k\,|\,R) \cdot P(R)}$$

or, using sigma notation,

$$P(u_k \mid v_k) = \frac{P(v_k \mid u_k) \cdot P(u_k)}{\sum_{x \in \{E,\ R\}} P(v_k \mid x) \cdot P(x)}.$$

However, most systems have more than two states. Thus, for other HMMs with S being the set of all (hidden) states, we have the following:

$$P(u_k \mid v_k) = \frac{P(v_k \mid u_k) \cdot P(u_k)}{\sum_{x \in S} P(v_k \mid x) \cdot P(x)}. \qquad (8)$$

Quick Review Question 6
For Holly's HMM, determine the following:

a. $P(R \mid M)$
b. $P(R \mid B)$
c. $P(E \mid B)$

Quick Review Question 7
Suppose $P(F) = 0.41$, $P(G) = 0.29$, $P(H) = 0.30$, $P(J \mid F) = 0.12$, $P(J \mid G) = 0.62$, $P(J \mid H) = 0.26$. Determine $P(F \mid J)$.

5   Probability of a Sequence of States Generating a Sequence of Observations

In this section, we consider a problem that would usually not occur but whose answer will help us in the solution of other more realistic problems: Knowing a sequence of states, u, what is the probability of a particular sequence of observations, v, or $P(v \mid u)$. For example, suppose we could spy on Holly in the jungle and discover that initially she rested ($u_1 = R$), but in the next 2 h she ate ($u_2 = E$ and $u_3 = E$), so that $u = REE$. Also, suppose we want to discover the probability that the monitoring equipment registers breathing ($v_1 = B$) followed by munching and then breathing noises in the next two readings ($v_2 = M$ and $v_3 = B$), or $v = BMB$. Thus, we are interested in $P(v = BMB \mid u = REE) = P(BMB \mid REE)$. Figure 5 presents a **trellis diagram** of the situation, with state circles unshaded, observation circles shaded, and arrows denoting conditional dependencies.

Fig. 5 Trellis diagram with state circles unshaded, observation circles shaded, and arrows denoting conditional dependencies

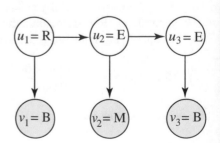

Because of the HMM property of output independence, the probability of each observation, v_i, only depends on the corresponding state, u_i. For example, the probability that the equipment's third reading is B, given that Holly is eating the third hour, $P(v_3 = B \mid u_3 = E)$, is the emission probability, $e(B \mid E) = 0.2$. Because of independence, to evaluate $P(\text{BMB} \mid \text{REE})$, we take the product of the three corresponding emission probabilities as follows:

$$
\begin{aligned}
P(\text{BMB} \mid \text{REE}) &= P(B \mid R) \cdot P(M \mid E) \cdot P(B \mid E) \\
&= e(B \mid R) \cdot e(M \mid E) \cdot e(B \mid E) \\
&= 0.9 \cdot 0.8 \cdot 0.2 \\
&= 0.144.
\end{aligned}
$$

In general, for sequences of length n, $u = u_1 u_2 \ldots u_n$ and $v = v_1 v_2 \ldots v_n$, we have

$$
P(v \mid u) = e(v_1 \mid u_1) \cdot e(v_2 \mid u_2) \ldots e(v_n \mid u_n).
$$

We can abbreviate the product with pi notation to yield the following equality for the probability of an observation sequence given a hidden sequence:

$$
P(v \mid u) = \prod_{i=1}^{n} e(v_i \mid u_i). \tag{9}
$$

Quick Review Question 8

a. Evaluate $P(\text{MMMB} \mid \text{ERER})$
b. Suppose a team catches up with Holly in the jungle and finds her sleeping for 6 h. Calculate the probability that the equipment in the base camp registers correctly records breathing over that period.

6 Probability of a State Sequence and an Observation Sequence

Another likelihood problem considers the probability of a particular observation sequence and a particular state sequence. There is an important difference between this and the problem in the previous section. For the previous problem, given a hidden sequence, u, we wanted to determine the conditional probability of a particular visible sequence, v, namely $P(v \mid u)$. In the current problem, we desire the joint probability that both a hidden sequence, u, and a visible sequence, v, occur, or $P(v, u)$.

For example, in the case of Holly, we might want to know the chance of the equipment registering a sequence of breathing (B), munching (M), and breathing (B), while she is actually resting (R), eating (E), and eating (E). By the joint probability equality (1), we can calculate this probability, $P(\text{BMB}, \text{REE})$, as follows:

$$P(\text{BMB}, \text{REE}) = P(\text{BMB} \mid \text{REE}) \cdot P(\text{REE}). \tag{10}$$

That is, the joint probability of both BMB and REE is the probability of BMB given REE and that REE really does occur.

We determined the conditional probability $P(\text{BMB} \mid \text{REE})$ in the previous section as 0.144, while Holly's Markov model enables us to calculate the chance of the hidden state sequence REE. As the trellis diagram in Fig. 5 illustrates, REE involves three independent events: Holly is initially resting; given that she is resting the first hour, she is eating the next hour; and she continues eating from the second hour to the next. Thus, the probability of the state sequence REE is as follows:

$$P(\text{REE}) = P(u_1 = \text{R}) \cdot P(u_2 = E \mid u_1 = \text{R}) \cdot P(u_3 = E \mid u_2 = \text{E}).$$

From Holly's HMM, we know $P(u_1 = \text{R}) = \pi(\text{R}) = 0.70$ and the transition probabilities $P(u_2 = \text{E} \mid u_1 = \text{R}) = t(\text{R}, \text{E}) = 0.2$ and $P(u_3 = \text{E} \mid u_2 = \text{E}) = t(\text{E}, \text{E}) = 0.6$. Thus, the probability of the state sequence REE is as follows:

$$P(\text{REE}) = \pi(\text{R}) \cdot t(\text{R}, \text{E}) \cdot t(\text{E}, \text{E}) = 0.70 \cdot 0.2 \cdot 0.6 = 0.084.$$

Substituting values into Eq. (10), we have the following:

$$P(\text{BMB}, \text{REE}) = P(\text{BMB} \mid \text{REE}) \cdot P(\text{REE}) = 0.144 \cdot 0.084 \approx 0.012.$$

Although given hidden sequence REE, there is approximately a 14% chance of observable sequence BMB, there is only about a 1% chance of BMB and REE simultaneously occurring.

Quick Review Question 9
Use the answers from Quick Review Question 8 as needed.

a. Evaluate $P(\text{MMMB},\text{ERER})$
b. Find the probability that Holly sleeps for 6 h and the monitoring equipment registers breathing over that same time period.

Generalizing, based on the Markov property, the probability of a sequence of observations, u, is as follows:

$$P(u) = \pi(u_1) \cdot \prod_{i=2}^{n} t(u_{i-1}, u_i). \tag{11}$$

Moreover, considering individual sequence elements, we have the following:

$$P(v \mid u) = \prod_{i=1}^{n} P(v_i \mid u_i) = \prod_{i=1}^{n} e(v_i \mid u_i). \tag{12}$$

Thus, using Eqs. (11) and (12), the generalized formula for the joint probability of particular observable and hidden sequences is as follows:

$$P(v, u) = P(v \mid u) \cdot P(u) = \left(\prod_{i=1}^{n} e(v_i \mid u_i) \right) \left(\pi(u_1) \cdot \prod_{i=2}^{n} t(u_{i-1}, u_i) \right). \tag{13}$$

A useful reorganization of Eq. (13), which follows, is evocative of the trellis diagram in Fig. 5:

$$P(v, u) = \pi(u_1) \cdot e(v_1 \mid u_1) \cdot \left(\prod_{i=2}^{n} [t(u_{i-1}, u_i) \cdot e(v_i \mid u_i)] \right)$$

$$= \pi(u_1) \cdot e(v_1 \mid u_1) \cdot t(u_1, u_2) \cdot e(v_2 \mid u_2) \cdot t(u_2, u_3) \cdot$$

$$e(v_3 \mid u_3) \cdots t(u_{n-1}, u_n) \cdot e(v_n \mid u_n).$$

Regrouping this expression, as follows, is revealing for the joint probability $P(v, u) = P(u, v)$:

$$P(v, u) = \left[\cdots \left[[[\pi(u_1) \cdot e(v_1 \mid u_1)] \cdot t(u_1, u_2) \cdot e(v_2 \mid u_2)] \cdot t(u_2, u_3) \cdot e(v_3 \mid u_3) \right] \cdots \right]$$

$$\cdot t(u_{n-1}, u_n) \cdot e(v_n \mid u_n).$$

Notice that the product of the first two terms, which are in the inner brackets, $[\pi(u_1) \cdot e(v_1 \mid u_1)]$, is $P(u_1, v_1)$, or $P(u, v)$, when u and v are one-element sequences. Moreover, the product of the terms in the next set of brackets is $P(u_{1,2}, v_{1,2})$, which is $P(u, v)$ when u and v are two-element sequences. The expression in the next set of brackets calculates $P(u_{1,3}, v_{1,3})$; and the expression in the outermost set of brackets is $P(u_{1,n-1}, v_{1,n-1})$, which is $P(u, v)$ for sequences of $n - 1$ elements. Thus, we can use recursion to define $P(u, v)$. A recursive task is one that calls itself. In this example, we employ the joint probability $P(u_{1,n-1}, v_{1,n-1})$ in the definition of the joint probability $P(u, v)$. The following recursive formula for $P(u, v)$ will be useful in further algorithms:

$$P(u, v) = P(u_{1,n}, v_{1,n}) = \begin{cases} \pi(u_1) \cdot e(v_1 \mid u_1), & \text{if } n = 1 \\ P(u_{1,n-1}, v_{1,n-1} \cdot t(u_{n-1}, u_n) \cdot e(v_n \mid u_n), & \text{if } n > 1 \end{cases}. \tag{14}$$

7 Probability of a Sequence of Observations: The Forward Algorithm

A more realistic likelihood problem is to determine the probability of an observed sequence, $P(v)$. For example, we might want to calculate the probability that the monitoring equipment registers breathing sounds and then munching noises the next 2 h, $P(v = \text{BMM})$.

7.1 Obvious Solution

In the last section, we determined the simultaneous probability of a visible and a hidden sequence. Thus, the obvious solution for this short sequence is to marginalize over all the possible three-element hidden state sequences. For $S = \{\text{E, R}\}$, there are two choices, E and R, for each of the three positions, yielding $2^3 = 8$ possible hidden sequences. Thus, marginalizing over this set of eight underlying sequences, $U = \{\text{RRR, RRE, RER, REE, ERR, ERE, EER, EEE}\}$, we can calculate $P(\text{BMM})$ as follows:

$$P(\text{BMM}) = \sum_{u \in U} P(\text{BMM}, u) = \sum_{u \in U} [P(\text{BMM} \mid u) \cdot P(u)].$$

Expanding the first sum, we have the following:

$$P(\text{BMM}) = P(\text{BMM, RRR}) + P(\text{BMM, RRE}) + P(\text{BMM, RER})$$
$$+ P(\text{BMM, REE}) + P(\text{BMM, ERR}) + P(\text{BMM, ERE})$$
$$+ P(\text{BMM, EER}) + P(\text{BMM, EEE}).$$

We can calculate each of these joint probabilities using Eqs. (12) or (13) from the last section. Such a solution seems feasible, although tedious. However, consider the situation where we have sequences of 10 elements. The number of 10-element sequences formed from $\{\text{E, R}\}$ is $2^{10} = 1024$, so that we would have 1024 summands. Doubling to 20 observations, the number of possible sequences, and, hence, the number of summands, is over a million, $2^{20} = 1,048,576$. Astoundingly, the number sequences of length 100 is over 10^{30}. As these calculations illustrate, for h hidden states and n observations, there are h^n number of possible hidden sequences; as n gets larger, h^n grows exponentially. Thus, this solution has an exponential growth rate on the order of h^n, written $O(h^n)$.

Quick Review Question 10
For each string length, find the number of strings formed from the bases A, C, T, and G.

a. 3
b. 10
c. 20
d. 21

7.2 Forward Algorithm

For realistic problems, often the number of states (h) and certainly the number of observations (n) are large, so that h^n is enormous, too great to compute by hand, and, in fact, intractable even for a computer. Clearly, we must find a better way to calculate $P(v)$, the probability of an observed sequence. Fortunately, the **forward algorithm** is much faster. The algorithm employs **dynamic programming**, which divides a problem into a collection of smaller problems and uses the solutions to these smaller problems to solve the larger problem. For the forward algorithm, we store answers to the smallest problems in the first column of a matrix. Repeatedly, we solve progressively larger problems, employing the answers in the previous column and storing the answers in a new column. Finally, the answer to the overall problem, the probability of an observed sequence, is the sum of the elements in the last column.

For $P(\text{BMM})$, we employ a matrix with the same number of rows as the number of states, h, and the same number of columns as the length of the observation sequence, n. For Holly's HMM, with $S = \{E, R\}$ and observed sequence $v = \text{BMM}$, $h = 2$ and $n = 3$, so we store values in a 2×3 matrix, F, with row and column headings from S and v, respectively:

$$F = \begin{array}{c} \\ E \\ R \end{array} \begin{array}{c} \text{B M M} \\ \begin{bmatrix} \square\ \square\ \square \\ \square\ \square\ \square \end{bmatrix} \end{array}.$$

7.3 Forward Algorithm Initialization

Initially, we solve the smaller problem of $P(\text{B})$, which by marginalizing Eq. (5) is as follows:

$$P(\text{B}) = \sum_{x \in S} P(\text{B}, x) = P(\text{B, E}) + P(\text{B, R}).$$

We will place the calculation of $P(\text{B, E})$ in f_{E1}, the element in the E row and first column of F, while f_{R1} will be $P(\text{B, R})$. The probability of $u_1 = \text{B}$ is the sum of these two values in the first column. For the calculation of the summands, we employ the step for $n = 1$ of the recursive formula (14) as follows:

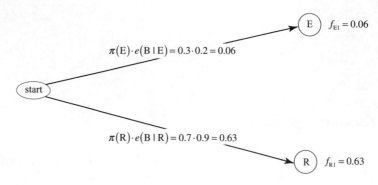

Fig. 6 Initialization step of the forward algorithm

$$P(B, E) = \pi(E) \cdot e(B \mid E)$$

and

$$P(B, R) = \pi(R) \cdot e(B \mid R).$$

The initial probabilities are $\pi(E) = 0.30$ and $\pi(R) = 0.70$, and the emission probabilities are $e(B \mid E) = 0.2$, $e(B \mid R) = 0.9$. Thus, as Fig. 6 illustrates, $P(B, E) = 0.06$ and $P(B, R) = 0.63$. Thus, placing these values from the initialization step in the first column, the matrix F is as follows:

$$F = \begin{array}{c} \\ E \\ R \end{array} \begin{array}{ccc} B & M & M \\ \left[\begin{array}{ccc} 0.06 & \square & \square \\ 0.63 & \square & \square \end{array} \right] \end{array}.$$

Quick Review Question 11

The HMM in Quick Review Question 1 and Fig. 3 contains the following information:

$S = \{A, B, C\}$ and $O = \{G, H\}$

$\pi(A) = 0.2$, $\pi(B) = 0.1$, $\pi(C) = 0.7$

$$\begin{array}{c} u_k/u_{k+1} \end{array} \begin{array}{ccc} A & B & C \end{array}$$

$$T = \begin{array}{c} A \\ B \\ C \end{array} \left[\begin{array}{ccc} 0.1 & 0.4 & 0.5 \\ 0.2 & 0.2 & 0.6 \\ 0.3 & 0.2 & 0.5 \end{array} \right]$$

$$\begin{array}{cc}
hidden/observable & \begin{array}{cc} G & H \end{array} \\
\begin{array}{c} A \\ B \\ C \end{array} & \left[\begin{array}{cc} 0.9 & 0.1 \\ 0.6 & 0.4 \\ 0.1 & 0.9 \end{array}\right].
\end{array}$$

Suppose we wish to use the forward algorithm to calculate $P(\text{HHGH})$.

a. Give the size of F, the forward algorithm matrix.
b. Calculate f_{A1}.
c. Write the conditional probability for f_{A1} symbolically; that is, write f_{A1} as $P(x \mid y)$ with the symbols for x and y for this example.
d. Calculate f_{B1}.
e. Write the conditional probability for f_{B1} symbolically.
f. Calculate f_{C1}.
g. Write the conditional probability for f_{C1} symbolically.
h. Calculate $P(\text{H})$.

7.4 Forward Algorithm Step 2

In the calculation of $P(\text{BMM})$, after initialization, which calculates the summands of $P(\text{B})$, we approach a little bit larger problem, $P(\text{BM})$. Fortunately, as we will see, we can use the values in the first column in the solution of this probability. As we did in the initialization step, we employ marginality to split the problem into a sum of joint probabilities. With marginality, we have

$$P(\text{BM}) = \sum_{x \in S^2} P(\text{BM}, x), \tag{15}$$

where S^2 represents the set of all two-element sequences with elements from $S = \{\text{E, R}\}$, $S^2 = \{\text{EE, ER, RE, RR}\}$; and, in general, S^n denotes the set of all n-element sequences over S. Thus, expanding Eq. (15) symbolically, we have the following sum:

$$P(\text{BM}) = P(\text{BM, EE}) + P(\text{BM, ER}) + P(\text{BM, RE}) + P(\text{BM, RR}). \tag{16}$$

Let us calculate one of these summands, $P(\text{BM, ER})$. Applying the second line of the recursive definition of $P(u, v)$, Eq. (14), we have the following:

$$P(\text{BM, ER}) = P(\text{B, E}) \cdot t(\text{E, R}) \cdot e(\text{M} \mid \text{R}).$$

Fortunately, we calculated $P(B,E)$ in the initialization step, and its value is matrix element $f_{E1} = 0.06$. Moreover, Holly's HMM gives $t(E, R)$ and $e(M \mid R)$ as 0.4 and 0.1, respectively. Thus, $P(BM, ER) = 0.06 \cdot 0.4 \cdot 0.1 = 0.0024$.

Quick Review Question 12
Calculate $P(BM, RR)$.

For f_{E2} we add the elements of Eq. (16) with sequences that end in E. Similarly, f_{R2} is the sum of elements with sequences that end in R. Rewriting Eq. (16) illustrates the process:

$$P(BM) = [P(BM, EE) + P(BM, RE)] + [P(BM, E\underline{R}) + P(BM, R\underline{R})] \tag{17}$$
$$= \qquad f_{E2} \qquad + \qquad f_{R2}.$$

With sigma notation, we have the following equations for the second-column elements:

$$f_{E2} = \sum_{x \in S} P(BM, xE)$$

and

$$f_{R2} = \sum_{x \in S} P(BM, xR).$$

From the calculations above and in Quick Review Question 12, we know that $P(BM, ER) = 0.0024$ and $P(BM, RR) = 0.0504$. Thus, $f_{R2} = P(BM, ER) + P(BM, RR) = 0.0024 + 0.0504 = 0.0528$.

Quick Review Question 13
Using $P(BM, EE) = 0.0288$ and $P(BM, RE) = 0.1008$, calculate f_{R2}.

Figure 7 illustrates the calculation of the second column of matrix F with $P(BM, EE) + P(BM, RE) = f_{E2}$ and $P(BM, E\underline{R}) + P(BM, R\underline{R}) = f_{R2}$. Thus, the developing F now is as follows:

$$F = \begin{array}{c} \\ E \\ R \end{array} \begin{array}{ccc} B & M & M \\ \left[\begin{array}{ccc} 0.06 & 0.1296 & \Box \\ 0.63 & 0.0528 & \Box \end{array} \right] \end{array}.$$

Moreover, the sum of the elements in the second column, 0.1824, is $P(BM)$; that is, for sequences of length two, there is an 18.24% chance of BM being an output sequence.

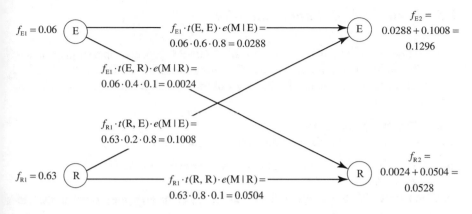

Fig. 7 Calculation of the second column of forward matrix, F

Quick Review Question 14

The HMM in Quick Review Questions 1 and 11 and Fig. 3 contains the following information:

$S = \{A, B, C\}$ and $O = \{G, H\}$

$\pi(A) = 0.2, \pi(B) = 0.1, \pi(C) = 0.7$

$$
\begin{array}{cc}
u_k/u_{k+1} & \begin{array}{ccc} A & B & C \end{array} \\
\begin{array}{c} A \\ T = B \\ C \end{array} &
\left[\begin{array}{ccc}
0.1 & 0.4 & 0.5 \\
0.2 & 0.2 & 0.6 \\
0.3 & 0.2 & 0.5
\end{array}\right]
\end{array}
$$

$$
\begin{array}{cc}
hidden/observable & \begin{array}{cc} G & H \end{array} \\
\begin{array}{c} A \\ B \\ C \end{array} &
\left[\begin{array}{cc}
0.9 & 0.1 \\
0.6 & 0.4 \\
0.1 & 0.9
\end{array}\right]
\end{array}.
$$

Suppose we wish to use the forward algorithm to calculate $P(\text{HHGH})$. As calculated in Quick Review Question 11, the first column of the forward algorithm matrix F contains $f_{A1} = 0.02$, $f_{B1} = 0.04$, and $f_{C1} = 0.63$.

a. Give the number of summands in the calculation of f_{B2}.
b. For f_{B2}, calculate the first summand, which employs f_{A1} as a factor.
c. For f_{B2}, calculate the second summand, which employs f_{B1} as a factor.
d. For f_{B2}, calculate the third summand.
e. Calculate f_{B2}.

7.5 Forward Algorithm Completion

The calculation of the third column of F for $P(BMM)$ proceeds in a similar manner to that of the second column. We use the small-problem answers in the previous column to calculate the values for the new column. Quick Review Question 15 steps through the process.

Quick Review Question 15
This question completes the forward algorithm for $P(BMM)$.

a. Write $P(BMM)$ using sigma notation similar to that in Eq. (15).
b. Write out S^3.
c. Enumerate $P(BMM)$ symbolically as in Eq. (17), grouping together the elements of f_{E3} and f_{R3}.
d. Notice in the calculation of f_{E3} that two probabilities have sequences that end in EE, $P(BMM, EEE)$ and $P(BMM, REE)$. Moreover, by Eq. (14), we have the following:

$$P(BMM, \text{EEE}) = P(BM, \text{EE}) \cdot t(\mathbf{E}, \mathbf{E}) \cdot e(\mathbf{M} \,|\, \mathbf{E})$$

$$P(BMM, \text{REE}) = P(BM, \text{RE}) \cdot t(\mathbf{E}, \mathbf{E}) \cdot e(\mathbf{M} \,|\, \mathbf{E}).$$

With the last two factors being identical, we can group the sum of the probabilities as follows:

$$P(BMM, \text{EEE}) + P(BMM, \text{REE}) = [P(BM, \text{EE}) + P(BM, \text{RE})] \cdot t(\mathbf{E}, \mathbf{E}) \cdot e(\mathbf{M}|\mathbf{E}).$$

In a similar manner, show the development to write the sum of the remaining two terms of f_{E3} using f_{R2}.
e. As in Part d, write the sum of two terms of f_{E3} using f_{E2}.
f. As in Part d, write the sum of two terms of f_{R3} using f_{E2}.
g. Evaluate $P(BMM, EEE) + P(BMM, REE)$, developed symbolically in Part d.
h. Referring to the answer to Part d, evaluate $P(BMM, ERE) + P(BMM, RRE)$.
i. Referring to the answer to Part e, evaluate $P(BMM, EER) + P(BMM, RER)$.
j. Referring to the answer to Part f, evaluate $P(BMM, ERR) + P(BMM, RRR)$.
k. Using the answers to Parts g and i, calculate f_{E3}.
l. Using the answers to Parts h and j, calculate f_{R3}.
m. Calculate $P(BMM)$

Figure 8 illustrates the development of the last column of the forward matrix, which Quick Review Question (15) solved. Moreover, with this and previous calculations, we can complete the forward matrix as follows:

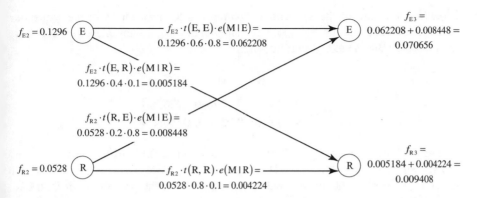

Fig. 8 Step 3 of forward algorithm in calculation of $P(\text{BMM})$

$$F = \begin{matrix} & \text{B} & \text{M} & \text{M} \\ \text{E} & \\ \text{R} & \end{matrix} \begin{bmatrix} 0.06 & 0.1296 & 0.070656 \\ 0.63 & 0.0528 & 0.094080 \end{bmatrix} \qquad (18)$$

Then, we can calculate $P(\text{BMM})$ as the sum of the elements in the last column, or $P(\text{BMM}) = 0.080064$. There is approximately an 8% chance of the output sequence BMM.

As the work for Quick Review Question 15 indicates, we can iterate through the matrix, using one column in the evaluation of the next. For $S = \{E, R\}$ and ith output symbol, v_i, we have.

$$f_{Ei} = f_{E(i-1)} \cdot t(E, E) \cdot e(v_i \mid E) + f_{R(i-1)} \cdot t(R, E) \cdot e(v_i \mid E)$$

$$= \sum_{y \in S} \left[f_{y(i-1)} \cdot t(y, E) \cdot e(v_i \mid E) \right]$$

and

$$f_{Ri} = f_{E(i-1)} \cdot t(E, R) \cdot e(v_i \mid R) + f_{R(i-1)} \cdot t(R, R) \cdot e(v_i \mid R)$$

$$= \sum_{y \in S} \left[f_{y(i-1)} \cdot t(y, R) \cdot e(v_i \mid R) \right].$$

In general, for state x,

$$f_{xi} = \sum_{y \in S} \left[f_{y(i-1)} \cdot t(y, x) \cdot e(v_i \mid x) \right].$$

Moreover, the sum of the elements in column i is the probability of the sequence of the first i observations, $P(v_{1,i})$. For example, in (18), we completed the forward matrix for $P(\text{BMM})$ to be the following:

$$F = \begin{array}{c} \\ E \\ R \end{array} \begin{array}{ccc} B & M & M \\ \left[\begin{array}{ccc} 0.06 & 0.1296 & 0.070656 \\ 0.63 & 0.0528 & 0.094080 \end{array}\right] \end{array} \cdot$$

The sum of the elements in the first column is $P(\text{B}) = 0.69$, while the sum of the second-column elements is $P(\text{BM}) = 0.1824$. The answer we desire, $P(\text{BMM})$, is the sum of the elements in the final column, 0.080064; the probability of BMM is approximately 8%. To summarize,

$$P(v) = \sum_{x \in S} f_{xn}, \tag{19}$$

where v is a sequence of n observations and f_{xn} is the element in row x and column n of the forward matrix F. The following quick review question considers the total number of calculations for such a probability using the forward algorithm.

Quick Review Question 16
This question analyzes the complexity of the forward algorithm.

a. Suppose a HMM has 5 hidden states. Give the number of rows in the forward matrix, F.
b. In Holly's HMM with 2 hidden states, the calculation of each of the two elements in the first column involved one product, for a total of 2 products. For 5 hidden states, give the total number of products in the evaluation of the first column.
c. For h states, give the total number of products in the evaluation of the first column.
d. Suppose x is one of these 5 hidden states. Not including the initialization step, give the number of terms having $f_{x(i-1)}$ as a factor, where $f_{x(i-1)}$ is the forward matrix value in row x and column $(i-1)$. This value is the same number of lines emanating from an f-value. For example, in Fig. 8, two terms had f_{2E} (or f_{2R}) as a factor and two lines emanating from f_{2E} (or f_{2R}).
e. For 2 hidden states, each of these terms involves two products. Give the number of products when we have 5 hidden states.
f. Give the number of these products when we have h hidden states. Note that the number of products does not depend on the number of hidden states.
g. Give the number of elements in the forward matrix for 5 hidden states and 20 observations, not including those in the first column.
h. Give the number of elements in the forward matrix for h hidden states and n observations, not including those in the first column.
i. Using your answers from Parts b, e, and g, for 5 hidden states and 20 observations, give the total number of products.

j. Using your answers from Parts c, f, and h, for h hidden states and n observations, give the total number of products. Simplify the result.

k. Give the total number of sums used in the calculations of the elements in the first column of a forward matrix, F.

l. For 5 hidden states and not including the initialization step, give the number of sums used in the calculation of f_{xi}, where x is a hidden state and i is a column. For example, in Fig. 8, f_{3E} (or f_{3R}) has two summands and, thus, one sum.

m. For h hidden states, give the number of sums used in the calculation of f_{xi}, where x is a hidden state and i is a column greater than 1.

n. For h hidden states and n observations, give the total number of sums used in the calculations of the elements of the forward matrix. Use answers from Parts h, k, and m. Simplify the result.

o. For h hidden states and n observations, give the total number of sums and products in the calculation of the elements of F. Use answers from Parts j and n.

p. Give the complexity of the forward algorithm.

As indicated in Quick Review Question 16, for h states and n observations, the forward matrix, F, has h rows and n columns, yielding hn number of elements. Most of these elements employ h summands of three factors each, although elements in the first column have fewer terms. Thus, the approximate amount of work involved in creating the forward matrix is on the order of nh^2. Moreover, to calculate the probability of the observed sequence, we add the h elements in the last column. In comparison to nh^2, h is inconsequential, so the forward algorithm is on the order of nh^2, written $O(nh^2)$. The complexity $O(nh^2)$ is a dramatic improvement over the exponential complexity, $O(h^n)$, for our original solution. For example, suppose our HMM has $h = 10$ hidden states and we wish to know the probability of a sequence of $n = 9$ observations. The original approach to the solution would involve $h^n = 10^9 = 1,000,000,000$, a billion calculations, while the forward algorithm would employ only about $nh^2 = 9 \cdot 102 = 900$ calculations.

8 Probability of a Genomics Sequence

In this section, we employ the forward algorithm to solving an important biological problem—locating genes.

8.1 Biology Background

Proteins, which are the basic molecules of life, perform many essential functions, such as forming the structural components of cells and, in the case of protein enzymes, catalyzing chemical reactions. Simple proteins are chains formed from 20 amino acids.

In the cell, the nucleic acid **DNA (deoxyribonucleic acid)** contains the encoded information for the building of all the proteins needed by a cell. DNA acts through an intermediary nucleic acid, **RNA (ribonucleic acid)**, to synthesize proteins. RNA has one long chain of molecules, called **nucleotides**, while two strands of nucleotides compose DNA. Each nucleotide is made up of a sugar, a phosphate, and a nitrogen base, which can be **adenine (A)**, **guanine (G)**, **cytosine (C)**, or **thymine (T)** in DNA or **uracil (U)** in RNA. In DNA, A in one strand bonds with T in the other strand, while C and G bond together. Each of these pairs of **complementary bases** is referred to as a **base pair (bp)**.

Virtually every cell (except red blood cells) in the human body contains **chromosomes**, or sequences of very long DNA molecules. The **genome** is the complete set of chromosomes in a cell and contains the organism's hereditary information. For example, a human genome has 23 pairs of chromosomes (46 total) with each pair having one chromosome from each parent. A **gene** is a subsequence of a chromosome that contains information for building a protein. Although lengths vary greatly, an average gene contains about 28,000 base pairs [1]. Some contiguous sections of each chromosome are not part of any gene, but some are important for the regulation of gene expression.

8.2 Locations of Genes

We can employ the forward algorithm to solve an important problem in genomics—the locations of genes—by helping us find areas of high CG concentration. In mammalian DNA, the sequence of bases CG appears more frequently upstream of, or before, a gene than in other parts of a DNA sequence, where CG occurs much less than expected from random occurrences of C and G. We call an area of greater concentration of CG a **CpG island**, where "p" in "CpG" represents a phosphate linking the bases C and G. The reason that CG occurs infrequently in much of a sequence of DNA is that often CG transforms to (methyl-C)G before mutating to TG. However, the transformation from CG to TG is suppressed in the CpG islands. A DNA segment of 200 bases is a CpG island if CG occurs at least 50% of the time and the ratio of observed-to-expected number of CpGs is greater than 0.6 [7].

Figure 9a and b contains transition matrices for sequences in and not in CpG islands, respectively. The matrices were derived from a database of human DNA

a		*xi*				b		*xi*			
	+	A	C	G	T		+	A	C	G	T
	A	0.180	.0274	0.426	0.120		A	0.300	0.205	0.285	0.210
	C	0.171	0.368	0.274	0.188		C	0.322	0.298	0.078	0.302
xi	G	0.161	0.339	0.375	0.125	*xi*	G	0.248	0.246	0.298	0.208
	T	0.079	0.355	0.384	0.182		T	0.177	0.239	0.292	0.292

Fig. 9 Possible transition matrix for (**a**) samples within CpG islands and for (**b**) samples not within CpG islands [4]

sequences using 48 accepted CpG islands. In the matrix for CpG islands, the probability of the pair CG (or the probability that G occurs, given that C has just appeared) is 0.274, written as $P(x_i = G \mid x_{i-1} = C) = P(G \mid C) = 0.274$, while in the CpG negative matrix (Fig. 9b), the probability of the sequence CG is much lower, $P(G \mid C) = 0.078$. Stanke [21] uses $\pi = (0.148, 0.334, 0.365, 0.154)$ as the initial state probability vector for CpG islands and $\pi = (0.260, 0.249, 0.241, 0.251)$ for non-CpG islands. For each situation, we assume the emission matrix is as follows:

$$
\begin{array}{cc}
hidden/observable & A\ C\ G\ T \\
\begin{array}{c} A \\ C \\ G \\ T \end{array} &
\begin{bmatrix}
1 & 0 & 0 & 0 \\
0 & 1 & 0 & 0 \\
0 & 0 & 1 & 0 \\
0 & 0 & 0 & 1
\end{bmatrix}
\end{array} \cdot
$$

Project 5 employs these hidden Markov models in accessing the probabilities of output sequences inside and outside CpG islands.

9 Parallel Forward Algorithm (Optional)

Although the forward algorithm has a big improvement in speed over the brute-force technique discussed initially, for a large number of observations and states, such as often encountered in genetics problems, the sequential forward algorithm can take a long time. To speed the task, we can employ parallel programming.

9.1 Communication

One obvious way to parallelize is to have different processes calculate different initial values for the first column and to have different processes compute separate summands, such as those in the middle sections of Figs. 7 and 8. In Holly's HMM with $s = 2$ states, we employ 2 processes for the initialization step; and in general, for s states, each of s processes could compute a different first-column element. In subsequent steps for Holly's HMM as in Figs. 7 and 8, each new state is a linear combination of the previous states, so each process calculates the new state by using the inputs from the arrows that point to that state. In theory, s processes are needed for s states to calculate the summands simultaneously. In reality, communication of appropriate forward matrix elements must occur for evaluation of the next column elements, and such communication limits the speedup of the parallel algorithm. Thus, it can often be more efficient to have less processes, so that each process does more work and less communication.

The heart of the forward algorithm is the matrix that holds the probabilities for the sequence of observations. From the perspective of the parallel program, each process could have its own copy of the matrix and would have to communicate the value of its calculated cell to all the other processes. Alternatively, the processes could share the same matrix, which would reduce the amount of needed communication. Such consideration of communication is important in deciding how to implement the parallel algorithm.

9.2 Implementation of the Parallel Forward Algorithm

The parallel forward algorithm, available to professors by request, is implemented using the OpenMP library. OpenMP (OMP) uses threads, which are like processes except that the threads share the same memory. Thus, threads communicate by reading and writing to the same matrix.

Communication limits speedup of the algorithm because when multiple threads have to use the same matrix cell, these lightweight processes must take turns—two threads cannot access a cell simultaneously. Thus, increasing the number of threads for any parallel program comes with a tradeoff: Each thread has less work to do, so the work can be done faster; but a thread must wait longer when multiple threads need to access the same matrix cell. Speedup of a parallel program depends on the number of threads used, and increasing the number of threads to its theoretical limit does not always improve speedup.

To illustrate, Table 1 gives timings and speedups and Fig. 10 depicts these timings for serial and parallel OMP implementations of the forward algorithm. One definition of **speedup** using n threads (or processes) is the length of time to execute the serial algorithm divided by the length of time to execute the corresponding parallel algorithm. Notice that the speedup of the parallel algorithm is a function of the number of states and the number of threads. For example, for 2048 states, the speedup is the greatest at about 4 threads:

$$\text{speedup} = 9.9255\,\text{s}/\,4.0269\,\text{s} = 2.4648.$$

However, with 4096 states, 8 threads are better:

$$\text{speedup} = 43.6465\,\text{s}/\,11.9184\,\text{s} = 3.6621.$$

Even with more states, the use of more threads eventually suppresses speedup because of the need for additional communication.

Table 1 Timings and speedup for forward algorithm

Time in seconds								
# States	Serial	1 thread	2 threads	3 threads	4 threads	8 threads	16 threads	32 threads
512	0.214	0.2490	0.1696	0.1722	0.2473	0.7697	1.3663	2.4829
1024	2.6585	2.7506	1.7230	1.1925	1.0848	1.8645	2.8908	4.9911
2048	**9.9255**	10.0194	7.2314	5.6453	**4.0269**	4.1924	5.9881	10.0612
4096	**43.6465**	44.2792	27.9575	20.2691	16.7255	**11.9184**	13.4140	20.6784

Speedup								
# States		1 thread	2 threads	3 threads	4 threads	8 threads	16 threads	32 threads
512		0.8593	1.2619	1.2424	0.8653	0.2780	0.1566	0.0862
1024		0.9665	1.5429	2.2293	2.4507	1.4258	0.9196	0.5326
2048		0.9906	1.3726	1.7582	**2.4648**	2.3675	1.6575	0.9865
4096		0.9857	1.5612	2.1534	2.6096	**3.6621**	3.2538	2.1107

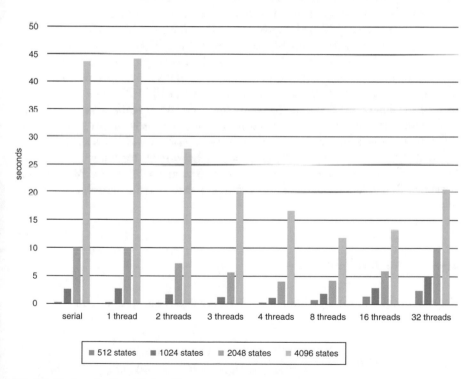

Fig. 10 Timings for forward algorithm

10 Decoding Problem

We employ the HMM forward algorithm to solve likelihood problems, such as the likelihood of monitoring equipment registering breathing sounds (B) and then munching noises (M) the next 2 h for Holly Howler, $P(v = \text{BMM})$. We use the **Viterbi algorithm** to solve another type of HMM problem, **decoding**. In this case, given a sequence of observations, such as $v = \text{BMM}$, we determine the most probable sequence of underlying states, u, to yield v. Thus, we wish to determine the u with maximum $P(u \mid v)$, such as the sequence of three states, u, that yields $\max(P(u \mid v = \text{BMM}))$.

10.1 Obvious Solution

In the development above, we learned by joint probability that $P(u, \text{BMM}) = P(u \mid \text{BMM}) \cdot P(\text{BMM})$. Thus, the problem of determining the state sequence u that maximizes $P(u \mid \text{BMM})$ is equivalent to the problem of determining the u that maximizes $P(u, \text{BMM})/P(\text{BMM})$. However, because $P(\text{BMM})$ is a constant, the problem simplifies to finding u where $P(u, \text{BMM})$ is maximum. The most obvious method to solve the problem is to determine $P(u, \text{BMM})$ for every possible three-element state sequence, $S^3 = \{\text{RRR, RRE, RER, REE, ERR, ERE, EER, EEE}\}$, using a version of the forward algorithm, and then to select the u with the largest probability. However, this solution is exponentially large because for a sequence of n observations with $h = 2$ states, the number of possible n-element state sequences, or the number of elements in S^n, is $h^n = 2^n$. In general, for h number of states, the number of n-element state sequences is h^n. The much faster **Viterbi algorithm** is another dynamic programming algorithm, which has the forward algorithm as its base.

10.2 Viterbi Algorithm

The key to the Viterbi algorithm is Eq. (14), repeated below, for calculating a joint probability:

$$P(u, v) = P(u_{1,n}, v_{1,n}) = \begin{cases} \pi(u_1) \cdot e(v_1|u_1), & \text{if } n = 1 \\ P(u_{1,n-1}, v_{1,n-1}) \cdot t(u_{n-1}, u_n) \cdot e(v_n|u_n), & \text{if } n > 1 \end{cases} \tag{14}$$

Beginning the Viterbi algorithm in the same way as the forward algorithm for observation sequence BMM in Holly's HMM, we employ a 2×3 matrix, G, with first-column elements being $g_{E1} = P(\text{B, E}) = \pi(\text{E}) \cdot e(\text{B} \mid \text{E})$ and $g_{R1} = P(\text{B, R}) = \pi(\text{R}) \cdot e(\text{B} \mid \text{R})$. The initialization step is identical to that of the forward algorithm (see Fig. 6), resulting in the following initial Viterbi matrix, G:

$$G = \begin{array}{c} \\ E \\ R \end{array} \begin{array}{cc} B & M\ M \\ \left[\begin{array}{ccc} 0.06 & \square & \square \\ 0.63 & \square & \square \end{array}\right] \end{array}.$$

Quick Review Question 17

Calculate the first-column elements of the Viterbi matrix to calculate u for $\max(P(u \mid HHGH))$ using the HMM in Fig. 11, which contains the following information:

$$S = \{A, B, C\} \text{ and } O = \{G, H\}$$

$$\pi(A) = 0.2, \quad \pi(B) = 0.1, \quad \pi(C) = 0.7$$

$$\begin{array}{cccc} u_k/u_{k+1} & A & B & C \\ A & \left[\begin{array}{ccc} 0.1 & 0.4 & 0.5 \\ \end{array}\right. \\ T = B & \left. 0.2 & 0.2 & 0.6 \right. \\ C & \left. 0.3 & 0.2 & 0.5 \right] \end{array}$$

$$\begin{array}{cc} hidden/observable & G \quad H \\ A & \left[\begin{array}{cc} 0.9 & 0.1 \\ \end{array}\right. \\ B & \left. 0.6 & 0.4 \right. \\ C & \left. 0.1 & 0.9 \right] \end{array}.$$

Computations of the second column of the Viterbi and the forward matrices also begin in the same way with calculating the product of a first-column element, a transition value, and an emission value. As in Fig. 7 for the forward algorithm, Fig. 12 of the current module for the Viterbi algorithm makes the following

Fig. 11 HMM diagram for Quick Review Question 17

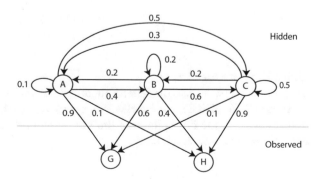

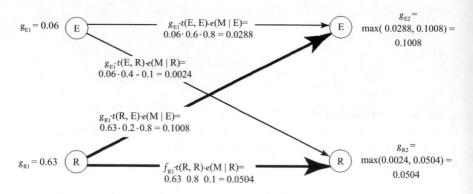

Fig. 12 Calculation of the second column of Viterbi matrix, G

computations using first-column values, g_{E1} and g_{R1}, that correspond to f_{E1} and f_{R1}, respectively, of the forward matrix:

$$g_{E1} \cdot t(E, E) \cdot e(M | E)$$
$$g_{E1} \cdot t(E, R) \cdot e(M | R)$$
$$g_{R1} \cdot t(R, E) \cdot e(M | E)$$
$$g_{R1} \cdot t(R, R) \cdot e(M | R)$$

However, instead of taking the sum of pairs of expressions (first and third transitioning to E, second and fourth transitioning to R) as we did with the forward algorithm, we take the maxima as follows:

$$g_{E2} = \max(g_{E1} \cdot t(E, E) \cdot e(M | E), \quad g_{R1} \cdot t(R, E) \cdot e(M | E))$$
$$g_{R2} = \max(g_{E1} \cdot t(E, R) \cdot e(M | R), \quad g_{R1} \cdot t(R, R) \cdot e(M | R)).$$

Figure 12 details these computations with boldface arrows indicating the maxima. The following displays the developing Viterbi matrix, G:

$$G = \begin{matrix} & B & M & M \\ E & \\ R & \end{matrix} \begin{bmatrix} 0.06 & 0.1008 & \square \\ 0.63 & 0.0504 & \square \end{bmatrix}.$$

Quick Review Question 18

Suppose we wish to use the Viterbi algorithm to find the state sequence, u, with maximum $P(u, \text{HHGH})$ for the HMM in Quick Review Question 17. As calculated in that question, the first column of the Viterbi algorithm matrix G contains $g_{A1} = 0.2$, $g_{B1} = 0.4$, and $g_{C1} = 0.63$.

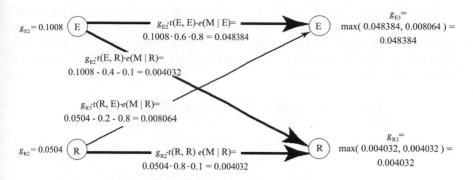

Fig. 13 Step 3 of Viterbi algorithm in calculation of P(BMM)

a. The calculation of g_{B2} involves three expressions whose values are 0.0032, 0.0032, and 0.0504. Calculate g_{B2}.

b. Calculate g_{A2}.

Calculations of subsequent Viterbi matrix elements for this example proceed in a similar fashion. With observation v_i, we employ the following evaluations for the elements of column i:

$$g_{Ei} = \max(g_{E(i-1)} \cdot t(E, E) \cdot e(v_i \mid E), \quad g_{R(i-1)} \cdot t(R, E) \cdot e(v_i \mid E))$$
$$g_{Ri} = \max(g_{E(i-1)} \cdot t(E, R) \cdot e(v_i \mid R), \quad g_{R(i-1)} \cdot t(R, R) \cdot e(v_i \mid R))$$

With boldface arrows indicating maximum values, Fig. 13 illustrates the calculation of the final column of the Viterbi matrix. Note that there are two paths to R that yield the maximum, 0.004032. The completed Viterbi matrix is as follows:

$$
\begin{array}{cc}
 & \begin{array}{ccc} B & M & M \end{array} \\
G = \begin{array}{c} E \\ R \end{array} & \left[\begin{array}{ccc} 0.06 & 0.1008 & 0.048384 \\ 0.63 & 0.0504 & 0.004032 \end{array} \right]
\end{array}.
$$

In general, using Viterbi's algorithm for any HMM; for state, x; observation, v_i; and set of states, S, we have the following calculation for a Viterbi matrix element in row x and column i:

$$g_{xi} = \max_{y \in S} \left(g_{y(i-1)} \cdot (y, x) \cdot e(v_i \mid x) \right).$$

Quick Review Question 19
Suppose we wish to use the Viterbi algorithm to find the state sequence, u, with maximum $P(u, \text{HHGH})$ for the HMM in Quick Review Question 17. Suppose, also, that the third column of the Viterbi matrix G contains $g_{A3} = 0.0765$, $g_{B3} = 0.0340$, and $g_{C3} = 0.0142$. Calculate g_{C4} to four decimal places.

To calculate the probability of the visible sequence with the forward algorithm, we added the probabilities in the final column. However, to calculate the maximum joint probability of a hidden sequence and a given visible sequence using the Viterbi algorithm, we find the maximum of the values in the final column. Thus, for Holly's HMM, we have the following:

$$\max_{u \in S^3}(u, \text{ BMM}) = \max(0.048384, 0.004032) = 0.048384.$$

However, we would like to calculate $\max(P(u|\text{BMM}))$ for this u. Recall that

$$P(u, \text{ BMM}) = P(u \mid \text{BMM}) \cdot P(\text{BMM}).$$

Dividing both sides by the factor $P(\text{BMM})$, we have

$$P(u \mid \text{BMM}) = P(u, \text{ BMM})/P(\text{BMM}).$$

Moreover, using the forward algorithm, we discovered $P(\text{BMM}) = 0.080064$. Thus, over all three-element hidden sequences, u,

$$\max(P(u \mid \text{BMM})) = 0.048384/0.080064 = 0.604317.$$

More important than finding this maximum probability, we would like to discover the particular state sequence that yields this maximum. Fortunately, by backtracking through the Viterbi matrix, we can determine this hidden sequence. Figure 14 summarizes results of Figs. 9 and 10 with arrows indicating the expressions generating the maxima. To calculate the state sequence, u, that results in $\max(P(u \mid \text{BMM})) = 0.604317$, we start by finding the maximum in the final column, 0.048384, which is in row E. Backtracking through the path indicated by the arrows, we then go to column 2, row E and finally to column 1, row R. Reading the row values from left to right, we obtain the state sequence $u = \text{REE}$. Thus, given observed sequence BMM, REE is the most likely state sequence, and $P(\text{REE} \mid \text{BMM}) = 0.604317$, which is over 60%.

Quick Review Question 20
Suppose we wish to use the Viterbi algorithm to find the state sequence, u, with maximum $P(u \mid \text{HHGH})$ for the HMM in Quick Review Question 17. Suppose, also, that $P(\text{HHGH}) = 0.1028$ and the Viterbi matrix, G, is in Fig. 15, with arrows indicating the direction from which maxima came.

Fig. 14 Final Viterbi matrix with arrows indicating the paths

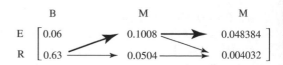

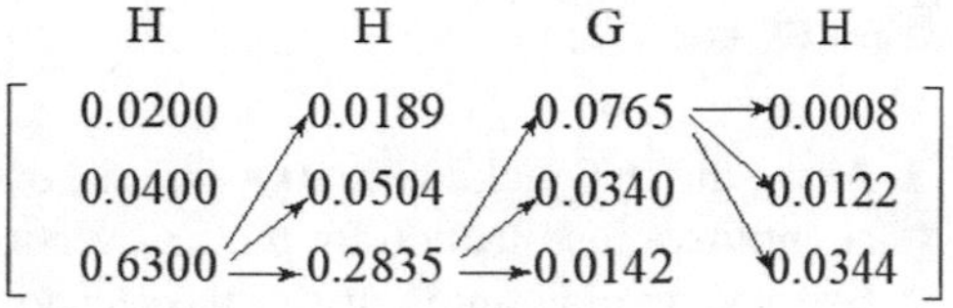

Fig. 15 Viterbi matrix for
Quick Review Question 20
with arrows indicating the
direction from which maxima
came

$$
\begin{array}{cccc}
H & H & G & H \\
\left[
\begin{array}{cccc}
0.0200 & 0.0189 & 0.0765 & 0.0008 \\
0.0400 & 0.0504 & 0.0340 & 0.0122 \\
0.6300 & 0.2835 & 0.0142 & 0.0344
\end{array}
\right]
\end{array}
$$

a. Calculate the maximum $P(u, \text{HHGH})$ for hidden state sequence u.
b. Calculate the maximum $P(u \mid \text{HHGH})$ for hidden state sequence u.
c. Give the u that achieves these maxima.

10.3 Parallel Viterbi Algorithm (Optional)

As with the forward algorithm, we can use high performance computing to achieve faster results when a decoding problem involves a large number of states and/or observations. Moreover, we can parallelize the Viterbi algorithm similarly to the forward algorithm with OpenMP and threads communicating by reading and writing to the same matrix. Project 6 calculates the speedup that can be achieved using HPC with this algorithm.

11 Detecting CpG Islands

One example of an HMM decoding problem with a solution employing the Viterbi algorithm involves detecting genes. As discussed in section "Probability of a Genomics Sequence," an area of greater frequency of the base sequence CG can be an indicator that a gene is to follow. The section presented initial probabilities, emission matrices, and possible transition matrices for samples within and not within such CpG islands, called positive and negative areas, respectively [4]. Suppose we also have transition probabilities from bases in positive areas (A^+, C^+, T^+, G^+) to bases in negative areas (A^-, C^-, T^-, G^-, respectively) and vice versa. Then, using the Viterbi algorithm, for a given observed sequence of bases from $\{A, C, T, G\}$, we can compute the most likely hidden sequence from the set of states, $S = \{\text{begin/end}, A^+, C^+, T^+, G^+, A^-, C^-, T^-, G^-\}$, where the sign indicates whether the base is probably in a CpG island or not. Project 7 considers such a decoding problem, where we can decode areas of high CpG concentration, containing bases A^+, C^+, T^+, and G^+.

Exercises

1 As we did in Quick Review Question 3 using the Venn diagram in Fig. 4, provide an intuitive justification for the first version of Bayes' Theorem in Eq. (6).

2 Use a Venn diagram to justify the second version of Bayes' Theorem in Eq. (7).

3 This exercise relates to the HMM in Quick Review Question 14.

   a. Calculate f_{A2}.
   b. Calculate f_{C2}.
   c. Calculate $P(HH)$.
   d. Calculate $P(HHG)$.

Projects

1 Write a sequential program to calculate the probability of a state given an observation using Eq. (8).

2 a. Develop a sequential program to calculate the probability of a sequence of states generating a sequence of observations using Eq. (9).

   b. Develop a parallel version of this program, having each thread responsible for a portion of the factors and gather the results into a final product.

   c. For large sequence length, time the parallel version for increasing numbers of threads. Produce a graph of the speedup versus the number of threads similar to Fig. 10.

3 a. Develop a sequential program to calculate the probability of a state sequence and an observation sequence using Eq. (13).

   b. Develop a parallel version of this program, having each thread responsible for a portion of the factors and gather the results into a final product.

   c. For large sequence length, time the parallel version for increasing numbers of threads. Produce a graph of the speedup versus the number of threads similar to Fig. 10.

4 a. Using the forward algorithm, develop a sequential program to calculate the probability of a sequence of observations.

   b. Develop a parallel version of this program.

   c. For large sequence length, time the parallel version for increasing numbers of threads. Produce a graph of the speedup versus the number of threads similar to Fig. 10.

5 Download *ProbabilitiesHumanPN.txt* [15], which stores the transition matrices from Fig. 9, and "Accessing Chromosome 19 Data" (*AccessingChr19Data.pdf*) [14], which describes how to access gene locations and subsequences from chromosome 19 of the human genome. Using the UCSC Genome Browser [23], select several subsequences of about 50 bases that occur in a CpG island and several that do not. Using a sequential or parallel forward algorithm program that you develop, determine the probability of each subsequence twice, once for each hidden Markov model in section "Probability of a Genomics Sequence," and calculate how many more times likely the subsequence is to be in a CpG island than not. Do your results concur with those of the UCSC Genome Browser?

6 a. Using the Viterbi algorithm, develop a sequential program to determine the most likely sequence of hidden states corresponding to a sequence of observations.

  b. Develop a parallel version of this program.

  c. For a large sequence length, time the parallel version for increasing numbers of threads. Produce a graph of the speedup versus the number of threads similar to Fig. 10.

7 Download *ProbabilitiesHumanV.txt* [16], which contains the transition matrix and other data described in the section "Detecting CpG Islands" from [8], and "Accessing Chromosome 19 Data" (*AccessingChr19Data.pdf*) [14], which describes how to access gene locations and subsequences from chromosome 19 of the human genome. Using the UCSC Genome Browser [23], select several subsequences of about 50 bases that overlap CpG islands. Using a sequential or parallel Viterbi algorithm program that you develop, for each downloaded sequence, determine the most likely hidden sequence of states from

$$S = \{\text{begin/end}, A^+, C^+, T^+, G^+, A^-, C^-, T^-, G^-\}$$

and its probability. Do your results concur with those of the UCSC Genome Browser?

8 Scientists have used telemetry data, such as from radio tags or collars, to monitor wildlife movement and HMMs to infer an animal's hidden behavior [9, 11]. Download *telemetry.nlogo* [17], a NetLogo program to simulate using telemetry to follow an animal and interpret its actions, and "Using a NetLogo Program" (*UsingNetLogo.pdf*) [18], a brief description of how to interact with a NetLogo simulation [12]. With NetLogo, which is free to download, we can generate agent-based models that have autonomous, decision-making agents, which have states and behaviors. In the simulation *telemetry.nlogo*, an animal wanders at random unless thirsty. When thirsty, the animal moves to water, in the center of the world, and stays in the water until satiated. The user can observe, through this simulated telemetry, if the animal is in water (W), facing water but not in water (F), or not facing water and not in water (A). The state of the animal is thirsty, either on way to water or in water (T); or not thirsty, not on way to water and not in water (N).

  a. Read the description of the model under the Info tab for the program. Estimate the parameters for an HMM to model the scenario as described in the section "Things to Try."

  b. Using this HMM, take part of an observation sequence from file output, say of about 50 observations; and with your forward algorithm program, determine the probability of that subsequence.

    c. Using your Viterbi algorithm program and the subsequence of observations from Part b, determine the most likely corresponding state sequence. Calculate the percentage of the sequence derived using the Viterbi algorithm that agrees with the actual state sequence generated by the NetLogo program.

    d. Revise the NetLogo program, *telemetry.nlogo*, to save to a file at least 100 observation sequences of length 50 and to store in another file the corresponding hidden state sequences. Revise your Viterbi program to read the file of observation sequences and to produce another file of derived state sequences. Write another program to read the latter output file of derived state sequences and the file of state sequences from your NetLogo program and to calculate the fraction of time the two files agree. That is, what is the probability that your Viterbi program will accurately generate the underlying state sequence.

    e. Extend the model in *telemetry.nlogo* as suggested under the Info tab in the section "Extending the Model." Tutorials on programming with NetLogo can be obtained from [20] and [12].

9 [9] described using telemetry to obtain movement data of bison, which could be in underlying, hidden states of "encamped" or "exploratory." Their paper discussed several types of HMMs to use observations to infer a bison's states. Observations were bivariant, involving step length and turning angles, or directions. In an encamped state, the animals were observed to have numerous long steps and few turns; while in an exploratory state, bison were seen having short steps with more frequent reversals.

    a. Similar to *telemetry.nlogo*, available on the website containing this module and discussed in the previous project, develop a NetLogo simulation of bison movement that outputs important totals and writes to one file a sequence of hidden states and to another file the corresponding sequence of observations [17]. In Part b of this project, we will use the output data to estimate an HMM. Then, using this HMM, we will take a sequence of observations and attempt to derive the underlying state sequence. We can compare our derivation with the state sequence from the NetLogo simulation. "Using a NetLogo Program" and tutorials on programming with NetLogo can be obtained from [18] and [20], respectively. Note that we could model the step lengths as being random numbers taken from different exponential distributions, each with step lengths ranging from 0.0 to 0.6 km/d. The turning angles for an exploring and for an encamping animal can be modeled as random numbers from normal distributions with means of 0 and 180 degrees/d, respectively. ([9] employed Weibull distributions for step lengths and wrapped Cauchy distributions for turning angles.)

    b. Estimate the parameters for an HMM to model the scenario.

    c. Using this HMM, take part of an observation sequence from file output, say of about 50 observations; and with your forward algorithm program, determine the probability of that subsequence.

d. Using this HMM, take part of an observation sequence from file output, say of about 50 observations; and with your Viterbi algorithm program, derive the most likely corresponding subsequence of observations. Calculate the percentage of the sequence derived using the Viterbi algorithm that agrees with the actual state sequence generated by the NetLogo program.

e. Revise your NetLogo program to save to a file at least 100 observation sequences of length 50 and to store in another file the corresponding hidden state sequences. Revise your Viterbi program to read the file of observation sequences and to produce a file of derived state sequences. Develop another program to take the latter output file of derived state sequences and the file of state sequences from your NetLogo program and to calculate the fraction of time the two files agree. That is, what is the probability that your Viterbi program will accurately generate the underlying state sequence.

10 a. In the programming language of your choice (such as Python, R, MAT-LAB, Mathematica, or C++), develop a program that uses the probabilities from Holly's HMM of this module to generate a given number of states and corresponding observations. Save the sequences to separate files. For example, the initial state should be selected at random using the initial state probabilities. Have the program select the corresponding observation at random with probabilities indicted by the emission matrix. The next state should be picked at random using probabilities from the transition (Markov) matrix, and so forth.

b. Repeat Parts c e from Project 9 using this program.

11 a. Repeat Project 10 a for a generalized situation. That is, develop a procedure with input of a vector of state names or their abbreviations, a vector of observation names or their abbreviations, an HMM (initial vector, transition matrix, and emission matrix), and a number (n) of states/observations for an output sequence. Then, using the HMM's probabilities, have the procedure generate and return a sequence of n states and corresponding sequence of n observations.

b. Develop the two HMMs from Project 5. Using your procedure from Part a, generate subsequences of length 50 for the positive and the negative models. Using a forward algorithm program that you develop, determine the probability of each subsequence twice, once for each hidden Markov model. For the positive subsequence, calculate how many more times likely the subsequence is to be in a CpG island than not. For the negative subsequence, calculate how many more times likely the subsequence is not to be in a CpG island than to be in a CpG island.

c. A gene in DNA and the corresponding subsequence of RNA have **exons**, which contain part of the encoding information, such as for proteins, separated by **introns**, or non-coding segments. **RNA splicing** removes the introns and reassembles the exons so that translation to the protein can eventually occur, and a **splice** is the location of a cut between an intron and an exon. In DNA and RNA, each non-terminal nucleotide attaches to the #5 carbon, written **5' carbon**, of a sugar at one end and the #3 carbon, **3' carbon**, of its neighbor at the other end. Thus, we consider a nucleotide chain to have a specific **5'-3' orientation**. Develop the "Toy HMM" for a 5' splice recognition problem in [5], and using your procedure from Part a, repeat Project 9, Parts c–e.

For each of the following HMMs, using your procedure from Part a, repeat Project 9, Parts c–e:

d. Project 7
e. The unfair-casino HMM in Fig. 1 of [3] or other sources
f. The weather HMM on the fourth slide of [10]
g. The weather HMM in Sections 1 and 2 of [6]
h. The light-dark chocolate candy HMM in Section 5, "Exercises," of [6]
i. The tree-growth rings HMM in Section 1, "A Simple Example," of [13].

12 a. Study the Baum-Welch (or forward-backward) algorithm for solving the training (or learning) problem mentioned in sections "Introduction" and "Example Model," and develop a program to solve the problem.

b. Using the NetLogo program *telemetry.nlogo*, available on the website containing this module and discussed in Project 9, generate a collection of state and corresponding observation sequences [17]. Using these and your program from Part a, determine the parameters for an HMM.

c. With your procedure from Project 10 a, generate a collection of state and corresponding observation sequences. Using these and your program from Part a, determine the parameters for a trained HMM. Compare this HMM to Holly's HMM.

Do the same process as described in Part c for the following HMMs:

d. The two HMMs of Project 11 b
e. Project 11 c
f. Project 11 d
g. Project 11 e
h. Project 11 f
i. Project 11 g
j. Project 11 h
k. Project 11 i

Answers to Quick Review Questions

1 a. {A, B, C}
  b. {G, H}
  c. 0.6
  d. 0.2
  e.

$$u_k/u_{k+1} \quad A \quad B \quad C$$

$$T = \begin{array}{c} A \\ B \\ C \end{array} \begin{bmatrix} 0.1 & 0.4 & 0.5 \\ 0.2 & 0.2 & 0.6 \\ 0.3 & 0.2 & 0.5 \end{bmatrix}$$

f. 0.2

g. $0.7 = 1 - \pi(A) - \pi(B)$

h. 0.6

i. 0.9

j.

$$hidden/observable \quad G \quad H$$

$$\begin{array}{c} A \\ B \\ C \end{array} \begin{bmatrix} 0.9 & 0.1 \\ 0.6 & 0.4 \\ 0.1 & 0.9 \end{bmatrix}$$

2 a. $P(E, B) = P(B \mid E) \cdot P(E) = 0.2 \cdot 0.3 = 0.06$

b. $P(R, M) = P(M \mid R) \cdot P(R) = 0.1 \cdot 0.7 = 0.07$

c. $P(R, B) = P(B \mid R) \cdot P(R) = 0.9 \cdot 0.7 = 0.63$

d. $1.00 = 0.24 + 0.06 + 0.07 + 0.63$

e. $0.30 = 0.24 + 0.06$

f. $0.70 = P(E, M) + P(R, B) = 0.07 \mid 0.63$

g. 1.00

h. $0.31 = P(E, M) + P(R, M) = 0.24 + 0.07$

i. $0.69 = P(E, B) + P(R, B) = 0.06 + 0.63$

j. 1.00

3

$$P(A, B) \stackrel{?}{=} P(B|A) \cdot P(A)$$

$$\frac{c}{a+b+c+d} \stackrel{?}{=} \frac{c}{\cancel{(a+c)}} \cdot \frac{\cancel{(a+c)}}{a+b+c+d}$$

$$\frac{c}{a+b+c+d} = \frac{c}{a+b+c+d}$$

4 a. $0.43 = 0.05 + 0.18 + 0.20$

b. $0.57 = 0.36 + 0.11 + 0.10$

c. {F, G, H}

d. 0.42 because

$$P(G, J) = P(G|J) \cdot P(J), \text{ so } P(G|J) = P(G, J)/P(J) = 0.18/0.43 \approx 0.42$$

e. 0.19 because

$P(G, K)=P(G | K) \cdot P(K)$, so $P(G | K)=P(G, K)/P(K)=0.11/0.57 \approx 0.19$

f. $\displaystyle\sum_{x \in \{E, G, H\}} P(K, x)$ or $\displaystyle\sum_{x \in \{E, G, H\}} P(x, K)$

g. $0.41 = 0.05 + 0.36$

h. 0.12 because

$P(J, F) = P(J | F) \cdot P(F)$, so $P(J | F) = P(J, F)/P(F) = 0.05/0.41 \approx 0.12$

i. 0.88 because

$P(K, F)=P(J | F) \cdot P(F)$, so $P(K | F)=P(K, F)/P(F)=0.36/0.41 \approx 0.88$

j. $0.29 = 0.18 + 0.11$

k. $\{J,K\}$

l. $\displaystyle\sum_{x \in \{J, K\}} P(G, x)$ or $\displaystyle\sum_{x \in \{J, K\}} P(x, G)$

m. $0.30 = 0.20 + 0.10$

5 a. 0.62 because $P(J | G) = P(G | J) \cdot P(J)/P(G) = 0.42 \cdot 0.43/0.29 \approx 0.62$

  b. 0.69 because $P(K | G) = P(G | K) \cdot P(K)/P(G) = 0.35 \cdot 0.57/0.29 \approx 0.69$

  c. 0.11 because $P(F | J) = P(J | F) \cdot P(F)/P(J) = 0.12 \cdot 0.41/0.43 \approx 0.11$

  d. 0.63 because $P(F | K) = P(K | F) \cdot P(F)/P(K) = 0.88 \cdot 0.41/0.57 \approx 0.63$

6 a. $0.23 \approx 1 - P(E | M) \approx 1 - 0.77$

  b. 0.91 because $P(R | B) = P(B | R) \cdot P(R)/(P(B | E) \cdot P(E) + P(B | R) \cdot P(R)) = 0.9 \cdot 0.7/(0.2 \cdot 0.3 + 0.9 \cdot 0.7) = 0.9 \cdot 0.7/0.69 \approx 0.91$

  c. $0.09 \approx 1 - P(R | B) \approx 1 - 0.91$; alternatively, $P(E | B) = P(B | E) \cdot P(E)/(P(B | E) \cdot P(E) + P(B | R) \cdot P(R)) = 0.2 \cdot 0.3/0.69 \approx 0.09$

7 $0.16 \approx P(F | J) = P(J | F) \cdot P(F)/[P(J | F) \cdot P(F) + P(J | G) \cdot P(G) + P(J | H) \cdot P(H)] = 0.12 \cdot 0.41/[0.12 \cdot 0.41 + 0.62 \cdot 0.29 + 0.26 \cdot 0.30]$

8 a. $0.0576 = P(MMMB | ERER) = e(M | E) \cdot e(M | R) \cdot e(M | E) \cdot e(B | R) = 0.8 \cdot 0.1 \cdot 0.8 \cdot 0.9$

  b. $0.53144 \approx P(BBBBBB | RRRRRR) = e(B | R)^6 = 0.9^6$

9 a. $0.00055 \approx P(MMMB,ERER) = P(MMMB | ERER) \cdot P(ERER) = 0.0576 \cdot \pi(E) \cdot t(E, R) \cdot t(R, E) \cdot t(E, R) = 0.0576 \cdot 0.3 \cdot 0.4 \cdot 0.2 \cdot 0.4$, where $P(MMMB | ERER) = 0.0576$ is from Quick Review Question 8 a.

  b. $0.1219 \approx P(BBBBBB, RRRRRR) = P(BBBBBB | RRRRRR) \cdot P(RRRRRR) = 0.53144 \cdot \pi(R) \cdot t(R, R)^5 = 0.53144 \cdot 0.70 \cdot 0.8^5$ where $P(BBBBBB | RRRRRR) \approx 0.53144$ is from Quick Review Question 8 b.

10 a. $64 = 4^3$

  b. $1,048,576 = 4^{10}$

  c. $1,099,511,627,776 = 4^{20}$

  d. $4,398,046,511,104 = 4^{21}$

11 a. 3×4

  b. $0.02 = \pi(H) \cdot e(H | A) = 0.2 \cdot 0.1$

  c. $P(A, H)$

d. $0.04 = \pi(\text{H}) \cdot e(\text{H} \mid \text{B}) = 0.1 \cdot 0.4$

e. $P(\text{B}, \text{H})$

f. $0.63 = \pi(\text{H}) \cdot e(\text{H} \mid \text{C}) = 0.7 \cdot 0.9$

g. $P(\text{C}, \text{H})$

h. $0.69 = P(\text{A}, \text{H}) + P(\text{B}, \text{H}) + P(\text{C}, \text{H}) = f_{A1} + f_{B1} + f_{C1} = 0.02 + 0.04 + 0.63$

12 $0.0504 = P(\text{BM}, \text{RR}) = P(\text{B}, \text{R}) \cdot t(\text{R}, \text{R}) \cdot e(\text{M} \mid \text{R}) = f_{R1} \cdot t(\text{R}, \text{R}) \cdot e(\text{M} \mid \text{R}) = 0.63 \cdot 0.8 \cdot 0.1$

13 $0.1296 = P(\text{BM}, \text{EE}) + P(\text{BM}, \text{RE}) = 0.0288 + 0.1008$

14 a. 3

b. $0.0032 = f_{A1} \cdot t(\text{A}, \text{B}) \cdot e(\text{H} \mid \text{B}) = 0.02 \cdot 0.4 \cdot 0.4$

c. $0.0032 = f_{B1} \cdot t(\text{B}, \text{B}) \cdot e(\text{H} \mid \text{B}) = 0.04 \cdot 0.2 \cdot 0.4$

d. $0.0504 = f_{C1} \cdot t(\text{C}, \text{B}) \cdot e(\text{H} \mid \text{B}) = 0.63 \cdot 0.2 \cdot 0.4$

e. $0.0568 = 0.0032 + 0.0032 + 0.0504$

15 a. $P(\text{BMM}) = \sum_{x \in S^3} P(\text{BMM}, x)$

b. {EEE, EER, ERE, ERR, REE, RER, RRE, RRR}

c. $P(\text{BMM}) = [P(\text{BMM}, \text{EEE}) + P(\text{BMM}, \text{ERE}) + P(\text{BMM}, \text{REE}) + P(\text{BMM}, \text{RRE})] + [P(\text{BMM}, \text{EER}) + P(\text{BMM}, \text{ERR}) + P(\text{BMM}, \text{RER}) + P(\text{BMM}, \text{RRR})] = f_{E3} + f_{R3}$

d. $P(\text{BMM}, \text{ERE}) + P(\text{BMM}, \text{RRE}) = P(\text{BM}, \text{ER}) \cdot t(\text{R}, \text{E}) \cdot e(\text{M} \mid \text{E}) + P(\text{BM}, \text{RR}) \cdot t(\text{R}, \text{E}) \cdot e(\text{M} \mid \text{E}) = [P(\text{BM}, \text{ER}) + P(\text{BM}, \text{RR})] \cdot t(\text{R}, \text{E}) \cdot e(\text{M} \mid \text{E}) = f_{R2} \cdot t(\text{R}, \text{E}) \cdot e(\text{M} \mid \text{E})$

e. $P(\text{BMM}, \text{EER}) + P(\text{BMM}, \text{RER}) = f_{E2} \cdot t(\text{E}, \text{R}) \cdot e(\text{M} \mid \text{R})$

f. $P(\text{BMM}, \text{ERR}) + P(\text{BMM}, \text{RRR}) = f_{R2} \cdot t(\text{R}, \text{R}) \cdot e(\text{M} \mid \text{R})$

g. $0.062208 = 0.1296 \cdot 0.6 \cdot 0.8$

h. $0.005184 = 0.1296 \cdot 0.4 \cdot 0.1$

i. $0.008448 = 0.0528 \cdot 0.2 \cdot 0.8$

j. $0.004224 = 0.0528 \cdot 0.8 \cdot 0.1$

k. $0.70656 = 0.062208 + 0.008448$

l. $0.009408 = 0.005184 + 0.004224$

m. $0.080064 = 0.070656 + 0.009408$

16 a. 5

b. 5

c. h

d. 5

e. 2

f. 2

g. $95 = 5 \cdot 19$

h. $h \cdot (n - 1)$

i. $195 = 5 + 95 \cdot 2$

j. $2hn - h$ because $h + h \cdot (n - 1) \cdot 2 = h + 2hn - 2h = -h + 2hn$

k. 0

l. 4

m. $h - 1$

n. $h^2(n - 1) - hn - h$ or $h^2 n - h^2 - hn - h$ because $0 + (h - 1) \cdot h \cdot (n - 1) = (h^2 - h)(n - 1) = h^2 n - hn - h^2 - h = h^2(n - 1) - hn - h$

o. $h^2n - h^2 + hn - 2h$ because $(2hn - h) + (h^2n - h^2 - hn - h) = h^2n - h^2 + hn - 2h$

p. $O(h^2n)$

17 $g_{A1} = 0.2$, $g_{B1} = 0.4$, and $g_{C1} = 0.63$

18 a. $0.0504 = \max(0.0032, 0.0032, 0.0504)$

b. $0.0189 = \max(0.0002, 0.0008, 0.0189)$

19 0.0344 because of the following: $g_{A3} \cdot t(A, C) \cdot e(H \mid C) = 0.034425$; $g_{B3} \cdot t(B, C) \cdot e(H \mid C) = 0.01836$; $g_{C3} \cdot t(C, C) \cdot e(H \mid C) = 0.00639$; and the maximum of these expressions is 0.034425.

20 a. 0.0344, the maximum in the final column

b. $0.33463 = 0.0344/0.1028$

c. CCAC

Further Reading

- Baldi, P., Chauvin, Y., Hunkapiller, T. and McClure, M.A.: Hidden Markov models of biological primary sequence information. Proc. of the Natl. Academy of Sciences, **91**(3), 1059–1063 (1994)
- Gales, M., Young, S.: The application of hidden Markov models in speech recognition. Foundations and Trends in Signal Processing, **1**(3), 195–304 (2007)
- Kamal, M.S., Chowdhury, L., Khan, M.I., Ashour, A.S., Tavares, J.M.R., Dey, N.: Hidden Markov model and Chapman Kolmogrov for protein structures prediction from images. Computational Biology and Chemistry, **68** 231–244 (2017)
- Krogh, A., Brown, M., Mian, I.S., Sjölander, K., Haussler, D.: Hidden Markov models in computational biology: Applications to protein modeling. J. of Molecular Biology, **235**(5), 1501–1531 (1994)
- Manogaran, G., Vijayakumar, V., Varatharajan, R., Kumar, P.M., Sundarasekar, R., Hsu, C.H.: Machine learning based big data processing framework for cancer diagnosis using hidden Markov model and GM clustering. Wireless Personal Communications, **102**(3), 2099–2116 (2018)
- McGibbon, R.T., Ramsundar, B., Sultan, M.M., Kiss, G., Pande, V.S.: Understanding protein dynamics with L1-regularized reversible hidden Markov models. arXiv preprint arXiv:1405.1444 (2014)
- Petersen, B.K., Mayhew, M.B., Ogbuefi, K.O., Liu, V.X., Greene, J.D., Ray, P.: Modeling sepsis disease progression using hidden Markov models (No. LLNL-CONF-740757). Lawrence Livermore National Lab.(LLNL), Livermore, CA (U. S.) (2017)
- Rabiner, L.R.: A tutorial on hidden Markov models and selected applications in speech recognition. Proc. of the IEEE **77**(2): 257–286 (1989)
- Ramanathan, N.: Applications of hidden Markov models. http://www.cs.umd.edu/~djacobs/CMSC828/ApplicationsHMMs.pdf

- Sharp, C., Bray, J., Housden, N.G., Maiden, M.C., Kleanthous, C.: Diversity and distribution of nuclease bacteriocins in bacterial genomes revealed using Hidden Markov Models. PLoS Computational Biology, **13**(7), p.e1005652 (2017)
- Williams, J.P., Storlie, C.B., Therneau, T.M., Jack Jr, C.R., Hannig, J.: A Bayesian approach to multi-state hidden Markov models: application to dementia progression. arXiv preprint arXiv:1802.02691 (2018)
- Yoon, B.: Hidden Markov models and their applications in biological sequence analysis. Curr. Genomics **10**(6), 402–415 (2009)

Acknowledgements Our thanks go to the Fulbright Specialist Program, University "Magna Græcia" of Catanzaro, and Wofford College for funding the Shiflets' visit to the university and to the National Computational Science Institute Blue Waters Student Internship Program for funding Dmitriy Kaplun's internship.

References

1. B10NUMB3R5: the database of useful biological numbers. http://bionumbers.hms.harvard.edu/bionumber.aspx?&id=105336&ver=2
2. Centers for Disease Control and Prevention: 2–20 years: girls stature-for-age and weight-for-age percentiles. https://www.cdc.gov/growthcharts/data/set2clinical/cj41c072.pdf
3. Cerulo, L., Ceccarelli, M., Di Penta, M., Canfora, G.: A hidden Markov model to detect coded information islands. In: Source Code Analysis and Manipulation (SCAM) IEEE 13th International Working Conference on Source Code Analysis and Manipulation: 157–166. https://doi.org/10.1109/SCAM.2013.6648197 (2013)
4. Durbin, R., Eddy, S.R., Krogh, A., Mitchison, G.: Biological Sequence Analysis. Cambridge University Press, Cambridge (1998)
5. Eddy, S.R.: What is a hidden Markov model? Nat Biotechnol. **22**(10), 1315–6 (2004) https://doi.org/10.1038/nbt1004-1315
6. Fosler-Lussier, E.: Markov models and hidden Markov models: a brief tutorial. International Computer Science Inst. http://di.ubi.pt/~jpaulo/competence/tutorials/hmm-tutorial-1.pdf
7. Gardiner-Garden, M., Frommer, M. CpG islands in vertebrate genomes. J. Mol. Biol. **196**, 261–282 (1987)
8. Huson, D.: Chapter 8: Markov chains and hidden Markov models. In course: Algorithms in Bioinformatics University of Tübingen. https://ab.inf.uni-tuebingen.de/teaching/ss08/gbi/script/chapter08-hmms.pdf
9. Langrock, R., King, R., Matthiopoulos, J., Thomas, L., Fortin, D., Morales, J.M.: Flexible and practical modeling of animal telemetry data: hidden Markov models and extensions. Ecology, **93**:2336–2342 (2012)
10. Lyngsø, R.: "Hidden Markov models." http://www.stats.ox.ac.uk/~mcvean/DTC/STAT/Lectures/Weds_wk2/hidden_markov_models.pdf
11. Morales, J.M., Haydon, D.T., Frair, J.L., Holsinger, K.E., Fryxell, J.M.: Extracting more out of relocation data: building movement models as mixtures of random walks. Ecology **85**:2436–2445 (2004)
12. NetLogo home page. https://ccl.northwestern.edu/netlogo/
13. Stamp, M.: A revealing introduction to hidden Markov models. https://www.cs.sjsu.edu/~stamp/RUA/HMM.pdf
14. Shiflet, A.: Accessing Chromosome 19 data. AccessingChr19Data.pdf in https://ics.wofford-ecs.org/files/HMM.zip

15. Shiflet, A.: Possible transition matrices for samples within and not within CpG islands. ProbabilitiesHumanPN.txt in https://ics.wofford-ecs.org/files/HMM.zip
16. Shiflet, A.: Transition matrix and other data from Hudson. ProbabilitiesHumanV.txt in https://ics.wofford-ecs.org/files/HMM.zip
17. Shiflet, A.: Telemetry program telemetry.nlogo in NetLogo. telemetry.nlogo in https://wofford-ecs.org/files/HMM.zip
18. Shiflet, A. Using a NetLogo program. UsingNetLogo.pdf in https://ics.wofford-ecs.org/files/HMM.zip
19. Shiflet, A., Shiflet, G. Introduction to Computational Science: Modeling and Simulation for the Sciences, 2nd ed., Princeton University Press (2014)
20. Shiflet, A., Shiflet, G. NetLogo agent-based files. https://ics.wofford-ecs.org/agent/NetLogo
21. Stanke, M.: Markov chains and hidden Markov models, Free University of Berlin. http://www.mi.fu-berlin.de/wiki/pub/ABI/HiddenMarkovModelsWS13/script.pdf
22. Sungkaworn, T., Jobin, M.L., Burnecki, K., Weron, A., Lohse, M.J., Calebiro, D.: Single-molecule imaging reveals receptor-G protein interactions at cell surface hot spots. Nature **550**(7677) 543–547 (2017)
23. UCSC Genome Browser. https://genome.ucsc.edu/cgi-bin/hgGateway

Using Neural Networks to Identify Bird Species from Birdsong Samples

Russell Houpt, Mark Pearson, Paul Pearson, Taylor Rink, Sarah Seckler, Darin Stephenson, and Allison VanderStoep

Abstract Modern data analysis techniques include the use of artificial neural networks for classification, estimation, and prediction. An area in which data science can be helpful is in species identification and enumeration from images or sound recordings. In this chapter, we develop the necessary tools for using machine learning to identify bird species from recordings of bird calls. This includes a basic introduction to wavelet transforms and scalograms and the construction of convolutional neural networks for solving classification problems. We give some ideas for extending our results on bird species classification, as well as ideas for using related neural networks for broader application to image and audio classification.

Suggested Prerequisites *Second year college mathematics including linear algebra and multivariate calculus. Programming experience in Python is helpful, but can be developed as needed through various online tutorials and examples.*

1 Introduction

Recent developments in the areas of machine learning and data science have given rise to new tools and processes to improve computer-aided recognition of images and sounds. While there are many highly sophisticated networks already in use,

Mark Pearson, **Paul Pearson**, and **Darin Stephenson** are faculty members in the Hope College Mathematics Department.

Russell Houpt and **Sarah Seckler** are 2018 Hope College graduates, and **Taylor Rink** and **Allison VanderStoep** are 2019 Hope College graduates.

R. Houpt · M. Pearson · P. Pearson · T. Rink · S. Seckler · D. Stephenson (✉) · A. VanderStoep
Department of Mathematics, Hope College, Holland, MI, USA
e-mail: pearson@hope.edu; pearsonp@hope.edu; stephenson@hope.edu

© Springer Nature Switzerland AG 2020
H. Callender Highlander et al. (eds.), *An Introduction to Undergraduate Research in Computational and Mathematical Biology*, Foundations for Undergraduate Research in Mathematics, https://doi.org/10.1007/978-3-030-33645-5_9

many interesting data models can be developed and understood with a minimal list of mathematical prerequisites. The wide-open nature of data science research and the multitude of important applications make this an ideal area in which to pursue research with undergraduate students.

This article describes undergraduate research projects that used signal processing and machine learning to identify bird species from recordings of birdsong. These projects were conducted at Hope College with four undergraduate students and three faculty supervisors during the summers of 2016 and 2017. Over the last several years, the faculty mentors have worked to educate themselves about data analytics and machine learning, but none of them had any prior training in those areas. The students who participated in these projects had mathematics backgrounds that included at least second year linear algebra and multivariate calculus, but they typically had little (or no) prior computer programming experience. Consequently, a considerable amount of time was devoted to teaching students how to program in Python. We found that good math students who had no prior programming experience could learn programming well enough in the time allotted for summer research as long as we provided them plenty of support.

During each summer research project, two students worked for 8 weeks learning to write code in Python and associated packages (NumPy, SciPy, pandas, Keras, ...), learning about sound signal processing using wavelets, constructing neural networks, and training neural networks to solve species classification problems. Students began learning Python through the tutorials at After Hours Programming [1], and then followed this by learning about the NumPy and pandas Python packages [16, 22]. Students were introduced to neural networks through several specific examples, such as constructing a network for handwriting analysis based on the MNIST dataset [10]. For reference, the students used online textbooks by Goodfellow et al., Hagan et al., and Nielsen [15, 19, 35], Stanford University's Online CS 231n notes [24], and an article by Zeiler and Fergus [55]. Early on, students were led through a construction of a basic neural network, and they computed a few iterations of back-propagation by hand using the chain rule and the appropriate gradients.

This article describes the signal processing and machine learning tools we used to analyze recordings of birdsong and identify the species of bird present. Our analysis of the recordings proceeded in two steps. The first step was to convert audio recordings (or signals) into grayscale images called scalograms via a process called the wavelet transform. In Sect. 2, we describe the wavelet transform we used to produce a scalogram that shows the pitch, time, and power information contained in a time signal. The second step in our analysis was to use artificial neural networks to assign to each scalogram a probability that it represents a particular species of bird. In Sect. 3, we describe several types of artificial neural networks, how they are constructed, and how they can be used to analyze scalograms. Together, Sects. 2 and 3 provide the mathematical framework for our analysis of birdsong. In Sects. 4 and 5, we provide some additional details about our dataset and the techniques we employed to pre-process the data. We highlight the results of our analyses in Sects. 6, and 7 offers some suggestions for further inquiry.

2 Creating Scalograms

We first describe WAV files and the signals arising from them. We introduce the Morlet wavelet and the continuous wavelet transform, and give details about their use in basic signal analysis. The overall goal is to turn each signal into a grayscale image called a *scalogram* containing time, pitch, and power information about the recorded sound.

2.1 Time Signals, the WAV Audio Format, and Scalograms

A continuous (or discrete) time signal is a function $f : \mathbb{R} \to \mathbb{R}$ (or $f : \mathbb{Z} \to \mathbb{R}$). In our research, we analyzed digital audio recordings (i.e., discrete time signals) in WAV file format consisting of amplitude measurements of birdsong taken at equally spaced time intervals. The WAV format is an uncompressed audio format which represents sound as a sequence of numbers. Because our samples were recorded using 16-bit mono recording, a WAV file is essentially a vector of integers between $-32{,}768$ and $32{,}767$. The recording was done at a sample rate of 44.1 kHz, which means that there will be 44,100 integers in the vector per second of recording time. This allows us to analyze sounds in the human-audible range 0–22.05 kHz. The Python package SciPy has both a scipy.signal module and a scipy.io.wavfile module that are helpful in working with WAV files [6].

Plotting the list of integers in a given WAV file as a time series will give us a graphical representation of the sound recording. For example, a WAV recording of a cuckoo bird's (*Cuculus canorus*) song might produce the graph in Fig. 1.

In Fig. 1, the four amplitude bursts show the times at which each of the four "notes" of the birdsong begins and ends, but the pitch of each note is not clear by visual inspection. One method of turning a sound recording into a more revealing

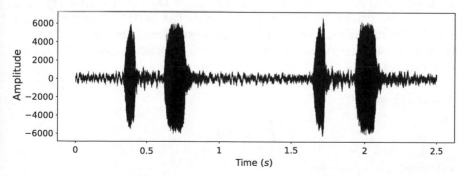

Fig. 1 A plot of 2.5 s of the WAV file of a cuckoo bird's (*Cuculus canorus*) song. Notice the four amplitude bursts of birdsong amid low-level background noise. Audio source: http://www.birdsongs.it/songs/cuculus_canorus/call1.wav

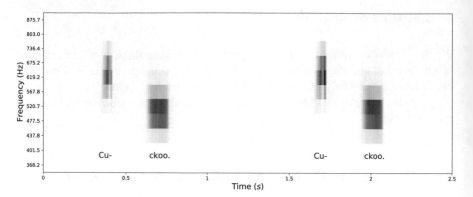

Fig. 2 A scalogram of 2.5 s of the cuckoo bird's song showing two repetitions of "cuckoo" with "cu-" at a higher frequency and shorter time duration than "ckoo." Grayscale shading shows the power values from lowest (white) to highest (black). On a standard piano keyboard the "cu-" is approximately the E♭ above middle C, and the "ckoo" is approximately the B below middle C

image is to use wavelet analysis to change the picture from a time-amplitude representation to a time-pitch-power representation called a scalogram. Figure 2 gives a scalogram for the cuckoo WAV file graphed in Fig. 1.

The scalogram gives a representation of the sound recording involving time (horizontal axis), pitch (vertical axis), and power (darkness of a pixel). In these regards, a scalogram is very similar to sheet music with dynamics markings. The notion of a wavelet and the process for producing the scalogram are detailed in the following subsections.

2.2   Wavelets

Roughly, a *wavelet* is a small wave-like function ψ that can be used to represent a part of another function, such as a time signal. The wavelet, along with certain scaled and translated versions of itself, should ideally form a basis of a certain space of functions, allowing for useful representations of a function in terms of the wavelet at various resolutions.

For our purposes, a wavelet is a function $\psi : \mathbb{R} \to \mathbb{R}$ (or $\psi : \mathbb{R} \to \mathbb{C}$) with the zero-average property

$$\int_{-\infty}^{\infty} \psi(t)\, dt = 0$$

and the unit energy property

$$\int_{-\infty}^{\infty} |\psi(t)|^2\, dt = 1.$$

There is also an *admissibility condition* for wavelet functions that is more complicated.[1]

Wavelets can be used to determine when certain frequencies occur in the
oscillations of a signal f by means of comparing scaled and shifted wavelets to
the signal. A wavelet ψ can be scaled (i.e., stretched or compressed) to produce
another wavelet of a different frequency (or pitch). By comparing wavelets of
different scales to a signal f, we can determine which wavelet scales "match"
the oscillations of a signal well, thereby identifying frequencies that are present in
the signal. Further, a wavelet can be shifted in time to produce another wavelet of the
same frequency. By comparing a time-shifted wavelet to a signal, we can determine
which time-shifted wavelets "match" the oscillations of a signal well, thereby
identifying *when* a signal has a particular frequency. The details of how to measure
the "match" between a wavelet and a signal are given below in the discussion of
the cross-correlation and the power (which is the square of the magnitude of the
cross-correlation). We chose to use wavelets because they have advantages over the
classical Fourier series methods for extracting frequencies. In particular, standard
Fourier analysis does not account for variability in the signal f over time, but
wavelets do.

Our main examples are wavelets that arise from a basic sine wave multiplied by
a Gaussian term that makes the amplitude very small outside of a narrow range of
values of t. Figure 3 shows the graph of a basic wavelet formed from a sine wave

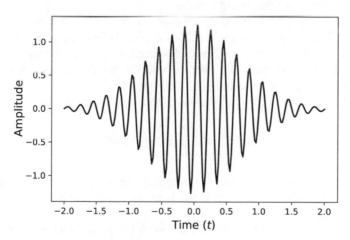

Fig. 3 A wavelet $\psi(t) = ke^{-t^2}\sin(10\pi t)$ based on damping a sinusoidal oscillation of
frequency 5

---

[1]This condition amounts to $\int_{-\infty}^{\infty} \dfrac{|\widehat{\psi}(\omega)|^2}{|\omega|}\, d\omega < \infty$, where $\widehat{\psi}(\omega) = \int_{-\infty}^{\infty} \psi(t)e^{-2\pi i\omega t}\, dt$ is the
Fourier transform of ψ. Full details can be found in many texts that cover continuous wavelet
transforms, such as [23].

of frequency 5 that has been damped by a Gaussian so that the amplitude of the wave becomes very small outside of $[-2, 2]$ and decays toward 0 as $t \to \pm\infty$. The formula for this function is

$$\psi(t) = ke^{-t^2} \sin(10\pi t),$$

where k is chosen so that

$$\int_{-\infty}^{\infty} |\psi(t)|^2 \, dt = 1.$$

This ensures that the wavelet has unit energy (i.e., L^2 norm of 1). We note also that

$$\int_{-\infty}^{\infty} \psi(t) \, dt = 0,$$

which means that the wavelet's average value is zero.

The wavelet ψ can then be compared with a signal f to see at which times f matches well with a frequency 5 oscillation. The overall process involves scaling and shifting the wavelet ψ and comparing it with f at many different scales (frequencies) and shifts (locations in time). By comparing the signal f to scaled and shifted versions of the wavelet ψ, we can discover the frequencies and times at which f oscillates like the scaled wavelet.

A signal f can be compared to a wavelet ψ through a process called *cross-correlation*. We will first describe the cross-correlation process in general, as if the functions f and ψ were both continuous functions. In practice, our signal f coming from a WAV file is discrete, and later we will discuss the discrete version of the cross-correlation for implementation on a computer.

The cross-correlation of two functions f and ψ is defined as the number

$$f * \psi = \int_{-\infty}^{\infty} f(t)\psi^*(t) \, dt,$$

where $\psi^*(t)$ represents the complex conjugate of $\psi(t)$. The magnitude of this integral will typically be large if f has behavior in common with ψ, and small if not.

To see this, suppose $\psi(t)$ is defined as above, and g is a function giving rise to a basic wave of frequency 5 that is in phase with ψ. For example, take

$$g(t) = \sin(10\pi t).$$

Then $g * \psi \approx 1.12$. In contrast, the function

$$h(t) = \sin(14\pi t)$$

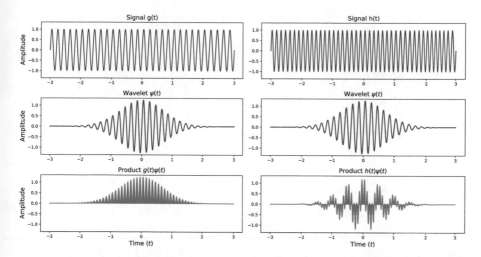

Fig. 4 (Top) Signals g and h of frequencies 5 and 7, respectively. (Middle) Wavelet ψ of frequency 5. (Bottom) Product of the signal with the wavelet. For the shaded regions in the bottom row, notice that the signed area $g * \psi \approx 1.12$ is large in comparison to $h * \psi \approx 10^{-17}$. This is because the frequency of oscillation for g matches that of ψ, but the frequency of h does not

oscillates at a frequency different from ψ, and consequently $h * \psi$ is much smaller; indeed, $h * \psi \approx 10^{-17}$. These results for $g * \psi$ and $h * \psi$ can be interpreted graphically as shown in Fig. 4.

In fact, if

$$j_n(t) = \sin(n\pi t),$$

and we plot $j_n * \psi$ for $n \in [0, 15]$, we obtain the response curve plotted in Fig. 5. This response curve compares the wavelet ψ to signals j_n of different frequencies to see which values of the frequency parameter n produce the largest values of the cross-correlation $j_n * \psi$. Notice that $j_n * \psi$ becomes large only when the frequency of the wave defined by j_n is nearly the same as that determined by ψ and is otherwise close to zero.

One issue that we have to confront when using a wavelet such as ψ to determine the frequency (or frequencies) present in a signal f is that ψ needs to be not only the same frequency as f but also in phase with f to give good results. For example, with ψ as defined above and $g_2(t) = \cos(10\pi t)$, we find that $g_2 * \psi = 0$, even though g_2 has the same underlying frequency as ψ.

Often, the cross-correlation is developed to be a function of a translation τ. For $\tau \in \mathbb{R}$, define

$$f * \psi(\tau) = \int_{-\infty}^{\infty} f(t)\psi^*(t - \tau)\, dt.$$

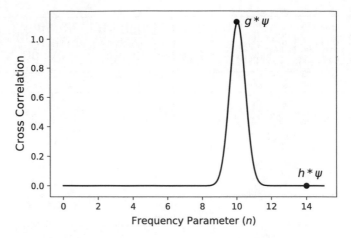

Fig. 5 Values of the cross-correlation $j_n * \psi$ for $n \in [0, 15]$. Notice that $j_n * \psi$ is largest at $n = 10$ and small or near zero away from $n = 10$

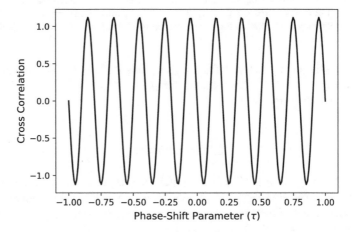

Fig. 6 The values of $g_2 * \psi(\tau)$ for $\tau \in [-1, 1]$. Notice that $g_2 * \psi(\tau)$ is small when g_2 and $\psi(\tau)$ are out of phase (e.g., $\tau = 0$), and large when they are in phase (e.g., $\tau = 0.15$)

For g_2 and ψ as defined above, a graph of $g_2 * \psi(\tau)$ vs. τ is given in Fig. 6. Here, we see that $g_2 * \psi(\tau)$ gets large in absolute value when the wavelet moves into phase with g_2, and becomes small in absolute value when the wavelet moves out of phase with g_2. We overcome this difficulty through the use of a complex-valued Morlet wavelet, described below, in which the real and imaginary parts both resemble the wavelet ψ defined above, but where one part is a phase shift of the other.

For more information on wavelets and their use in signal and image processing, the reader may see [23, 30, 52].

2.3   The Continuous Wavelet Transform

As seen in the previous subsection, the wavelet ψ can be shifted by the time-shift parameter $\tau \in \mathbb{R}$ to determine when it is in phase with a signal f. The wavelet ψ can also be scaled by a frequency-adjustment (or scale-adjustment) parameter $s > 0$ to determine which frequencies are present in the signal f. The scale parameter $s \in \mathbb{R}^+$ compresses (when $0 < s < 1$) or stretches (when $s > 1$) the wavelength of ψ, thereby increasing or decreasing its frequency. This leads to functions of the form

$$\psi_{s,\tau}(t) = \frac{1}{\sqrt{s}} \psi\left(\frac{t-\tau}{s}\right).$$

The factor $1/\sqrt{s}$ occurs in this definition to ensure that

$$\int_{-\infty}^{\infty} |\psi_{s,\tau}(t)|^2 \, dt = 1,$$

regardless of the values of s and τ, so that $\psi_{s,\tau}$ has unit energy.

In order to compare the shifted, translated wavelet with our signal $f(t)$, we compute the cross-correlation

$$C_f(s, \tau) = f * \psi_{s,\tau} = \int_{-\infty}^{\infty} f(t)\psi_{s,\tau}^*(t) \, dt.$$

The function $C_f(s, \tau)$ is called the *continuous wavelet transform* of f with respect to the wavelet ψ. For a signal $f : \mathbb{R} \to \mathbb{R}$ with $\int_{-\infty}^{\infty} |f|^2 \, dt < \infty$ and a function ψ satisfying the admissibility condition, the function f can be recovered from its continuous wavelet transform $C_f(s, \tau)$. See, for example, [23, Section 3.2].

To ensure that a constant-amplitude, periodic signal f does not appear as if it has periodic amplitude (as in the graph of $g_2 * \psi(\tau)$ vs. τ in Fig. 6), we employ a complex-valued wavelet when computing the continuous wavelet transform. This will provide better information about the frequencies that occur in the signal, the times at which they occur, and effectively deal with phase issues.

The basic Morlet wavelet, which we will use from this point forward, is a function $\psi : \mathbb{R} \to \mathbb{C}$ with formula

$$\begin{aligned}
\psi(t) &= K e^{it\omega_0 - t^2/2} \\
&= K e^{-t^2/2}(\cos(\omega_0 t) + i \sin(\omega_0 t)),
\end{aligned}$$

where K and ω_0 are positive real constants. Here, K is chosen so that

$$\int_{-\infty}^{\infty} |\psi(t)|^2\, dt = 1,$$

again ensuring that the wavelet has unit energy. The term "wavelet" is typically used for the Morlet wavelet, which performs well in practice even though ψ only approximately satisfies the admissibility condition. See [17, 29] for further details.

The real and imaginary parts of $\psi(t)$ are

$$\mathrm{Re}\,\psi(t) = K e^{-t^2/2} \cos(\omega_0 t), \qquad\qquad \mathrm{Im}\,\psi(t) = K e^{-t^2/2} \sin(\omega_0 t).$$

Therefore, the Morlet wavelet encompasses two real oscillatory wavelets of the same frequency, each of which is similar to the real wavelet introduced at the start of Sect. 2.2. The real and imaginary parts of ψ are out of phase with one another, allowing us to do a better job of determining the amplitude and duration of a periodic signal. A basic plot of the real and imaginary components of the Morlet wavelet is given in Fig. 7.

Because the signal f is real-valued, we have

$$\mathrm{Re}(f * \psi_{s,\tau}) = f * (\mathrm{Re}\,\psi_{s,\tau})$$

and

$$\mathrm{Im}(f * \psi_{s,\tau}) = -f * (\mathrm{Im}\,\psi_{s,\tau}).$$

We determine the strength of similarity between the scaled wavelet and the signal f by the computing the *power*, defined as

$$|C_f(s,\tau)|^2 = [f * (\mathrm{Re}\,\psi_{s,\tau})]^2 + [f * (\mathrm{Im}\,\psi_{s,\tau})]^2.$$

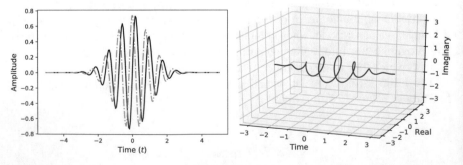

Fig. 7 (Left) A graph of the real (black solid) and imaginary (gray dashed) components of the Morlet wavelet for $\omega_0 = 8$. Notice that the real and imaginary components are out of phase with each other by $\pi/16$ radians. (Right) A graph of the Morlet wavelet for $\omega_0 = 8$ in $\mathbb{R} \times \mathbb{C}$. Viewed from above it is $\mathrm{Re}(\psi)$, and viewed from the side it is $\mathrm{Im}(\psi)$

Morlet wavelets are used to determine parts of an audio signal oscillating at a particular pitch, as well as to locate the beginning time and ending time of a pitch. To see an example of this, we consider a signal that is defined by a sine wave of frequency 5 over the interval $[0, 2]$ and is zero elsewhere:

$$f(t) = \begin{cases} \sin(10\pi t) & \text{if } 0 \le t \le 2, \\ 0 & \text{otherwise.} \end{cases}$$

Suppose that $\psi(t)$ is a Morlet wavelet with $\omega_0 = 2\pi$, so that ψ is based on a wave with underlying frequency 1. We plot $|C_f(s, \tau)|^2$ vs. τ for a variety of scales s in Fig. 8. The scales are chosen from the set

$$\mathscr{S} = \left\{ \frac{2^n}{5} : n = -2, -1, 0, 1, 2, 3 \right\}.$$

The cross-correlation is largest when the underlying frequency of the Morlet components matches that of the signal. In this example, the cross-correlation is largest when $s = 1/5$. The magnitudes of the cross-correlations with the wavelet at other scales are negligible compared to the one for which the frequencies match.

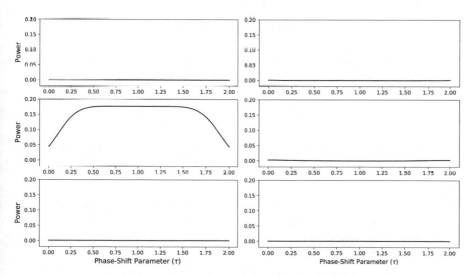

Fig. 8 Graphs of the phase-shift parameter τ versus power $|C_f(s, \tau)|^2$ for scales $s = 1/20$ (top left), $s = 1/10$ (top right), $s = 1/5$ (middle left), $s = 2/5$ (middle right), $s = 4/5$ (bottom left), and $s = 8/5$ (bottom right). Notice that $s = 1/5$ produces a graph with the largest power response for a broad range of τ values. This is because the signal has frequency 5. When $s = 1/5$, notice the "edge effects" near $\tau = 0$ and $\tau = 2$ that occur because near these values the wavelet overlaps with the part of the signal that is zero, whereas when $\tau \in [0.5, 1.5]$ the wavelet overlaps with the nonzero part of the signal

Figure 8 gives plots of power for only six scales. If we were interested in a large number of scales, then the graphical approach in Fig. 8 would become unwieldy because of the sheer number of plots to interpret. To simplify interpreting a large number of plots, we introduce a scalogram matrix to give an easily understood representation of the values $|C_f(s, \tau)|^2$ at many different times τ and scales s. Scalograms are matrices containing the values of $\frac{1}{s}|C_f(s, \tau)|^2$ for a finite set of values of s and τ. The factor $1/s$ here is needed to assure comparability of the power across different scales. This is discussed in more detail in Sect. 2.5. Under these assumptions, scalograms are viewable as grayscale images, and are also ideal for using as matrix inputs for a neural network.

2.4   Transforms on Discrete Signals

Before we get down to the business of explaining the scalogram generation process, we first need to understand how relevant scales can be chosen and how cross-correlations can be approximated using a computer. In general, it is helpful to assume that units on the t axis are $1/44,100$ s, so that our discrete signal f, which was sampled every $1/44,100$ s, can be evaluated at every integer time t. We continue to assume that the Morlet wavelet has formula

$$\psi(t) = Ke^{it\omega_0 - t^2/2},$$

but we note that, with this change to the t axis, the underlying frequencies of the wavelets $\psi(t)$ and $\psi_{s,\tau}(t)$ are, respectively, in cycles per second,

$$\frac{44,100\omega_0}{2\pi} \quad \text{and} \quad \frac{44,100\omega_0}{2\pi s}.$$

We chose a Morlet wavelet with $\omega_0 = 8$ and a list of 64 scales from the set

$$\mathscr{S} = \{2^{(j+11)/8} : 0 \le j \le 63\}.$$

This gives 8 scales per octave with an overall frequency range of about 92 Hz to about 21,650 Hz, which covers much of the human-audible range (20–20,000 Hz).

For a given discrete signal f, we are estimating the magnitude of the integral

$$f * \psi_{s,\tau} = \int_{-\infty}^{\infty} \frac{1}{\sqrt{s}} f(t)\psi^* \left(\frac{t-\tau}{s}\right) dt.$$

Using a basic linear substitution which replaces t by $st + \tau$, we find

$$f * \psi_{s,\tau} = \int_{-\infty}^{\infty} \sqrt{s} f(st + \tau)\psi^*(t) dt.$$

Because ψ is a continuous function, we can evaluate ψ^* wherever we like. As stated above, f can be evaluated at any integer time. We can think of f as a vector $\mathbf{x} = \langle x_0, x_1, \ldots, x_{N-1} \rangle$, where each $x_j = f(j)$ is an integer.

Choose a real number B such that the values of $\psi(t)$ are of negligible magnitude outside of the interval $[-B, B]$. (The value $B = 5$ works well for the Morlet wavelet as defined above.) We approximate the integral over the whole real number line by the integral over an interval slightly larger than $[-B, B]$, and we estimate this integral using the midpoint rule.

Given a scale s, we define M to be the first odd integer larger than $2Bs$. We approximate the integral on the interval $\left[-\dfrac{M}{2s}, \dfrac{M}{2s} \right]$. The midpoints t_j needed for the midpoint rule with $\Delta t = 1/s$ are

$$\mathscr{T} = \left\{ t_k = \frac{-M + 2k + 1}{2s} \ : \ k = 0, \ldots, M - 1 \right\}.$$

Let τ be an integer with $0 \leq \tau \leq N - 1$. Assume that f is zero before the signal starts and after it ends; that is, $f(j) = 0$ if $j < 0$ or $j > N - 1$. In order to approximate the integral for a given s and τ, we create a vector by evaluating ψ^* at all of the points in $\mathscr{T}$. We create a second vector by evaluating f at points in the following set of consecutive integers

$$\{ st_k + \tau \ : \ k = 0, \ldots, M - 1 \}.$$

Then

$$\int_{-\infty}^{\infty} \sqrt{s} f(st + \tau) \psi^*(t) \, dt$$

$$\approx \sqrt{s} \langle x_{st_0+\tau}, x_{st_1+\tau}, \ldots x_{st_{M-1}+\tau} \rangle \cdot \langle \psi^*(t_0), \psi^*(t_1), \ldots, \psi^*(t_{M-1}) \rangle \frac{1}{s}$$

$$= \frac{1}{\sqrt{s}} \langle x_{st_0+\tau}, x_{st_1+\tau}, \ldots x_{st_{M-1}+\tau} \rangle \cdot \langle \psi^*(t_0), \psi^*(t_1), \ldots, \psi^*(t_{M-1}) \rangle.$$

There are NumPy and SciPy routines to compute a "vector cross-correlation" like the one above. We used scipy.signal.fftconvolve due to its speed advantages over other routines. One must take care because the fftconvolve routine does not automatically conjugate and also has a time reversal in the second vector. Therefore, to be precise, we need to call fftconvolve with the signal $\langle x_0, x_1, \ldots, x_{N-1} \rangle$ in the first entry and the vector $\langle \psi^*(t_{M-1}), \ldots, \psi^*(t_1), \psi^*(t_0) \rangle$ in the second.[2]

We used the Morlet wavelet transform in our work for several reasons. First, the Morlet wavelet transform is a continuous wavelet transform which, unlike a discrete

---

[2]Ultimately, neither the reversal nor the conjugation makes a difference in practice, due to the essential symmetry of the Morlet wavelet and the fact that we are eventually only interested in the square of the magnitude of the integral.

wavelet transform, allows a signal to be resolved into as many different scales (or pitches) as desired. Fine pitch resolution allows us to capture small variations of pitch commonly found in birdsongs. Second, because the real and imaginary parts of the Morlet wavelet are out of phase with each other, the complex Morlet wavelet transform measures all of the important phase information in a signal. Third, Morlet wavelets and windowed Fourier transforms are conceptually similar, except that the former uses damping and the latter uses periodic extension, so it would be possible to compare the results of one to the other. For more details about the Morlet wavelet transform and a comparison to the windowed Fourier transform, see the article by Torrence and Compo [52]. To learn more about discrete wavelet transforms, see the book by Van Fleet [54], or the articles by Bénéteau and Van Fleet [9] and Mulcahy [32]. One avenue for future birdsong research could be to use mel-frequency cepstral coefficients, which have been used successfully for speech recognition, or a windowed Fourier transform instead of a Morlet wavelet transform.

2.5 Creating Scalograms

Using a Morlet wavelet ψ with $\omega_0 = 8$ and scale factors $s \in \mathscr{S}$ and $\tau \in \{0, 1, \ldots, N - 1\}$ as outlined above, we can use the technique of the last subsection to approximate the power $|C_f(s, \tau)|^2$ of a discrete signal f at each scale and time-shift. We store the values

$$\frac{1}{s}|C_f(s, \tau)|^2$$

in a matrix, where the rows indicate increasing scales and each column indicates a time-shift. Because frequency and scale are inversely proportional, as the scale index j increases from 0 (top row) to 63 (bottom row) in the power matrix, the frequency decreases. The presence of the extra factor $1/s$ here is due to the fact that we would like the output defined by signals of the same amplitude at different scales to give comparable power.

To explain the reason for this extra factor of $1/s$, we give the following example. The basic Morlet wavelet with $\omega_0 = 8$ has formula

$$\psi(t) = Ke^{-t^2/2}(\cos 8t + i \sin 8t).$$

Suppose $f(t) = A \sin 8t$ is a wave of the same frequency. Then

$$|C_f(1, 0)|^2 = \left|\int_{-\infty}^{\infty} f(t)\psi^*(t)\, dt\right|^2.$$

Given a scale $s \neq 0, 1$, the wave $g(t) = f(t/s)$ has the same amplitude as $f(t)$ but a different frequency. In order for power values in the scalogram to be comparable, we would like the entry for $\psi_{s,0}$ with $g(t)$ to be the same as that for $\psi(t)$ with $f(t)$. However

$$|C_g(s,0)|^2 = \left|\int_{-\infty}^{\infty} f(t/s)\frac{1}{\sqrt{s}}\psi^*(t/s)\,dt\right|^2 = \left|\int_{-\infty}^{\infty} \sqrt{s}f(u)\psi^*(u)\,du\right|^2 = s|C_f(1,0)|^2.$$

For more on the necessity of $1/s$, see [52, Section 3.c.] for further details.

The matrix consisting of the power values $\frac{1}{s}|C_f(s,\tau)|^2$ for specified values of s and τ is called a *scalogram*. In processing the WAV files, we chose $\omega_0 = 8$ for the underlying Morlet wavelet. Because we used the 64 scales in $\mathscr{S} = \{s_0, s_1, \ldots s_{63}\}$ and the integer time-shifts $0 \le \tau \le N - 1$ in creating the scalogram, it turns out to be a $64 \times N$ matrix of the form:

$$\begin{pmatrix} \frac{1}{s_0}|C_f(s_0,0)|^2 & \frac{1}{s_0}|C_f(s_0,1)|^2 & \frac{1}{s_0}|C_f(s_0,2)|^2 & \cdots & \frac{1}{s_0}|C_f(s_0,N-1)|^2 \\ \frac{1}{s_1}|C_f(s_1,0)|^2 & \frac{1}{s_1}|C_f(s_1,1)|^2 & \frac{1}{s_1}|C_f(s_1,2)|^2 & \cdots & \frac{1}{s_1}|C_f(s_1,N-1)|^2 \\ \vdots & \vdots & \vdots & \ddots & \vdots \\ \frac{1}{s_{63}}|C_f(s_{63},0)|^2 & \frac{1}{s_{63}}|C_f(s_{63},1)|^2 & \frac{1}{s_{63}}|C_f(s_{63},2)|^2 & \cdots & \frac{1}{s_{63}}|C_f(s_{63},N-1)|^2 \end{pmatrix}.$$

Because the entries of the scalogram are non-negative real numbers, the scalogram can be viewed as a $64 \times N$ grayscale image where values near 0 give white pixels and values near the maximum value in the entire matrix give black pixels as in Fig. 9. This allows us to use neural networks that work well for image classification to help classify bird species from scalograms. The process of constructing convolutional neural networks to accomplish this task is described in the next section.

3 Neural Networks

We now introduce artificial neural networks and the process of training them to accomplish a classification task. We then go on to define convolutional neural networks which are particularly useful for image classification.

3.1 *Densely Connected Neural Networks*

Roughly, an artificial neural network is a function with many parameters that is constructed to perform a classification or regression task with minimal error. The parameters are usually "learned" from a set of data, which provide a set of known input and target output value pairs for the function to try to associate with each other.

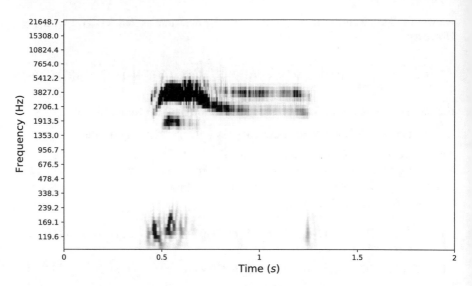

Fig. 9 Scalogram of a Roadside Hawk call (*Rupornis magnirostris*) recording from xeno-canto [2]. Each scale s on the scalogram was converted to a frequency for labeling the vertical axis using the formula $44,100\omega_0/(2\pi s)$. Audio source: https://www.xeno-canto.org/428440

A *densely connected neural network* is a composition of functions that alternate between affine functions and activation functions. A *layer* in a neural network is an affine function followed by an activation function.

An example of a densely connected neural network with one layer is the function $f : \mathbb{R}^2 \to \mathbb{R}^3$ defined by

$$f(x_1, x_2) = \text{softmax}(-3x_1 - 2x_2 + 6, \ -x_1 + 3x_2 - 6, \ 3x_1 - x_2 - 3),$$

where the activation function softmax $: \mathbb{R}^3 \to \mathbb{R}^3$ is defined by

$$\text{softmax}(u_1, u_2, u_3) = \left(\frac{e^{u_1}}{e^{u_1} + e^{u_2} + e^{u_3}}, \ \frac{e^{u_2}}{e^{u_1} + e^{u_2} + e^{u_3}}, \ \frac{e^{u_3}}{e^{u_1} + e^{u_2} + e^{u_3}} \right).$$

This neural network can be written as $f(x) = \text{softmax}(W_1 x + b_1)$, where the affine function $x \mapsto W_1 x + b_1$ has weight matrix W_1 and bias vector b_1 given by

$$W_1 = \begin{pmatrix} -3 & -2 \\ -1 & 3 \\ 3 & -1 \end{pmatrix}, \qquad b_1 = \begin{pmatrix} 6 \\ -6 \\ -3 \end{pmatrix}.$$

Notice that each component of the output of the softmax function is a number between 0 and 1, and that the sum of the output components is always 1. This allows

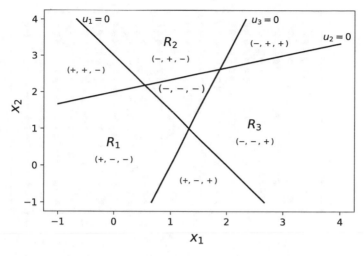

Fig. 10 Regions defined by the very simple neural network defined by f. The 3-tuples of positive and negative signs show the signs of (u_1, u_2, u_3) in each of the seven regions, where $u - W_1 x + b_1$

each component of the softmax output to be interpreted as a probability that an input point belongs to a particular output class.

The neural network f partitions the input space $\mathbb{R}^2$ into regions that have signature output values. For instance, the equation $u = W_1 x + b_1 = 0$ yields equations for three lines $3x_1 + 2x_2 = 6$ (or $u_1 = 0$), $x_1 - 3x_2 = -6$ (or $u_2 = 0$), and $3x_1 - x_2 = 3$ (or $u_3 = 0$) that divide $\mathbb{R}^2$ into seven disjoint regions, three of which are labeled R_1, R_2, and R_3 in Fig. 10. Specifically,

$$R_1 = \{(x_1, x_2) : -3x_1 - 2x_2 + 6 > 0, \ -x_1 + 3x_2 - 6 < 0, \ 3x_1 - x_2 - 3 < 0\},$$
$$R_2 = \{(x_1, x_2) : -3x_1 - 2x_2 + 6 < 0, \ -x_1 + 3x_2 - 6 > 0, \ 3x_1 - x_2 - 3 < 0\},$$
$$R_3 = \{(x_1, x_2) : -3x_1 - 2x_2 + 6 < 0, \ -x_1 + 3x_2 - 6 < 0, \ 3x_1 - x_2 - 3 > 0\}.$$

Under the function f defining the neural network, points in R_1 map to output values near $(1, 0, 0)$, points in R_2 map to outputs near $(0, 1, 0)$, and points in R_3 map to outputs near $(0, 0, 1)$. For example, $f(1, 3) = \text{softmax}(-3, 2, -3) \approx (0.0066, \ 0.9867, \ 0.0066)$, which is near $(0, 1, 0)$. Even if we did not know what the input point was, we could still interpret the output as saying that there is a very high probability (98.67%) the input point comes from region R_2, and a very low probability (0.66% + 0.66%) the input comes from either region R_1 or R_3. Thus, the output values of the neural network can tell us how to classify input points.

We can also use this network to classify points outside of the regions R_j depending on their proximity to the regions in question. For example, $f(-1, 3) \approx (0.268941, 0.731057, 0.000002)$, indicating that the point $(-1, 3)$ is nearest to R_2 (and quite far from R_3). Obviously, the point $(-1, 3)$ is not in any R_j here, and so the probabilities should really be thought of heuristically; in practice, they give

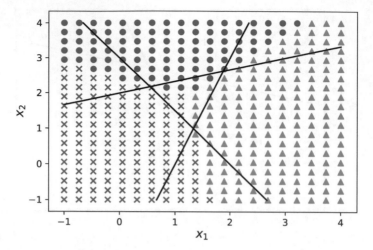

Fig. 11 Partition of the plane given by the neural network f. An x, filled circle, or filled triangle at a point (a, b) in the plane indicates that the output value of the neural network, $f(a, b)$, was closest to $(1, 0, 0)$, $(0, 1, 0)$, or $(0, 0, 1)$, respectively

a way to interpret the output of a softmax activation function in cases where the regions are defined by data and thus may be "fuzzy." This allows the network to classify each point in the plane as "most likely" nearest to one of the three regions (see Fig. 11).

A layer of a neural network is often depicted as consisting of nodes (neurons) with connections corresponding to the weights. The network above might be depicted as in Fig. 12. The diagram in Fig. 12 represents the composition f : $\mathbb{R}^2 \xrightarrow{W_1 x + b_1} \mathbb{R}^3 \xrightarrow{\text{softmax}} \mathbb{R}^3$. Each dot ($\bullet$) represents a node, which corresponds to a standard unit vector in $\mathbb{R}^2$ or $\mathbb{R}^3$. The "densely connected" nature of the network is seen in the arrows connecting the input layer to the middle column of nodes. Each arrow in that part of the diagram is labeled with a weight from the matrix W, and each node in the middle layer is labeled with the appropriate bias from the vector b_1. In neural network parlance, we think of each node as a neuron. The incoming weights and the bias determine the strength of the effect of input from the previous layer on that neuron, and the activation function determines the extent to which that neuron "fires" (i.e., passes information on to the next layer) based on its value relative to the other neurons in its layer.

In general, a densely connected neural network is a function $f : \mathbb{R}^m \to \mathbb{R}^n$, with

$$f = (f_k \circ L_k) \circ (f_{k-1} \circ L_{k-1}) \circ \cdots \circ (f_2 \circ L_2) \circ (f_1 \circ L_1),$$

where each L_j is an affine function of the form $L_j(x) = W_j x + b_j$ for some weight matrix W_j and bias vector b_j, and each f_j is an activation function. Each composition of the form $f_j \circ L_j$ is called layer j in this network. The functions

Fig. 12 Visual depiction of
the 1-layer neural network f.
Note: we consider the
composition of an affine
function followed by an
activation function in a
densely connected neural
network to be one layer, not
two

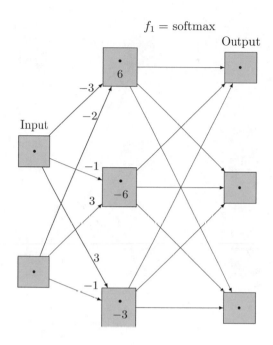

$f_1 = \text{softmax}$

f_j do not all have to be softmax functions; there are a number of functions that are typically used to achieve the desired network behavior. Another common activation function, which we will use later in our neural network construction, is the rectified linear unit (often abbreviated as ReLU). The basic ReLU function $\text{relu} : \mathbb{R} \to \mathbb{R}$ is

$$\text{relu}(x) = \begin{cases} x & \text{if } x \geq 0, \\ 0 & \text{otherwise.} \end{cases}$$

Like the softmax function, this function introduces non-linearity into the neural network. As an activation function in a network, the ReLU function is applied element-wise to vectors, matrices, or higher-rank tensors. In a neural network such as the one in Fig. 13, it is common to use ReLU activation functions everywhere except for the last layer, where the softmax is used. Note that if $f_1 = \text{relu}$ in Fig. 13, then only the horizontal arrows between nodes in columns two and three would be needed because ReLU is a function from $\mathbb{R}$ to $\mathbb{R}$ that is applied entry-wise to vectors.

The diagram in Fig. 13 is a relatively simple 2-layer network. In practice, each layer can have hundreds or thousands of nodes, which leads to a huge number of weights in a densely connected layer. For some purposes, it is helpful to define some layers of the network to be certain kinds of sparsely connected layers. Later, we will define the *convolutional layer*, a kind of sparsely connected layer that is often helpful in image classification problems.

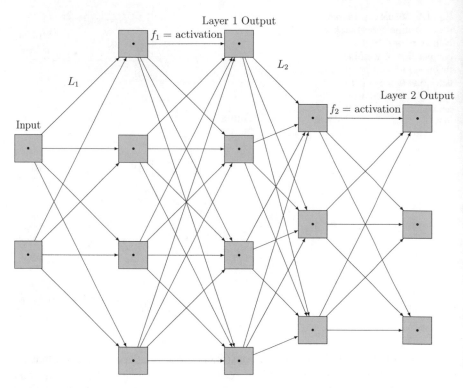

Fig. 13 A 2-layer densely connected neural network. The input layer has 2 nodes, layer 1 consists of $f_1 \circ L_1$ and has 4 nodes, while layer 3 (the output layer) consists of $f_2 \circ L_2$ and has 3 nodes. The affine function L_1 divides the input space ($\mathbb{R}^2$) into regions defined by 4 lines. Similarly, the affine function L_2 divides the output space of layer 1 ($\mathbb{R}^4$) into regions defined by 3 hyperplanes

3.2 Training Network Parameters

The 1-layer example f in the previous subsection could be somewhat misleading, in that we define the specific weights and biases of the neural network at the start. In general, the weights and biases of a neural network are the *parameters* of the model and are unknown values at the start. The basic shape ("architecture") of the network is predefined, but the values of the weights and biases must be determined ("learned") by training on the available data.

In order to illustrate the idea of training, we will look at a network similar to the one in the previous example from the perspective of training with data. Suppose we know that points in the plane are divided into three types (Type 1, Type 2, and Type 3), but we do not know which points are of which type. Suppose we have, however, gathered some data (in Table 1) that tells us the types of 21 points in the plane. The 1-hot encoding of Type j is just the standard unit vector $\mathbf{e}_j \in \mathbb{R}^3$.

Table 1 Data categorizing some points in the plane into one of three types

Point x	Type	Type 1-Hot encoding $T(x)$
$(0, 0)$	1	$(1, 0, 0)$
$(2, 2)$	3	$(0, 0, 1)$
$(1, 3)$	2	$(0, 1, 0)$
$(2, 4)$	2	$(0, 1, 0)$
$(0, 1)$	1	$(1, 0, 0)$
$(3, 1)$	3	$(0, 0, 1)$
$(0, 4)$	2	$(0, 1, 0)$
$(1, 1)$	1	$(1, 0, 0)$
$(3, 2)$	3	$(0, 0, 1)$
$(2, 1)$	1	$(1, 0, 0)$
$(0, 2)$	1	$(1, 0, 0)$
$(3, 3)$	2	$(0, 1, 0)$
$(2, 0)$	1	$(1, 0, 0)$
$(3, 0)$	1	$(1, 0, 0)$
$(1, 4)$	2	$(0, 1, 0)$
$(1, 0)$	1	$(1, 0, 0)$
$(0, 3)$	1	$(1, 0, 0)$
$(4, 2)$	3	$(0, 0, 1)$
$(4, 0)$	1	$(1, 0, 0)$
$(4, 1)$	3	$(0, 0, 1)$
$(1, 2)$	3	$(0, 0, 1)$

The association of points to the type of 1-hot encoding defines the "target" function T : $X \rightarrow Y$, where $X \subset \mathbb{R}^2$ is the set of all 21 points in the point column and $Y = \{(1, 0, 0), (0, 1, 0), (0, 0, 1)\} \subset \mathbb{R}^3$

Suppose we want to construct a single-layer neural network model $f : \mathbb{R}^2 \rightarrow \mathbb{R}^3$ of the form $f = f_1 \circ L$, where $f_1 : \mathbb{R}^3 \rightarrow \mathbb{R}^3$ is the softmax function. We assume that $L : \mathbb{R}^2 \rightarrow \mathbb{R}^3$ is defined by $L(x) = Wx + b$, where

$$W = \begin{pmatrix} w_{11} & w_{12} \\ w_{21} & w_{22} \\ w_{31} & w_{32} \end{pmatrix}, \quad \text{and} \quad b = \begin{pmatrix} b_1 \\ b_2 \\ b_3 \end{pmatrix}$$

are unknown matrices.

Now, we only know the general shape of the affine part of the model, the activation function being used, and the data from Table 1 (depicted in Fig. 14). The main task is choosing the weight matrix W and the bias vector b in order to accurately model this data.

We begin by defining an *error function*. Ideally, we would like a model that would take each of our data points in $\mathbb{R}^2$ as input and produce the 1-hot encoded version of the associated type as output. Thus, in an ideal world, our model f would map

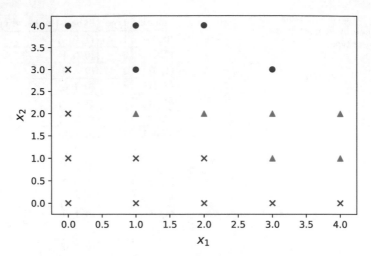

Fig. 14 A scatter plot of the training data given in Table 1. An x is class 1, a filled circle is class 2, and a filled triangle is class 3

$(0, 0)$ to $(1, 0, 0)$, $(2, 2)$ to $(0, 0, 1)$, $(1, 3)$ to $(0, 1, 0)$, and so forth. Such an exact correspondence is generally not feasible given the model constraints, and so we would instead like to choose the weights and biases to minimize the error function.

Here, we will use the standard sum of squares error. Given a data point x, we will use $T(x)$ to represent the 1-hot encoded type of x. We define the error function E as the sum of the squares of the entries of the vector $T(x) - f(x)$:

$$E(x, W, b) = ||T(x) - f(x)||^2 = ||T(x) - f_1(Wx + b)||^2.$$

Finding matrices of W and b for which the average error over all 21 data points is minimized is now a multivariate calculus problem. Ideally, we would like to minimize the average overall error, which we can think of as a function of the matrices W and b:

$$E(W, b) = \frac{1}{21} \sum_x E(x, W, b) = \frac{1}{21} \sum_x ||T(x) - f(x)||^2.$$

In practice, we typically find approximate solutions to this minimization problem via a method such as *gradient descent*. We will illustrate this method here. The simplest form of gradient descent involves incremental training; that is, we consider randomly chosen input data points x, one at a time, and take small steps in the parameter space that reduce the associated component of the error.

We begin by initializing the weights and biases to some randomly chosen values. Then, we use the data to train the network by gradient descent by the process described below.

First, we compute the relevant gradients. Given a fixed data point x, let $E = E(x, W, b)$. We need to find

$$\nabla E = \left\langle \frac{\partial E}{\partial w_{11}}, \frac{\partial E}{\partial w_{12}}, \frac{\partial E}{\partial w_{21}}, \frac{\partial E}{\partial w_{22}}, \frac{\partial E}{\partial w_{31}}, \frac{\partial E}{\partial w_{32}}, \frac{\partial E}{\partial b_1}, \frac{\partial E}{\partial b_2}, \frac{\partial E}{\partial b_3} \right\rangle.$$

In order to do this, let $u = L(x) = Wx + b$ and $y = f_1(u)$, where f_1 is the softmax function defined above. Here, $L : \mathbb{R}^9 \to \mathbb{R}^3$ is a function of nine variables (the six weights in W and the three biases in b are "variables," but x is a fixed data point and not a variable), and its derivative matrix is

$$D(L) = \begin{pmatrix} x_1 & x_2 & 0 & 0 & 0 & 0 & 1 & 0 & 0 \\ 0 & 0 & x_1 & x_2 & 0 & 0 & 0 & 1 & 0 \\ 0 & 0 & 0 & 0 & x_1 & x_2 & 0 & 0 & 1 \end{pmatrix}.$$

The derivative matrix of the softmax function $f_1 : \mathbb{R}^3 \to \mathbb{R}^3$ has a nice structure if we write it in terms of the dependent variable y:

$$D(f_1) = \begin{pmatrix} y_1 - y_1^2 & -y_1 y_2 & -y_1 y_3 \\ -y_1 y_2 & y_2 - y_2^2 & -y_2 y_3 \\ -y_1 y_3 & -y_2 y_3 & y_3 - y_3^2 \end{pmatrix}.$$

Finally, the derivative matrix for E as a function of (y_1, y_2, y_3) is

$$D(E) = \left(\begin{matrix} 2(y_1 - t_1) & 2(y_2 - t_2) & 2(y_3 - t_3) \end{matrix} \right),$$

where $T(x) = (t_1, t_2, t_3)$. We can now use the chain rule to compute

$$\nabla E = D(E)D(f_1)D(L)$$

given any input point x, its target output $T(x)$, and current parameter choices W and b.

We will perform a number of steps, each of which runs as follows:

1. Choose a learning rate α. (This value can vary with the step, but we will choose $\alpha = 0.05$ for each step for simplicity.)
2. Randomly choose one of the inputs x and find its target $T(x)$.
3. Compute ∇E based on the chain rule computations above. Reshape the first 6 entries of ∇E into a 3×2 matrix d_W, and the final 3 entries of ∇E into a vector d_b.
4. Compute $W_{new} = W - \alpha d_W$ and $b_{new} = b - \alpha d_b$.

After one step has been completed, we begin the next set with a new random data point, and with W_{new} and B_{new} in place of W and B.

We now illustrate one step of this process. Other steps will be similar, but with different starting data. For the purposes of this example, let us suppose the initial random weights and biases are

$$
W = \begin{pmatrix} 0.5 & -0.5 \\ -0.5 & -0.5 \\ 0.5 & 0.5 \end{pmatrix}, \qquad b = \begin{pmatrix} 0 \\ 0 \\ 1 \end{pmatrix}.
$$

Suppose that our first randomly chosen training point is $x = (2, 4)$ with $T(2, 4) = (t_1, t_2, t_3) = (0, 1, 0)$. Then, we compute

$$
\begin{pmatrix} u_1 \\ u_2 \\ u_3 \end{pmatrix} = L(2, 4) = \begin{pmatrix} 0.5 & -0.5 \\ -0.5 & -0.5 \\ 0.5 & 0.5 \end{pmatrix} \begin{pmatrix} 2 \\ 4 \end{pmatrix} + \begin{pmatrix} 0 \\ 0 \\ 1 \end{pmatrix} = \begin{pmatrix} -1 \\ -3 \\ 4 \end{pmatrix}.
$$

$$
\begin{pmatrix} y_1 \\ y_2 \\ y_3 \end{pmatrix} = f_1 \begin{pmatrix} u_1 \\ u_2 \\ u_3 \end{pmatrix} \approx \begin{pmatrix} 0.00668679 \\ 0.00090496 \\ 0.99240825 \end{pmatrix}.
$$

The associated derivative matrices are

$$
D(L) = \begin{pmatrix} 2 & 4 & 0 & 0 & 0 & 0 & 1 & 0 & 0 \\ 0 & 0 & 2 & 4 & 0 & 0 & 0 & 1 & 0 \\ 0 & 0 & 0 & 0 & 2 & 4 & 0 & 0 & 1 \end{pmatrix},
$$

$$
D(f_1) \approx \begin{pmatrix} 0.00664208 & -0.00000605 & -0.00663603 \\ -0.00000605 & 0.00090414 & -0.00089809 \\ -0.00663603 & -0.00089809 & 0.00753412 \end{pmatrix},
$$

$$
D(E) \approx \begin{pmatrix} 0.01337359 & -1.99819008 & 1.98481649 \end{pmatrix}.
$$

Then, the chain rule yields

$$
\nabla E \approx \begin{pmatrix} -0.02614076 \\ -0.05228152 \\ -0.00717853 \\ -0.01435706 \\ 0.03331929 \\ 0.06663859 \\ -0.01307038 \\ -0.00358926 \\ 0.01665964 \end{pmatrix}^T.
$$

This yields

$$dw = \begin{pmatrix} -0.02614076 & -0.05228152 \\ -0.00717853 & -0.01435706 \\ 0.03331929 & 0.06663859 \end{pmatrix}, \qquad d_b = \begin{pmatrix} -0.01307038 \\ -0.00358926 \\ 0.01665964 \end{pmatrix},$$

which gives

$$W_{new} = \begin{pmatrix} 0.50130703 & -0.49738592 \\ -0.49964107 & -0.49928214 \\ 0.49833403 & 0.49666807 \end{pmatrix}, \qquad b_{new} = \begin{pmatrix} 0.00065351 \\ 0.00017946 \\ 0.99916701 \end{pmatrix}.$$

If we continue to train the model by repeatedly choosing a random point and updating the weights W and bias vector b, the model will often converge to a point of minimum error, resulting in a partition of points in the plane as shown in Fig. 15. Because the network in this case only has one affine layer, the resulting model is essentially a simple linear regression model followed by a normalization. Therefore, we should not be surprised that this 1-layer model resulted in linear decision boundaries for the three types.

With more layers, one can train in the same manner to obtain more complicated decision boundaries. The result of the trained 2-layer model given in Fig. 13 with activation functions $f_1 = $ relu and $f_2 = $ softmax on the data from Table 1 is shown

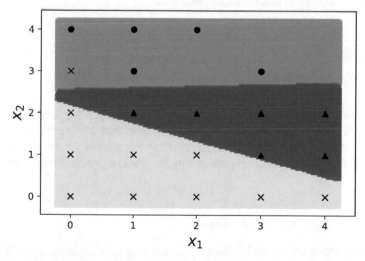

Fig. 15 Partition of the plane given by training the 1-layer network on the data from Table 1. Shading of the plane indicates whether the neural network output is largest in the first component (class 1 = light gray), second component (class 2 = medium gray), or third component (class 3 = dark gray). Training data are displayed with an x for class 1, a filled circle for class 2, and a filled triangle for class 3. Notice that some of the training data are misclassified by the neural network (e.g., the point (0, 3)) while others are quite close to the decision boundaries (e.g., the point (3, 1))

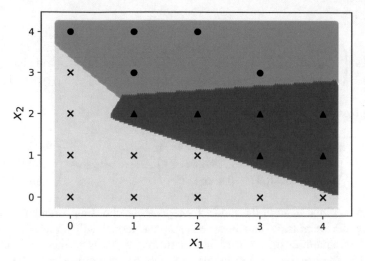

Fig. 16 Partition of the plane given by training a 2-layer network on the data from Table 1. Notice that this 2-layer neural network partitions the plane via non-linear decision boundaries and makes fewer classification errors on the training data than the 1-layer neural network in Fig. 15

in Fig. 16. The non-linear decision boundaries in Fig. 16 are the result of composing two non-linear functions $f_1 \circ L_1$ and $f_2 \circ L_2$ that divide $\mathbb{R}^2$, which is the input to layer 1, into regions defined by 4 lines and $\mathbb{R}^4$, which is the input to layer 2, into regions defined by 3 hyperplanes, respectively.

3.3 Other Types of Layers (Convolutional, Pooling, and Reshaping)

Networks with only dense layers have some disadvantages for image analysis. In particular, a dense layer in a network with n inputs and m outputs has $(n + 1)m$ parameters, which can be a very large number. As a result, a densely connected network with L layers, vector inputs of length n_0, and n_k nodes at layer k will have

$$\sum_{k=0}^{L-1} (n_k + 1)n_{k+1}$$

parameters to be determined. For instance, analyzing medium-sized 200×200 pixel images via a densely connected network with 50 nodes in layer 1 and 3 nodes in layer 2 yields a total of $(200^2 + 1) \cdot 50 + (50 + 1) \cdot 3 = 2{,}000{,}203$ parameters in the neural network! The large number of parameters and the size of vector inputs into each layer of the network can cause the network to train very slowly and even have unrealistically large training times. Also, networks with too many parameters can overtrain to match the data in the sample, rather than building a model that correctly classifies key characteristics and patterns that may be present in out-of-sample data. Moreover, the fact that each vector input is a "flat" vector fails to take advantage of the fact that the original inputs are images, and therefore have a rectangular array structure. For all of these reasons, densely connected neural networks might not be the ideal tool for image classification.

One way to overcome these difficulties is to insert certain types of sparsely connected layers at the start of the network. The sparse nature of these layers will reduce the total number of parameters to be determined, and the layers will also be structured in order to take advantage of the rectangular array structure of the inputs. In this subsection, we introduce a few of the types of layers available. One of the main types, the convolutional layers, will be based on a rectangular kernel of parameters which will sweep across the entire input image and train themselves to extract key features from the images. For a full description of the layer types discussed here and how they fit into a network designed for image analysis, see [24, Module 2].

2-Dimensional Convolutional Layers We first describe the basic 2-dimensional convolution operation, which is a 2-dimensional analogue of the 1-dimensional cross-correlation we used in the wavelet transform. A 2-dimensional convolution can be thought of as a process for identifying regions within a larger matrix (M) that are similar to features in a smaller matrix (K). This is done by moving the smaller matrix over contiguous regions of the larger one and measuring similarity at each step via a weighted sum of entries from the larger matrix with weights from the smaller matrix. For the purposes of giving a manageable example, we will work with a fairly small input matrix M. Suppose, for example, that the input to a layer is the 6×8 matrix

$$M = \begin{pmatrix} 2 & 3 & -1 & -3 & 0 & 4 & 1 & 2 \\ 1 & -1 & -2 & 2 & 3 & 0 & -1 & 3 \\ -2 & 1 & 0 & -2 & 3 & 1 & -1 & 0 \\ -1 & 3 & -1 & 1 & 2 & 2 & -4 & 1 \\ 4 & -2 & 1 & 0 & -3 & 1 & 0 & 1 \\ 1 & 2 & -1 & -1 & 3 & -1 & 2 & 2 \end{pmatrix}.$$

Convolution kernels are smaller matrices which are typically trained based on the data. For this example, suppose that the kernel is

$$K = \begin{pmatrix} 1 & 0 & -1 \\ 0 & 2 & 0 \\ -1 & 0 & 1 \end{pmatrix},$$

which will be used to identify features parallel to the main diagonal because the entries on the main diagonal are positive weights, the sub- and super-diagonal entries are neutral (zero), and the remaining entries are negative. The result of applying the convolution C_K defined by K to the matrix M is the 4×6 matrix given by passing K over each contiguous 3×3 submatrix of M and then recording the sum of the resulting products of elements. (This corresponds to the dot product of the two vectors obtained by reshaping K and the 3×3 submatrix of M into row vectors of length 9.) In the following diagram, the kernel K is dotted with the first 3×3 submatrix of M

$$K = \begin{pmatrix} 1 & 0 & -1 \\ 0 & 2 & 0 \\ -1 & 0 & 1 \end{pmatrix}, \qquad M = \begin{pmatrix} 2 & 3 & -1 & -3 & 0 & 4 & 1 & 2 \\ 1 & -1 & -2 & 2 & 3 & 0 & -1 & 3 \\ -2 & 1 & 0 & -2 & 3 & 1 & -1 & 0 \\ -1 & 3 & -1 & 1 & 2 & 2 & -4 & 1 \\ 4 & -2 & 1 & 0 & -3 & 1 & 0 & 1 \\ 1 & 2 & -1 & -1 & 3 & -1 & 2 & 2 \end{pmatrix},$$

to produce the first entry of $C_K(M)$:

$$1 \cdot 2 + 0 \cdot 3 + (-1) \cdot (-1) + 0 \cdot 1 + 2 \cdot (-1) + 0 \cdot (-2) + (-1) \cdot (-2) + 0 \cdot 1 + 1 \cdot 0 = 3.$$

Next, we would move the kernel one column to the right to compute the next entry of $C_K(M)$.

$$K = \begin{pmatrix} 1 & 0 & -1 \\ 0 & 2 & 0 \\ -1 & 0 & 1 \end{pmatrix}, \qquad M = \begin{pmatrix} 2 & 3 & -1 & -3 & 0 & 4 & 1 & 2 \\ 1 & -1 & -2 & 2 & 3 & 0 & -1 & 3 \\ -2 & 1 & 0 & -2 & 3 & 1 & -1 & 0 \\ -1 & 3 & -1 & 1 & 2 & 2 & -4 & 1 \\ 4 & -2 & 1 & 0 & -3 & 1 & 0 & 1 \\ 1 & 2 & -1 & -1 & 3 & -1 & 2 & 2 \end{pmatrix}.$$

The next entry of $C_K(M)$ is therefore

$$1 \cdot 3 + 0 \cdot (-1) + (-1) \cdot (-3) + 0 \cdot (-1) + 2 \cdot (-2) + 0 \cdot 2 + (-1) \cdot 1 + 0 \cdot 0 + 1 \cdot (-2) = -1.$$

We continue this process over all 3×3 contiguous submatrices of M to produce

$$C_K(M) = \begin{pmatrix} 3 & -1 & 6 & 2 & -5 & -1 \\ 5 & -5 & -6 & 9 & 0 & -6 \\ 1 & 3 & -5 & 2 & 11 & -7 \\ -6 & 1 & 1 & -7 & 7 & 4 \end{pmatrix}.$$

It is possible to pad the matrix M with rows and columns of zeros around the outside edges so that M and $C_K(M)$ have the same dimensions. It is also possible to move the kernel K across M with a stride length of more than one cell. For more details on padding and striding with 2d-convolution, please see [19, 35].

Given the basic operation of 2-d convolution, we now describe how to use this to make a layer in a neural network. We choose a certain size for our convolution kernel (say 3×3) and also a number of kernels to use in the layer (say 16). This number is sometimes called the number of channels. Using more than one channel allows us to apply several different kernels of the same size to the input at the same time. In this situation, $K_1, K_2, \ldots, K_{16}$ would each represent 3×3 convolution kernels which start off as unknown parameters (i.e., they are initialized to random values in some way). The training of the model determines the entries in each of these kernels in order to extract key features from the data. If the input to this convolutional layer is a $m \times n$ array, the output would consist of 16 different $(m - 1) \times (n - 1)$ arrays, which is usually considered as a single $(m - 1) \times (n - 1) \times 16$ array. Such a multidimensional array is called a rank-3 tensor, where standard matrices are examples of rank-2 tensors. Modern data modeling software such as Google's TensorFlow takes advantage of training data using tensors of rank higher than 2. We would refer to the convolutional layer described above as a "3×3 convolutional layer with 16 channels."

2-Dimensional Pooling Layers A convolutional layer is likely to generate several large matrices based on a single matrix as input. In order to simplify and shrink this data while retaining its dominant features, we can use pooling layers. These layers have no parameters to be trained. Instead, we divide the input matrix into rectangular arrays of the same size, and retain only some information (for example, the maximum value or the average value). In our networks, we used max-pooling, which keeps only the maximum value in each submatrix.

For example, a 2×2 max-pooling layer with input given by the output matrix $C_K(M)$ from above would yield the following breakdown:

$$
C_K(M) = \begin{pmatrix}
\begin{array}{cc|cc|cc}
3 & -1 & 6 & 2 & -5 & -1 \\
5 & -5 & -6 & 9 & 0 & -6 \\
\hline
1 & 3 & -5 & 2 & 11 & 7 \\
-6 & 1 & 1 & -7 & 7 & 4
\end{array}
\end{pmatrix}
$$

leading to the following output matrix

$$
\text{maxpool}_{2\times 2}(C_K(M)) = \begin{pmatrix} 5 & 9 & 0 \\ 3 & 2 & 11 \end{pmatrix}.
$$

Reshaping Layers Sometimes layers that reshape inputs are also helpful. For example, a common type of layer is the flatten layer, which will take a rank-k tensor as input (where $k > 1$) and flatten it into a vector (a rank-1 tensor). For example, if the input to a flatten layer is an $m \times n$ matrix, the output would have the same entries, but would be reshaped to a vector of length mn.

3.4 Convolutional Neural Networks

In constructing a convolutional neural network (CNN) to solve the bird species classification problem, we explored several different network architecture types common in image recognition applications (see [24], for example). The top performing models that we created for our data were arranged as in the diagram in Fig. 17. The example in Fig. 17 has 4 kernels that pass over the scalogram to produce each of the 4 channels of the first convolutional layer. Shading in the scalogram shows one place where convolution with a kernel is used to produce a number in one channel of the convolutional layer. Similarly, the shading in the lower-left corner of one of the channels of the first convolutional layer shows one place where max-pooling is used to produce a number in the max-pool layer. The flatten operation rearranges all

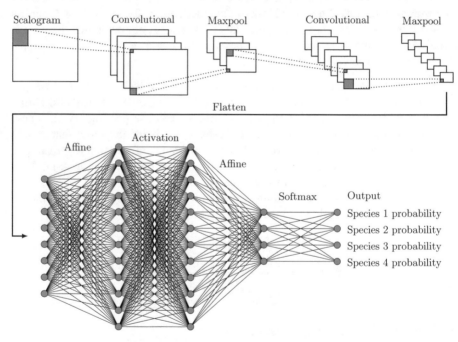

Fig. 17 Example of a schematic diagram of a convolutional neural network where the input is a scalogram of a birdsong and the output is vector of probabilities for four different types of bird species. The top row shows convolutional and max-pooling layers, which are rank-3 tensors. The convolutional and max-pooling layers on the left have 4 channels, while the ones on the right have 7 channels. Each convolutional layer scans the previous layer for features that are similar to the kernels. Each max-pooling layer compresses the previous layer while retaining the dominant features. The top row produces an abstract feature vector from the scalogram. The densely connected network in the bottom row decides which of these abstract features are important for classifying which bird species is singing. Note: The densely connected network is intentionally drawn with too few inputs and only two layers to save space

of the numerical entries from the last max-pool layer (a rank-3 tensor) into a vector (a rank-1 tensor). The bottom row is a densely connected neural network that takes this vector as input and returns a vector of probabilities as output.

4 Data Description

Each year since 2014, ImageCLEF has conducted a machine learning competition specifically dedicated to understanding more about biodiversity and habitats of living species [4]. It is hoped that increased knowledge in these areas will lead to better policies relating to sustainability and conservation. One task within this "LifeCLEF" contest is known as "BirdCLEF," which focuses on the identification of bird species from birdsong samples.

The BirdCLEF dataset is an annually updated, curated set of WAV files containing birdsong. The data for this task comes from the collection of birdsong data maintained by the xeno-canto Foundation [2]. Each WAV file is associated with an XML file which gives the date and location at which the recording was made, the name of the person who made the recording, and the bird species represented in the recording. This labeled and curated dataset provides an ideal framework for learning about audio signal processing and machine learning classification protocols.

For this project, we worked with the BirdCLEF 2016 dataset. This is a set of 24,607 WAV files, each of which is labeled with one of 999 different bird species native to South America. The files range from under 1 s to over 45 min in length, with a median length of around 25 s. The number of WAV files per species ranges from a minimum of 10 (for 7 different species) to a maximum of 160. On average, a species has around 25 associated WAV files.

In most cases, only short parts of a longer WAV file contain birdsong. Therefore, once we have generated a scalogram, we need to select appropriate short sections from each scalogram, called "snippets," to use as input for the neural network. Ideally, the selected snippets will contain clear examples of birdsong and be reasonably free of other noise. For entry into the neural network, each of the snippets must have the same size. The process of programming the computer to choose relevant snippets is detailed in the following section.

5 Pre-processing and Snippet Selection

In this section, we will describe the pre-processing steps involved in preparing the scalograms for a neural network. All of us started this project as novices at audio signal processing. Therefore, the procedures outlined in this section are not intended to be put forth as "best possible" pre-processing steps, but only as the most successful of the many different things we have tried. The following should be considered as a rough guide as to the sorts of processes that can be attempted.

Interested undergraduate students can easily get into this area by altering existing processes in various ways to see if greater classification accuracy can be attained.

We began our work with just 4 of the 999 species in the overall BirdCLEF dataset: Bananaquit (*Coereba flaveola*), Roadside Hawk (*Rupornis magnirostris*), Green Violetear (*Colibri cyanotus*), and Buff-breasted Wren (*Cantorchilus leucotis*). Using a small number of species dramatically reduced the computation time when trying different pre-processing steps and neural network model configurations, which helped us to quickly discover what works well and what does not. Once we had achieved some success in classification, we worked to extend our results to larger collections of species. We give details of our processes and results for four species in this section and the next.

5.1 Making Scalograms

We began by making a scalogram for every WAV file labeled with one of the four chosen species. Each scalogram was stored in a dataset together with its WAV file identifier and species label. Each scalogram was essentially a $64 \times (44100 \cdot T)$ matrix of non-negative real numbers, where T is the length of the recording in seconds.

5.2 Max-Pooling

Because the time dimension of the scalogram arrays was typically very large, we found that it was difficult to store and process data from all of the complete scalograms. However, having 44,100 observations of the continuous wavelet transform per second was not entirely necessary, because there was quite a bit of redundant information. We therefore chose to "max-pool" the scalograms along the time dimension. Each scalogram was broken into disjoint 1×50 sections, and only the maximum value from each section was retained. This maximum value became a single entry in the max-pooled scalogram. The new scalograms had smaller dimensions of $64 \times (882 \cdot T)$, and still contained the relevant information from the original scalograms.

5.3 Training–Validation–Testing Split

We split the collection of WAV files into three sets: a training set on which we trained our neural network model, a validation set on which we frequently evaluated our model, and a testing set on which we performed one final evaluation of our model. We opted for a small validation set that is separate from the testing set in

Table 2 Distribution of snippets and WAV files in the training, validation, and testing sets

	Training set		Validation set		Testing set		All three sets	
	Snippets	WAVs	Snippets	WAVs	Snippets	WAVs	Snippets	WAVs
Bananaquit	3322	62	615	9	2024	36	5961	107
Roadside Hawk	6792	61	1134	11	1558	28	9484	100
Green Violetear	8382	57	1139	11	2641	33	12,162	101
Buff-breasted Wren	4298	66	558	10	2044	26	6900	102
All four species	22,794	246	3446	41	8267	123	34,507	410

order to help us choose the best network architecture from among all the networks we constructed. The validation set also helped us evaluate how well the network performs on data it has not learned, thereby giving us a means to evaluate how well the neural network model generalizes from the sample (training data) to the population (data outside the training set). A network generalizes well when it performs well on the training and the validation sets, and poorly when it does significantly better on training data than on validation data. There were a total of 410 WAV files, and each of the 4 bird species was represented almost equally. Among the WAV files, we randomly took 60% (246 files) for the training set, 10% (41 files) for the validation set, and used the remaining 30% (123 files) for testing. Because the four bird species were almost equally represented in each of these three sets, we say that the split is stratified. The distribution of WAV files among the four bird species within each of the training, validation, and testing sets is given in Table 2.

5.4 Selecting Snippets from Scalograms

Our next step was to select short "snippets" from each scalogram to use as input to our neural network. This was necessary because our neural network requires every input matrix to have the same size. Snippets were created by sliding a window across each scalogram, keeping pieces that likely have birdsong and discarding ones that likely do not have birdsong. In particular, starting from the beginning of each scalogram we created one 1.5 s long snippet every quarter of a second. Each of these snippets, some of which overlap each other, was a 64×1323 contiguous submatrix of non-negative real numbers from the scalogram. For each snippet, we computed the total power, which is the sum of all the entries in the snippet. If the total power of a snippet was greater than the median of the total power for all the snippets from a WAV file, we kept the snippet as it likely had birdsong or other high-amplitude sounds in it. Otherwise, we discarded the snippet because it likely had no discernible signal in it and was mostly low-amplitude background noise.

The snippets that we kept from each scalogram became inputs to the convolutional neural network. For the 4 species under consideration, this resulted in 22,794 snippets from 246 WAV files in the training set, 3446 snippets from 41 WAV files in the validation set, and 8267 snippets from 123 WAV files in the testing set. The distribution of these snippets among the four bird species in each of the training, validation, and testing sets is given in Table 2. Note that although the number of WAV files for each of the four bird species is approximately equal within the training, validation, and testing sets, the number of snippets is not close to being equal within each set. This is largely the result of some bird species having more snippets because of longer recordings.

5.5 Denoising

One issue we struggled with was removing noise from the recordings. Here, "noise" means any sound in the recording that is not birdsong from the primary species. Some of the noise was due to low-volume noise, which we could typically remove from the scalogram via a thresholding technique. In fact, during the first summer of this project we denoised scalograms using wavelet-based thresholding techniques such as VisuShrink and SureShrink, which are good at removing low-power background noise from scalograms [54]. In the second summer of research, we discontinued wavelet-based thresholding and instead tried other means to remove noise due to human voices, other birdsongs or animal noises, and weather-related effects. The volume of this noise often was on par with or higher than the volume of the birdsong that we were extracting, and we had some difficulty getting the computer to distinguish between actual birdsong and these other sorts of noise. Our efforts to separate birdsong from noise, which focused on identifying noisy snippets and then throwing them away, were largely unsuccessful because they failed to automatically identify and discard high-amplitude noise. Interested readers could almost certainly improve upon our results by implementing effective denoising techniques, such as wavelet-based methods or band-pass filtering.

6 Results

Once we had developed some working pre-processing steps for our audio data, we worked to design convolutional neural networks to address the bird species classification problem. The neural network takes 64×1323 matrices S (snippets) as input, and the target output is equal to the one-hot encoded species identifier for that species. This means that snippets from a Bananaquit have a target output of $(1, 0, 0, 0)$, snippets from a Roadside Hawk have a target output of $(0, 1, 0, 0)$, snippets from a Green Violetear have a target output of $(0, 0, 1, 0)$, and snippets from a Buff-breasted Wren have a target output of $(0, 0, 0, 1)$. The output $f(S)$ of our network should be a vector in $\mathbb{R}^4$ with positive entries that sum to 1. To measure the error between the target output $T(S) = (t_1, t_2, t_3, t_4)$ and the neural

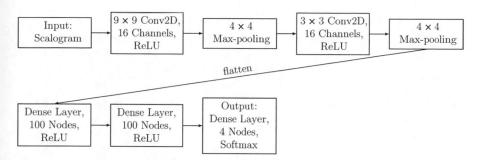

Fig. 18 The structure of the primary convolutional neural network used to address the species classification problem

network output $f(S) = (a_1, a_2, a_3, a_4)$, we used the "categorical cross-entropy" loss function

$$\text{Loss}_f(S) = -\sum_{i=1}^{4} t_i \log a_i.$$

This cross-entropy sum is only nonzero on the term corresponding to the correct species. Notice that it is built to penalize a low probability a_i in that position with a high positive loss.

Our most successful networks thus far have been structured as in Fig. 18. Each convolutional or dense layer in Fig. 18 is followed by an activation function. For the output layer, the activation function is the softmax function defined earlier. The other convolutional and dense layers use a rectified linear unit activation function. Using the ReLU activation function tends to speed up network training.

Each WAV file has many snippets associated with it. For each snippet, the neural network output is a probability vector with four entries whose largest entry indicates which bird species is most likely singing in the snippet. Thus, the bird species predicted to be in the WAV file by the neural network is the one that receives the most votes from among all of its snippets.

We trained our networks on snippets chosen from training set WAV files associated with the four species. Once acceptable training accuracy was reached, we validated our model on snippets chosen from the validation WAV files. This led to two sorts of accuracy: the percentage of chosen validation snippets for which the species is correctly identified, and the percentage of validation WAV files for which the species is correctly identified by combining the results of the associated snippets. The training process was stochastic, and so different training runs produced different levels of accuracy.

The results in Table 3 show that the ten neural networks constructed had average accuracy of 84.5% with standard deviation of 4.2% when all four bird species are considered. Table 3 also shows that the accuracy on WAV files was significantly higher for the training set than on the validation and testing sets. This may indicate that the model is overfitting, which means that it has learned the detailed features

Table 3 Average accuracy across ten neural network classifiers

	Training set		Validation set		Testing set	
	Accuracy		Accuracy		Accuracy	
	Snippets	WAVs	Snippets	WAVs	Snippets	WAVs
Bananaquit	92.6%	98.4%	90.8%	90.0%	69.1%	86.9%
Roadside Hawk	94.6%	99.5%	84.4%	89.1%	71.5%	83.6%
Green Violetear	97.0%	99.7%	67.3%	75.5%	79.2%	85.5%
Buff-breasted Wren	84.0%	97.1%	83.9%	98.0%	68.4%	80.8%
All four species	93.2%	98.6%	79.8%	87.8%	72.6%	84.5%
Standard deviation	2.1%	2.1%	4.1%	4.2%	4.2%	4.2%

Table 4 Some neural network classification results for the WAV files in the testing set

Predicted					Predicted					Predicted			
Actual	BQ	RH	GV	BW	Actual	BQ	RH	GV	BW	Actual	BQ	RH	GV	BW
BQ	32	1	3	0	BQ	31.3	1.1	2.6	1.0	BQ	86.9%	3.1%	7.2%	2.7%
RH	1	22	2	3	RH	1.8	23.4	0.7	2.1	RH	6.4%	83.6%	2.5%	7.5%
GV	0	3	30	0	GV	2.6	2.1	28.2	0.1	GV	7.9%	6.4%	85.5%	0.3%
BW	0	5	1	20	BW	0.1	4.3	0.6	21.0	BW	0.4%	16.5%	2.3%	80.8%

BQ Bannanaquit, *RH* Roadside Hawk, *GV* Green Violetear, *BW* Buff-Breasted Wren. (Left) A typical confusion matrix for one neural network. The entry in row i and column j is the number of WAV files for which the birdsong was actually species i but was predicted to be species j in the testing set. Each row sum is the number of WAV files for the bird species in that row. (Center) Average confusion matrix for all ten neural networks. (Right) Average confusion matrix for all ten neural networks converted to percentages (i.e., each row entry of the average confusion matrix is divided by the row sum and then converted to a percent)

of the training data too well, resulting in a model that performs sub-optimally on data that does not have features identical to those in the training set. Overfitting can be prevented by doing fewer weight updates or by techniques such as regularization and dropout [19]. We reduced overfitting by only learning from the training data by cycling through it three times (i.e., three epochs). We also tried regularization and dropout, but did not see significant improvements from them, so we omitted them from our neural network model.

The results in Table 4 highlight some classification results for our ten neural networks for the WAV files in the testing set. Each matrix in this table is called a confusion matrix, in which diagonal entries indicate correct classification and entries not on the main diagonal indicate incorrect classification.

7 Suggested Future Research Projects

Experiments in signal processing, neural network design, and data modeling provide great opportunities to engage students in statistical research projects. In addition to learning about neural network design and implementation, students develop useful skills in data processing, computer programming, and computer-aided data modeling. In this section, we discuss potential research avenues that might be undertaken by interested groups of undergraduate faculty and students.

Students who are interested in studying bird species classification can start from scratch or extend our work in this area.[3] Ideas for extending or improving bird species classification using birdsong include:

Research Project 1: Optimizing Signal Processing Explore different types of signal transforms for making power spectra, such as scalograms, and identify their best parameter values for birdsong analysis.

For this research project you may want to:

(a) Modify the Morlet wavelet transform by adjusting its parameters, changing the number of frequency bands and their widths to yield better results when analyzing birdsong. To speed up processing, try to find the minimal number of frequency bands that yield good results when analyzing birdsong.
(b) Instead of using a Morlet wavelet transform, create scalograms via a Daubechies, discrete Meyer, or another discrete wavelet transform.
(c) Instead of using a Morlet wavelet transform, create spectrograms via short-term or windowed Fourier transform methods.
(d) Instead of using a Morlet wavelet transform, create a short-term power spectrum representation using mel-frequency cepstral coefficients.

Research Project 2: Denoising Explore methods of denoising birdsong in order to improve the signal-to-noise ratio.

For this research project you may want to:

(a) Reduce low-power background noise in scalograms by applying wavelet-based thresholding techniques such as VisuShrink and SureShrink, and BayesShrink, which typically apply a single, universal threshold to all wavelet coefficients.
(b) Reduce high-power foreground noise in scalograms by applying band-pass filtering that leaves wavelet coefficients in some frequency bands alone while reducing some wavelet coefficients in other bands (possibly to zero). Such filtering could be particularly useful for reducing noise due to human voices, other birds or animals, and weather. It is challenging to reduce this kind of noise without reducing the birdsong signal, and to do so in a systematic way that is effective, universally applicable, and not ad hoc.

---

[3]Our Python computer code is available at https://github.com/paultpearson/BirdsongPaper.

Research Project 3: Selecting Relevant Snippets Explore methods for selecting the most relevant short segments (snippets) of birdsong from a long recording.

For this research project you may want to:

(a) Examine the scalograms for many bird species closely and develop your own criteria for snippet generation and selection. (In other words, get to know your data well and then devise a method for snippet selection that is tailored to your data and will generalize well to data you have not seen yet.) For snippet generation from scalograms, we used a sliding window approach with an arbitrarily chosen fixed window width and stride length, and we retained only the snippets that had above median power. Our snippet generation and selection methods could be improved upon. You may want to explore the birdsong literature for other ideas for effective detection and segmentation practices.
(b) Use object detection (or recognition) methods from machine learning to create a more effective method for generating and choosing segments (snippets) of birdsong in a scalogram from a long recording. Sliding window approaches that use machine learning to select snippets might perform better than those that use statistics, like our approach did.

Research Project 4: Design a Better Neural Network Create a neural network that achieves greater classification accuracy for birdsong scalograms.

For this research project, you may want to:

(a) Modify the existing neural network architecture (e.g., the number and size of convolutional layers, max-pooling layers, dense layers, etc.) or add dropout layers to improve classification accuracy.
(b) Use normalization on scalograms to make them have the same power relative to the other scalograms. In the neural network, batch normalization layers for the CNN or regularization for the dense neural network may also help improve performance, especially on testing and out-of-sample data.
(c) Explore other types of neural networks (e.g., recurrent neural networks, radial basis function neural networks, capsule networks, etc.).
(d) Explore ensemble methods, which combine the results of several different networks into a single classifier, to improve overall classification accuracy.

Research Project 5 Scale up the project by building a classifier that works for more bird species with a high degree of accuracy, and works on devices deployable where birds live.

For this research project, you may want to:

(a) Take what you learned from analyzing a small number of bird species and try to build a classifier that can correctly classify thousands of bird species and work well on new recordings. This may involve optimizing code for improved speed and data compression, and running the code on computers with sufficient hardware resources (e.g., memory, hard drive space, and graphical processing units).
(b) Build a remote sensing device (e.g., Arduino or Raspberry Pi) that can gather audio recordings, analyze the data in near real time using a pre-built classifier, and then report which bird species are present in a region either by transmitting them via the internet or allowing results to be downloaded directly from the device.

The research projects listed above build upon and extend a growing body of research in birdsong classification. Students interested in learning more about birdsong classification have many resources to draw on because in recent years Kaggle, NIPS, and BirdCLEF competitions have popularized birdsong analysis and resulted in much research. A recent review of the field of birdsong classification can be found in Priyadarshani et al. [37], and a thorough and recent literature review in appendix 1 of Knight et al. [25]. Many birdsong recognition contestants have also written about their work for the Kaggle 2013 contest (winner) [28], NIPS4B [48], BirdCLEF2014 [27], and BirdCLEF2016 (winner) [47]. Birdsong classification has recently been the subject of several undergraduate, masters, and Ph.D. theses [7, 31, 36, 46]. A non-exhaustive list of techniques employed in birdsong classification research includes wavelets [36, 38, 50], mel-frequency cepstral coefficients [7, 13, 14, 18, 20, 31, 48, 49], Fourier transform based methods [8, 12, 28, 41, 42, 44, 47], convolutional neural networks [25, 26, 31, 33, 43, 47, 53], densely connected neural networks [8, 27, 45], extreme learning machine [40], sparse-instance-based active learning [39], Gaussian mixture models [7], hidden Markov models [7, 14, 25], Bayes risk-minimizing classifier [13], multi-instance multi-label learning [11, 12], random forests [11, 28, 48, 49], support vector machines [14, 18, 20, 34, 51], and linear predictive coefficients [14, 21]. We encourage students to contact researchers and competition contestants in birdsong recognition as they dig into the literature.

Learning wavelet transforms and neural networks through birdsong research develops many skills that can be applied to analyze other types of data. For instance, interested students can also use scalograms and neural networks to classify other kinds of sound files. This could include sound-based recognition of other species types or more general sound recognition (e.g., human voice recognition). Students could also explore the idea of automated music recognition or transcription using wavelets such as the Morlet wavelet. Convolutional neural networks have a wide variety of important applications to image recognition and classification problems. Machine learning competitions on websites such as Kaggle [5], ImageCLEF [4], and BirdCLEF [3] regularly post large datasets, and are a great way to develop skills and learn from the other competitors. The availability of a vast number of large datasets online gives a wide range of approachable data modeling opportunities for students with diverse mathematical and statistical backgrounds.

References

1. After hours programming python tutorials. https://www.afterhoursprogramming.com/tutorial/python/python-overview/. Accessed: 2019-02-04.
2. A xeno-canto foundation. http://www.xeno-canto.org/. Accessed: 2019-02-04.
3. BirdCLEF 2016 competition, website. http://www.imageclef.org/lifeclef/2016/bird. Accessed: 2019-02-04.
4. ImageCLEF website. https://www.imageclef.org/. Accessed: 2019-02-04.
5. Kaggle. http://www.kaggle.com/. Accessed: 2019-02-04.
6. Scipy.org. https://www.scipy.org/, https://docs.scipy.org/doc/scipy/reference/signal.html https://docs.scipy.org/doc/scipy-0.19.0/reference/io.html#module-scipy.io.wavfile.
7. Fredrik Fløttum Aagaard. Modeling and confidence in a system for automatic classification of birdsong. Master's thesis, NTNU, 2015.
8. Anuradha Abewardana and Upul Sonnadara. Classification of birds using FFT and artificial neural networks. 2012.
9. Catherine Bénéteau and Patrick J Van Fleet. Discrete wavelet transformations and undergraduate education. *Notices of the AMS*, 58(05), 2011.
10. Aryal Bibek. Mnist handwritten digits classification using Keras. https://www.pytorials.com/mnist-handwritten-digits-classification-using-keras/. Accessed: 2019-02-04.
11. Forrest Briggs, Xiaoli Z Fern, and Jed Irvine. Multi-label classifier chains for bird sound. *arXiv preprint arXiv:1304.5862*, 2013.
12. Forrest Briggs, Balaji Lakshminarayanan, Lawrence Neal, Xiaoli Z Fern, Raviv Raich, Sarah JK Hadley, Adam S Hadley, and Matthew G Betts. Acoustic classification of multiple simultaneous bird species: A multi-instance multi-label approach. *The Journal of the Acoustical Society of America*, 131(6):4640–4650, 2012.
13. Forrest Briggs, Raviv Raich, and Xiaoli Z Fern. Audio classification of bird species: a statistical manifold approach. In *2009 Ninth IEEE International Conference on Data Mining*, pages 51–60. IEEE, 2009.
14. Jinkui Cheng, Bengui Xie, Congtian Lin, and Liqiang Ji. A comparative study in birds: call-type-independent species and individual recognition using four machine-learning methods and two acoustic features. *Bioacoustics*, 21(2):157–171, 2012.
15. Howard B Demuth, Mark H Beale, Orlando De Jesús, and Martin T Hagan. *Neural network design 2nd edition: http://hagan.okstate.edu/nnd.html*. Martin Hagan, 2014. http://hagan.okstate.edu/nnd.html.
16. Julia Evans. Pandas cookbook. https://github.com/jvns/pandas-cookbook. Accessed: 2019-02-04.
17. Marie Farge. Wavelet transforms and their applications to turbulence. *Annual review of fluid mechanics*, 24(1):395–458, 1992.
18. Nicolas Figueiredo, Felipe Felix, Carolina Brum Medeiros, and Marcelo Queiroz. A comparative study on filtering and classification of bird songs. http://doi.org/10.5281/zenodo.1422609. Accessed: 2019-02-04.

19. Ian Goodfellow, Yoshua Bengio, Aaron Courville, and Yoshua Bengio. *Deep learning:* *http://www.deeplearningbook.org/*, volume 1. MIT press Cambridge, 2016. http://www. deeplearningbook.org/.
20. Xin Guo and Qing-Zhong Liu. A comparison study to identify birds species based on bird song signals. In *ITM Web of Conferences*, volume 12, page 02002. EDP Sciences, 2017.
21. Michihiro Jinnai, Neil J Boucher, Jeremy Robertson, and Sonia Kleindorfer. Design considerations in an automatic classification system for bird vocalisations using the two dimensional geometric distance and cluster analysis. In *International Congress of Acoustics*, 2010.
22. Justin Johnson. Python numpy tutorial. http://cs231n.github.io/python-numpy-tutorial/. Accessed: 2019-02-04.
23. Gerald Kaiser. *A friendly guide to wavelets*. Springer Science & Business Media, 2010.
24. Andrej Karpathy and F. Li. Course notes: Cs231n at Stanford University. http://cs231n.github. io/. Accessed: 2019-02-04.
25. Elly Knight, Kevin Hannah, Gabriel Foley, Chris Scott, R Brigham, and Erin Bayne. Recommendations for acoustic recognizer performance assessment with application to five common automated signal recognition programs. *Avian Conservation and Ecology*, 12(2), 2017.
26. Qiuqiang Kong, Yong Xu, and Mark D Plumbley. Joint detection and classification convolutional neural network on weakly labelled bird audio detection. In *Signal Processing Conference (EUSIPCO), 2017 25th European*, pages 1749–1753. IEEE, 2017.
27. Hendrik Vincent Koops, Jan Van Balen, and Frans Wiering. Automatic segmentation and deep learning of bird sounds. In *International Conference of the Cross-Language Evaluation Forum for European Languages*, pages 261–267. Springer, 2015.
28. Mario Lasseck. Bird song classification in field recordings: winning solution for nips4b 2013 competition. In *Proc. of int. symp. Neural Information Scaled for Bioacoustics, sabiod.org/nips4b, joint to NIPS, Nevada*, pages 176–181, 2013. http:// www.animalsoundarchive.org/RefSys/Nips4b2013NotesAndSourceCode/WorkingNotes_ Mario.pdf.
29. Daniel TL Lee and Akio Yamamoto. Wavelet analysis: theory and applications. *Hewlett Packard journal*, 45:44–44, 1994.
30. Stéphane Mallat. *A wavelet tour of signal processing*. Elsevier, 2009.
31. John Martinsson. Bird species identification using convolutional neural networks. Master's thesis, Chalmers University of Technology, University of Gothenburg, Sweden, 2017.
32. Colm Mulcahy. Plotting and scheming with wavelets. *Mathematics Magazine*, 69(5):323–343, 1996.
33. Revathy Narasimhan, Xiaoli Z Fern, and Raviv Raich. Simultaneous segmentation and classification of bird song using cnn. In *Acoustics, Speech and Signal Processing (ICASSP), 2017 IEEE International Conference on*, pages 146–150. IEEE, 2017.
34. David Nicholson. Comparison of machine learning methods applied to birdsong element classification. In *Proceedings of the 15th Python in Science Conference*, pages 11–17, 2016.
35. Michael Nielsen. Neural networks and deep learning. http://neuralnetworksanddeeplearning. com/. Accessed: 2019-02-04.
36. Nirosha Priyadarshani. *Wavelet-based birdsong recognition for conservation: a thesis presented in partial fulfilment of the requirements for the degree of Doctor of Philosophy in Computer Science at Massey University, Palmerston North, New Zealand*. PhD thesis, Massey University, 2017.
37. Nirosha Priyadarshani, Stephen Marsland, and Isabel Castro. Automated birdsong recognition in complex acoustic environments: a review. *Journal of Avian Biology*, 49(5):jav–01447, 2018.
38. Nirosha Priyadarshani, Stephen Marsland, Isabel Castro, and Amal Punchihewa. Birdsong denoising using wavelets. *PloS one*, 11(1):e0146790, 2016.
39. Kun Qian, Zixing Zhang, Alice Baird, and Björn Schuller. Active learning for bird sounds classification. *Acta Acustica united with Acustica*, 103(3):361–364, 2017.
40. Kun Qian, Zixing Zhang, Fabien Ringeval, and Björn Schuller. Bird sounds classification by large scale acoustic features and extreme learning machine. In *Signal and Information Processing (GlobalSIP), 2015 IEEE Global Conference on*, pages 1317–1321. IEEE, 2015.

41. Louis Ranjard and Howard A Ross. Unsupervised bird song syllable classification using evolving neural networks. *The Journal of the Acoustical Society of America*, 123(6):4358–4368, 2008.
42. José Francisco Ruiz-Munoz, Mauricio Orozco-Alzate, and Germán Castellanos-Domínguez. Multiple instance learning-based birdsong classification using unsupervised recording segmentation. In *IJCAI*, pages 2632–2638, 2015.
43. Justin Salamon, Juan Pablo Bello, Andrew Farnsworth, and Steve Kelling. Fusing shallow and deep learning for bioacoustic bird species classification. In *Acoustics, Speech and Signal Processing (ICASSP), 2017 IEEE International Conference on*, pages 141–145. IEEE, 2017.
44. Maria Sandsten, Mareile Große Ruse, and Martin Jönsson. Robust feature representation for classification of bird song syllables. *EURASIP Journal on Advances in Signal Processing*, 2016(1):68, 2016.
45. Arja Selin, Jari Turunen, and Juha T Tanttu. Wavelets in recognition of bird sounds. *EURASIP Journal on Advances in Signal Processing*, 2007(1):051806, 2006.
46. Sinduran Sivarajan, Björn Schuller, and Eduardo Coutinho. Bird sound classification, 2016. Imperial College London, Bachelor Thesis.
47. Elias Sprengel, Martin Jaggi, Yannic Kilcher, and Thomas Hofmann. Audio based bird species identification using deep learning techniques. In *LifeCLEF 2016*, number EPFL-CONF-229232, pages 547–559, 2016.
48. Dan Stowell and Mark D Plumbley. Feature design for multilabel bird song classification in noise (nips4b challenge). *Proceedings of NIPS4b: neural information processing scaled for bioacoustics, from neurons to big data*, 2013.
49. Dan Stowell and Mark D Plumbley. Automatic large-scale classification of bird sounds is strongly improved by unsupervised feature learning. *PeerJ*, 2:e488, 2014.
50. Rong Sun, Yihenew Wondie Marye, and Hua-An Zhao. Wavelet transform digital sound processing to identify wild bird species. In *Wavelet Analysis and Pattern Recognition (ICWAPR), 2013 International Conference on*, pages 306–309. IEEE, 2013.
51. Ryosuke O Tachibana, Naoya Oosugi, and Kazuo Okanoya. Semi-automatic classification of birdsong elements using a linear support vector machine. *PloS one*, 9(3):e92584, 2014.
52. Christopher Torrence and Gilbert P Compo. A practical guide to wavelet analysis. *Bulletin of the American Meteorological society*, 79(1):61–78, 1998.
53. Bálint Pál Tóth and Bálint Czeba. Convolutional neural networks for large-scale bird song classification in noisy environment. In *CLEF (Working Notes)*, pages 560–568, 2016. http://ceur-ws.org/Vol-1609/16090560.pdf.
54. Patrick J Van Fleet. *Discrete wavelet transformations: An elementary approach with applications*. John Wiley & Sons, 2011.
55. Matthew D Zeiler and Rob Fergus. Visualizing and understanding convolutional networks. In *European conference on computer vision*, pages 818–833. Springer, 2014.

Using Regularized Singularities to Model Stokes Flow: A Study of Fluid Dynamics Induced by Metachronal Ciliary Waves

Elizabeth L. Bouzarth, Kevin R. Hutson, Zachary L. Miller, and Mary Elizabeth Saine

Abstract Pulmonary cilia are instrumental in the mucociliary clearance process that propels contaminants and bacteria out of the lungs. Cilia are small, slender fibers found in dense, carpet-like patches that impart a force on the surrounding fluid through coordinated motion. The motion of a single cilium in the lung is comprised of an upright power stroke relative to the cell surface from which it protrudes and an out-of-plane recovery stroke relative to the power stroke. A phase delay between neighboring cilia creates a metachronal wave that moves across the top of the cilia. Due to the small size of the cilia in this situation, simplifications to the governing fluid dynamics equations (the incompressible Navier–Stokes equations) allow us to use the linear, time-independent quasi-steady Stokes equations to model this scenario. We utilize the method of regularized Stokeslets to discretize the continuous body of each cilium and compute resulting fluid flow. In addition to introducing this problem, we also include results from a recent undergraduate project studying the effects of metachronal waves on fluid transport and provide possible future project directions this work can lead.

Suggested Prerequisites *This topic has elements of vector calculus, differential equations, scientific computation, computational fluid dynamics, physics, and biology in it. However, depending on the depth of the intended project, a student need not have experience in all of these things. A strong foundation in vector calculus and experience with programming are essential.*

E. L. Bouzarth (✉) · K. R. Hutson
Department of Mathematics, Furman University, Greenville, SC, USA
e-mail: liz.bouzarth@furman.edu; kevin.hutson@furman.edu

Z. L. Miller
Department of Aeronautics and Astronautics, Stanford University, Stanford, CA, USA
e-mail: zack1@stanford.edu

M. E. Saine
School of Mathematical and Statistical Sciences, Clemson University, Clemson, SC, USA
e-mail: msaine@g.clemson.edu

© Springer Nature Switzerland AG 2020
H. Callender Highlander et al. (eds.), *An Introduction to Undergraduate Research in Computational and Mathematical Biology*, Foundations for Undergraduate Research in Mathematics, https://doi.org/10.1007/978-3-030-33645-5_10

443

1   Biological Introduction

Cilia are generally small, slender fibers that serve a variety of biological functions. In some instances, cilia are immotile, fairly rigid, and may serve a sensory role. In others, cilia move and their primary function is to create fluid flow. For example, primary nodal cilia are rigid cilia that move in a conical fashion about a tilted axis. Primary nodal cilia are fairly sparsely packed and are critical in developing asymmetry in embryos before organ development [7, 8, 20]. Put another way, the conical motion of primary nodal cilia starts the biological signaling that puts your heart on the left side of your body rather than the right. When primary nodal cilia do not function properly, there is a chance that the signaling will get mixed up and the heart will form on the right side of the body in a condition known as situs inversus.

The pulmonary cilia that we are studying in this project are much more flexible and move in a different way than the previously described primary nodal cilia. Their primary function is fluid transport. Pulmonary cilia are flexible, densely packed, hair-like fibers that move with a coordinated whip-like motion that propels contaminants out of the lungs. The surface of the lung is lined with cells called *epithelial cells*. Patches of pulmonary cilia extend out from these epithelial cells into the airway surface liquid (ASL). The ASL is composed of two fluids: a periciliary liquid layer (PCL) and a mucus layer. The PCL is a water-like substance adjacent to the epithelial cells. The mucus layer is a thick, sticky fluid that sits on top of the PCL. The depth of the PCL is approximately the length of a cilium, but the thickness of the mucus layer can vary dramatically within a lung [9]. The mucus layer is responsible for trapping harmful contaminants from the airway above [19]. A sketch of the fluid surrounding the lung is shown in Fig. 1 (not to scale). Pulmonary cilia form a densely packed carpet on the surface of the lungs and the bulk effect of the cilia's coordinated motion is a *metachronal wave* that travels along the carpet's surface [19]. Recent studies have explored the formation of such metachronal waves [15, 18] as well as the efficiency of different wave configurations [9, 10, 24]. The metachronal wave helps propel the PCL and the contaminants trapped in the

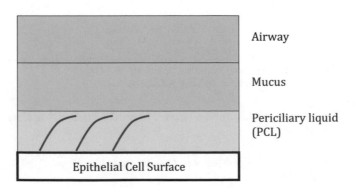

Fig. 1 Fluid layers above the pulmonary surface with cilia shown in red

adjacent mucus layer above it out of the lungs. This process is known as *mucociliary clearance*, clearing contaminants and protecting the lungs from the effects of these pollutants [19].

Our research draws motivation from a current subject of study in the biological community. Cystic fibrosis and primary ciliary dyskinesia (PCD) are two diseases with connections to the movement of cilia in the airway surface liquid in the lungs. Abnormalities caused by the cystic fibrosis gene lead to the depletion of the periciliary liquid and a thicker layer of mucus adhering to the airway surface [3]. Therefore, the coordinated beating of the cilia is inhibited in patients with cystic fibrosis because of the presence of unusually sticky mucus and the lack of periciliary liquid. This provides a favorable environment for infection [3]. A patient with primary ciliary dyskinesia suffers from impaired movement and possible immotility of the cilia, characterized by abnormalities in the beat, stroke, or coordination of cilia. This ultimately leads to reduced or absent mucociliary clearance [22]. Similar to the consequences of cystic fibrosis, without properly functioning cilia, the bacteria and contaminants remain in the mucus, leaving the lungs more susceptible to infection. Investigating the transport and activity of fluid tracers as a result of their interaction with healthy cilia will help further the biological study of cystic fibrosis, PCD, and other pulmonary or cilia-related diseases [4]. Research concerning properties of the PCL and mucus layers and their interaction with cilia is a subject of intense research because the mechanism by which the cilia cause clearance of the infected mucus remains uncertain.

There are competing theories as to the interaction between the cilia and the mucus layer. It is not known for certain whether or not the tips of the cilia penetrate the mucus layer during their beat cycle [16]. Given the variety and nature of questions regarding this physical scenario, creating a mathematical model of this fluid–structure interaction can provide insight into questions that might be difficult to explore with biological experiments. In our model, we choose to focus on the effects of a metachronal wave traveling through a patch of cilia on the surrounding periciliary liquid. As such, we make a simplifying assumption that there is no mucus layer for the cilia to directly interact with, which is consistent with the assumption that the cilia do not penetrate into the mucus and the entire length of the cilia is covered by the PCL. This decision is supported in [28]. Some models in the literature consider both the mucus and the PCL using different fluid modeling techniques and assumptions [9, 10].

The beat form of a single cilium is divided into two basic motions, shown in Fig. 2. The first is the *power* or *effective stroke*, where the cilium is nearly straight and in a plane perpendicular to the cell surface. Let us assume for discussion purposes in $\mathbb{R}^3$ that the cilium is moving in the xz-plane during its power stroke, so in the view of Fig. 2, consider the plane of the page to be the xz-plane where $y = 0$ and the power stroke is in the positive x-direction when moving to the right. During the power stroke the cilium will move quickly, in an attempt to propel the mucus and PCL. This is followed by the *recovery stroke*, which propagates a bend from the base to the tip of the cilium, returning it to its original form and starting position [4].

Fig. 2 Side view of the power stroke (black) and recovery stroke (gray) of a pulmonary cilium. Reproduced with permission of the ©ERS 2019: *European Respiratory Journal* 44 (6) 1579–1588; DOI: 10.1183/09031936.00052014 Published 30 November 2014 [27]

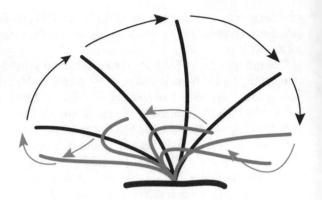

For the recovery stroke, the cilium moves out of the xz-plane and has position values with $y < 0$ (out of the page) when the cilium is moving in the negative x-direction (to the left) as it returns to its original position. Different descriptions of the cilium beat form exist in the literature. Some are parameterizations of a flexible cilium [16] while others make simplifications by making the cilium more rigid [17]. As discussed further in Sect. 2.3, our model uses the latter approach.

During a full beat cycle, the cilium quickly moves through its power stroke, propelling fluid forward and then makes a whip-like motion as it moves through its recovery stroke, back to the beginning of the next power stroke. The ratio of the recovery stroke duration to the power stroke duration is referred to as the temporal asymmetry, Rt. In pulmonary cilia, $Rt > 1$ because the power stroke is always faster than the recovery stroke [17]. Much research concerning pulmonary cilia has included a recovery stroke that is in the same plane as the effective stroke [14]. Our research introduces a model with an out-of-plane recovery stroke, which appears in biological studies [17], but is not present in all cilia instances [11].

Each cilium in a patch moves in coordination with the others, with neighboring cilia having a slight phase difference. The delay between beating cilia creates a metachronal wave affecting the propulsion of the PCL and mucus above. A metachronal wave is defined by its direction of movement relative to the power stroke direction. In our model, if a cilium's power stroke is in the positive x-direction, the metachronal wave travels generally in the negative x-direction. A wave that is propagating somewhat parallel or antiparallel to the power stroke direction is labeled as *symplectic* or *antiplectic*, respectively [28]. In fact, our antiplectic metachronal wave travels backwards and to the left of the effective stroke direction, which is similar to Gheber and Priel's model [17]. Sanderson and Sleigh observe an antiplectic wave in patches of rabbit tracheal cilia [28], so we implement various antiplectic metachronal waves in our model. Recent works explore the formation of the metachronal wave and explore properties of various configurations [9, 10, 24]. We are not exploring the formation of the metachronal waves in this project, rather focusing on the effects of the coordinated motion.

In the remainder of this chapter, we will introduce some mathematical foundations of computational fluid dynamics, specifically the method of regularized

Stokeslets, that can be used to model biological fluids. We will also include an undergraduate project completed on this topic as well as suggestions for future studies.

2   Mathematical Model

We are interested in the factors influencing fluid flow of the PCL in healthy lungs, especially in the presence of a ciliary metachronal wave. To study this, we will focus on a technique in computational fluid dynamics referred to as the *method of regularized Stokeslets* (MRS). Our model approximates the body of a pulmonary cilium by placing a collection of 30 regularized Stokeslets along the centerline of that cilium, discretizing the continuous body. As will be discussed further in Sect. 2.2, a regularized Stokeslet is a regularized singularity that mathematically converts a force exerted on the fluid to a resulting velocity anywhere in the fluid. We then prescribe a beat form by specifying the location of the cilium at discrete time steps and implement that beat form for a patch of cilia representative of patches in the lungs. The MRS provides a linear relationship between velocities and forces that can be used to both calculate the forces the cilia are exerting on the fluid necessary to maintain the prescribed velocities of the beat form as well as the velocity anywhere in the fluid resulting from the ciliary motion. The regularization is necessary to ensure finite fluid velocities at the location of the forces, which is not the case when using singular, non-regularized Stokeslets.

Other studies of metachronal waves use adaptations of the immersed boundary method to model a one- or two-phase fluid flow, where both the mucus and PCL are considered [9, 10]. The MRS used in this work is a Lagrangian method, meaning that we calculate the fluid velocities only at the points in the fluid we are interested in studying and follow those points along their trajectories. Eulerian methods such as the immersed boundary method require the calculation of the fluid velocity at a grid that spans the entire fluid domain, which could be quite computationally expensive depending on the desired resolution. Imagine having one particular location that you want to watch evolve over time. The Lagrangian approach would take the fluid at the initial location of interest and follow it through its motion as it likely moves away from its initial position. The Eulerian approach would keep the initial position fixed and explore the fluid that flows past its fixed location. In this approach, to get a sense of the global behavior of the fluid, one would need to calculate fluid velocities at many locations throughout the fluid whereas with the Lagrangian approach, one would only need to calculate the velocity at one point at each time step.

The Lagrangian nature of the MRS saves in computational time when only interested in a few points in the fluid domain. The MRS also has flexibility in modeling problems with complicated geometries whereas some other methods struggle with that. For instance, boundary integral methods require integration over a closed surface, so the time-varying geometry of a beating cilium is unlikely to produce integrable surface integrals because there is no simple mathematical description of the surface. Section 2.1 provides an introduction to the Navier–Stokes equations, the governing equations of fluid dynamics. Section 2.2 introduces the method of regularized Stokeslets while Sects. 2.3 and 2.4 further describe the cilium beat form and patches, respectively.

2.1   Fluid Dynamics

While fluids in general encompass a variety of substances including liquids, gases, and plasmas, we will be focusing on the behavior of liquids in this project. The *Navier–Stokes equations* are a collection of nonlinear partial differential equations that describe the relationship between a fluid's velocity, pressure, density, and viscosity. *Viscosity* is a fluid's resistance to flow, or its thickness in the sense that honey is "thicker" and more resistant to flow than water, hence honey has a higher viscosity than water. Liquids are often categorized as *Newtonian* or *non-Newtonian*, depending on the relationship between the *shear stress* and *local strain rate*. The shear stress describes internal forces within the fluid while the local strain rate described the rate of deformation of the fluid. For Newtonian fluids, this relationship is linear while non-Newtonian fluids have a nonlinear relationship. Non-Newtonian fluids are often referred to as viscoelastic fluids because the nonlinear dependence between stress and strain rates provides some memory to the fluid, creating an elastic effect. Non-Newtonian fluids can also have shear-thinning or shear-thickening properties as a result of the nonlinearity. For instance, ketchup is shear-thinning in that it does not flow very readily unless the shear force of a knife is applied when spreading it on a hamburger bun. On the other hand, silly putty is shear-thickening. If you leave it in its storage container overnight, it will take the shape of its container indicating that it is indeed a fluid. However, if you roll it into a ball and bounce it on the ground, it will act like a solid. During the impact with the ground, the instantaneous forces from the collision cause a quick change in the properties of the fluid. The water-like PCL in the lung is considered a Newtonian fluid, so modeling Newtonian fluids will be the focus of this chapter. If one wants to expand consideration to include the mucus layer, then non-Newtonian effects would need to be taken into consideration.

While more general forms exist, the incompressible Navier–Stokes equations describe Newtonian fluids:

$$\rho \frac{\partial \mathbf{u}}{\partial t} + \rho \mathbf{u} \cdot \nabla \mathbf{u} = -\nabla p + \mu \nabla^2 \mathbf{u} + \mathbf{F} \tag{1}$$

$$\nabla \cdot \mathbf{u} = 0, \tag{2}$$

where $\mathbf{u}$ represents the fluid velocity, ρ fluid density, p pressure, μ dynamic viscosity, and $\mathbf{F}$ external force. Note that bold variables represent vector quantities in $\mathbb{R}^3$. When modeling fluid flow in three dimensions as we are here, Eq. (1) represents three equations due to the vector nature of velocity and these equations account for conservation of momentum in the system. Sometimes the left-hand side of Eq. (1) is represented as $\rho\frac{D\mathbf{u}}{Dt}$, where $\frac{D\mathbf{u}}{Dt}$ is a *material derivative* and encapsulates both the local acceleration term $\frac{\partial \mathbf{u}}{\partial t}$ and the convective acceleration term $\mathbf{u} \cdot \nabla\mathbf{u}$ that accounts for the fact that the fluid of interest is flowing to other areas of the fluid domain. The right-hand side of Eq. (1) brings in the effects of pressure, viscosity, and external forces. Equation (2) is sometimes referred to as the *continuity equation* and represents conservation of mass for an incompressible fluid whose density remains constant, which often means there are no large temperature variations within the fluid domain. Textbooks exploring fluid dynamics principles at the advanced undergraduate level or graduate school level include [2, 23, 25] and provide more detailed derivations and analysis of the incompressible Navier–Stokes equations, including further discussion of the physical interpretations of these equations as well as extensions to other fluid scenarios and models.

It is worth noting that while the Navier–Stokes equations are used to model many fluids, the proofs of existence and uniqueness of solutions are open mathematical problems, and are among the set of Millennium Prizes offered by the Clay Mathematics Institute [1]. Solving the Navier–Stokes equations exactly is challenging, so scholars often resort to either making simplifying assumptions that change the nature of the PDEs or using numerical methods to approximate solutions to Eqs. (1) and (2). In this work, we recognize that we can use characteristics of the biological system to make assumptions that will simplify the nonlinear time-dependent Navier–Stokes equations into the linear quasi-steady Stokes equations (namely the small length scale of cilia and low fluid velocities contribute to a low Reynolds number). The *Reynolds number* provides insight into the balance of inertial and viscous forces in a fluid and is given by

$$Re = \frac{UL\rho}{\mu}, \tag{3}$$

where U is a characteristic velocity, L is a characteristic length scale, ρ is the fluid density, and μ is the dynamic viscosity. The Reynolds number gives a sense of how the inertial forces and viscous forces balance in a given fluid flow. Notice that the Reynolds number is not a property of the fluid alone, but also includes information about the length scales and speed of the motion as well. Fluid flow that has very high speeds, large length scales, and low densities would have a high Reynolds number and inertial forces would dominate viscous forces. Flow in this regime is turbulent, so the flow around a flying airplane or water rushing out of a fire hydrant would be examples of high Reynolds number flow. In our case, the small size of the cilium coupled with the density and viscosity of water contributes to a low Reynolds number situation where flow is laminar rather than turbulent and inertial forces are negligible relative to viscous forces. You can often see the transition from high Reynolds number to low Reynolds number intuitively at some kitchen sink faucets. If you turn the water all the way on, the water flow often looks turbulent and bumpy, however if you reduce the amount of water coming out, there is often a point where the water becomes calm, smooth, and transparent, a sign of laminar flow. Note that the low Reynolds number assumption in this work ($Re \ll 1$) is at a much smaller Reynolds number than the onset of laminar flow (on the order of 10^3).

The average velocity of our cilium over its beat cycle is $U = 0.122\,\mu\text{m/s}$ and the length scale we use is the length of a cilium, $L = 6\,\mu\text{m}$. Discussion on cilia parameters can be found in [16, 29, 30]. We assume the density and viscosity of the PCL are the same as water [16], so $\rho = 917\,\text{kg/m}^2$ and $\mu = 1\,\text{kg/ms}$. This gives us a Reynolds number of $Re = 6.75 \times 10^{-5}$, which is sufficiently small so that the incompressible Navier–Stokes equations (1) and (2) can be simplified to the quasi-steady Stokes equations after nondimensionalization

$$-\nabla p + \mu \nabla^2 \mathbf{u} + \mathbf{F} = 0 \tag{4}$$

$$\nabla \cdot \mathbf{u} = 0. \tag{5}$$

Notice the terms in the left-hand side of Eq. (1) become negligible in a low Reynolds number regime. Those terms represent acceleration terms in the Navier–Stokes equation, providing evidence that the inertial terms are no longer present in the quasi-steady Stokes equations. An example of the nondimensionalization and transition from the incompressible Navier–Stokes equations to the quasi-steady Stokes equations in low Reynolds flow can be found in [4].

Notice that the Navier–Stokes equations (1) and (2) are nonlinear PDEs with a time derivative (inertial term). The quasi-steady Stokes equations remove both the nonlinearity and the inertial term, meaning solutions to Eqs. (4) and (5) can be computed as a result of only considering the current force configuration since inertial effects are negligible. The previous force configuration is not needed to calculate fluid velocities. Since time is not immediately visible in Eqs. (4) and (5), any fluid flow in a low Reynolds number regime is reversible. That is, if one imposes a force on a fluid and calculates the resulting fluid flow at that instant in time, its motion will not persist to the next moment in time (because inertial effects have been ignored in the equations). Additionally, the fluid flow can be reversed by applying an equal and opposite force in the same location at another instant in time. Thus, small organisms that either want to generate fluid flow or to move themselves in a fluid environment need to move with motions that are not symmetric in the sense that if a cilium moves forward in a certain way and returns to its initial position in the same way but in the reverse direction, the fluid would have moved from the original motion, but would move back to its original state when the cilium returns to its original position. This is an instance of the scallop theorem introduced by Purcell [26] in reference to the idea that a scallop can open its shell slowly but close it quickly to move by forcing water out of it, but this does not occur in a low Reynolds number fluid regime. If the scallop were in a low Reynolds regime, the opening and closing of the shell (regardless of speed as long as the motion was low Reynolds number) will result in the scallop ending up where it started because the motion of the shell opening and closing have reciprocal motions. For cilia to generate fluid transport in a low Reynolds number regime, they must undergo a motion, described previously in this section, that breaks symmetry. The lack of time information in Eqs. (4) and (5) does not restrict our study only to steady flows, just flows whose parameters fall in a low Reynolds number regime.

2.2 Method of Regularized Stokeslets

Fundamental solutions play a role in solving linear partial differential equations. A *fundamental solution* or *Green's function* F satisfies $\mathscr{L}F = \delta(x)$, where $\mathscr{L}$ is a linear differential operator and $\delta(x)$ is the Dirac delta function, whose value is 0 everywhere except when $x = 0$ where it is undefined. Solutions to a non-homogeneous differential equation $\mathscr{L}f = g(x)$ can then be found through convolutions of F and g. Further discussions of fundamental solutions and convolutions can be found in various textbooks discussing differential equations, e.g., [21, 25]. When solving a linear differential equation, we can take advantage of superposition of solutions to find new solutions. In the case of solving the quasi-steady Stokes equations (4) and (5), boundary integral techniques rely on superimposing fundamental solutions. However, the fundamental solutions, or *Stokeslets*, are singular, meaning that they are not defined everywhere in the fluid domain. For some boundary integral calculations where you are integrating around a closed boundary and the singularities are separated from the fluid domain by the boundary, the singularity does not cause much concern. However, when using fundamental solutions to communicate forces directly on a fluid, there is not always a separation between the fluid domain and where the singularities exist, causing computational issues. This motivates the modification of the singular fundamental solutions (Stokeslets) to become regularized fundamental solutions (regularized Stokeslets).

The method of regularized Stokeslets (MRS), as developed by Cortez et al. [12, 13], uses Green's functions to produce solutions to the quasi-steady Stokes equations (4) and (5) in free-space. The MRS computes the fluid velocity $\mathbf{u}$ at a location $\mathbf{x}$ resulting from a force $\mathbf{F}$ exerted on the fluid at a location $\mathbf{x_0}$. We define the distance r between where the velocity is being calculated and a force is being exerted as $r = \|\hat{\mathbf{x}}\|$, where $\hat{\mathbf{x}} = \mathbf{x} - \mathbf{x_0}$ and $\| \cdot \|$ represents the L^2-norm. A Stokeslet is a fundamental solution to the singularly forced Stokes equations (4) and (5), where $\mathbf{F} = \delta(\hat{\mathbf{x}})\mathbf{f}$ (e.g., see [25]). Alternatively, a *regularized Stokeslet* is a Green's function to the Stokes equations when $\mathbf{F} = \phi_\epsilon(\hat{\mathbf{x}})\mathbf{f}$. Here ϕ_ϵ is a *cutoff function* that approximates a delta distribution as the regularization parameter $\epsilon \to 0$ and $\mathbf{f} \in \mathbb{R}^3$ controls the direction and magnitude of the force. The Dirac delta function is radially symmetric and integrates to 1, so we typically use cutoff functions that have the same properties. In building an intuitive difference between these two fundamental solutions, imagine exerting a localized force by pushing on a fluid with a pencil. In the singular case, you are using the sharp point of the pencil to push the fluid and calculate the resulting flow. In the regularized case, you use the rounded eraser of the pencil to push the fluid and calculate the resulting flow. You might imagine that the flow locally will be different near the site of the force, but away from the site of the force, the difference will be negligible. The flow will be most different at the location of the force because in the regularized case, you can calculate the velocity there, but in the singular case you cannot. This feature is quite important in establishing the usefulness of the MRS.

In general, the choice of cutoff function is not unique, but rather is often motivated by the calculation details needed in a given problem. For instance,

$$\phi_\epsilon(r) = \frac{15\epsilon^4}{8\pi(r^2 + \epsilon^2)^{3/2}} \tag{6}$$

works well for our purposes [13]. This cutoff function produces the regularized Stokeslet S^{ϕ_ϵ}:

$$S^{\phi_\epsilon}_{ij}(\hat{\mathbf{x}}) = \delta_{ij}\frac{r^2 + 2\epsilon^2}{(r^2 + \epsilon^2)^{3/2}} + \frac{\hat{x}_i\hat{x}_j}{(r^2 + \epsilon^2)^{3/2}}. \tag{7}$$

As $\epsilon \to 0$, we recover the singular Stokeslet [25],

$$S_{ij}(\hat{x}) = \frac{\delta_{ij}}{r} + \frac{\hat{x}_i\hat{x}_j}{r^3}, \tag{8}$$

which is undefined when $r = 0$ (when evaluating the velocity at the location of a force). Notice that $S^{\phi_\epsilon}_{ij}$ in Eq. (7) does not have the same property when $\epsilon > 0$. Velocities due to a regularized Stokeslet are found by

$$\mathbf{u}^{\phi_\epsilon}(\hat{\mathbf{x}}) = \frac{1}{8\pi\mu}S^{\phi_\epsilon}(\hat{\mathbf{x}})\mathbf{f}, \tag{9}$$

which is sometimes helpful to express in terms of the ith component of the velocity

$$u^{\phi_\epsilon}_i(\hat{\mathbf{x}}) = \sum_{j=1}^{3n}\frac{1}{8\pi\mu}S^{\phi_\epsilon}_{ij}(\hat{\mathbf{x}})f_j, \tag{10}$$

where n is the number of forces exerted on the fluid. Notice Eqs. (9) and (10) display a linear relationship between forces exerted on a fluid and resulting velocities calculated anywhere in the fluid. Similar relationships hold when replacing the regularized Stokeslet S^{ϕ_ϵ} with the singular Stokeslet S:

$$\mathbf{u}(\hat{\mathbf{x}}) = \frac{1}{8\pi\mu}S(\hat{\mathbf{x}})\mathbf{f}. \tag{11}$$

With this comes a natural question of the errors associated with calculating the same quantity with two different methods. There are errors due to regularization and quadrature (numerical approximation of integrals in both space and time), which are discussed in more detail in sources such as [4–6, 13], but will not be the focus of this discussion.

We use regularized Stokeslets to model a solid body (cilium) and do so by placing a collection of regularized Stokeslets along the centerline of the cilium. We select a value for the spreading parameter ϵ that spreads the force to approximately the radius of a pulmonary cilium [6, 29]. In our models we use the value $\epsilon = 0.1\,\mu m$. Using this value, we also calculate the number of regularized Stokeslets necessary to discretize the body of our cilia. Since each regularized Stokeslet has a diameter of $0.2\,\mu m$, it takes 30 to model a cilium at a length $L = 6\,\mu m$.

The regularized Stokeslets we are considering here are free-space solutions to the quasi-steady Stokes equations in three dimensions. However, the pulmonary cilia protrude from epithelial cells, and we are interested in modeling the PCL in three dimensions above the cell surface. To model the epithelial cells we implement a system of regularized image singularities that create a zero-velocity plane that emulates the no-slip cell surface. This creates a no-slip plane at $z = 0$ in $\mathbb{R}^3$, and we only consider the fluid motion in $\mathbb{R}^3$ where $z > 0$. To create a zero-velocity plane mathematically, we place a regularized Stokeslet as well as a collection of other regularized singularities (doublet, dipole, and two rotlets) at the image point reflected from the original regularized Stokeslet across the desired zero-velocity plane. These particular singularities were chosen because their net effect cancels that of the original regularized Stokeslet to create a plane within the fluid domain with no velocity that simulates the presence of a physical stationary boundary as the epithelial cells (at $z = 0$). The specific cutoff functions and form of the other singularities can be found in [4, 14] and a discussion of the singular ($\epsilon = 0$) versions of these can be found in [25]. Matlab code for the regularized singularities in both the free-space case and semi-infinite case (image singularities) can be found at https://github.com/ELBou21/FURM.

Figure 3 shows an example resulting from this combination of regularized singularities. Imagine two forces exerted in the xz-plane (shown in red). The resulting fluid flow is shown at a variety of tracking points, or tracer points, chosen in a grid pattern for velocity visualization purposes here. These are infinitesimally small, massless particles that move with the fluid, but do not affect the fluid itself. The blue arrows represent the fluid velocity in the xz-plane at the given locations resulting from the two red forces. Notice in the free-space solution on the left in Fig. 3a, there are visible blue velocity vectors where $z = 0$ indicating that there is fluid flow in that plane (the xy-plane that is perpendicular to the page). Contrast this with the velocity field shown in Fig. 3b that implements the system of image singularities. Notice that the net effect provides similar flow near the red forces, but now there are no visible blue arrows where $z = 0$, indicating that the velocities in the xy-plane are zero vectors. Since we are creating this zero-velocity plane with the intent of mimicking a no-slip plane that represents the cell surface, we would only consider the fluid domain above the no-slip plane where $z > 0$. Notice there is fluid flow generated below the plane in Fig. 3b, however, we disregard this in

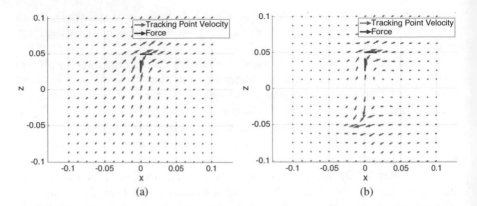

Fig. 3 Fluid flow resulting from two regularized forces (shown in red) (**a**) in free-space and (**b**) with image singularities (not shown) to create a no-slip plane at $z = 0$ where the velocities vanish. In practice, only the fluid above the cell surface (where $z > 0$) would be studied

practice. It is shown here for the purposes of demonstrating the effects of the system of image singularities, but when using the MRS, one would only put tracking points at desired locations (not necessarily in a grid) in the feasible fluid domain where $z > 0$.

The MRS provides a linear way to relate external forces exerted on the fluid into velocities anywhere in the fluid. Since the motion of the cilia is providing these external forces on the fluid, we must translate the beat form information into external forces exerted on the fluid. Often the beat form is known as a series of snapshots of the cilium's position, which we use to calculate velocity. While the specifics of the beat form we use will be discussed in Sect. 2.3, consider $\mathbf{x}^k$ to be the location of a regularized Stokeslet on a cilium's centerline at time k. Using the discretized position information from the beat form in time intervals of length h, we implement a second order finite difference scheme with respect to time to calculate the velocity of each regularized Stokeslet at time k using information about the discretized position at four different (known) instances in time:

$$\mathbf{u}^k = \frac{\mathbf{x}^{k-2h} - 8\mathbf{x}^{k-h} + 8\mathbf{x}^{k+h} - \mathbf{x}^{k+2h}}{12h}. \tag{12}$$

Note that because the cilium beat form is repeated and the position is known at each instance in time, there is no concern regarding how to initialize this velocity calculation. We arrange the information from each regularized singularity into a system of linear equations for each point of interest. We exploit the linear, homogeneous properties of the quasi-steady Stokes equations to create the system $\mathbf{u} = A\mathbf{f}$, where $\mathbf{u}$ is a $3m \times 1$ vector of velocities, $\mathbf{f}$ is a $3n \times 1$ vector of forces, and A is a $3m \times 3n$ matrix containing the information of each regularized Stokeslet as well as the collection of regularized image singularities.[1] Here m is the number of velocity calculation points and n is the number of regularized Stokeslets or forcing points. Thus, the full solution superimposes solutions for each individual regularized Stokeslet and the collection of image singularities. To find the forces $\mathbf{f}$ that would recreate the desired ciliary motion, we iteratively solve the linear system created by imposing the velocities at each regularized Stokeslet given by the beat cycle using the generalized minimum residual method (GMRES) in MATLAB. In this case, we have the same number of tracking points, m, and forcing points, n, so A is a square matrix that we invert to solve. Note that this matrix becomes singular for large enough values of ϵ given a fixed spatial distribution [6].

After calculating the forces required to impose a cilium's beat form, we can move on to calculating velocities at other points of interest anywhere in the fluid. We again use $\mathbf{u} = A\mathbf{f}$, but now $m \neq n$ and A is no longer a square matrix. The forces are already known from the previous calculation, but A and $\mathbf{u}$ are different in size because of the change in where the velocity is being calculated. Thus, we can solve for the fluid velocity at a given point through matrix multiplication as the net effect of each force we place in the fluid domain in the semi-infinite system. Once we have the velocity, we use Euler's method to update the tracer's new position.

2.3 Cilium Beat Form

Literature on modeling pulmonary cilia shows different approaches to modeling cilia and defining the beat form. Fulford and Blake parameterized a curved cilium that features a recovery stroke in the same plane as the power stroke [16]. The length of the cilium in their model's recovery stroke is shorter than in the power stroke and patches of cilia are synchronized [16]. Gheber and Priel simplify a cilium into a straight line in the power stroke that sweeps out an arc of a circle and two lines connected at a bend point in an out-of-plane recovery stroke [17]. This model was created to study the transmission of light through a field of cilia. Consequently, their model's recovery stroke does not build a full cilium, only the projection of the out-of-plane stroke onto the cell surface. Smith, Gaffney, and Blake model a line of cilia using the Fulford and Blake parameterization to represent each step in the beat form, but do not model the patches with metachronal waves we are interested in or an out-of-plane recovery stroke [30].

---

[1] If one wanted to modify the system of regularized singularities included in the model, this change would affect the matrix A, but the bigger picture of utilizing this framework remains the same. For instance, if you did not want to implement the system of regularized image singularities to create a no-slip plane, your matrix A would only consist of the regularized Stokeslet information in creating the free-space solution, so $A = S_{ij}^{\phi_\epsilon}$. When incorporating other regularized singularities, A is formed by taking a cleverly weighted sum of a regularize Stokeslet, doublet, dipole, and two rotlets as discussed previously in this section.

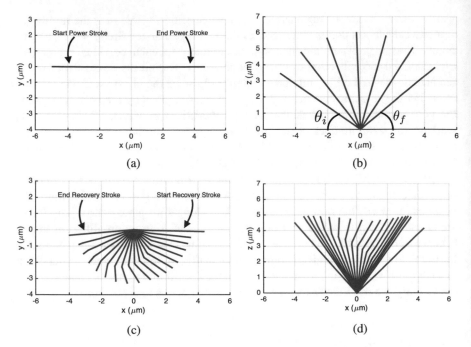

Fig. 4 Stages of the (**a, b**) power and (**c, d**) recovery strokes. The power stroke moves from left to right while the recovery stroke moves from right to left. The same time step is used in all plots showing faster speeds during power stroke than during recovery stroke. (**a**) Top view of our cilium's power stroke. (**b**) Side view of the power stroke beginning at θ_i and ending at θ_f. The effective stroke angle, θ_e, is the angle between the most extreme cilium locations so that $\theta_i + \theta_e + \theta_f = 180°$. (**c**) Top view of our cilium's recovery stroke. (**d**) Side view of the recovery stroke

Taking beat parameters and inspiration for how to build a cilium in its recovery stroke from the model created by Gheber and Priel [17], we create and test a cilium beat cycle featuring an out-of-plane recovery stroke with a motile bend, motile bend angle, and a varying recovery angle. Top and side views of our power stroke are shown in Fig. 4a, b and the same views of our recovery stroke are shown in Fig. 4c, d. These features were selected to mimic the description of cilia in rabbit trachea given by Sanderson and Sleigh [28]. Our cilium's beat cycle closely resembles Gheber and Priel's [17], but we are concerned with more than the projection of the cilium onto the epithelial cells. We define a recovery stroke that has a similar projection to Gheber and Priel's model but is comprised of the full length cilium moving in a three-dimensional fluid domain.

To model a cilium of length L, we utilize parameters of a cilium's temporal asymmetry and the arc of radius L of the power stroke from Gheber and Priel's model [17]. They model the power stroke of a cilium by sweeping out an arc of radius L that begins at angle θ_i and ends at angle θ_f, both of which are measured from the cell surface, as shown in Fig. 4b. These angles can be calculated from the effective stroke angle θ_e

$$\theta_i = \theta_f = (180° - \theta_e)/2, \tag{13}$$

as seen in [28]. We use $\theta_e = 110°$, which makes $\theta_i = \theta_f = 35°$. We define temporal asymmetry in the same manner as Gheber and Priel and choose the ratio of the recovery stroke to effective stroke to be $Rt = 3$ [17].

During the recovery stroke, we model the cilium as two lines connected at a bend point. This bend point propagates from $L/4$ to L over the entire duration of the recovery stroke. The minimum bend angle of $135°$ is achieved halfway through the recovery stroke, where the angle is measured between the two lines of regularized Stokeslets. Thus, a smaller angle corresponds to a larger bend. The angle of the bend necessarily starts and ends at $180°$ to match the straight cilium at the beginning and end of the power stroke.

To match the projection described by Gheber and Priel [17], we instituted a parameter of the recovery stroke defining the elevation angle at the midpoint of the recovery stroke. This angle dictates how close the cilium gets to the xy-plane. An elevation angle of $0°$ corresponds to the recovery stroke sweeping through the xy- plane and an elevation angle of $90°$ would be completely vertical. To maintain the length of our cilium and match the projection given by Gheber and Priel [17], we elevate the cilium in the recovery stroke. Measured up from the xy-plane, our cilium progresses from the elevation angle of $35°$ at the end of the power stroke up to $55°$ at the midpoint of the recovery stroke when it is tilted over the negative y-axis (out of the page in Fig. 4d) and back down to $35°$ for the beginning of the next power stroke.

2.4 Patches of Cilia

After modeling a single cilium that includes characteristics that mimic the behavior of realistic cilia noted by Gheber and Priel [17], we copy an individual cilium to build a patch of cilia, spaced 1 μm apart in both the x- and y-directions. The power stroke moves in the positive x-direction and the recovery stroke circulates out in the negative y-direction, as shown in Fig. 4. To implement a phase difference as outlined by Gheber and Priel [17], we varied the starting stages of each cilium in the patch. While we refer to the phase difference as a delay, we actually implement the phase changes as a forward phase difference. Thus, the leading cilium that is farthest away from the initial cilium (located at the origin) is ahead of the initial cilium, as shown in Fig. 5. The initial cilium began at the first step of the power stroke, with each subsequent cilium in the x- or y-direction starting one step ahead of the previous one. Thus, in a 30×5 patch, with one time step "delay" in both the x- and y-directions, the leading cilium starts 33 steps ahead of the initial cilium. In our model, we use a time step of 50 steps per stroke with a temporal asymmetry of $Rt = 3$, meaning that the effective stroke completes in 13 steps and the recovery stroke

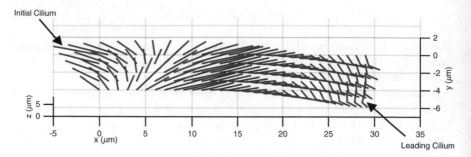

Fig. 5 Starting stages of the initial cilium and the leading cilium

completes in 37 steps. Therefore, the leading cilium and much of the surrounding cilia begin in a recovery stroke step. This observation will resurface when we begin our fluid flow analysis as it is central to the explanation of the trajectories of tracer points.

3 Results

With a focus on fluid transport as a result of the metachronal wave, our research looked at the trajectories of a specific set of tracer points subject to different metachronal waves. Tracer points have no impact on the fluid and are only used for visualizing flow. They are massless infinitely small particles that move with the local fluid velocity. By following their position over a sequence of time snapshots, we can visualize their trajectories. We chose a set of 42 points to serve as our basis for comparison across the different cilia patches. Points were placed at a variety of locations with respect to the cilia. Locations below the cilia tips were less desirable to analyze because of near-field interactions with the cilia. However, points at $z = 7\,\mu m$ provided a good basis for analysis as they sat right above the tips of the cilia and did not experience the interactions with other cilia as significantly. As tracer points approached the edge of the patch, their motion changed as a result of moving away from the patch. For this project, we focused on the behavior due to the metachronal wave and were mainly interested in the behavior over the patch itself when edge effects were negligible. For future work, patches could be constructed with different dimensions depending on the scope of the problem of interest, but this does increase computational cost. Realistically, a patch in the lungs would be neighbored by other patches, but our model only focuses on a single patch.

We focused on several patches with different wave characteristics, including a synchronized patch, a patch with a "regular" metachronal wave, and two patches with compressed metachronal waves, outlined in Sect. 3.2. The size of the patches varies from a rectangular patch of 150 cilia (30 × 5) to a square patch of 100 cilia (10 × 10). These choices were made to balance computational cost and the ability to see the effects of the metachronal wave on particle trajectories before encountering edge effects. Modeling larger patches of cilia increases computational time, but depending on a particular project's goals, this may be worth implementing.

3.1 No Metachronal Wave

As a first step in analyzing the effects of a metachronal wave, we modeled a synchronized patch of cilia without any phase delay, as seen in Fig. 6a, b. This facilitates analysis by emphasizing the influence of a delay on the trajectories of tracer points and ultimately the transport of fluid. Regardless of the tracer points' starting positions, our synchronized patch, of size 30×5, caused only slight movement in the effective stroke direction, as shown by the blue paths in Fig. 7.

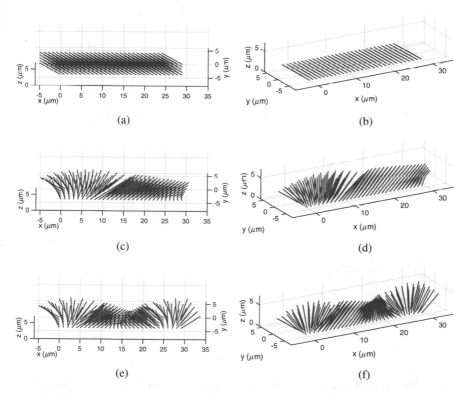

Fig. 6 Synchronized (**a, b**), regular wave (**c, d**), and compressed wave (**e, f**) at the initial time step on a 30×5 patch. (**a**) Synchronized 30×5 patch of cilia. (**b**) Synchronized 30×5 patch of cilia. (**c**) Regular wave on a 30×5 patch of cilia. (**d**) Regular wave on a 30×5 patch of cilia. (**e**) Compressed wave on a 30×5 patch of cilia. (**f**) Compressed wave on a 30×5 patch of cilia

3.2 Various Metachronal Waves

Using the same tracer points discussed above, we examine the effects of a regular metachronal wave, beginning with a delay of one time step in both the x- and y-directions as discussed in Sect. 2.4 with the goal of finding the characteristics of a wave that influence fluid transport. This wave is shown in Fig. 6c, d. In comparison to the synchronized patch, the tracer points have much larger transport in the effective stroke direction. Looking at Fig. 7, we see that in the synchronized 30×5 patch the tracer point (blue) makes very little forward progress in the positive x-direction and instead loops back nearly to its starting location. In contrast, the tracer point from a patch with a one time-step delay in both directions (magenta) makes

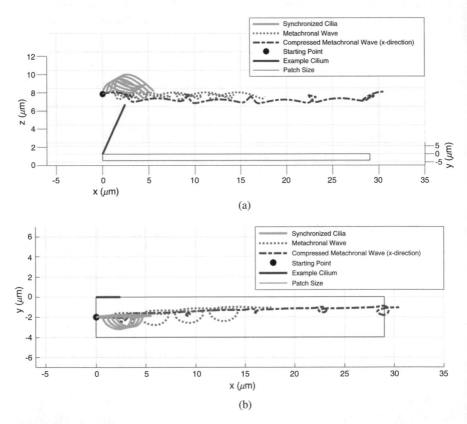

Fig. 7 Two views of trajectory comparisons between a synchronized patch of cilia and two metachronal waves for the tracer point beginning at $(0, -2, 7)$. A single cilium is shown for perspective, but the red rectangular region shows the footprint of the 30×5 cilia patch. The implementation of a metachronal wave increases transport (magenta) over the synchronized case (blue) and the compressed wave (green) greatly reduces the loops caused by the recovery stroke, increasing transport. (**a**) Synchronized 30×5 patch of cilia with fluid trajectories shown. (**b**) Top view

significant forward progress in the same time interval. From the top view in Fig. 7b, we see that the tracer's path during the recovery stroke sweeps out a smaller arc in the metachronal wave case than in the synchronized case. Our interpretation of this movement is that the wave mitigates the backward motion of the cilia in the recovery stroke and increases the forward transport of the PCL.

The next patch of cilia we examine is a 30×5 field with a metachronal wave with a delay of one time step in the y-direction and two time steps in the x-direction. We will refer to this metachronal wave as a compressed wave because it decreases the wavelength of the metachronal wave as seen in Fig. 6e, f. The ability of the compressed wave to mitigate the effects of the recovery stroke is evident when viewing the trajectory of fluid tracers over the compressed wave in Fig. 7 (green). Since we have shortened the wavelength of the metachronal wave, we have more cilia in their power stroke in the same 30×5 patch in our model. Therefore, as the recovery stroke passes tracer points in the compressed patch, tracers are closer to the previous and upcoming power strokes than they were in the patch with the regular wave.

We also investigate compressing the wave in the y-direction by testing our model on a 10×10 patch with a delay of one time step in the x-direction and two in the y-direction. We chose to use a 10×10 patch to investigate y-compression so that there was a sufficient number of cilia in that direction to demonstrate the effects of that compression. Comparing tracer points over a synchronized patch, patch with a regular wave, and patches with compressed waves in the x- and y-directions shows that compression in either direction works to remove the effects of the recovery stroke, as shown in Fig. 8b. In the 10×10 patch we see that compression has less of an impact on transport in the x-direction, but this may be a factor of patch length. However, loops seen in the tracer trajectories are greatly reduced in size with both implementations of a compressed wave relative to the synchronized and regular wave patches, demonstrating its effectiveness at decreasing motion due to the recovery stroke.

4 Conclusion

We sought to create a model of the carpet-like patches of pulmonary cilia found on the surface of the lungs, essential to the mucociliary clearance process. Using the method of regularized Stokeslets, we developed a model for pulmonary cilia that features an out-of-plane recovery stroke. By instituting a system of regularized singularities, we modeled the effects of the epithelial cells which act as a no-slip boundary for our fluid.

Our model uses many of the beat form parameters defined by Gheber and Priel. We also use their recovery stroke's projection on the epithelial cells as a reference from which we develop our full length, out-of-plane recovery stroke [17]. In our

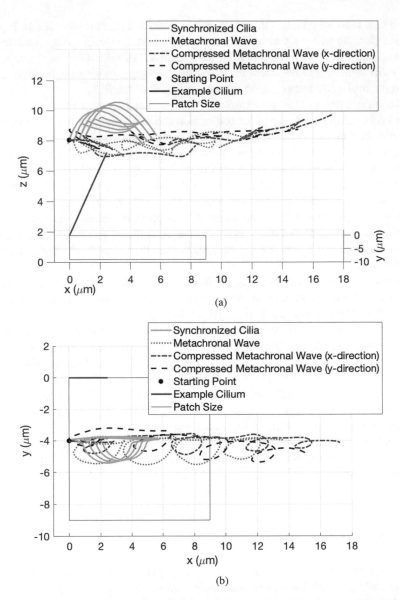

Fig. 8 Two views of trajectory comparisons between a synchronized patch of cilia and three metachronal waves for the tracer point beginning at $(0, -4, 7)$. A single cilium is shown for perspective, but the rectangular region shows the footprint of the 10×10 cilia patch. The implementation of a metachronal wave increases transport and a compression in the x-direction reduces the loops caused by the recovery stroke, thus increasing transport. (**a**) Trajectory comparisons over a 10×10 patch of cilia with various metachronal waves. (**b**) Top view

models, we look at patches of pulmonary cilia and institute a metachronal wave moving in an antiplectic direction relative to the power stroke. We impose the beat form rather than letting it evolve from hydrodynamic interactions as other studies do to study the effects of the metachronal wave [15, 18]. By varying the time-step delay used to create these waves, we are able to investigate various compressions of the wave. These variations serve to shorten the wavelength of the metachronal wave and we show that this has a substantial effect on fluid transport over the patch of cilia. Tracer points over a 30×5 patch with a metachronal wave compressed in the x-direction travel farther than tracer points starting at the same location travel over a patch with a regular wave or the synchronized patch.

In the case where we compressed the wave in the y-direction, we observe less contribution by the recovery stroke, demonstrated through smaller loops in the tracer point trajectories over the 10×10 patches than in the synchronized and regular wave patches. The tracer points do not demonstrate a large difference in x-transport, however, we believe this is due to a smaller patch size in the x-direction.

5 Suggested Projects

There are many extensions of ideas in this project that can be investigated by undergraduates. The interests and experiences of a particular student might make some of these projects more suitable than others. As with the nature of academic research, this topic of biologically inspired computational fluid dynamics is a current topic of research and the scope of cutting edge papers is often larger and more involved than would be appropriate for an undergraduate project, however, the tools presented here are very valuable to use in a creative way to find new problems that are appropriate relative to the experiences and interests of a student.

Future research ideas from the project described here focus on improvements upon our current model. While we currently use two lines connected at a bend point to model the body of the cilium during the recovery stroke, our model would benefit from the creation of a smooth cilium body that includes the same, current out-of-plane recovery stroke.

Research Project 1 Create a model of a cilium that more accurately captures the features of the pulmonary cilium beat form. That is, create a smooth curve that has an out-of-plane recovery stroke and matches the physiological properties of the beat form [16, 27]. This may involve using parametric curves, interpolation, and/or splines in both time and space. Does the new beat form produce significantly different results in terms of transport when modeled with the method of regularized Stokeslets?

Since a regularized Stokeslet is a way to transmit local forces onto a fluid, we used collections of regularized Stokeslets to model slender solid objects (cilia) here, but that can be expanded to model objects or boundaries of any shape. The flexibility of the method allows a user to customize the regularized Stokeslet placement to model any desired geometry, as long as the flow conditions fall into the low Reynolds number regime. We used the idea of regularized image singularities to create a no-slip (zero-velocity) plane that emulates the epithelial cell surface, however other fluid geometries, either in biological contexts or in a laboratory setting, could use other techniques to model the flow. The flexibility of the MRS being able to adapt to irregular geometries is one of its main strengths over other computational methods, some of which require a mathematical formulation of the potentially complicated boundary. The fundamental idea here is that you can enforce a velocity condition anywhere you place a regularized Stokeslet (and use the linear relationship to translate that to the appropriate external forces as discussed in Sect. 2.2). In modeling our pulmonary cilia, we were enforcing non-zero velocities, but regularized Stokeslets can also be used to enforce stationary objects with zero velocity, like no-slip boundaries of irregular geometries.

Research Project 2 Use the method of regularized Stokeslets to model the geometry of primary nodal cilia. These cilia are more rigid, less densely spaced, and have a different motion than pulmonary cilia. The geometry of the biological setting is also different in that the cilia sit in a more enclosed region. Sources in the literature such as [7, 8, 20, 30] give more details about the flow induced by primary nodal cilia.

The system of regularized image singularities mentioned in Sect. 2.2 creates a no-slip plane at $z = 0$ in a three-dimensional fluid domain to create a semi-infinite fluid domain as shown in Fig. 3. In a laboratory setting, one may be interested in studying fluid flow in a tank where wall effects are not negligible. The tank floor would be modeled by the existing system of image singularities, but what if other tank walls need to be taken into account?

Research Project 3 Modify the system of image singularities to create the walls of a fluid tank. That is, for each regularized Stokeslet you place in the fluid domain, what collection of regularized singularities do you need to place outside the fluid domain to create five no-slip walls (the floor and four vertical sides)? Where do these singularities need to be placed?

Research Project 4 Can you attempt to achieve the same effect of creating a fluid tank with five walls without using a system of image singularities? Perhaps utilize the fact that regularized Stokeslets can be used to implement velocity conditions on the fluid. In the aforementioned work in this chapter, we used the known velocity of the cilium to solve for the forces needed to mimic that effect. Could you use the known velocity of a tank wall to mimic those wall effects without image singularities?

There are other boundary conditions that people are interested in beyond no-slip where velocity vanishes on the boundary (particularly a solid boundary). However, if you were looking at a multi-fluid model, you might have an interface between two fluids with different properties.

Research Project 5 Can you formulate conditions that would be appropriate to create a no-stress boundary condition that might appear at the interface of two fluids?

The model described in this work assumes the cilia are immersed in a semi-infinite water-like fluid to model the PCL. However, the biological system has a mucus layer sitting atop the PCL. Can you modify the MRS to incorporate more than one fluid? One step would be to explore adding a second Newtonian fluid layer. If you make the assumption that the second layer is non-Newtonian, the MRS may no longer be helpful because the linearity of the Stokes equations that we rely on breaks down.

Research Project 6 Explore the idea of implementing a two-phase fluid with the MRS. What challenges do you face? Can you come up with a creative way to approximate the effects of the second fluid that rests atop the first?

The literature involving metachronal waves includes work on the formation of the metachronal waves. Our discussion here imposes the waves and studies resulting flow, but modifications could be made to the technique to create the wave formation.

Research Project 7 Modify the techniques in this chapter to explore the formation of the metachronal waves instead of imposing them. Perhaps this involves treating different segments of the cilia differently so that adjacent regularized Stokeslets react to each other and the fluid instead of having their positions all enforced through the beat form. Since regularized Stokeslets transmit forces to the fluid, one might consider using external spring forces between nearby Stokeslets to create a flexible cilium that responds to the surrounding fluid.

When studying the motion of "particles" in the fluid in this project, the particles are actually infinitesimal passive fluid tracers that have no impact on the fluid. However, in the real world, the contaminants trapped in the mucus have physical dimensions and their presence affects the fluid around them. Their density affects their buoyancy; their size affects their drag forces. In mucociliary clearance, their size may impact their transport. In a different application, if one is trying to deliver drugs through an inhaled medication (like an asthma inhaler), the drug particles need to penetrate the fluid flow and get absorbed by the epithelial cells. Different drugs have different particle sizes that may present challenges to transport and absorption.

Research Project 8 Do the properties of the particles affect flow induced by a metachronal wave in a significant way? Can you model individual particles of a given radius or density to explore the effects of the particle's physical presence on the flow? What particle properties are desirable for the case of contaminant transport out of the lung? What about in the case of a drug delivery system?

Since turbulence is not present in low Reynolds number systems, mixing can be an interesting topic. Some organisms use their motion to create mixing opportunities to find nutrients.

Research Project 9 Can you modify the cilia model presented here to study questions related to mixing of tracer particles in the fluid? If your goal was

(continued)

Research Project 9 (continued)
to create a low Reynolds fluid environment that is favorable to mixing near cilia, would you propose the cilia move in a metachronal wave or would you propose a different motion to enhance mixing? Can you find evidence that different motions are more adept at mixing the surrounding fluids?

The sizes of the cilia patches in this project were chosen to reduce computational time while still observing desired metachronal wave effects on fluid trajectories.

Research Project 10 Explore properties of the matrices involved in this system $\mathbf{u} = A\mathbf{f}$ (both when A is square and when it is not). Do you have the same matrix characterization if A only contains regularized Stokeslets (in a free-space model) or if it also contains the systems of regularized image singularities (in a semi-infinite model)? Can you leverage any of these properties to explore ways of more efficiently calculating the forces when velocities are known or increasing computational speed when calculating velocities are calculated at tracer points from known forces?

While cilia are the foundation of the biological application focused on in this work, there is much research interest in biological fluid flows relating to transport of small flagellated organisms like bacteria and spermatozoa. Like cilia, flagella are slender objects immersed in a fluid, but they often serve the role of propelling a microorganism through a fluid.

Research Project 11 Explore other biological fluid flow scenarios to find an application of the MRS to suit your interests.

Acknowledgements The authors thank the Furman University Summer Research Fellowship Program for funding support.

References

1. URL http://www.claymath.org/millennium-problems/navier-stokes-equation
2. Batchelor, G.: An Introduction to Fluid Dynamics. Cambridge University Press (1967)
3. Boucher, R.C.: New concepts of the pathogenesis of cystic fibrosis lung disease. European Respiratory Journal **23**(1), 146–158 (2004)

4. Bouzarth, E.L.: Regularized singularities and spectral deferred correction methods: A mathematical study of numerically modeling Stokes fluid flow. Ph.D. thesis, The University of North Carolina at Chapel Hill (2008)
5. Bouzarth, E.L., Minion, M.L.: A multirate time integrator for regularized Stokeslets. Journal of Computational Physics **229**, 4208–4224 (2010)
6. Bouzarth, E.L., Minion, M.L.: Modeling slender bodies with the method of regularized Stokeslets. Journal of Computational Physics **230**(10), 3929–3947 (2011)
7. Cartwright, J., Piro, N., Piro, O., Tuval, I.: Embryonic nodal flow and the dynamics of nodal vesicular parcels. J. R. Soc. Interface **4**, 49–55 (2007)
8. Cartwright, J., Piro, O., Tuval, I.: Fluid-dynamical basis of the embryonic development of left-right asymmetry in vertebrates. PNAS **101**(19), 7234–7239 (2004)
9. Chateau, S., D'Ortona, U., Poncet, S., Favier, J.: Transport and mixing induced by beating cilia in human airways. Frontiers in Physiology **9**(161) (2018)
10. Chateau, S., Favier, J., D'Ortona, U., Poncet, S.: Transport efficiency of metachronal waves in 3D cilia arrays immersed in a two-phase flow. Journal of Fluid Mechanics **824**, 931–961 (2017)
11. Chilvers, M.A., O'Callaghan, C.: Analysis of ciliary beat pattern and beat frequency using digital high speed imaging: Comparison with the photomultiplier and photodiode methods. Thorax **53**, 314–317 (2000)
12. Cortez, R.: The method of regularized Stokeslets. SIAM J. Sci. Comput. **23**(4), 1204–1225 (2001)
13. Cortez, R., Fauci, L., Medovikov, A.: The method of regularized Stokeslets in three dimensions: Analysis, validation, and application to helical swimming. Physics of Fluids **17**(3), 031,504 (2005)
14. Cortez, R., Varela, D.: A general system of images for regularized Stokeslets and other elements near a plane wall. Journal of Computational Physics **285**, 41–54 (2015)
15. Elegeti, J., Gompper, G.: Emergence of metachronal waves in cilia arrays. PNAS **110**(12), 4470–4475 (2013)
16. Fulford, G.R., Blake, J.R.: Muco-ciliary transport in the lung. Journal of Theoretical Biology **121**(4), 381–402 (1986)
17. Gheber, L., Priel, Z.: Extraction of cilium beat parameters by the combined application of photoelectric measurements and computer simulation. Biophysical Journal **72**(1), 449–462 (1997)
18. Guirao, B., Joanny, J.F.: Spontaneous creation of macroscopic flow and metachronal waves in an array of cilia. Biophysical Journal **92**, 1900–1917 (2007)
19. Healthcare, F..P.: Introduction to mucociliary transport video microscopy (2009)
20. Hirokawa, N., Tanaka, Y., Okada, Y., Takeda, S.: Nodal flow and the generation of left-right asymmetry. Cell **125**, 33–45 (2006)
21. Humi, M., Miller, W.B.: Boundary Value Problems and Partial Differential Equations. PWS-Kent (1992)
22. Knowles, M.R., Boucher, R.C.: Mucus clearance as a primary innate defense mechanism for mammalian airways. The Journal of Clinical Investigation **109**(5), 571 (2002)
23. Kundu, P.K., Cohen, I.M.: Fluid Mechanics. Elsevier Academic Press (2004)
24. Osterman, N., Vilfan, A.: Finding the ciliary beating pattern with optimal efficiency. PNAS **108**(38), 15,727–15,732 (2011)
25. Pozrikidis, C.: Introduction to Theoretical and Computational Fluid Dynamics. Oxford University Press (1997)
26. Purcell, E.M.: Life at low Reynolds number. American Journal of Physics **45**(1) (1977)
27. Raidt, J., Wallmeier, J., Hjeij, R., Onnebrink, J.G., Pennekamp, P., Loges, N., Olbrich, H., Haffner, K., Dougherty, G., Omran, H., Werner, C.: Ciliary beat pattern and frequency in genetic variants of primary ciliary dyskinesia ciliary beat pattern and frequency in genetic variants of primary ciliary dyskinesia. European Respiratory Journal **44**, 1579–1588 (2014)
28. Sanderson, M.J., Sleigh, M.A.: Ciliary activity of cultured rabbit tracheal epithelium: beat pattern and metachrony. Journal of Cell Science **47**(1), 331–347 (1981)

29. Smith, D.J.: A boundary element regularized Stokeslet method applied to cilia- and flagella-driven flow. Proceedings of the Royal Society of London A: Mathematical, Physical and Engineering Sciences **465**(2112), 3605–3626 (2009)
30. Smith, D.J., Gaffney, E.A., Blake, J.R.: Discrete cilia modelling with singularity distributions: application to the embryonic node and the airway surface liquid. Bulletin of Mathematical Biology **69**(5), 1477–1510 (2007)

Printed in the United States
By Bookmasters